Springer Proceedings in Mathematics & Statistics

Volume 82

For further volumes:
http://www.springer.com/series/10533

This book series features volumes composed of select contributions from workshops and conferences in all areas of current research in mathematics and statistics, including OR and optimization. In addition to an overall evaluation of the interest, scientific quality, and timeliness of each proposal at the hands of the publisher, individual contributions are all refereed to the high quality standards of leading journals in the field. Thus, this series provides the research community with well-edited, authoritative reports on developments in the most exciting areas of mathematical and statistical research today.

N. S. Narasimha Sastry
Editor

Groups of Exceptional Type, Coxeter Groups and Related Geometries

Springer

Editor
N. S. Narasimha Sastry
Statistics and Mathematics Unit
Indian Statistical Institute
Bangalore
Karnataka
India

ISSN 2194-1009 ISSN 2194-1017 (electronic)
ISBN 978-81-322-1813-5 ISBN 978-81-322-1814-2 (eBook)
DOI 10.1007/978-81-322-1814-2
Springer New Delhi Heidelberg New York Dordrecht London

Library of Congress Control Number: 2014935062

Printed on acid-free paper

Springer is part of Springer Science+Business Media (www.springer.com)

Preface

This is a collection of invited articles by some very active researchers in the interface of the theory of finite groups and their representations, Coxeter and Artin groups, incidence geometries and the theory of buildings. This includes articles by some of the invited speakers at the International Conference on "Groups and Geometries" organized at the Indian Statistical Institute, Bangalore Centre, during December 10–21, 2012. The request to the authors was to discuss the latest developments in the above topics with an emphasis on the basic questions of current interest. This was also the objective of the conference.

I would like to thank all the authors for kindly writing the articles and many colleagues who carefully refereed the articles in time. I also wish to thank the speakers and the participants at the conference.

This conference was a follow-up of the ICM Satellite Conference on "Buildings, Finite geometries and Groups" held at the Indian Statistical Institute, Bangalore Centre, during August 29–31, 2010. Springer has kindly published a volume comprising of articles by the participants in that conference and other authors who could not attend the conference (*Buildings, Finite Geometries and Groups*, Springer Proceedings in Mathematics Series, Vol. 10).

The year 2012 marked the 125th birth anniversary of the legendary mathematician of Indian origin, Srinivasa Ramanujan. There had been year-long celebrations throughout India, starting from December 22, 2011. This conference was organized as part of these celebrations.

During 2012, the Indian mathematical community lost Prof. Somesh C. Bagchi, who worked at the Indian Statistical Institute, Kolkata, India, and Prof. D. N. Verma, who worked at the Tata Institute of Fundamental Research, Mumbai, India. Both of them contributed very substantially during the last 3–4 decades to the growth of mathematics in India in terms of teaching and training, organization, and research. This programme was dedicated to the memory of these two colleagues.

The conference was supported by the Indian Statistical Institute. Partial support was also received from the National Board for Higher Mathematics, India. I thank these organizations for their generous support. I also thank Springer for kindly publishing this volume, Dr. Hans J. Koelsch for his encouragement, and Mr. Shamim Ahmad for his friendly and efficient cooperation in bringing out this volume in time. Finally, I record my indebtedness to Ms. Asha Lata of Statistics and Mathematics Unit, Indian Statistical Institute, Bangalore Centre, for her help through all stages of organizing the conference and the preparation of this volume.

September 30, 2013 N. S. Narasimha Sastry

Talks Given in the Conference

Bhaskar Bagchi	The non-existent projective plane of order ten revisited
Bart De Bruyn	The uniqueness of the generalized octagon of order (2, 4) containing a suboctagon of order (2, 1)
Ilaria Cardinali	An outline of polar spaces: basics and advances
Bruce Cooperstein	Witt-type theorems for subspaces of exceptional geometries and Generating Sets for Lie Geometries
Sudhir Ghorpade	Automorphisms of Grassmann varieties
A. A. Ivanov	Monster group and Sporadic geometries
S. Kannan	Schubert Varieties, Coxeter elements, GIT quotients
Ramazan Koc	A Mathematica Program for Projecting A4 Lattice Vectors on the Coxeter Plane
Mehmet Koca	Rank 3 Coxeter groups, quaternions, polyhedra, chiral polyhedra and their duals; Rank 4 Coxeter groups, quaternions, 4D polytopes and their dual polytopes and Quasicrystallography as the projections of the Lie algebra lattices
Tom De Medts	Moufang sets and exceptional groups
Koji Nuida	On the isomorphism problem for Coxeter groups and related topics
Luis Paris	The $K(\pi, 1)$ conjecture for Artin groups
N. S. N. Sastry	Codes from some rank 2 geometries
Peter Sin	On the structure of Classical Permutation Modules
David Stewart	The archaeology of unipotent radicals
Koen Struyve	Affine buildings: metric properties and rigidity
B. Sury	What is Tits index and how to work with it
Donna Testerman	Exceptional groups of Lie type: subgroup structure and unipotent elements
Maneesh Thakur	Exceptional groups
T. N. Venkataramana	Arithmeticity of Certain Symplectic Hypergeometric Groups

Participants

Bhaskar Bagchi, ISI, Bangalore, India (bbagchi@isibang.ac.in)
Sarjick Bakshi, CMI, Chennai, India (sarjick91@gmail.com)
Sonia Balagopalan, NUI Maynooth, Ireland (s.balagopalan@gmail.com)
Lien Boelaert, Ghent University, Belgium (lboelaer@cage.ugent.be)
Anirban Bose, ISI, Delhi (anirban.math@gmail.com)
Bart De Bruyn, University of Ghent, Belgium (bdb@cage.ugent.be)
Ilaria Cardinali, University of Siena, Italy (ilaria.cardinali@unisi.it)
B. Narasimha Chary, CMI, Chennai, India (chary@cmi.ac.in)
Bruce Cooperstein, University of California, USA (coop@ucsc.edu)
Shripad M. Garge, IIT, Bombay, India (smgarge@gmail.com)
Sudhir R. Ghorpade, IIT, Mumbai, India (srg@math.iitb.ac.in)
Neha Hooda, ISI, Delhi, India (neha.hooda2@gmail.com)
Alexander A. Ivanov, Imperial College, London (a.ivanov@imperial.ac.uk)
Anuj Jakhar, IISER, Mohali, India (ms09020@iisermohali.ac.in)
Upendra Kulkarni, CMI, Chennai, India (upendra@cmi.ac.in)
Ramazan Koc, Gaziantep University, Turkey (koc@gantep.edu.tr)
Mehmet Koca, Sultan Qaboos University, Oman (kocam@squ.edu.om)
Tom De Medts, University of Ghent, Belgium (tdemedts@java.ugent.be)
Anubhav Mukerjee, ISI, Bangalore, India (bmat1103@ms.isibang.ac.in)
Apurv Nakade, CMI, Chennai, India (apurv@cmi.ac.in)
Koji Nuida, AIST, Japan (k.nuida@aist.go.jp)
Luis Paris, Institute de Mathematiques de Bourgogne, France (Luis.Paris@u-bourgogne.fr)
Binod Kumar Sahoo, NISER, Bhubaneswar, India (bksahoo@niser.ac.in)
Preena Samuel, CMI, Chennai, India (preena@cmi.ac.in)
N. S. N. Sastry, ISI, Bangalore, India (nsastry@isibang.ac.in)
S. Senthamarai Kannan, CMI, Chennai, India (kannan@cmi.ac.in)
Brajesh Sharma, University of Allahabad, Allahabad, India (brajeshsharma72@gmail.com)
Ravindra Shukla, University of Allahabad, Allahabad, India (shuklarp@gmail.com)
Peter Sin, University of Florida, USA (sin@ufl.edu)
Anupam Singh, IISER, Pune, India (anupamk18@gmail.com)
David Stewart, University of Oxford, UK (david.stewart@new.ox.ac.uk)

Koen Struyve, University of Ghent, Belgium (kstruy@gmail.com)
B. Sury, ISI, Bangalore, India (sury@isibang.ac.in)
Donna Testerman, Ecole Polytechnique Federale de Lausanne, Switzerland (donna.testerman@epfl.ch)
Maneesh Thakur, ISI, Delhi, India (maneesh.thakur@gmail.com)
T. N. Venkataramana, TIFR, Mumbai, India (venky@math.tifr.res.in)

Contents

Contributors

Rieuwert J. Blok Department of Mathematics and Statistics, Bowling Green State University, Ohio, USA

Ilaria Cardinali Department of Information Engineering and Mathematics, University of Siena, Siena, Italy

A. Castillo-Ramirez Department of Mathematics, Imperial College London, London, UK

Bruce N. Cooperstein Department of Mathematics, University of California, Santa Cruz, USA

Hans Cuypers Department of Mathematics and Computer Science, Technische Universiteit Eindhoven, Eindhoven, The Netherlands

Bart De Bruyn Department of Mathematics, Ghent University, Ghent, Belgium

Corneliu G. Hoffman University of Birmingham, Edgbaston, UK

Jos in 't panhuis Department of Mathematics and Computer Science, Technische Universiteit Eindhoven, Eindhoven, The Netherlands

A. A. Ivanov Department of Mathematics, Imperial College London, London, UK

S. Senthamarai Kannan Chennai Mathematical Institute, Siruseri, Kelambakkam, Tamil Nadu, India

Koji Nuida Innovative Security Research Group, Research Institute for Secure Systems, National Institute of Advanced Industrial Science and Technology Ibaraki, Japan

Luis Paris Institut de Mathématiques de Bourgogne, Université de Bourgogne, Dijon cedex, France

A. Pasini Department of Information Engineering and Mathematics, University of Siena, Siena, Italy

N. S. Narasimha Sastry Division of Theoretical Statistics and Mathematics, Indian Statistical Institute, Bangalore, India

Iulian I. Simion École Polytechnique Fédérale de Lausanne, Lausanne, Switzerland

Peter Sin Department of Mathematics, University of Florida, Gainesville, FL, USA

Donna M. Testerman École Polytechnique Fédérale de Lausanne, Lausanne, Switzerland

Hendrik Van Maldeghem Department of Mathematics, Ghent University, Ghent, Belgium

Chapter 1
A Classification of Curtis-Tits Amalgams

Rieuwert J. Blok and Corneliu G. Hoffman

Abstract A celebrated theorem of Curtis and Tits on groups with finite BN-pair shows that these groups are determined by the local structure arising from their fundamental subgroups of ranks 1 and 2. This result was later extended to Kac-Moody groups by P. Abramenko and B. Mühlherr and Caprace. Their theorem states that a Kac-Moody group G is the universal completion of an amalgam of rank two (Levi) subgroups, as they are arranged inside G itself. Taking this result as a starting point, we define a *Curtis-Tits structure* over a given diagram to be an amalgam of groups such that the sub-amalgam corresponding to a two-vertex sub-diagram is the Curtis-Tits amalgam of some rank-2 group of Lie type. There is no a priori reference to an ambient group, nor to the existence of an associated (twin-) building. Indeed, there is no a priori guarantee that the amalgam will not collapse. We then classify these amalgams up to isomorphism. In the present paper we consider triangle-free simply-laced diagrams. Instead of using Goldschmidt's lemma, we introduce a new approach by applying Bass and Serre's theory of graphs of groups, not to the amalgams themselves but to a graph of groups consisting of certain automorphism groups. The classification reveals a natural division into two main types: "orientable" and "non-orientable" Curtis-Tits structures. Our classification of orientable Curtis-Tits structures naturally fits with the classification of all locally split Kac-Moody groups over fields with at least four elements using Moufang foundations. In particular, our classification yields a simple criterion for recognizing when Curtis-Tits structures

---

Subject Classification: [2010] Primary 20G35; Secondary 51E24, 20E42
The authors wish to thank the Banff International Research Station, Banff, Canada for a very productive visit as part of their Research In Teams program (08rit130).

R. J. Blok
Department of Mathematics and Statistics, Bowling Green State University, Bowling Green, OH 43403, USA
e-mail: blokr@member.ams.org

C. G. Hoffman (✉)
University of Birmingham, B15 2TT, Edgbaston, UK
e-mail: C.G.Hoffman@bham.ac.uk

N. S. N. Sastry (ed.), *Groups of Exceptional Type, Coxeter Groups and Related Geometries*, Springer Proceedings in Mathematics & Statistics 82,
DOI: 10.1007/978-81-322-1814-2_1,

give rise to Kac-Moody groups. The class of non-orientable Curtis-Tits structures is in some sense much larger. Many of these amalgams turn out to have non-trivial interesting completions inviting further study.

Keywords Twin buildings · Amalgams · Kac-Moody groups · Bass-Serre theory

1 Introduction

Kac-Moody Lie algebras are infinite dimensional Lie algebras defined by relations analogous to the Serre relations for finite dimensional semi-simple Lie algebras. They have been introduced in the mid 1960s by V. Kac and R. Moody. The affine Kac-Moody and generalized Kac-Moody Lie algebras have extensive applications to theoretical physics, especially conformal field theory, monstrous moonshine and more.

Finite dimensional semi-simple Lie algebras admit Chevalley bases which allow the construction of Chevalley groups, Lie-type groups over arbitrary fields. By analogy, J. Tits gave a new interpretation of Kac-Moody groups as groups having a twin-root datum, which implies that they are symmetry groups of Moufang twin-buildings (see [27, 28]). In the case that the corresponding diagram is spherical, the corresponding group is a Chevalley group. These and other similar groups play a very important role in various aspects of geometric group theory. In particular, they provide examples of infinite simple groups (see for example [10, 11, 13, 16]).

A celebrated theorem of Curtis and Tits [14, 26] (later extended by Timmesfeld (see [22–25] for spherical groups), by Abramenko and Mühlherr [1] and Caprace [12] to two-spherical Kac-Moody groups) on groups with finite BN-pair states that the Kac-Moody groups are the universal completion of the concrete amalgam of the Levi components of the parabolic subgroups with respect to a given (twin-) BN-pair.

In case that the amalgam is unique, this suffices to recognize the group. In general however, this is an inconvenience since it is usually easy to recognize isomorphism classes of subgroups but perhaps not so easy to globally manage their embedding. This is the reason that one often restricts to the so called "split" Kac-Moody groups. However "twisted" versions of Kac-Moody groups do exist, as constructed in [16, 19] and they in turn give Curtis-Tits amalgams.

A natural question is therefore the following: how can one recognize these amalgams as abstract group amalgams? More generally one would like to classify all amalgams that are "locally" isomorphic to the usual Curtis-Tits ones and identify their universal completions. In this paper we use a variation of Bass-Serre theory to classify all Curtis-Tits structures over a field with at least four elements and with connected simply laced triangle-free Dynkin diagrams. As a by-product we obtain a description of all Kac-Moody groups in this case.

Throughout the paper k will be a commutative field of order at least four. We need the restriction on the order for the classification of the amalgams. Precise definitions will be given in Sect. 3. Let Γ be a Dynkin diagram over a set I. A *Curtis-Tits (CT)*

structure over k with Dynkin diagram Γ is an amalgam $\mathscr{G} = \{G_i, G_{i,j} \mid i, j \in I\}$ such that, for every $i, j \in I$, $\{G_i, G_j, G_{i,j}\}$ is a Curtis-Tits amalgam of type $\Gamma_{i,j}$. In case Γ is simply-laced, this means that the rank-1 groups G_i are isomorphic to $SL_2(k)$, where $G_{i,j} = \langle G_i, G_j \rangle$, that G_i and G_j commute if $\{i, j\}$ is a non-edge in Γ and their images in $G_{i,j} \cong SL_3(k)$ form a standard pair (see Definition 3) if $\{i, j\}$ is an edge in Γ. Note that we do not impose any further conditions on the embedding maps $G_i \hookrightarrow G_{i,j}$.

We are only interested in CT structures that admit a non-trivial completion. The universal completion of a (non-collapsing) Curtis-Tits structure is called a *Curtis-Tits group*.

In fact, a slight extension of our methods allows to classify Curtis-Tits structures for a larger class of diagrams, including for instance all three-spherical Dynkin diagrams and, more general amalgams such as Phan type amalgams. However, in order to present these new methods and new results in a transparent manner, we chose to restrict to all simply-laced diagrams without triangles, just as Tits did in his classification of Moufang foundations for these diagrams in [28].

Curtis-Tits groups that are not Kac-Moody groups do exist. As an application in [5] we give constructions of all possible Curtis-Tits structures with diagram $\widetilde{A}_n$, realizing them as concrete amalgams inside their respective non-trivial completions, and giving sufficient conditions for these completions to be universal. In addition, we furnish geometric objects on which these groups act naturally. This leads us to describe two very interesting collections of groups. The first is a collection of twisted versions of the Kac-Moody group $SL_n(k[t, t^{-1}])$ whose natural quotients are labeled by the cyclic algebras of center k. The corresponding twin-building is related to Drinfeld's vector bundles over a non-commutative projective line. The second is a collection of Curtis-Tits groups that are not Kac-Moody groups. One of these maps surjectively to $Sp_{2n}(q)$, $\Omega^+_{2n}(q)$, and $SU_{2n}(q^l)$, for all $l \geq 1$, making this family of unitary groups into a family of expanders [8].

The main result of the present paper is the following.

Theorem 1.1. *Let Γ be a connected simply laced Dynkin diagram with no triangles and k a field with at least four elements. There is a natural bijection between isomorphism classes of CT-structures over the field k on a graph Γ of a given type and group homomorphisms $\Phi\colon \pi(\Gamma, i_0) \to \mathbb{Z}_2 \times \mathrm{Aut}(k)$.*

Here, $\pi(\Gamma, i_0)$ denotes the fundamental group of the graph Γ with base point i_0. As mentioned above, the motivation for the work came from the Curtis-Tits amalgam presentations for Kac-Moody groups. In fact in the spherical case these were proved to be the only such amalgams. In general the proportion of orientable amalgams among all Curtis-Tits amalgams decreases with the rank of $\pi(\Gamma, i_0)$. More precisely they are those amalgams in the theorem corresponding to maps Φ so that $\mathrm{Im}(\Phi) \leq \mathrm{Aut}(k)$. We call such amalgams *orientable*.

Corollary 1.2. *Let Γ be a connected simply laced Dynkin diagram with no triangles and k a field with at least four elements. The universal completion of a Curtis-Tits structure over a commutative field k and diagram Γ is a central extension of a locally* *split Kac-Moody group over k with Dynkin diagram Γ (and $\mathcal{G}$ is the Curtis-Tits amalgam for this group) if and only if $\mathcal{G}$ is orientable.*

Note that for example in [1, 12, 28] the amalgam is a priori required to live in the corresponding Kac-Moody group. This is rather inconvenient since it gives no intrinsic description of the amalgam. Our result above defines Kac-Moody groups as universal completions of certain abstract amalgams hence giving concrete presentations for those groups.

Note that Theorem 1.1 together with Corollary 1.2 provides an alternative proof of the classification of sound Moufang foundations with connected simply-laced triangle-free Dynkin diagram as given in [28, Sect. 6.5].

We can refine Corollary 1.2 as follows. See Sect. 5.2 for the exact definitions.

Corollary 1.3. *Let Γ be a connected simply laced Dynkin diagram with no triangles and k a field with at least 4 elements. Any locally split Kac-Moody group over k with diagram Γ is a central quotient of a twist $G^{\delta}_{\Gamma}(k)$ of the Curtis-Tits amalgam inside the corresponding split Kac-Moody group. Moreover any two twists are equivalent if and only if they correspond to the same homomorphism $\Phi: \pi(\Gamma, i_0) \to \mathbb{Z}_2 \times \mathrm{Aut}(k)$.*

For a definition of the twist $G^{\delta}_{\Gamma}(k)$ see Sect. 6.

Conceivably Corollary 3 above can be proved directly from results of Tits and Mühlherr in [28] and [17]. Indeed in [17] a bijective correspondence is established between Moufang foundations and twin-buildings and in [28] whereas [28, Sect. 6.5] contains the correspondence between Moufang foundations and homomorphisms $\pi_1(\Gamma, i_0) \to \mathrm{Aut}(k)$. However, to our knowledge no explicit correspondence between Curtis-Tits amalgams and Moufang foundations exists in the literature. As for Corollary 1.2, it provides a group theoretic definition of Kac-Moody groups that does not make a reference to Moufang foundations or other data that are a priori associated to twin-buildings. Indeed, in the absence of Theorem 1.1, it is not immediately obvious that different choices of an orientable CT amalgam would give different foundations. See also Corollary 2 for a more precise construction of the amalgams in the spirit of [12] (see the application to Theorem A in loc. cit.).

The paper is organized as follows. In Sect. 2 we review our modification of Bass-Serre theory from [6] and add a few improvements. In Sect. 3 we define Curtis-Tits structures and their morphisms, and prove some general technical lemmas regarding Property (D). In Sect. 4 we prove Theorem 1.1. In Sect. 5 we prove Corollary 1.2 and in Sect. 6 we prove Corollary 1.3.

Note. In the original version of the present paper we developed a technique for classifying amalgams and applied this to the classification of Curtis-Tits amalgams with simply laced Dynkin diagram. By coincidence, the paper [6], which contains a generalization of that classification technique, was published first. In the interest of self-containedness and efficiency, in the present paper we will quickly review the results from loc. cit., add a few improvements, and then focus on their application in the classification of Curtis-Tits structures with simply laced Dynkin diagram.

Acknowledgement. The original version of this paper was written during some wonderful, if very claustrophobic and accident prone three weeks in Birmingham. We thank Irina and Karin for putting up with it all.

The authors would also like to thank the two referees for their excellent comments, which improved the final version and eliminated some ambiguities in the manuscript.

2 Classification of Amalgams Using Graphs of Groups

In this section we review some definitions and results from [6] that are essential to our main results and furnish some additional details. For proofs, examples and applications, we refer to loc. cit.

2.1 Amalgams, Graphs of Groups, and Pointings

Throughout the paper we fix an oriented graph $\overrightarrow{\Gamma} = (I, \overrightarrow{E})$ where for each edge $e \in \overrightarrow{E}$ there is a distinct reverse edge denoted $\bar{e}$. Moreover we denote by $d_0(e)$ the starting node of the oriented edge e and by $d_1(e)$ the end vertex of e. Thus $d_0(e) = d_1(\bar{e})$. Morphisms of graphs must preserve these properties.

We shall call the graph *admissible* if it is connected and has no circuits of length ≤ 3 and we will only consider such graphs.

Definition 2.1. *An* amalgam *over the graph* $\overrightarrow{\Gamma} = (I, \overrightarrow{E})$ *is a collection* $\mathscr{G} = \{G_i, G_e, \varphi_e \mid i \in I, e \in \overrightarrow{E}\}$*, where* G_i *and* $G_e = G_{\bar{e}}$ *are groups and, for each* e *we have a monomorphism* $\varphi_e \colon G_{d_0(e)} \hookrightarrow G_e$*, called an* inclusion map. *A* completion *of* $\mathscr{G}$ *is a group* G *together with a collection* $\phi = \{\phi_e, \phi_i \mid i \in I, e \in \overrightarrow{E}\}$ *of homomorphisms* $\phi_e \colon G_e \to G$, $\phi_i \colon G_i \to G$ *such that for any* e *we have* $\phi_e \circ \varphi_e = \phi_{d_0(e)}$. *For simplicity we shall write* $\bar{G}_e = \varphi_e(G_{d_0(e)}) \leq G_e$. *The amalgam* $\mathscr{G}$ *is* non-collapsing *if it has a non-trivial completion. A completion* $(\widetilde{G}, \widetilde{\phi})$ *is called* universal *if for any completion* (G, ϕ) *there is a (necessarily unique) surjective group homomorphism* $\pi \colon \widetilde{G} \to G$ *such that* $\phi = \pi \circ \widetilde{\phi}$.

Definition 2.2. *A* special homomorphism *between the two amalgams*

$$\mathscr{G}^{(1)}(\overrightarrow{\Gamma}) = \{G_i^{(1)}, G_e^{(1)}, \varphi_e^{(1)} \mid i \in I, e \in \overrightarrow{E}\} \text{ and}$$
$$\mathscr{G}^{(2)}(\overrightarrow{\Gamma}) = \{G_i^{(2)}, G_e^{(2)}, \varphi_e^{(2)} \mid i \in I, e \in \overrightarrow{E}\}$$

is a map $\phi = \{\phi_i, \phi_e \mid i \in I, e \in \overrightarrow{E}\}$ *such that, for all* $e \in \overrightarrow{E}$

$$\begin{array}{ccc} G_e^{(1)} & \xrightarrow{\varphi_e} & G_e^{(2)} \\ {\scriptstyle \varphi_e^{(2)}} \uparrow & & \uparrow {\scriptstyle \varphi_e^{(2)}} \\ G_{d_0(e)}^{(1)} & \xrightarrow{\phi_{d_0(e)}} & G_{d_0(e)}^{(2)} \end{array}$$

is a commutative diagram of group homomorphisms. We call ϕ an isomorphism *of amalgams if ϕ_i and ϕ_e are bijective for all $i, j \in I$, and ϕ^{-1} is a homomorphism of amalgams.*

Remark 2.3. One could define a more general notion of homomorphism of amalgams involving automorphisms of the graph Γ. We have chosen not to do this and therefore fix a labeling I of the vertices of Γ and the corresponding groups in $\mathscr{G}$ throughout.

Definition 2.4. *Consider an amalgam $\mathscr{G}_0 = \{G_i, G_e, \psi_e \mid i \in I, e \in \vec{E}\}$ over $\vec{\Gamma} = (I, \vec{E})$. An* amalgam of type $\mathscr{G}_0$ *is an amalgam over $\vec{\Gamma}$ with the same groups G_i, G_e and where, for every $e \in \vec{E}$, the inclusion map φ_e is possibly different from ψ_e except that it has the same image $\bar{G}_e$.*

For the rest of this section, $\mathscr{G}_0$ will be a fixed amalgam over $\vec{\Gamma} = (I, \vec{E})$. We shall therefore simply write $\mathscr{G}_0 = \{G_i, G_e, \psi_e\}$ tacitly understanding that i is taken over all elements in I and e is taken over all elements of $\vec{E}$. We shall adopt a similar shorthand for amalgams $\mathscr{G}$ of type $\mathscr{G}_0$.

Given a collection of subgroups $G_1, \dots, G_k$ of the group G, we define $\mathrm{Aut}_G(G_1, \dots, G_k)$ to be the subgroup of $\mathrm{Aut}(G)$ stabilizing each G_i. Given a monomorphism of groups $\phi\colon G \to H$, there is a corresponding homomorphism $\mathrm{ad}(\phi)\colon \mathrm{Aut}_H(\phi(G)) \to \mathrm{Aut}(G)$ such that for any $a \in \mathrm{Aut}_H(\phi(G))$ we have $\mathrm{ad}(\phi)(a) = \phi^{-1} \circ a \circ \phi$.

Lemma 2.5. *Let $f \in \vec{E}$ and suppose that the set map*

$$\mathrm{Aut}_{G_f}(\bar{G}_f, \bar{G}_{\bar{f}}) \to \mathrm{Aut}(\bar{G}_f) \times \mathrm{Aut}(\bar{G}_{\bar{f}})$$

given by restriction is onto. Then the isomorphism class of any amalgam $\mathscr{G} = \{G_i, G_e, \varphi_e\}$ of type $\mathscr{G}_0$ does not depend on φ_f or $\varphi_{\bar{f}}$.

Proof. For $i = 1, 2$, let $\mathscr{G}^{(i)} = (G_i, G_e, \varphi_e^{(i)}\}$ be an amalgam of type $\mathscr{G}_0$. Suppose that $\varphi_e^{(1)} = \varphi_e^{(2)}$ for all $e \in \vec{E} - \{f, \bar{f}\}$. Then since the groups $\bar{G}_i$ and $\bar{G}_e$ are the same in both amalgams, the compositions $a_{d_0(f)} = \varphi_f^{(2)} \circ \mathrm{id}_{G_{d_0(f)}} \circ (\varphi_f^{(1)})^{-1}$ and $a_{d_1(f)} = \varphi_{\bar{f}}^{(2)} \circ \mathrm{id}_{G_{d_0(\bar{f})}} \circ (\varphi_{\bar{f}}^{(1)})^{-1}$ are automorphisms of $\bar{G}_f$ and $\bar{G}_{\bar{f}}$ respectively. By assumption there exists $a_f \in \mathrm{Aut}_{G_f}(\bar{G}_f, \bar{G}_{\bar{f}})$ that simultaneously induces $a_{d_0(f)}$ and $a_{d_1(f)}$. It follows that the set $\phi = \{\phi_i, \phi_e \mid i \in I, e \in \vec{E}\}$ where

$$\phi_i = \mathrm{id}_{G_i} \text{ for all } i \in I \text{ and } \qquad \phi_e = \begin{cases} a_f & \text{if } e = f \\ \mathrm{id}_{G_e} & \text{else} \end{cases}$$

is an isomorphism $\mathscr{G}^{(1)} \to \mathscr{G}^{(2)}$.

Remark 2.6.

(a) The situation of Lemma 2.5 arises for instance if G_f equals $\bar{G}_f \oplus \bar{G}_{\bar{f}}$ (internal direct sum) or $\bar{G}_f * \bar{G}_{\bar{f}}$ (free product).

(b) The situation of Lemma 2.5 also arises if $G_f \cong \bar{G}_f \circ \bar{G}_{\bar{f}}$ (central product), and $\bar{G}_f \cong \bar{G}_{\bar{f}} \cong \mathrm{SL}_2(k)$ for some field k. Namely, the center of $\mathrm{SL}_2(k)$ equals $\{\pm 1\}$ and any automorphism of $\mathrm{SL}_2(k)$ will fix this. Therefore $\mathrm{Aut}_{G_f}(\bar{G}_f, \bar{G}_{\bar{f}}) \cong \mathrm{Aut}_{\bar{G}_f \times \bar{G}_{\bar{f}}}(\bar{G}_f, \bar{G}_{\bar{f}})$.

(c) As a consequence of Lemma 2.5, for the purposes of classifying amalgams of type $\mathscr{G}_0$ up to isomorphism, we can remove any edges f as in that lemma from $\vec{E}$ and assume that $\varphi_f = \psi_f$.

Definition 2.7. *A* weak graph of groups *is a pair* $(\mathbb{A}, \vec{\Gamma})$ *where* $\vec{\Gamma}$ *is a graph as above and* $\mathbb{A}$ *associates to each* $i \in I$ *a group* A_i *and to each directed edge* $e \in \vec{E}$ *a group* $A_e = A_{\bar{e}}$. *Moreover, for each edge we have a homomorphism* $\alpha_e \colon A_e \to A_{d_0(e)}$. *We call the weak graph of groups a* graph of groups *if each* α_e *is a monomorphism; these are graphs of groups in the sense of* [2] *and* [21]. *Since we shall work with a fixed graph* $\vec{\Gamma} = (I, \vec{E})$, *if we want to specify the groups of* $(\mathbb{A}, \vec{\Gamma})$ *we shall write* $\mathbb{A} = \{A_i, A_e, \alpha_e\}$ *rather than* $\mathbb{A} = \{A_i, A_e, \alpha_e \mid i \in I, e \in \vec{E}\}$.

Definition 2.8. *Consider the reference amalgam* $\mathscr{G}_0 = \{G_i, G_e, \psi_e\}$. *Recall that every amalgam of type* $\mathscr{G}_0$ *has the same target subgroups* $\bar{G}_e$. *Define a weak graph of groups as follows. For each* $e \in \vec{E}$ *we take* $A_{d_0(e)} = \mathrm{Aut}(G_{d_0(e)})$, $A_e = \mathrm{Aut}_{G_e}(\bar{G}_e, \bar{G}_{\bar{e}})$ *and* $\alpha_e = \mathrm{ad}(\psi_e)$. *The resulting weak graph of groups will be denoted by* $(\mathbb{A}_0, \vec{\Gamma})$ *and will be called the* reference graph of groups.

Remark 2.9. Since the maps α_e in Definition 2.8 are not necessarily monomorphic, so $(\mathbb{A}_0, \vec{\Gamma})$ is in general not a proper graph of groups in the sense of [2, 21] or in the more general language of complexes of groups as described and developed in [9, Part III.C].

Definition 2.10. *Given graphs of groups* $(\mathbb{A}^{(k)}, \vec{\Gamma})$ *for* $k = 1, 2$, *an* inner morphism *is a collection* $\phi = \{\phi_i, \phi_e \mid i \in I, e \in \vec{E}\}$ *of group homomorphisms* $\phi_i \colon A_i^{(1)} \to A_i^{(2)}$ *and* $\phi_e \colon A_e^{(1)} \to A_e^{(2)}$ *so that for each* $e \in \vec{E}$ *there exists an element* $\delta_e \in A_{d_0(e)}^{(2)}$ *so that*

$$\phi_{d_0(e)} \circ \alpha_e^{(1)} = \mathrm{ad}(\delta_e^{-1}) \circ \alpha_e^{(2)} \circ \phi_e.$$

Here $\mathrm{ad}(x)(y) = x^{-1}yx$. *We call an inner morphism* central *if* $\delta_e = 1$ *for all* $e \in \vec{E}$.

Definition 2.11. *Let* $(\mathbb{A}_0, \vec{\Gamma})$ *be a weak graph of groups, a* pointing *is a pair* $((\mathbb{A}, \vec{\Gamma}), \delta)$, *where* $\delta = \{\delta_e \mid e \in \vec{E}\}$ *is a collection of elements* $\delta_e \in A_{d_0(e)}$ *and* $(\mathbb{A}, \vec{\Gamma})$ *is a weak graph of groups obtained from* $(\mathbb{A}_0, \vec{\Gamma})$ *by setting* $\alpha'_e = \mathrm{ad}(\delta_e^{-1}) \circ \alpha_e$, *for each* $e \in \vec{E}$.

Lemma 2.12. *For any pointing $((\mathbb{A}, \vec{\Gamma}), \delta)$ of $(\mathbb{A}_0, \vec{\Gamma})$ the weak graphs of groups $(\mathbb{A}, \vec{\Gamma})$ and $(\mathbb{A}_0, \vec{\Gamma})$ are isomorphic.*

Definition 2.13. *An isomorphism between pointings $((\mathbb{A}, \vec{\Gamma}), \delta^{(k)})$ $(k = 1, 2)$ of a weak graph of groups $(\mathbb{A}_0, \vec{\Gamma})$ is an inner isomorphism ϕ of $(\mathbb{A}_0, \vec{\Gamma})$ such that there exist $m_i \in A_i$ and $m_e \in A_e$ so that $m_e = m_{\bar{e}}$ and with $\phi_i = \mathrm{ad}(m_i)$ and $\phi_e = \mathrm{ad}(m_e)$ for each $i \in I, e \in \vec{E}$ and we have*

$$\delta_e^{(1)}\alpha_e(m_e) = m_{d_0(e)}\delta_e^{(2)}. \tag{1}$$

We will say that $\{m_e, m_i \mid i \in I, e \in \vec{E}\}$ induces *the isomorphism.*

Definition 2.14. *If $\mathscr{G} = \{G_i, G_e, \varphi_e\}$ is an amalgam of type $\mathscr{G}_0$, we define an associated pointing $((\mathbb{A}, \vec{\Gamma}), \delta^{\mathscr{G}})$ of $\mathbb{A}_0$ as follows. If $e \in \vec{E}$ then $\delta_e^{\mathscr{G}} = \varphi_e^{-1} \circ \psi_e$ where of course by φ_e^{-1} we mean the inverse of the map $\varphi_e \colon G_{d_0(e)} \to \bar{G}_e$.*

Conversely if $((\mathbb{A}, \vec{\Gamma}), \delta)$ is a pointing of $\mathbb{A}_0$ then we define an amalgam $\mathscr{G}^\delta = \{G_i, G_e, \varphi_e^\delta\}$ of type $\mathscr{G}_0$ via $\varphi_e^\delta = \psi_e \delta_e^{-1}$.

Theorem 2.15. [6, Theorem 1] *The correspondence $\mathscr{G} \mapsto ((\mathbb{A}, \vec{\Gamma}), \delta^{\mathscr{G}})$ with inverse $((\mathbb{A}, \vec{\Gamma}), \delta) \mapsto \mathscr{G}^\delta$ yields a bijection between special isomorphism classes of amalgams of type $\mathscr{G}_0$ and isomorphism classes of pointings of $\mathbb{A}_0$.*

2.2 Amalgams and Fundamental Groups

Definition 2.16. *Call the weak graph of groups $(\mathbb{A}, \vec{\Gamma})$* bijective *if each α_e is bijective. Note that this means in particular that $(\mathbb{A}, \vec{\Gamma})$ is a graph of groups in the sense of* [2, 21].

In our classification of Curtis-Tits structures with admissible Dynkin diagram we shall demonstrate that we can restrict ourselves to pointings of bijective graphs of groups. In anticipation of this, in this subsection we make the following assumption:

(B) $\mathbb{A}_0$ is bijective.

Since $\mathbb{A}_0$ is a graph of groups in the more restricted sense of [2] we can use Definitions 2.17 and 2.18 introduced there.

Definition 2.17. *For a given graph of groups $(\mathbb{A}, \vec{\Gamma})$ we define the* path group *as follows*

$$\pi(\mathbb{A}) = ((*_{i \in I} A_i) * F(\vec{E}))/R$$

where $F(\vec{E})$ is the free group on the set $\vec{E}$, $*$ denotes free product and R is the following set of relations: for any $e \in \vec{E}$, we have

$$\begin{aligned} e\bar{e} &= \text{id} && \textit{and} \\ e \cdot \alpha_{\bar{e}}(a) \cdot \bar{e} &= \alpha_e(a) && \textit{for any } a \in A_e. \end{aligned} \tag{2}$$

Definition 2.18. *Given a graph of groups* $(\mathbb{A}, \vec{\Gamma})$, *a* path of length n *from* i_1 *to* i_{n+1} *in* $\mathbb{A}$ *is a sequence* $\gamma = (a_1, e_1, a_2, \dots, e_n, a_{n+1})$, *where* $e_1, \dots, e_n$ *is an edge path in* $\vec{\Gamma}$ *with vertex sequence* $i_1, \dots, i_{n+1}$ *and* $a_k \in A_{i_k}$ *for each* $k = 1, \dots, n+1$. *We call* γ reduced *if it has no returns. Since* $(\mathbb{A}, \vec{\Gamma})$ *is bijective this is equivalent to saying that* $e_{i+1} \neq \bar{e}_i$ *for any* $i = 1, \dots, n-1$. *The path* γ *defines an element* $|\gamma| = a_1 \cdot e_1 \cdot a_2 \cdots e_n \cdot a_{n+1} \in \pi(\mathbb{A})$. *We denote by* $\pi[i, j]$ *the collection of elements* $|\gamma|$, *where* γ *runs through all paths from* i *to* j *in* $\mathbb{A}$. *Concatenation induces a group operation on* $\pi(\mathbb{A}, i_0) = \pi[i_0, i_0]$ *and we call this group* the fundamental group of $\mathbb{A}$ with base point i_0.

Remark 2.19. One verifies easily that we can naturally identify

$$\pi_1(\vec{\Gamma}, i_0) = \pi(\mathbb{A}, i_0) \cap \langle e \mid e \in \vec{E} \rangle,$$

where the former denotes the fundamental group of the graph $\vec{\Gamma}$ with base point i_0.

Consider the bijective graph of groups $(\mathbb{A}_0, \vec{\Gamma})$ and let $\eta = e_1, \dots, e_n$ be a path from l to m in $\vec{\Gamma}$. Define $\beta_\eta \colon A_l \to A_m$ by setting $\beta_\eta(a) = \alpha_{\bar{e}_n} \circ \alpha_{e_n}^{-1} \circ \cdots \circ \alpha_{\bar{e}_1} \circ \alpha_{e_1}^{-1}(a)$, for each $a \in A_l$.

Lemma 2.20. *Any element* $|\gamma| \in \pi(\mathbb{A}_0, i_0)$ *can be uniquely realized as* $e_1 e_2 \cdots e_n g$ *for some edge path* $\eta = e_1, \dots, e_n$ *with* $d_0(e_1) = i_0 = d_1(e_n)$ *and* $g \in A_{i_0}$. *More precisely, if* $\gamma = (d_1, e_1, d_2, \dots, d_n, e_n, d_{n+1})$ *with* $d_k \in \mathbf{A}_{d_0(e_k)}$ *for* $k = 1, \dots, n$, *and* $d_{n+1} \in \mathbf{A}_{i_0}$, *then we have*

$$g = \beta_{\bar{\eta}}(d_1) \cdots \beta_{\bar{e}_n \bar{e}_{n-1}}(d_{n-1}) \cdot \beta_{\bar{e}_n}(d_n) d_{n+1}.$$

Proof. The first part is a special case of Corollary 1.13 in [2] since all maps α_e are surjective. The second part follows from the relations in Definition 2.17 and the definition of β_η above.

Definition 2.21. *If* $((\mathbb{A}, \vec{\Gamma}), \delta)$ *is a pointing of the graph of groups* $(\mathbb{A}_0, \vec{\Gamma})$ *then any edge path* $\eta = e_1, \dots, e_n$ *in* $\vec{\Gamma}$ *gives rise to a path in* $\mathbb{A}$ *via* $\eta \mapsto \eta_\delta = \delta_{e_1} e_1 \delta_{\bar{e}_1}^{-1} \delta_{e_2} \cdots e_{n-1} \delta_{\bar{e}_{n-1}}^{-1} \delta_{e_n} e_n \delta_{\bar{e}_n}^{-1}$. *Fixing a base point* i_0, *the map* $\eta \mapsto |\eta_\delta|$ *restricts to a monomorphism*

$$\Phi_\delta \colon \pi_1(\Gamma, i_0) \to \pi(\mathbb{A}_0, i_0),$$

where $\pi_1(\Gamma, i_0)$ *denotes the fundamental group of* Γ *with base point* i_0. *The image of this map is called* the fundamental group of the pointing *and denoted by* $\pi(\mathbb{A}, i_0, \delta)$.

Define the map $\Pi_\delta \colon \pi_1(\vec{\Gamma}, i_0) \to A_{i_0}$ *as follows: given an edge path* $\eta = e_1, \dots, e_n$ *with* $|\eta| \in \pi_1(\vec{\Gamma}, i_0)$, *let* $\gamma = \eta_\delta$ *and let* $\Pi_\delta(\eta) = g$, *where* γ *and* g *are related as in Lemma 2.20. Concretely:*

$$\Pi_\delta(\eta) = \beta_{\bar{\eta}}(\delta_{e_1}) \cdots \beta_{\bar{e}_n \bar{e}_{n-1}}(\delta_{\bar{e}_{n-2}}^{-1} \delta_{e_{n-1}}) \cdot \beta_{\bar{e}_n}(\delta_{\bar{e}_{n-1}}^{-1} \delta_{e_n}) \cdot \delta_{\bar{e}_n}^{-1}.$$

Lemma 2.22. *We have $\pi(\mathbb{A}_0, i_0) \cong \pi(\overrightarrow{\Gamma}, i_0) \ltimes A_{i_0}$. Let $H = \{\eta_j\}_{j \in J}$ be a base for the free group $\pi_1(\overrightarrow{\Gamma}, i_0)$. Then, both Φ_δ and its image $\pi_1(\mathbb{A}, i_0, \delta)$ are uniquely determined by the assignment $\eta \mapsto \Pi_\delta(\eta)$.*

Proof. Clearly as subgroups of $\pi_1(\mathbb{A}, i_0)$ we have $A_{i_0} \cap \pi_1(\overrightarrow{\Gamma}, i_0) = \{1\}$. From Lemma 2.20 it follows in particular, that $\pi_1(\overrightarrow{\Gamma}, i_0)$ normalizes A_{i_0}. This proves the first statement. Now given the base H, and the assignment $\eta \mapsto \Pi_\delta(\eta)$, using Lemma 2.20 we see that $\Phi_\delta(\eta) = \eta \Pi_\delta(\eta)$ and $\pi_1(\mathbb{A}, i_0, \delta) = \langle \eta \Pi_\delta(\eta) \mid \eta \in H \rangle \leq \pi_1(\mathbb{A}, i_0)$. Conversely, given $\eta \in H$ since the composition $\pi_1(\mathbb{A}, i_0, \delta) \hookrightarrow \pi_1(\overrightarrow{\Gamma}, i_0) \ltimes A_{i_0} \to \pi_1(\overrightarrow{\Gamma}, i_0)$ yields an isomorphism, there is a unique $a \in A_{i_0}$ such that $\eta a \in \pi_1(\mathbb{A}, i_0, \delta)$. Thus, we must have $\eta \mapsto \Pi_\delta(\eta) = a$.

Corollary 2.23. *The fundamental groups of pointings are bijectively parametrized by the homomorphisms $\pi_1(\overrightarrow{\Gamma}, i_0) \to A_{i_0}$.*

Proof. The data $\{(\eta, \Pi_\delta(\eta)) \mid \eta \in \pi_1(\overrightarrow{\Gamma}, i_0)\}$ from Lemma 2.22 bijectively parametrizes such homomorphisms.

The following theorem appears as Theorem 3.7 in [6].

Theorem 2.24. *Two pointings of $\mathbb{A}_0$ are isomorphic if and only if they have the same fundamental group.*

Remark 2.25. Note that all such fundamental groups are isomorphic to the fundamental group $\pi_1(\Gamma, i_0)$, but not all of them are the same.

Corollary 2.26. *The isomorphism classes of amalgams of type $\mathscr{G}_0$ are in bijective correspondence with the homomorphisms $\pi_1(\overrightarrow{\Gamma}, i_0) \to A_{i_0}$.*

Proof. This follows from Theorems 2.15 and 2.24, and Corollary 2.23.

3 CT-Structures

In this section we introduce the notion of a CT-structure over a commutative field and define its category. Throughout the paper k will be a commutative field, which after Lemma 3.5 will be assumed to have at least four elements.

3.1 CT-Structures and Standard Pairs

Definition 3.1. *Let Γ be a Dynkin diagram over a set I and let $K(I)$ be the complete graph on I. Let $\overrightarrow{K(I)} = (I, \overrightarrow{F})$ be the oriented graph associated to $K(I)$ and let*

$\overrightarrow{\Gamma} = (I, \overrightarrow{E})$ *be the oriented graph associated to* Γ*; that is,* $e \in \overrightarrow{E}$ *if and only if the subdiagram of* Γ *induced on* e *is not of type* $A_1 \times A_1$.

A Curtis-Tits (CT) structure *over* k *with Dynkin diagram* Γ *is an amalgam* $\mathscr{G} = \{G_i, G_e, \varphi_e \mid i \in I, e \in \overrightarrow{F}\}$ *over* $\overrightarrow{K(I)}$ *such that, for every* $e \in \overrightarrow{F}$, $\{\bar{G}_e, G_e, \bar{G}_{\bar{e}}\}$ *is a Curtis-Tits amalgam with respect to a BN-pair of type* Γ_e *for* G_e*. Here* Γ_e *denotes the subdiagram of* Γ *induced on the edge* e*.*

Definition 3.2. *A* simply laced Dynkin diagram *over the set* I *is a simple graph* $\Gamma = (I, E)$*. That is,* Γ *has vertex set* I*, and an edge set* E *that contains no loops or double edges.*

In the present paper we shall restrict ourselves to the case where Γ is simply-laced. To specify the Curtis-Tits amalgam in each rank-2 group of a CT-structure over Γ, we make use of standard pairs.

Definition 3.3. *Let* V *be a vector space of dimension* 3 *over* k*. We call* (S_1, S_2) *a* standard pair *for* $S = \mathrm{SL}(V)$ *if there are decompositions* $V = U_i \oplus V_i$, $i = 1, 2$, *with* $\dim(U_i) = 1$ *and* $\dim(V_i) = 2$ *such that* $U_1 \subseteq V_2$ *and* $U_2 \subseteq V_1$ *and* S_i *is the subgroup of elements in* S *that centralize* U_i *and preserve* V_i*.*

One also calls S_1 *a* standard complement *of* S_2 *and vice-versa. We set* $D_1 = N_{S_1}(S_2)$ *and* $D_2 = N_{S_2}(S_1)$*. A simple calculation shows that* D_i *is a maximal torus in* S_i*, for* $i = 1, 2$*. In general if* $G \cong \mathrm{SL}_3(k)$*, then* (G_1, G_2) *is a standard pair for* G *if there is an isomorphism* $\psi\colon G \to S$ *such that* $\psi(G_i) = S_i$ *for* $i = 1, 2$*.*

Definition 3.4. *Given a standard pair* (S_1, S_2)*, a* standard basis *for* (S_1, S_2) *is an ordered basis* $\mathcal{V} = (v_1, v_2, v_3)$ *of* V *such that* $V_1 = \langle v_1, v_2\rangle$, $U_1 = \langle v_3\rangle$, $U_2 = \langle v_1\rangle$, *and* $V_2 = \langle v_2, v_3\rangle$*.*

Identifying S with $\mathrm{SL}_3(k)$ via its left action on V with respect to $\mathcal{V}$, yields

$$S_1 = \left\{ \begin{pmatrix} A & 0 \\ 0 & 1 \end{pmatrix} \middle| \; A \in \mathrm{SL}_2(k) \right\} \text{and} \qquad S_2 = \left\{ \begin{pmatrix} 1 & 0 \\ 0 & A \end{pmatrix} \middle| \; A \in \mathrm{SL}_2(k) \right\}$$

so that

$$D_1 = \left\{ \begin{pmatrix} a & 0 & 0 \\ 0 & a^{-1} & 0 \\ 0 & 0 & 1 \end{pmatrix} \middle| \; a \in k^* \right\} \text{ and } \qquad D_2 = \left\{ \begin{pmatrix} 1 & 0 & 0 \\ 0 & a & 0 \\ 0 & 0 & a^{-1} \end{pmatrix} \middle| \; a \in k^* \right\}.$$

Lemma 3.5. *Let* S_1 *and* S_2 *be a standard pair for* $S = \mathrm{SL}_3(k)$*, where* k *has at least four elements.*

(a) The group D_1 *determines a standard basis* $\mathcal{V} = (v_1, v_2, v_3)$ *up to a permutation of* $\{v_1, v_2\}$ *and a linear diagonal transformation with respect to* $\mathcal{V}$*.*
(b) The pair (D_1, D_2) *and hence, a fortiori, the pair* (S_1, S_2) *determines a standard basis* $\mathcal{V}$ *for* (S_1, S_2) *up to a linear diagonal transformation with respect to* $\mathcal{V}$*.*
(c) The group S_1 *has exactly one standard complement* $S_2' \neq S_2$ *normalized by* D_1*.*

Proof. Since k has at least four elements, D_1 uniquely determines three 1-dimensional eigenspaces and S_1 fixes all vectors in exactly one of these eigenspaces, say E_1. In the notation above, these are $E_1 = U_1$, U_2 and $V_1 \cap V_2$. Any standard basis $\mathcal{V} = (v_1, v_2, v_3)$ must satisfy $U_2 = \langle v_1 \rangle$, $V_1 \cap V_2 = \langle v_2 \rangle$, and $U_1 = \langle v_3 \rangle$. Thus (3.5) and hence (3.5) follow. To see (3.5), note that any standard complement S_2 to S_1 that is normalized by D_1 is completely determined by the eigenspace $E \neq E_1$ that it fixes vector-wise. As we saw, there are two choices.

We will need the following lemma.

Lemma 3.6. *With the notations above, $D_1 = C_{S_1}(D_2)$ and $D_2 = C_{S_2}(D_1)$. Moreover, D_2 is the only torus in S_2 that is normalized by D_1.*

Proof. Note that if T is a torus in S_2 then $N_S(T)$ is the set of monomial matrices so $N_{S_1}(T)$ only contains one torus which is $C_{S_1}(T)$. The conclusion follows.

Definition 3.7. *Let $\Gamma = (I, E)$ be a simply laced Dynkin diagram. A* Curtis-Tits structure *over Γ is a non-collapsing amalgam $\mathcal{G}(\Gamma) = \{G_i, G_e, \varphi_e \mid i \in I, e \in \overrightarrow{F}\}$ over $\overrightarrow{K(I)}$ such that*

(CT1) for each vertex i, $G_i = \mathrm{SL}_2(k) = \mathrm{SL}(W_i)$, for some 2-dimensional vector space W_i over k, and for each edge $e \in \overrightarrow{F}$,

$$G_e \cong \begin{cases} \mathrm{SL}(V_e) & \text{if } e \in \overrightarrow{E} \\ \bar{G}_e \circ \bar{G}_{\bar{e}} & \text{if } e \notin \overrightarrow{E}, \end{cases}$$

where $V_e = V_{\bar{e}}$ is a 3-dimensional vector space over k, $\bar{G}_e = \varphi_e(G_{d_0(e)})$, and $\circ$ denotes central product,

(CT2) if $e \in \overrightarrow{E}$ then $(\bar{G}_e, \bar{G}_{\bar{e}})$ is a standard pair in G_e.

Remark 3.8. In view of Remark 2.6 for the purposes of classifying CT-structures with simply-laced Dynkin diagram, we may in fact classify the subamalgams defined over $\overrightarrow{\Gamma}$ only. In practice this means that, for $f \in \overrightarrow{F} - \overrightarrow{E}$, we can let φ_f be the same in every amalgam of this type.

Recall that we shall only consider admissible Dynkin diagrams, that is, graphs that are connected and have no circuits of length ≤ 3.

From now on $\Gamma = (I, E)$ will be an admissible Dynkin diagram and our reference amalgam $\mathcal{G}_0 = \mathcal{G}(\Gamma) = \{G_i, G_e, \psi_e \mid i \in I, e \in \overrightarrow{F}\}$ will be a non-collapsing Curtis-Tits structure over Γ.

We now focus on the following subgroups:

Definition 3.9. *For any $e \in \overrightarrow{E}$, let*

$$D_e = N_{G_e}(\bar{G}_{\bar{e}}) \cap \bar{G}_e$$

As noted in Definition 3.3, D_e is a torus in $\bar{G}_e \leq G_e$. By Lemma 3.6 D_e is the only torus in $\bar{G}_e$ normalized by $D_{\bar{e}}$.

Lemma 3.10. *If $e \in \overrightarrow{E}$, then D_e and $D_{\bar{e}}$ are contained in a unique common maximal torus $D_{e,\bar{e}}$ of G_e.*

Proof. Clearly in any completion of the amalgam, both D_e and $D_{\bar{e}}$ normalize $\bar{G}_e$ and $\bar{G}_{\bar{e}}$ so we have $D_e, D_{\bar{e}} \leq N_{G_e}(\bar{G}_e) \cap N_{G_e}(\bar{G}_{\bar{e}}) = D_{e,\bar{e}}$, which is the required maximal torus.

Definition 3.11. (property (D)) *Let $\mathscr{G} = \{G_i, G_e, \varphi_e \mid i \in I, e \in \overrightarrow{F}\}$ be a Curtis-Tits structure over Γ. We say that $\mathscr{G}$ has property (D) if, for each $i \in I$ there is a subgroup $D_i \leq G_i$ such that for any pair of edges $e, f \in \overrightarrow{E}$ with $d_0(f) = i = d_0(e)$ and any completion (G, ϕ), we have a commutative diagram of isomorphisms*

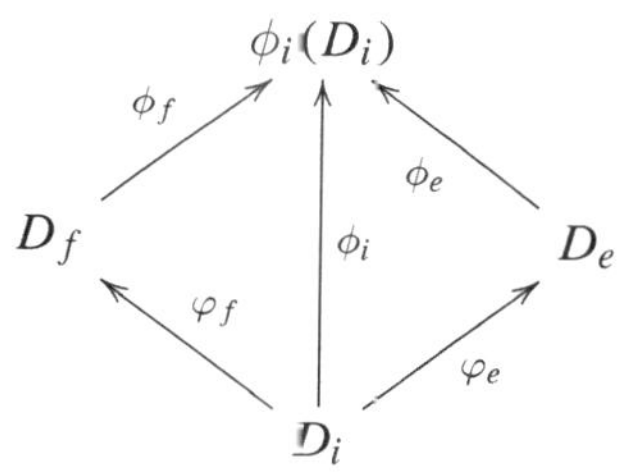

Property (D) is equivalent to requiring that for all edges $e, f \in \overrightarrow{E}$ with $d_0(f) = d_0(e)$ we have an isomorphism $\varphi_f \circ \varphi_e^{-1} \colon D_e \to D_f$.

That $\mathscr{G}_0$ is non-collapsing has the following consequence.

Lemma 3.12. *Let $\mathscr{G} = \{G_i, G_e, \varphi_e\}$ be a CT-structure with admissible Dynkin diagram Γ. Suppose that φ_e is injective for every $e \in \overrightarrow{E}$. Then, $\mathscr{G}$ has property (D).*

Proof. (See also [15]) Let $e, f \in \overrightarrow{E}$ and $i \in I$ be any two edges with $d_0(f) = i = d_0(e)$. By Definition 3.3 $\phi_{\bar{f}}(D_{\bar{f}})$ normalizes $\phi_f(\bar{G}_f) = \phi_i(G_i) = \phi_e(\bar{G}_e)$. It follows from the fact that the nodes $d_0(\bar{f})$ and $d_0(\bar{e})$ are not connected in Γ that $\phi_{\bar{f}}(D_{\bar{f}})$ commutes with $\phi_{\bar{e}}(D_{\bar{e}})$. Thus, $\phi_{\bar{f}}(D_{\bar{f}})$ normalizes $\phi_e(\bar{G}_e)$, and hence $\phi_e(G_e)$, while centralizing $\phi_{\bar{e}}(\bar{G}_{\bar{e}})$. Hence, by Lemma 3.6, $\phi_{\bar{f}}(D_{\bar{f}})$ normalizes the torus $\phi_e(D_e) = C_{\phi_e(\bar{G}_e)}(\phi_{\bar{e}}(D_{\bar{e}}))$ of $\phi_e(\bar{G}_e) = \phi_f(\bar{G}_f)$. By the second part of Lemma 3.6 $\phi_f(D_f)$ is the only torus in $\phi_f(G_f)$ normalized by $\phi_{\bar{f}}(D_{\bar{f}})$ and so $\phi_e(D_e) = \phi_f(D_f)$. The diagram is now commutative since (G, ϕ) is a completion of $\mathscr{G}_0$.

From now on we shall only consider Curtis-Tits amalgams with property (D). This allows us to make the following definition.

Definition 3.13. *For any $i \in I$, let $e \in \overrightarrow{E}$ be such that $d_0(e) = i$ (e exists since Γ is connected). Let D_i be the torus of G_i such that $\varphi_e(D_i) = D_e$ (as in Definition 3.9). Since $\mathscr{G}$ has property (D), D_i does not depend on the choice of e.*

Moreover, for $e \in \overrightarrow{E}$, we let $D_{e,\bar{e}}$ be the maximal torus generated by D_e and $D_{\bar{e}}$ in G_e (cf. Lemma 3.10).

In case $f \in \overrightarrow{F} - \overrightarrow{E}$, we set

$$D_f = \varphi_f(D_{d_0(f)}) \qquad \text{and} \qquad D_{f,\bar{f}} = D_f D_{\bar{f}} \le G_f.$$

We collect these groups in a set

$$\mathscr{D}(\mathscr{G}) = \{D_i, D_f, D_{f,\bar{f}} \mid i \in I, f \in \overrightarrow{F}\}.$$

Note that D_e and $D_{e,\bar{e}}$ ($e \in \overrightarrow{E}$) only depend on the type of $\mathscr{G}$, D_i ($i \in I$), and hence D_f and $D_{f,\bar{f}}$ ($f \in \overrightarrow{E} - \overrightarrow{F}$) depend, in addition, on the particular connecting maps φ_e.

Definition 3.14. *Note that a torus in $\mathrm{SL}_2(k)$ uniquely determines a pair of opposite root groups X^+ and X^-. We now choose one root group X_i^+ normalized by the torus D_i of G_i for each i. We call the CT-structure $\mathscr{G}_0$ having property (D) an orientable Curtis-Tits structure if it admits a system $\{X_i^+ \mid i \in I\}$ of root groups as above such that for any $f \in \overrightarrow{F}$ with $d_0(f) = i$ and $d_1(f) = j$, the groups $\varphi_f(X_i^+)$ and $\varphi_{\bar{f}}(X_j^+)$ are contained in a common Borel subgroup $B_f^+ = B_{\bar{f}}^+$ of $G_f = G_{\bar{f}}$.*

Lemma 3.15. *With the notation of Definition 3.14, assume that $\mathscr{G}$ is an orientable CT-structure. Suppose that $\{X_i^+ \mid i \in I\}$ is chosen. For each $i \in I$ select the negative root group X_i^-. Then, for any $e \in \overrightarrow{E}$, $X_{d_0(e)}^-$ and $X_{d_1(e)}^-$ are contained in a common Borel subgroup B_e^- that is opposite to B_e^+.*

Proof. Since $\mathscr{G}$ is orientable, it has property (D). The tori D_e and $D_{\bar{e}}$ determine a unique torus in G_e hence a unique apartment of the building associated to $\Delta_e = (G_e, B_e^+, N_{G_e}(D_{e,\bar{e}}))$. The torus D_e (resp. $D_{\bar{e}}$) determines a pair of opposite panels of Δ_e. These pairs of opposite panels intersect in a pair of opposite chambers. One corresponds to B_e^+ and the other to $B_{\bar{e}}^+ = B_e^-$. The conclusion follows.

3.2 Automorphisms of $\mathscr{G}(A_2)$

Let V be a (left) vector space of dimension n over k. Let $G = \mathrm{SL}(V)$ act on V as the matrix group $\mathrm{SL}_n(k)$ with respect to some fixed basis $\mathcal{V} = \{v_i \mid i = 1, 2, \ldots, n\}$. Let $\omega \in \mathrm{Aut}(\mathrm{SL}_n(k))$ be the automorphism given by

$$A \mapsto {}^tA^{-1}$$

where ${}^{t}A$ denotes the transpose of A.

Let $\Phi = \{(i, j) \mid 1 \le i \ne j \le n\}$. For any $(i, j) \in \Phi$ and $\lambda \in k$, we define the *root group* $X_{i,j} = \{X_{i,j}(\lambda) \mid \lambda \in k\}$, where $X_{i,j}(\lambda)$ acts as

$$\begin{array}{ll} v_j \mapsto v_j + \lambda v_i & \text{and} \\ v_k \mapsto v_k & \text{for all } k \ne j. \end{array}$$

Let $\Phi_+ = \{(i, j) \in \Phi \mid i < j\}$ and $\Phi_- = \{(i, j) \in \Phi \mid j < i\}$. We call $X_{i,j}$ *positive* if $(i, j) \in \Phi_+$ and *negative* otherwise. Let H be the torus of diagonal matrices in $\mathrm{SL}_n(k)$ and for $\varepsilon \in \{+, -\}$, let $X_\varepsilon = \langle X_{i,j} \mid (i, j) \in \Phi_\varepsilon\rangle$ and $B_\varepsilon = H \ltimes X_\varepsilon$.

Let $\Gamma L_n(k)$ be the group of all semilinear automorphisms of the vector space V and let $P\Gamma L_n(k) = \Gamma L_n(k)/Z(\Gamma L_n(k))$. Then $\Gamma L_n(k) \cong \mathrm{GL}_n(k) \rtimes \mathrm{Aut}(k)$, where we view $t \in \mathrm{Aut}(k)$ as an element of $\Gamma L_n(k)$ by setting $((a_{i,j})_{i,j=1}^n)^t = (a_{i,j}^t)_{i,j=1}^n$. The automorphism group of $\mathrm{SL}_n(k)$ can be expressed using $P\Gamma L_n(k)$ and ω as follows [20].

Lemma 3.16.

$$\mathrm{Aut}(\mathrm{SL}_n(k)) = \begin{cases} P\Gamma L_n(k) & \textit{if } n = 2; \\ P\Gamma L_n(k) \rtimes \langle\omega\rangle & \textit{if } n \ge 3. \end{cases}$$

Remark 3.17. Note that the actions of $\mathrm{Aut}(k)$ and ω are given with respect to the basis $\mathcal{V}$ of V.

4 Classification of CT-Structures with Simply-Laced Diagram

Let $\mathscr{G}$ be a Curtis-Tits structure with property (D) over a simply-laced diagram $\Gamma = (I, E)$. This is really an amalgam $\mathscr{G} = \{G_i, G_f, \varphi_f \mid i \in I, f \in \vec{F}\}$ over the complete graph $\overrightarrow{K(I)} = (I, \vec{F})$. In Sect. 2 we saw that amalgams of type $\mathscr{G}$ are classified by pointings of an associated weak graph of groups $(\mathbb{A}, \overrightarrow{K(I)})$. However, by Remark 2.6, for the purposes of classifying CT-structures of type $\mathscr{G}$ we may discard all edges that do not belong to the Dynkin diagram and classify the amalgams whose type is the subamalgam of $\mathscr{G}$ over the Dynkin diagram $\vec{\Gamma}$ instead.

In this section we shall take advantage of the special nature of CT-structures with simply laced diagram to define a graph of groups that is equally effective in classifying CT-structures, but whose vertex and edge groups are smaller than those of the weak graph of groups $\mathbb{A}$.

We shall only wish to classify those amalgams $\mathscr{G}$ of type $\mathscr{G}_0$ over Γ that have property (D). Using Lemma 3.16 one finds that, for each $i \in I$ and each $e \in \vec{E}$,

Corollary 4.1.

$$\mathrm{Aut}_{G_i}(D_i) \cong T_i \rtimes (\langle\omega\rangle \times \mathrm{Aut}(k))$$
$$\mathrm{Aut}_{G_e}(\bar{G}_e, \bar{G}_{\bar{e}}) \cong T_{e,\bar{e}} \rtimes (\langle\omega\rangle \times \mathrm{Aut}(k)).$$

where $T_i \trianglelefteq \mathrm{Aut}_{G_i}(D_i)$ and $T_{e,\bar{e}} \trianglelefteq \mathrm{Aut}_{G_e}(\bar{G}_e, \bar{G}_{\bar{e}})$ are the normal subgroups of diagonal automorphisms in the respective groups.

Note that the complements to T_i and $T_{e,\bar{e}}$ are both isomorphic to $\mathbb{Z}_2 \times \mathrm{Aut}(k)$, and are unique up to conjugation by elements in T_i and $T_{e,\bar{e}}$ respectively.

Definition 4.2. *Let $\mathscr{G}$ be a CT-structure with simply-laced diagram Γ that has property (D). As in Definition 3.13 $\mathscr{G}$ determines a unique collection of tori $\mathscr{D}(\mathscr{G}) = \{D_i, D_f, D_{f,\bar{f}} \mid i \in I, f \in \overrightarrow{F}\}$. With the notation from Corollary 4.1, for $f \in \overrightarrow{F} - \overrightarrow{E}$, we let*

$$T_f = \mathrm{ad}(\varphi_f^{-1})(T_{d_0(f)}) \qquad \text{and} \qquad T_{f,\bar{f}} = T_f \times T_{\bar{f}} \leq \mathrm{Aut}_{G_f}(D_{f,\bar{f}}).$$

From Corollary 4.1 it is clear that the set

$$\mathscr{T}(\mathscr{G}) = \{T_i, T_f, T_{f,\bar{f}} \mid i \in I, f \in \overrightarrow{F}\},$$

is also uniquely determined by $\mathscr{G}$.

Lemma 4.3. *Given any collection $\{\tau_i \in T_i \mid i \in I\}$, there exist unique $\tau_{f,\bar{f}} \in T_{f,\bar{f}}$ such that $\tau = \{\tau_i, \tau_{f,\bar{f}} \mid i \in I, f \in \overrightarrow{F}\}$ induces an automorphism of $\mathscr{G}$.*

Proof. First we note that, for each $f \in \overrightarrow{F}$, $\bar{\tau}_f = \mathrm{ad}(\varphi_f^{-1})(\tau_{d_0(f)})$ is a diagonal (linear) automorphism of $\bar{G}_f \leq G_f$. By definition, $\bar{\tau}_f \circ \varphi_f = \tau_{d_0(f)} \circ \varphi_f$.

If $f \in \overrightarrow{F} - \overrightarrow{E}$, then we define $\tau_{f,\bar{f}}$ to be the product $\bar{\tau}_f \times \bar{\tau}_{\bar{f}}\colon \bar{G}_f \times \bar{G}_{\bar{f}} \to \bar{G}_f \times \bar{G}_{\bar{f}}$. If $f \in \overrightarrow{E}$, suppose that with respect to some basis $\{v_1, v_2, v_3\}$ of eigenvectors for D_f and $D_{\bar{f}}$ we have $\bar{\tau}_f = \mathrm{diag}\{a, b, 1\}$ and $\bar{\tau}_{\bar{f}} = \mathrm{diag}\{1, c, d\}$, then let $\tau_{f,\bar{f}} = \mathrm{diag}\{ac, bc, bd\}$. In either case $\tau_{f,\bar{f}} \in \mathrm{Aut}_{G_f}(D_{f,\bar{f}})$. That τ commutes with the connecting maps follows easily.

Definition 4.4. *(Basis for $\mathscr{G}$) Let $\mathscr{G} = \{G_i, G_f, \varphi_f \mid i \in I, f \in \overrightarrow{F}\}$ be a CT structure with simply-laced diagram $\Gamma = (I, E)$ that has property (D). A* basis *of $\mathscr{G}$ is a collection $\mathcal{V} = \{\mathcal{V}_e, \mathcal{V}_{\bar{e}} \mid e \in \overrightarrow{E}\}$ so that $\mathcal{V}_e = (v_1^e, v_2^e, v_3^e)$ is an (ordered) standard basis in V_e for the standard pair $(\bar{G}_e, \bar{G}_{\bar{e}})$ of G_e (see Definition 3.4) and $\mathcal{V}_{\bar{e}} = (v_1^{\bar{e}} = v_3^e, v_2^{\bar{e}} = v_2^e, v_3^{\bar{e}} = v_1^e)$ is the same basis but the ordering is reversed. Note that each 1-space spanned by an element of $\mathcal{V}_e$ is stabilized by D_e and $D_{\bar{e}}$. The* edge reversal map *is the element ρ_e of $\mathrm{GL}(V_e)$ defined by $v_i^e \leftrightarrow v_i^{\bar{e}}$, for $i = 1, 2, 3$.*

Definition 4.5. *(Concrete graph of groups) Let $\mathscr{G}$ be a CT-structure with simply-laced Dynkin diagram $\Gamma = (I, E)$ that has property (D). Let $\mathcal{V}$ be a basis for $\mathscr{G}$ as in Definition 4.4. For each $i \in I$ let $\mathcal{W}_i = (w_1^i, w_2^i)$ be a basis for W_i identifying $G_i = \mathrm{SL}_2(k)$ with $\mathrm{SL}(W_i)$ such that D_i is diagonal with respect to $\mathcal{W}_i$.*

Let $\psi_e\colon G_{d_0(e)} \to \bar{G}_e \le G_e$ be the isomorphism induced by the linear map that takes the ordered basis $(w_1^{d_0(e)}, w_2^{d_0(e)})$ to (v_1^e, v_2^e). Explicitly, if $G_{d_0(e)}$ (resp. G_e) are identified with $\mathrm{SL}_2(k)$ (resp. $\mathrm{SL}_3(k)$) via the bases $\mathcal{W}_{d_0(e)}$ (resp. $\mathcal{V}_e$), then ψ_e is given by

$$\psi_e\colon G_{d_0(e)} = \mathrm{SL}_2(k) \to \mathrm{SL}_3(k) = G_e$$
$$A \mapsto \begin{pmatrix} A & 0 \\ 0 & 1 \end{pmatrix}.$$

This defines a graph of groups $(\mathbb{K}, \vec{\Gamma})$ in the following way. We let $\mathbb{K} = \{K_i, K_e, \psi_e \mid i \in I, e \in \vec{E}\}$ where K_i is the complement in $\mathrm{Aut}_{G_i}(D_i)$ to T_i defined with respect to $\mathcal{W}_i = (w_1^i, w_2^i)$ and K_e is the complement in $\mathrm{Aut}_{G_e}(\bar{G}_e, \bar{G}_{\bar{e}})$ to $T_{e,\bar{e}}$, defined by $\mathcal{V}_e$ (See Lemma 3.16). Note that this yields $K_e = K_{\bar{e}}$, as desired. Finally, we may define the map $\kappa_e\colon K_e \to K_{d_0(e)}$ as given by the restriction of $\mathrm{ad}(\psi_e)$ to K_e, as the following lemma shows. We call $(\mathbb{K}, \vec{\Gamma})$ the concrete graph of groups *for $\mathscr{G}$ associated with $\mathcal{V}$.*

Remark 4.6. (on Definition 4.4) Note that by Lemma 3.5 each $\mathcal{V}_e$ is uniquely determined by the type of $\mathscr{G}$ up to a linear transformation that is diagonal with respect to $\mathcal{V}_e$; it is independent of the connecting maps φ_e.

Remark 4.7. (on Definition 4.5):

(a) Note that as in the proof of Lemma 3.5, since k has at least four elements, $\mathcal{W}_i$ is uniquely determined up to a permutation and a diagonal linear transformation.
(b) The choice of $\mathcal{V}$ and $\mathcal{W}$ does not affect the maps φ_f ($f \in \vec{F}$); it only affects the way they are represented with respect to these bases. In particular, $\varphi_e^{-1}\psi_e \in \mathrm{Aut}_{G_{d_0(e)}}(D_{d_0(e)}) = T_{d_0(e)} \rtimes K_{d_0(e)}$.
(c) The concrete graph of groups $(\mathbb{K}, \vec{\Gamma})$ in Definition 4.5 depends on $\mathscr{G}$, $\mathcal{V}$ and $\mathcal{W}$, but is not related to the particular connecting maps φ_f ($f \in \vec{E}$). Lemma 4.9 below shows that $\mathbb{K}$ is uniquely determined by the *type of $\mathscr{G}$* up to central isomorphism.

Remark 4.8. Note that if $(\mathbb{A}, \vec{\Gamma})$ is the weak graph of groups associated to $\mathscr{G}$ as in Sect. 2, where we have removed all edges $f \in \vec{E} - \vec{F}$ using Lemma 2.5 then, for all $i \in I$ and $e \in \vec{E}$ we have

$$K_i \le A_i, \qquad K_e \le A_e, \text{ and } \qquad \kappa_e = \alpha_e|_{K_e}.$$

Lemma 4.9. *Let $\mathcal{G}$ and $\mathcal{G}'$ be CT-structures over Γ with property (D) and let $(\mathbb{K}, \vec{\Gamma})$ and $(\mathbb{K}', \vec{\Gamma})$ be the concrete graphs of groups associated as in Definition 4.5. If $\mathcal{G}'$ is of type $\mathcal{G}$, then there is a central isomorphism $(\mathbb{K}, \vec{\Gamma}) \to (\mathbb{K}', \vec{\Gamma})$ that only depends on a base change $(\mathcal{V}, \mathcal{W}) \mapsto (\mathcal{V}', \mathcal{W}')$.*

Proof. First note that $\mathcal{G}$ and $\mathcal{G}'$ have the same groups G_i and G_e for $i \in I$ and $e \in \vec{E}$. Now the construction of $(\mathbb{K}, \vec{\Gamma})$ does not depend on the connecting maps φ_f and φ'_f, but only involves the maps ψ_e, which in turn depend uniquely on the basis $\mathcal{V}$ for $\mathcal{G}$ and the collection $\mathcal{W} = \{\mathcal{W}_i = (w_1^i, w_2^i) \mid i \in I\}$ of bases chosen for the W_i. We now show that any other choice of $\mathcal{V}$ and $\mathcal{W}$ merely induces a central isomorphism between the resulting graphs of groups. Let $\mathcal{V}'$ and $\mathcal{W}'$ be another choice of a basis for $\mathcal{G}$ and the W_i's and let $(\mathbb{K}', \vec{\Gamma})$ be the resulting concrete graph of groups. For each $i \in I$, let $t_i \in \mathrm{Aut}(G_i)$ be induced by the linear map sending the ordered basis $\mathcal{W}_i$ to $\mathcal{W}'_i$ and for each $e \in \vec{E}$, let $t_e \in \mathrm{Aut}(G_e)$ be induced by the linear map sending the ordered basis $\mathcal{V}_e$ to $\mathcal{V}'_e$. Then the following diagram is commutative.

$$\begin{array}{ccc} G_e & \xrightarrow{\ t_e\ } & G_e \\ {\scriptstyle \psi_e}\uparrow & & \uparrow{\scriptstyle \psi'_e} \\ G_{d_0(e)} & \xrightarrow{\ t_{d_0(e)}\ } & G_{d_0(e)} \end{array}$$

Since the bases defining the complements K_e, K'_e, $K_{d_0(e)}$ and $K'_{d_0(e)}$ all correspond via the maps in this diagram, also these complements themselves correspond to each other via the adjoint maps. This shows that the map $\phi\colon (\mathbb{K}', \vec{\Gamma}) \to (\mathbb{K}, \vec{\Gamma})$ given by $\phi = \{\phi_e = \mathrm{ad}(t_e)\colon K'_e \to K_e, \phi_i = \mathrm{ad}(t_i)\colon K'_i \to K_i \mid i \in I, e \in \vec{E}\}$ is an isomorphism which is central because all maps are given as conjugation by a (linear) base-change.

It is our aim to classify all CT structures of type $\mathcal{G}_0$ with property (D) up to isomorphism. As a consequence of Remarks 4.6 and 4.7 and Lemma 4.9, we can then fix the following:

$$\begin{aligned}
\vec{\Gamma} &= (I, \vec{E}),\\
\mathcal{G}_0 &= \{G_i, G_e, \psi_e \mid i \in I, e \in \vec{E}\},\\
\mathcal{D}_0 &= \{D_i, D_f, D_{f,\bar{f}} \mid i \in I, f \in \vec{F}\},\\
\mathcal{T}_0 &= \{T_i, T_f, T_{f,\bar{f}} \mid i \in I, f \in \vec{F}\},\\
\mathcal{V}_0 &= \{\mathcal{V}_e, \mathcal{V}_{\bar{e}} \mid e \in \vec{E}\},\\
\mathcal{W}_0 &= \{\mathcal{W}_i \mid i \in I\},\\
\mathbb{K}_0 &= \{K_i, K_e, \kappa_e \mid i \in I, e \in \vec{E}\}.
\end{aligned}$$

where $\mathscr{G}_0$ is a Curtis-Tits structure having property (D) with simply-laced diagram Γ, $\mathscr{D}_0$ is the collection of groups uniquely determined by $\mathscr{G}_0$ as in Definition 3.13, $\mathscr{T}$ is the collection of groups uniquely determined by $\mathscr{G}_0$ as in Definition 4.2, $\mathcal{V}_0$ is a basis for $\mathscr{G}_0$, and $\mathcal{W}_0$, ψ_e, and $\mathbb{K}_0$ are as defined in Definition 4.5.

We note that in order to describe all possible CT structures $\mathscr{G}$ of type $\mathscr{G}_0$ with property (D) up to isomorphism, in view of Remarks 2.6, 4.6 and 4.7, it suffices to make a choice for the maps $\{\varphi_e \mid e \in \overrightarrow{E}\}$ in such a way that $\varphi_e^{-1}\psi_e \in \mathrm{Aut}_{G_{d_0(e)}}(D_{d_0(e)}) = T_{d_0(e)} \rtimes K_{d_0(e)}$, for all $e \in \overrightarrow{E}$. Our next aim is to show that modification of φ_e by elements in $T_{d_0(e)}$ and $T_{e,\bar{e}}$ results in isomorphic amalgams and that these modifications can be used to ensure that $\varphi_e^{-1}\psi_e \in K_{d_0(e)}$.

Definition 4.10. *Let $\mathscr{G}$ be a CT-structure of type $\mathscr{G}_0$ given by a family of connecting maps $\{\varphi_e \mid e \in \overrightarrow{E}\}$. We say that $\mathscr{G}$ is a* concrete CT structure associated to $(\mathcal{V}, \mathcal{W})$ *if its inclusion maps φ_e satisfy* $\mathrm{ad}(\varphi_e)(K_e) = K_{d_0(e)}$, *where $K_e, K_{d_0(e)} \in \mathbb{K}_0$.*

Note that, by definition, $\mathscr{G}_0$ itself is a concrete CT-structure associated to $(\mathcal{V}_0, \mathcal{W}_0)$.

Lemma 4.11. *Let $\mathscr{G}' = \{G_i, G_f, \varphi'_f \mid i \in I, f \in \overrightarrow{F}\}$ be a CT structure of type $\mathscr{G}_0$ that has property (D). Then, given a collection $\{K'_i \le \mathrm{Aut}_{G_i}(D_i) \mid i \in I\}$ of complements to the groups of diagonal automorphisms T_i, there exists a basis $\mathcal{V}' = \{\mathcal{V}'_e, \mathcal{V}'_{\bar{e}} \mid e \in \overrightarrow{E}\}$ and a collection $\{K'_e \mid e \in \overrightarrow{E}\}$ of complements to $T_{e,\bar{e}}$ such that, for each $e \in \overrightarrow{E}$, K'_e corresponds to $\mathcal{V}'_e$ and* $\mathrm{ad}(\varphi'_e)(K'_e) = K'_{d_0(e)}$. *Given $\{K'_i \mid i \in I\}$, the collection $\mathbb{K}' = \{K'_i, K'_e \mid i \in I, e \in \overrightarrow{E}\}$ is unique and the bases $\mathcal{V}'_e$ are unique up to multiplication by a scalar in Fix*$(\mathrm{Aut}(k))$.

Proof. We shall use an automorphism $\tau = \{\tau_{d_0(e)}, \tau_{e,\bar{e}} \mid f \in \overrightarrow{F}\}$ of $\mathscr{G}_0$ as in Lemma 4.3 to establish the existence of $\mathcal{V}'$ and the collection $\mathbb{K}'$.

Recall that the groups D_e and T_e $(e \in \overrightarrow{E})$ are uniquely determined by the type of $\mathscr{G}_0$ and are therefore shared by $\mathscr{G}'$. The group T_e acts—via $\mathrm{ad}(\cdot)$—transitively on the set of its complements, while acting on the corresponding bases. Two bases correspond to the same complement if and only if one is obtained from the other by permuting their elements and multiplying them by a scalar that is fixed by $\mathrm{Aut}(k)$. This proves the uniqueness part of the theorem.

For the existence we first pick a random base $\mathcal{V}''$ and modify it as follows. If $e \in \overrightarrow{E}$ then $\mathcal{V}''_e$ determines K''_e, a complement to $T_{e,\bar{e}}$. Restriction to $G_{d_0(e)}$ and $G_{d_0(\bar{e})}$ determines complements $K'_{d_0(e)}$ and $K'_{d_0(\bar{e})}$ to $T_{d_0(e)}$ and $T_{d_0(\bar{e})}$. These are conjugates of $K_{d_0(e)}$ and $K_{d_0(\bar{e})}$ under diagonal automorphisms $\tau_{d_0(e)} \in T_{d_0(e)}$ and $\tau_{d_0(\bar{e})} \in T_{d_0(\bar{e})}$. As in the proof of Lemma 4.3 there exists an automorphism $\tau = \{\tau_{d_0(e)}, \tau_{e,\bar{e}} \mid f \in \overrightarrow{F}\}$ of $\mathscr{G}_0$, where, for each $e \in \overrightarrow{E}$, $\tau_{e,\bar{e}} \in T_{e,\bar{e}}$ restricts to $\tau_{d_0(e)}$ and $\tau_{d_0(\bar{e})}$. The adjoint map $\mathrm{ad}(\tau_{e,\bar{e}})$ sends K''_e to a complement K'_e satisfying the statement of the lemma for the edge e, while the underlying linear map transforms the basis $\mathcal{V}''_e$ to the desired basis $\mathcal{V}'_e$.

Corollary 4.12. *Let $\mathcal{G}' = \{G_i, G_f, \varphi'_f \mid i \in I, f \in \overrightarrow{F}\}$ be a CT-structure of type $\mathcal{G}_0$ that has property (D). Then, $\mathcal{G}'$ is isomorphic to a concrete amalgam $\mathcal{G}$ of type $\mathcal{G}_0$. Moreover, the isomorphism can be taken to be diagonal.*

Proof. By Lemma 4.11, given the collection $\{K_i \mid i \in I\} \subseteq \mathbb{K}_0$, there exists a basis $\mathcal{V}'$ and a collection $\mathbb{K}' = \{K'_i, K'_e \mid i \in I, e \in \overrightarrow{E}\}$ that satisfies $K'_i = K_i$ for all $i \in I$, and $\mathrm{ad}(\varphi'_e)(K'_e) = K'_i$ for all $e \in \overrightarrow{E}$.

We now define an isomorphism $\phi = \{\phi_i, \phi_e \mid i \in I, e \in \overrightarrow{E}\}$ from a concrete amalgam $\mathcal{G}$ to $\mathcal{G}'$. Recall that $\mathcal{V}$ is the basis corresponding to the complements K_e. Now, for $e \in \overrightarrow{E}$, let $\phi_e \colon G_e \to G_e$ be the isomorphism induced by the (diagonal) linear map sending $\mathcal{V}_e$ to $\mathcal{V}'_e$ and let $\phi_i = \mathrm{id}_{G_i}$ for all $i \in I$. Now define $\mathcal{G}$ by setting $\varphi_e = \phi_e^{-1} \circ \varphi'_e \circ \phi_{d_0(e)}$. Note that $\mathcal{G} = \{G_i, G_e, \varphi_e \mid i \in I, e \in \overrightarrow{E}\}$ is concrete since $\mathrm{ad}(\varphi_e)(K_e) = \mathrm{ad}(\varphi'_e) \circ \mathrm{ad}({\phi_e}^{-1})(K_e) = K_{d_0(e)}$. Clearly ϕ defines an isomorphism between $\mathcal{G}$ and $\mathcal{G}'$.

As a consequence of Corollary 4.12, we only have to classify—up to isomorphism—those CT-structures of type $\mathcal{G}_0$ with property (D) that are concrete, which means that we can assume that $\varphi_e^{-1}\psi_e \in K_{d_0(e)} \in \mathbb{K}_0 \leq \mathbb{A}_0$ for all $e \in \overrightarrow{E}$.

Theorem 4.13. *For any simply-laced Dynkin diagram $\overrightarrow{\Gamma}$, there is a natural bijection between the set of isomorphism classes of concrete CT-structures of type $\mathcal{G}_0$ with property (D) over $\overrightarrow{\Gamma}$ and the set of isomorphism classes of pointings of the concrete graph of groups $(\mathbb{K}_0, \overrightarrow{\Gamma})$.*

Proof. Let $\mathcal{G}$ be a concrete CT structure of type $\mathcal{G}_0$ with property (D) over $\overrightarrow{\Gamma}$. Then $\mathcal{G}$ defines a pointing of $\mathbb{K}_0$ by setting $\delta_e = \varphi_e^{-1} \circ \psi_e$, for each $e \in \overrightarrow{E}$.

Conversely, given a pointing $((\mathbb{K}_0, \overrightarrow{\Gamma}), \delta)$ we define a CT-structure $\mathcal{G}$ over $\overrightarrow{\Gamma}$ setting $\varphi_e = \psi_e \circ \delta_e^{-1}$ for each $e \in \overrightarrow{E}$. In view of Remark 3.8 for $f \in \overrightarrow{F} - \overrightarrow{E}$ we can let $\varphi_f = \psi_f$.

The fact that the collection $\{\varphi_f \mid f \in \overrightarrow{F}\}$ defines a concrete CT-structure is immediate since, for $e \in \overrightarrow{E}$, $\mathrm{ad}(\varphi_e) = \mathrm{ad}(\delta_e^{-1}) \circ \mathrm{ad}(\psi_e)$, where $\mathrm{ad}(\psi_e)$ takes K_e to $K_{d_0(e)}$ and $\mathrm{ad}(\delta_e^{-1})$ preserves $K_{d_0(e)}$.

It now suffices to show that the correspondence preserves isomorphism classes. In view of Remark 4.8, which says that $\mathbb{K}_0$ is obtained from $\mathbb{A}_0$ by selecting subgroups and restricting the connecting maps, and the observation that the correspondence described here is as in Definition 2.14, this follows from (the proof of) Theorem 2.15.

Corollary 4.14. *For simply-laced Dynkin diagram $\overrightarrow{\Gamma}$, there is a natural bijection between the set of isomorphism classes of (not necessarily concrete) CT-structures of type $\mathcal{G}_0$ with property (D) over $\overrightarrow{\Gamma}$ and the set of isomorphism classes of pointings of the concrete graph of groups $(\mathbb{K}_0, \overrightarrow{\Gamma})$.*

Proof. This follows from Corollary 4.12 and Theorem 4.13.

Theorem 1.1 is now a consequence of Lemma 3.12, Corollary 4.14, Theorem 2.24 and Corollary 2.23.

5 The Curtis-Tits Theorem

5.1 Amalgams Resulting from the Curtis-Tits Theorem

Let G be a simply connected Kac-Moody group that is locally split over a field k (in the sense of [17], that is, the rank-1 and 2 groups are split algebraic groups) with an admissible simply laced Dynkin diagram Γ over some finite index set I. We shall prove that G possesses an amalgam that is an orientable Curtis-Tits structure. Let $(W, \{r_i\}_{i\in I})$ be the Coxeter system of type Γ. Then G has a twin BN-pair (B^+, N, B^-) of type Γ, which gives rise to a Moufang twin-building $\Delta = (\Delta_+, \Delta_-, \delta_+, \delta_-, \delta_*)$ of type Γ, where, for $\varepsilon = \pm$ we have $\Delta_\varepsilon = G/B^\varepsilon$ and

$$\begin{aligned} \delta_\varepsilon(gB^\varepsilon, hB^\varepsilon) &= w \in W & &\text{whenever } B^\varepsilon g^{-1} h B^\varepsilon = B^\varepsilon w B^\varepsilon, \\ \delta_*(gB^+, hB^-) &= w \in W & &\text{whenever } B^+ g^{-1} h B^- = B^+ w B^-. \end{aligned}$$

Two chambers C and D are called *opposite* if $\delta_*(C, D) = 1$. Fix two opposite chambers $C^+ = B^+$ and $C^- = B^-$.

Since Γ is simply laced, by [18] the local structure of Δ determines the global structure. The generalization of the Curtis-Tits theorem to 2-spherical diagrams given in [1] yields G as the universal completion of an amalgam of rank 1 and rank 2 Levi complements.

Since Γ is simply-laced, by [12], G is a central quotient of the universal completion of the following amalgam:

$$\mathscr{R} = \{R_i, R_f, \rho_f \mid i \in I, f \in \vec{F}\},$$

where $R_i = \langle U_i^+, U_i^- \rangle$, $R_f = \langle R_{d_0(f)}, R_{d_0(\bar{f})} \rangle$, and $\{U_i^+ \mid i \in I\}$ is a selection of positive root groups corresponding to a fundamental system of positive roots, and $\rho_f \colon R_{d_0(f)} \hookrightarrow R_f$ is given by inclusion of subgroups in G.

Remark 5.1. Note that an automorphism of Γ corresponds to a renaming of the types of Δ, and a corresponding renaming of the groups in the amalgam $\mathscr{R}$. Naturally the buildings corresponding to these respectively labeled amalgams are isomorphic via a type permuting isomorphism. In accordance with our convention for amalgams in Definition 2.1 and Remark 2.3 we shall regard such buildings and amalgams as being different. Note that Tits uses the same convention in [28].

The fact that G is locally split means that whenever $e \in \vec{E}$, then the e-residue on C^+ is isomorphic to the building associated to the group $\mathrm{SL}_3(k)$. A priori it is possible that the actual groups R_i and R_f are proper quotients of $\mathrm{SL}_2(k)$ and $\mathrm{SL}_3(k)$. However, the following is true.

Lemma 5.2. *Assume that the Dynkin diagram Γ is simply laced and admissible and that $|I| \geq 4$. Then, for all $i \in I$ and $f \in \vec{F}$,*

$$R_i \cong \mathrm{SL}_2(k)$$

$$R_f \cong \begin{cases} \mathrm{SL}_3(k) & \text{if } f \in \overrightarrow{E} \\ \mathrm{SL}_2(k) \circ \mathrm{SL}_2(k) & \text{else.} \end{cases}$$

This follows from the general result in [12] and the observation that on any subdiagram of type A_3, the rank 1 and 2 groups intersect the center of the corresponding rank 3 group trivially.

The fact that G has a torus generated by a coherent selection of tori in each R_i $(i \in I)$ implies that $\mathscr{R}$ has property (D). The fact that $\mathscr{R}$ is oriented follows from the observation that for each i, the root group U_i^+ of the fundamental positive root α_i belongs to R_i and $U_i^+ \subseteq B^+$. In particular, for any $f \in \overrightarrow{F}$, $U^+_{d_0(f)}$ and $U^+_{d_0(\bar{f})}$ belong to a common Borel group of R_f. In sum, $\mathscr{R}$ is an oriented Curtis-Tits structure (in particular it has property (D)) with Dynkin diagram Γ for G.

In the remainder of this section, we shall prove that every oriented CT-structure with admissible Dynkin diagram can be obtained as the Curtis-Tits amalgam of some simply connected Kac-Moody group that is locally split over k.

Let Γ be an admissible Dynkin diagram and $\mathscr{G}(\Gamma)$ an oriented CT-structure over some field k. The fact that $\mathscr{G}(\Gamma)$ is oriented allows us to define a Moufang foundation, which by a result of Mühlherr is integrable to a twin-building Δ.

5.2 Moufang Foundations and Orientable CT-Amalgams

We shall make use of the following definition of a foundation [17], which is equivalent to the definition in [27]:

Definition 5.3. *Let Γ be an admissible Dynkin diagram over I. A* foundation *of* type *Γ is a triple*

$$(\{\Delta_e \mid e \in \overrightarrow{E}\}, \{C_e \mid e \in \overrightarrow{E}\}, \{\theta_e^f \mid e, f \in \overrightarrow{E} \text{ with } d_0(e) = d_0(f)\}),$$

satisfying the following conditions:

(Fo1) $\Delta_e = \Delta_{\bar{e}}$ *is a building of type* A_2 *for each* $e \in \overrightarrow{E}$*;*
(Fo2) $C_e = C_{\bar{e}}$ *is a chamber of* Δ_e *for each* $e \in \overrightarrow{E}$*;*
(Fo3) θ_e^f *is a bijection between the* $d_0(e)$*-panel on* C_e *in* Δ_e *and the* $d_0(f)$*-panel on* C_f *in* Δ_f *such that* $\theta_e^f(C_e) = C_f$ *and if* $e, f, g \in \overrightarrow{E}$ *are such that* $d_0(e) = d_0(f) = d_0(g)$, *then* $\theta_f^g \circ \theta_e^f = \theta_e^g$.

This foundation is said to be of Moufang *type if* Δ_e *is a Moufang building for each* $e \in \overrightarrow{E}$ *and if in (Fo3) the map* θ_e^f *induces an isomorphism between the Moufang set induced by* Δ_e *on the* $d_0(e)$*-panel of* C_e *and the Moufang set induced by* Δ_f *on the* $d_0(f)$*-panel of* C_f.

We shall now describe how to obtain a Moufang foundation from a given orientable CT-structure $(\mathscr{G}, \overrightarrow{\Gamma})$. Let $\{X_i^+, \mid i \in I\}$ be the collection of root groups as in Definition 3.14 and let $\{B_e^+ \mid e \in \overrightarrow{E}\}$ be the collection of Borel groups in G_e such that $\varphi_e(X_{d_0(e)}^+)$ and $\varphi_{\bar{e}}(X_{d_0(\bar{e})}^+)$ are contained in B_e^+ for any $e \in \overrightarrow{E}$ (note that this in fact determines B_e^+ uniquely). For each $e \in \overrightarrow{E}$, let Δ_e be the Moufang building of type A_2 obtained from G_e via the BN-pair $(B_e^+, N_{G_e}(D_e))$ and let C_e be the chamber given by B_e^+. Now let $e, f \in \overrightarrow{E}$ be such that $d_0(e) = d_0(f)$. Let n_e be the element of $N_{G_e}(D_{e,\bar{e}})$ given by

$$\begin{pmatrix} 0 & -1 & 0 \\ 1 & 0 & 0 \\ 0 & 0 & 1 \end{pmatrix}$$

with respect to the ordered basis $\mathcal{V}_e$. Now, for every $e \in \overrightarrow{E}$, let $\bar{X}_e^+ = \varphi_e(X_{d_0(e)}^+)$. Then, the $d_0(e)$-panel of Δ_e containing C_e equals

$$\bar{\Delta}_e = \{C_e\} \cup \{\lambda n_e C_e \mid \lambda \in \bar{X}_e^+\}.$$

We now define

$$\begin{aligned} \theta_e^f : \bar{\Delta}_e &\to \bar{\Delta}_f \\ C_e &\mapsto C_f \\ \lambda n_e C_e &\mapsto \varphi_f \circ \varphi_e^{-1}(\lambda) n_f C_f \end{aligned}$$

Note that since the structure of the $d_0(e)$-panel of Δ_e on C_e (resp. of Δ_f on C_f) as a Moufang set is entirely determined by $G_{d_0(e)}$ the map θ_e^f, induced by the group isomorphism $\varphi_f \circ \varphi_e^{-1}$ preserves this structure. Clearly, if $g \in \overrightarrow{E}$ is such that $d_0(e) = d_0(f) = d_0(g)$, then $\theta_f^g \circ \theta_e^f = \theta_e^g$. To prove soundness, we observe that since Γ is admissible, the connected subdiagrams with three vertices are of type A_3 and the corresponding amalgam is the unique Curtis-Tits amalgam for $\mathrm{SL}_4(k)$.

By Lemma 3.15, for each $e \in \overrightarrow{E}$, the selection $\{X_{d_0(e)}^\pm, X_{d_0(\bar{e})}^\pm\}$ determines a pair of opposite chambers, hence an apartment Σ_e in Δ_e. Moreover, Σ_e intersects $\bar{\Delta}_e$ in the panel $\{C_e, n_e C_e\}$. Clearly then we have $\theta_e^f(n_e C_e) = n_f C_f$. This proves that the system $\mathscr{S} = \{\Sigma_e = \Sigma_{\bar{e}} \mid e \in \overrightarrow{E}\}$ is an apartment of $\mathbf{F}$.

This proves the following.

Lemma 5.4. *The triple*

$$\mathbf{F} = (\{\Delta_e \mid e \in \overrightarrow{E}\}, \{C_e \mid e \in \overrightarrow{E}\}, \{\theta_e^f \mid e, f \in \overrightarrow{E} \text{ with } d_0(e) = d_0(f)\})$$

obtained from the CT-structure $(\mathscr{G}, \overrightarrow{\Gamma})$ as above, is a sound Moufang foundation. Moreover, the selection $\{X_i^+, X_i^- \mid i \in I\}$ determines a unique apartment in $\mathbf{F}$.

Proof. of Corollary 1.2. As proved in the beginning of Sect. 5, the amalgam $\mathscr{R}$ arising from [12] is an orientable CT-structure with diagram $\vec{\Gamma}$.

Conversely, let $\mathscr{G}$ be a concrete orientable CT structure with diagram $\vec{\Gamma}$ and select a system of root groups $\mathscr{X} = \{X_i^{\pm} \mid i \in I\}$ and let $\mathbf{F}$ be the Moufang foundation with apartment $\mathscr{S}$ constructed from $\mathscr{G}$ as in Lemma 5.4. By the main result of [17] this sound Moufang foundation is integrable to a twin-building Δ which is unique up to isomorphism. Let (G, B^+, B^-, N) be the twin-BN-pair for the universal Kac-Moody group G giving rise to Δ.

Identify $\mathbf{F}$ with its image $E_2(C^+)$ in Δ where C^+ is the chamber represented by B^+. Then $\mathscr{S}$ determines a chamber C^- opposite to C^+ (see e.g. [17, Sect. 8]). By transitivity of G on pairs of opposite chambers, we may assume C^- corresponds to B^-. Pick any $e \in \vec{E}$. Then, $\bar{\Delta}_e$ is identified with the $d_0(e)$-panel on C^+. Let $\Sigma = \Sigma(C^+, C^-)$ be the twin-apartment that is the coconvex hull of C^+ and C^-. Then, $\Sigma_e = \Sigma \cap \Delta_e$. Let $U^{\pm}_{d_0(e)}$ be the root groups in G stabilizing the roots of Σ determined by the panel $\bar{\Delta}_e$ on C^+. By [17, Theorem 8.1], we can extend the action of $X^{\pm}_{d_0(e)}$ on Δ_e to Δ. This defines an isomorphism $X^{\pm}_{d_0(e)} \to U^{\pm}_{d_0(e)}$ that extends to a homomorphism $\phi_e \colon G_e \to R_e = \langle U^{\pm}_{d_0(e)}, U^{\pm}_{d_0(\bar{e})}\rangle_G$. From Lemma 5.2 it follows that this homomorphism is in fact an isomorphism yielding a commutative diagram:

$$\begin{array}{ccc} G_e & \xrightarrow{\phi_e} & R_e \\ \uparrow{\scriptstyle \varphi_e} & & \uparrow{\scriptstyle \iota_e} \\ G_{d_0(e)} & \xrightarrow{\phi_{d_0(e)}} & R_{d_0(e)} \end{array}$$

6 Twists of Split Kac-Moody Groups

Let $G_\Gamma(k)$ be the split Kac-Moody group with Dynkin diagram Γ over k and let $\mathscr{G}_0 = \{G_i, G_f, \psi_f \mid i \in I, f \in \vec{F}\}$ be the CT-structure associated to $G_\Gamma(k)$ as in Sect. 5.1. We may assume that this amalgam is concrete by Corollary 4.12 and let $(\mathbb{K}_0, \vec{\Gamma})$ be the associated concrete graph of groups. This suggests the following definition.

Definition 6.1. *For any pointing $((\mathbb{K}_0, \vec{\Gamma}), \delta)$ of the concrete graph of groups $(\mathbb{K}_0, \vec{\Gamma})$, the* δ-twist *is the universal completion $G^\delta_\Gamma(k)$ of the Curtis-Tits amalgam corresponding to δ as in Theorem 4.13. More precisely, for any $e \in \vec{E}$ and $i \in I$ we get a copy $G_i = \mathrm{SL}_2(k)$ and $G_e = G_{\bar{e}} = \mathrm{SL}_3(k)$ and the relations given by those in G_i and G_e together with the following:*

(i) if $e \in \vec{E}$ then $\varphi_e = \psi_e \circ \delta_e^{-1} \colon G_{d_0(e)} \hookrightarrow G_e$ identifies $G_{d_0(e)}$ with a subgroup of G_e;
(ii) if $f \in \vec{F} - \vec{E}$, then $[G_{d_0(f)}, G_{d_0(\bar{f})}] = 1$.

Two twists are considered equivalent *if the corresponding amalgams, and hence pointings, are isomorphic. As for amalgams and buildings, two twists that differ by a non-trivial graph automorphism are considered non-equivalent.*

We can now make an efficient selection of twists δ. To that end, let us fix a spanning tree $\overrightarrow{T}$ of Γ, together with a set of directed edges $\overrightarrow{H}$ that does not intersect $\overleftarrow{H} = \{\bar{e} \mid e \in \overrightarrow{H}\}$ and $\overrightarrow{F} = \overrightarrow{T} \cup \overrightarrow{H} \cup \overleftarrow{H}$. We will construct an amalgam as follows. For each $e \in \overrightarrow{H}$ we take $\delta_e \in K_{d_0(e)}$. Now let

$$\varphi_e = \begin{cases} \psi_e \circ \delta_e^{-1} & \text{if } e \in \overrightarrow{H}, \\ \psi_e & \text{else.} \end{cases}$$

The resulting amalgam is denoted by $\mathscr{G}^\delta$.

Corollary 6.2. *Let Γ be a connected simply laced Dynkin diagram with no triangles and k a field with at least four elements and let G be a universal Kac-Moody group with diagram Γ that is locally split over k. Then, the amalgam arising from the Curtis-Tits theorem for G is isomorphic to a unique Curtis-Tits amalgam $\mathscr{G}^\delta$. Moreover, in that case G is a central quotient of the universal completion of $\mathscr{G}^\delta$.*

Proof. Since the set $\overrightarrow{H}$ corresponds to a unique set of generators for the fundamental group of Γ, there is a natural bijection between sets $\{\delta_e \mid e \in \overrightarrow{H}\}$ and homomorphisms $\Phi\colon \pi(\overrightarrow{\Gamma}, i_0) \to \mathrm{Aut}(k)$. The result now follows from Corollary 1.2 and the main result from [12].

Remark 6.3. Note that Corollary 6.2 does not say that different twists necessarily correspond to non-isomorphic Kac-Moody groups. In fact this is false. For instance, if in Corrollary 6.2 one changes the labeling of Γ via a graph automorphism, the resulting twists and Curtis-Tits amalgams will be considered to be non-equivalent, even though the corresponding groups and buildings are isomorphic via a type permuting isomorphism.

We would like to thank a careful referee for drawing our attention to this fact.

Corollary 1.3 follows immediately from Corollary 6.2.

References

1. Abramenko, P., Mühlherr, B.: Présentations de certaines BN-paires jumelées comme sommes amalgamées. C. R. Acad. Sci. Paris Sér. I Math. **325**(7), 701–706 (1997)
2. Bass, H.: Covering theory for graphs of groups. J. Pure Appl. Algebra **89**(1–2), 3–47 (1993)
3. Blok, R.J., Gagola III, M.S.: Coxeter-chein loops, Commun Algebra (2013)
4. Blok, R.J., Hoffman, C.: A Curtis-Tits-Phan theorem for the twin-building of type $\widetilde{A}_{n-1}$. J. Algebra **321**(4), 1196–1224 (2009)

5. Blok R.J., Hoffman C.: Curtis-Tits groups generalizing Kac-Moody groups of type $\widetilde{A}_n$. Submitted, Jan 2013
6. Blok, R.J., Hoffman, C.: Bass-Serre theory and counting rank two amalgams. J. Group Theor. **14**(3), 389–400 (2011)
7. Blok R.J., Hoffman C.G.: 1-cohomology of simplicial amalgams of groups. J. Alg. Comb. (2012). doi:10.1007/s10801-012-0374-0
8. Blok, R.J., Hoffman, C.G., Vdovina, A.: Expander graphs from Curtis-Tits groups. J. Combin. Theor. Ser. A **119**(3), 521–525 (2012)
9. Bridson M.R., Haefliger A.: Metric spaces of non-positive curvature, In: Grundlehren der Mathematischen Wissenschaften [Fundamental Principles of Mathematical Sciences]. vol. 319, Springer-Verlag, Berlin (1999)
10. Burger M., Mozes S.: Lattices in product of trees. Inst. Hautes Études Sci. Publ. Math. **92**(2000), 151–194 (2001)
11. Burger M., Mozes S.: Groups acting on trees: from local to global structure. Inst. Hautes Études Sci. Publ. Math. **92**(2000), 113–150 (2001)
12. Caprace, P.-E.: On 2-spherical Kac-Moody groups and their central extensions. Forum Math. **19**(5), 763–781 (2007)
13. Caprace, P.-E., Rémy, B.: Simplicity and superrigidity of twin building lattices. Invent. Math. **176**(1), 169–221 (2009)
14. Curtis, C.W.: Central extensions of groups of Lie type. J. Reine Angew. Math. **220**, 174–185 (1965)
15. Dunlap J.: Uniqueness of Curtis-Tits amalgams. Ph.D. thesis, Bowling Green State University (2005)
16. Hée, J.-I.: Systèmes de racines sur un anneau commutatif totalement ordonné. Geometriae Dedicata **37**(1), 65–102 (1991)
17. Mühlherr, B.: Locally split and locally finite twin buildings of 2-spherical type. J. Reine Angew. Math. **511**, 119–143 (1999)
18. Mühlherr, B., Ronan, M.: Local to global structure in twin buildings. Invent. Math. **122**, 71–81 (1995)
19. Ramagge, J.: On certain fixed point subgroups of affine Kac-Moody groups. J. Algebra **171**(2), 473–514 (1995)
20. Schreier, O., Van der Waerden, B.: Die automorphismen der projektiven gruppen. Abhandlungen aus dem Mathematischen Seminar der Universität Hamburg **6**(1), 303–322 (1928)
21. J.-P. Serre.: Trees. Springer-Verlag, Berlin, (1980) Translated from the French by John Stillwell
22. Timmesfeld, F.G.: Presentations for certain Chevalley groups. Geom. Dedicata **73**(1), 85–117 (1998)
23. Timmesfeld, F.G.: On the Steinberg-presentation for Lie-type groups. Forum Math. **15**(5), 645–663 (2003)
24. Timmesfeld, F.G.: The Curtis-Tits-presentation. Adv. Math. **189**(1), 38–67 (2004)
25. Timmesfeld, F.G.: Steinberg-type presentation for Lie-type groups. J. Algebra **300**(2), 806–819 (2006)
26. Tits, J.: Buildings of Spherical Type and Finite BN-Pairs. Springer-Verlag, Berlin (1974)
27. Tits, J.: Uniqueness and presentation of Kac-Moody groups over fields. J. Algebra **105**(2), 806–819 (2006)
28. Tits J.: Twin Buildings and Groups of Kac-Moody Type. In: Groups, combinatorics and geometry, Durham, 1990. London Mathematical Society Lecture Note, vol. 165, pp. 249–286. Cambridge University Press, Cambridge (1992)

Chapter 2
The Use of Valuations for Classifying Point-Line Geometries

Bart De Bruyn

Abstract A valuation is a map from the point set of a point-line geometry $\mathcal{S}$ to the set $\mathbb{N}$ of nonnegative integers satisfying a number of well-chosen axioms. These axioms are chosen in such a way that these objects are (potentially) useful for classifying certain point-line geometries that contain an isomorphic copy of $\mathcal{S}$ as a full subgeometry. Depending on the list of chosen axioms, different "valuation theories" can be developed. The importance of each such theory depends on the successes that can be achieved in classification problems. Valuations have been successfully used for classifying certain near polygons (in particular generalized polygons). They were also useful tools for obtaining a number of other results. The aim of this paper is to survey some of these results.

Keywords Valuation · Generalized polygon · Near polygon

1 Introduction

A point-line geometry $\mathcal{S} = (\mathcal{P}, \mathcal{L}, \mathrm{I})$ with nonempty point set $\mathcal{P}$, line set $\mathcal{L}$ and incidence relation $\mathrm{I} \subseteq \mathcal{P} \times \mathcal{L}$ is called a *partial linear space* if every two distinct points are incident with at most one line In this paper, we will always assume that the point set $\mathcal{P}$ and the line set $\mathcal{L}$ of a point-line geometry $\mathcal{S}$ are disjoint. The distance $\mathrm{d}(x_1, x_2)$ between two points x_1 and x_2 of $\mathcal{S}$ is measured in its collinearity graph. If $x \in \mathcal{P}$ and $\emptyset \neq X \subseteq \mathcal{P}$, then $\mathrm{d}(x, X)$ denotes the minimal distance between x and a point of X. If $x \in \mathcal{P}$ and $i \in \mathbb{N}$, then $\Gamma_i(x)$ denotes the set of points at distance

MSC2000: 51E12, 05B25.

B. De Bruyn (✉)
Department of Mathematics, Ghent University, Gent, Belgium
e-mail: bdb@cage.ugent.be

N. S. N. Sastry (ed.), *Groups of Exceptional Type, Coxeter Groups and Related Geometries,* Springer Proceedings in Mathematics & Statistics 82,
DOI: 10.1007/978-81-322-1814-2_2,

i from x. A point-line geometry is said to have *order* (s, t) if every line is incident with precisely $s + 1$ points and if every point is incident with precisely $t + 1$ lines.

A point-line geometry $\mathcal{S} = (\mathcal{P}, \mathcal{L}, \mathrm{I})$ is called a *subgeometry* of a point-line geometry $\mathcal{S}' = (\mathcal{P}', \mathcal{L}', \mathrm{I}')$ if $\mathcal{P} \subseteq \mathcal{P}'$, $\mathcal{L} \subseteq \mathcal{L}'$ and $\mathrm{I} = \mathrm{I}' \cap (\mathcal{P} \times \mathcal{L})$. If $\{x \in \mathcal{P} \mid x \mathrm{I} L\} = \{x \in \mathcal{P}' \mid x \mathrm{I}' L\}$ for every line L of $\mathcal{S}$, then the subgeometry $\mathcal{S}$ of $\mathcal{S}'$ is called *full*. Consider the following problem.

> Given a family $\mathcal{F}$ of partial linear spaces and a partial linear space $\mathcal{S} = (\mathcal{P}, \mathcal{L}, \mathrm{I})$, determine those members $\mathcal{S}'$ of $\mathcal{F}$ that contain an isomorphic copy of $\mathcal{S}$ as a full (isometrically embedded) subgeometry.

A possible way to approach this problem is trying to reconstruct $\mathcal{S}'$ from its subgeometry $\mathcal{S}$. Such a reconstruction can be realized by labeling the points and lines of $\mathcal{S}'$ by certain objects of $\mathcal{S}$ and indicating which points are incident with what lines. The objects of $\mathcal{S}$ suitable for that purposes might depend on the specific situation. The approach followed in the present paper aims to label the points of $\mathcal{S}'$ by certain maps from $\mathcal{P}$ to $\mathbb{N}$, and the lines of $\mathcal{S}'$ by certain sets of those maps.

Suppose $\mathcal{V}$ is a family of maps from $\mathcal{P}$ to $\mathbb{N}$, defined by a certain list of axioms. Suppose the following can be shown: for every point-line geometry $\mathcal{S}' = (\mathcal{P}', \mathcal{L}', \mathrm{I}')$ belonging to $\mathcal{F}$ and having $\mathcal{S}$ as a full subgeometry and for every point $x \in \mathcal{P}'$, the map $f_x : \mathcal{P} \to \mathbb{N}; y \mapsto \mathrm{d}(x, y) - \min\{\mathrm{d}(x, z) \mid z \in \mathcal{P}\}$ belongs to $\mathcal{V}$. Then information regarding the elements of $\mathcal{V}$ (like for instance a whole classification) can sometimes be turned into information on how $\mathcal{S}$ can be embedded as a full subgeometry into a member of $\mathcal{F}$. For classification purposes, it might often also be worthwhile to study the following problem: for two maps $f_1, f_2 \in \mathcal{V}$, determine a list of conditions for pairs $(f_1, f_2) \in \mathcal{V} \times \mathcal{V}$ such that if x_1 and x_2 are distinct collinear points of $\mathcal{S}'$ then the pair (f_{x_1}, f_{x_2}) satisfies all these conditions. Observe that these conditions may only involve properties of $\mathcal{S}$ not of $\mathcal{S}'$. Indeed, in the classification problem the geometry $\mathcal{S}$ is known while $\mathcal{S}'$ is not ($\mathcal{S}'$ needs to be determined).

If $\mathcal{S} = (\mathcal{P}, \mathcal{L}, \mathrm{I})$ is a partial linear space, then a *valuation* of $\mathcal{S}$ can be understood as a certain map from $\mathcal{P}$ to $\mathbb{N}$ satisfying a number of axioms, chosen in such a way that the resulting "valuation theory" is interesting enough to be valuable in actual classifications. Concrete examples will be given later.

Valuations have effectively been used to obtain new classification results for point-line geometries. In the present paper we will give an overview of some of these results. All of them concern a class of point-line geometries called *near polygons*. Valuations have found some other applications as well, and a few of these will also be mentioned here.

2 The Basics of Valuation Theory

A partial linear space is called a *near polygon* if for every point x and every line L there exists a unique point y on L nearest to x. If $d \in \mathbb{N}$ is the maximal distance between two points, then the near polygon is called a *near 2d-gon*. Near polygons were introduced in [33]. The study of near polygons is often restricted to some interesting classes.

(1) A near polygon is called *dense* if every line is incident with at least three points and if every two points at distance two have at least two common neighbors. An important class of dense near polygons are the so-called dual polar spaces with at least three points on each line. These are related to polar spaces. With every polar space Π of rank $n \geq 1$ there is associated a *dual polar space* of rank n. This is the point-line geometry whose points are the maximal singular subspaces of Π (those of dimension $n-1$) and whose lines are the next-to-maximal singular subspaces of Π (those of dimension $n-2$), with incidence being reverse containment. By convention, a dual polar space of rank 0 is a point (no lines). The automorphism groups of finite thick (dual) polar spaces provide examples of so-called groups of Lie type.

(2) A near $2d$-gon with $d \in \mathbb{N} \setminus \{0, 1\}$ is called a *generalized $2d$-gon* if every point is incident with at least two lines and if for every two points x and y at distance $i \in \{1, 2, \ldots, d-1\}$ from each other there exists a unique point collinear with y at distance $i-1$ from x. Usually, generalized $2d$-gons are defined differently, namely as partial linear spaces $\mathcal{S} = (\mathcal{P}, \mathcal{L}, \mathrm{I})$ that have no subgeometries that are ordinary k-gons with $k \in \{3, 4, \ldots, 2d-1\}$, but "sufficiently many" subgeometries that are ordinary $2d$-gons. With sufficiently many we mean that for every two elements $A, B \in \mathcal{P} \cup \mathcal{L}$ there should exist a subgeometry $\mathcal{S}' = (\mathcal{P}', \mathcal{L}', \mathrm{I}')$ with $A, B \in \mathcal{P}' \cup \mathcal{L}'$ that is an ordinary $2d$-gon.

(3) A finite near $2d$-gon with $d \in \mathbb{N} \setminus \{0, 1\}$ is called *regular* if it has some order (s, t) and if there exist constants t_i, $i \in \{0, 1, \ldots, d\}$, such that for every two points x and y at distance i from each other there are $t_i + 1$ lines through y containing a (necessarily unique) point at distance $i-1$ from x. If this is the case, then $t_0 = -1$, $t_1 = 0$ and $t_d = t$. Finite thick dual polar spaces of rank d and finite generalized $2d$-gons are examples of regular near $2d$-gons. The regular near $2d$-gons are precisely those near $2d$-gons whose collinearity graph is a so-called *distance-regular graph* [2].

A *semi-valuation* of a point-line geometry $\mathcal{S} = (\mathcal{P}, \mathcal{L}, \mathrm{I})$ is a map $f : \mathcal{P} \to \mathbb{Z}$ satisfying the property that every line L contains a (unique) point x_L^* such that $f(x) = f(x_L^*) + 1$ for every point x of L distinct from x_L^*. The number $f(x)$ is called the *f-value of* x, or shortly the *value of* x if no confusion is possible. If f is a semi-valuation attaining a maximal value, then the set of all points with non-maximal value is a *hyperplane* H_f of $\mathcal{S}$, that is, a set of points distinct from $\mathcal{P}$ intersecting (the point set of) each line in either a singleton or the whole line. We call H_f the *hyperplane of* $\mathcal{S}$ *associated with* f. Two semi-valuations f_1 and f_2 of $\mathcal{S}$ are called *neighboring* if there exists an $\epsilon \in \mathbb{Z}$ such that $|f_1(x) - f_2(x) + \epsilon| \leq 1$ for every point $x \in \mathcal{P}$. The notion of (neighboring) semi-valuation(s) was introduced in [11].

Suppose $\mathcal{S}' = (\mathcal{P}', \mathcal{L}', \mathrm{I}')$ is a near $2d$-gon and $\mathcal{S} = (\mathcal{P}, \mathcal{L}, \mathrm{I})$ is a full subgeometry of $\mathcal{S}'$. For every point x of $\mathcal{S}'$ and every point y of $\mathcal{S}$, we define $f_x(y) := \mathrm{d}(x, y) - m_x$, where $m_x = \mathrm{d}(x, \mathcal{P})$. Then the map f_x satisfies the following properties.

(V1) There exists a point y of $\mathcal{S}$ for which $f_x(y) = 0$.
(V2) f_x is a semi-valuation of $\mathcal{S}$.

Moreover, the following can be said.

Proposition 2.1. *If x_1 and x_2 are two distinct collinear points of $\mathcal{S}'$, then the semi-valuations f_{x_1} and f_{x_2} are neighboring.*

Proof. Put $\epsilon = m_{x_1} - m_{x_2}$. Then for every point x of $\mathcal{S}$, we have $|f_{x_1}(x) - f_{x_2}(x) + \epsilon| = |(f_{x_1}(x) + m_{x_1}) - (f_{x_2}(x) + m_{x_2})| = |\mathrm{d}(x, x_1) - \mathrm{d}(x, x_2)| \leq d(x_1, x_2) = 1$.

We have seen that with every semi-valuation f attaining a maximal value there is associated a hyperplane H_f. One can now wonder whether the semi-valuation f can be reconstructed from its associated hyperplane H_f. The following proposition gives sufficient conditions for that to be the case.

Proposition 2.2. *Suppose $\mathcal{S} = (\mathcal{P}, \mathcal{L}, \mathrm{I})$ is a connected partial linear space with finite diameter. Let $f : \mathcal{P} \to \mathbb{N}$ be a map satisfying the properties $(V1)$ and $(V2)$ mentioned above. Suppose that every point with non-maximal f-value is collinear with a point whose f-value is one more. Let H_f denote the hyperplane associated with f and let M denote the maximal distance from a point of $\mathcal{S}$ to $\mathcal{P} \setminus H_f$. Then $f(x) = M - d(x, \mathcal{P} \setminus H_f)$ for every point x of $\mathcal{S}$.*

Proof. By Property (V2), we must have that $M - f(x) \leq \mathrm{d}(x, \mathcal{P} \setminus H_f)$. The fact that every point with non-maximal f-value is collinear with a point whose f-value is one more implies that $\mathrm{d}(x, \mathcal{P} \setminus H_f) \leq M - f(x)$.

Suppose $\mathcal{S}$ is a partial linear space. Let $f_i, i \in I$, be a collection of $|I| \geq 2$ mutually distinct maps from $\mathcal{P}$ to $\mathbb{N}$ satisfying the properties (V1) and (V2) mentioned above. We also assume that every f_i, $i \in I$, attains a maximal value M_i. Then we say that the set $\mathcal{F} = \{f_i \mid i \in I\}$ is an *L-set* if for every point x of $\mathcal{S}$, there exists a (necessary unique) $i \in I$ such that $f_j(x) - M_j = f_i(x) - M_i + 1$ for every $j \in I \setminus \{i\}$. The notion of an L-set was introduced in [14].

Proposition 2.3. *Let $\mathcal{S} = (\mathcal{P}, \mathcal{L}, \mathrm{I})$ be a partial linear space. Let f_i, $i \in I$, be a collection of mutually distinct maps from $\mathcal{P}$ to $\mathbb{N}$ satisfying the properties $(V1)$ and $(V2)$ mentioned above. Suppose that each f_i $(i \in I)$ attains a maximal value M_i and let H_i denote its associated hyperplane. If $\{f_i \mid i \in I\}$ is an L-set, then each point of $\mathcal{S}$ is contained in either one or all hyperplanes of the collection H_i, $i \in I$.*

Proof. Let $x \in \mathcal{P}$. Since the numbers $f_i(x) - M_i$, $i \in I$, are nonpositive, the fact that $\{f_i \mid i \in I\}$ is an L-set implies that either one or all these numbers are (strictly) negative, implying that x belongs to either one or all the hyperplanes H_i, $i \in I$.

Proposition 2.4. *Let $\mathcal{S}' = (\mathcal{P}', \mathcal{L}', \mathrm{I}')$ be a near 2d-gon having the partial linear space $\mathcal{S} = (\mathcal{P}, \mathcal{L}, \mathrm{I})$ as a full subgeometry, and let L' be a line of $\mathcal{S}'$. Suppose there exists a constant δ such that for every point x of L' the maximal distance from a point of $\mathcal{S}$ to x is equal to δ. For every point x of L' and every point y of $\mathcal{S}$, we put $f_x(y) = d(x, y) - m_x$ where $m_x = d(x, \mathcal{P})$. Then the following hold.*

(1) If the maps f_x, $x \in L'$, are mutually distinct, then $\{f_x \mid x \in L'\}$ is an L-set.
(2) If there exists a line L in $\mathcal{S}$ at distance $\delta - 1$ from L', then the maps f_x, $x \in L'$, are mutually distinct and $\{f_x \mid x \in L'\}$ is an L-set.

Proof. (1) Suppose the maps f_x, $x \in L'$, are mutually distinct. For every point x of L', the maximal value that f_x attains is equal to $\delta - m_x$. Now, let y be an arbitrary point of $\mathcal{S}$. If x denotes the unique point of L' nearest to y, then for every point $z \neq x$ on L', we have $f_z(y) - M_z = \mathrm{d}(y,z) - m_z - M_z = \mathrm{d}(y,z) - \delta = \mathrm{d}(y,x) + 1 - \delta = \mathrm{d}(y,x) - m_x - M_x + 1 = f_x(y) - M_x + 1$. This implies that $\{f_x \mid x \in L'\}$ is an L-set.

(2) Suppose there exists a line L in $\mathcal{S}$ at distance $\delta - 1$ from L'. In view of (1), it suffices to show that the maps f_x, $x \in L'$, are mutually distinct. So, consider two distinct points x_1' and x_2' of L' and let x_i, $i \in \{1,2\}$, denote the unique point of L nearest to x_i'. Since $\mathrm{d}(x_i', L) \geq \delta - 1$ and $\mathrm{d}(x_i', y) \leq \delta$ for every point y of L, we necessary have $\mathrm{d}(x_i, x_i') = \delta - 1$. Now, $x_1 \neq x_2$, otherwise the point $x_1 = x_2$ would lie at distance at most $\delta - 2$ from a point of L' since it lies at distance $\delta - 1$ from the points x_1' and x_2'. Suppose now that $f_{x_1'} = f_{x_2'}$. Since $\mathrm{d}(x_1', x_2) = \delta$ and $\mathrm{d}(x_2', x_2) = \delta - 1$, the fact that $f_{x_1'}(x_2) = f_{x_2'}(x_2)$ implies that $m_{x_1'} = m_{x_2'} + 1$. On the other hand, since $\mathrm{d}(x_2', x_1) = \delta$ and $\mathrm{d}(x_1', x_1) = \delta - 1$, we would also have $m_{x_2'} = m_{x_1'} + 1$. As this is impossible, we should have $f_{x_1'} \neq f_{x_2'}$ as we needed to prove.

3 Valuations of Dense Near Polygons

3.1 Basic Definitions

Let $\mathcal{S}$ be a dense near polygon. A set X of points of $\mathcal{S}$ is called a *subspace* if every line of $\mathcal{S}$ having two of its points in X has all its points in X. A *subspace* X of $\mathcal{S}$ is called *convex* if every point on a shortest path between two points of X is also contained in X. If X is a convex subspace of $\mathcal{S}$, then the lines of $\mathcal{S}$ that have all their points in X define a point-line geometry $\widetilde{X}$ on the set X which itself is also a near polygon. The maximal distance between two points of a convex subspace X is called the *diameter* of X. By results of [33, Proposition 2.5] and [3, Theorem 4], every two points of a dense near polygon at distance δ from each other are contained in a unique convex subspace of diameter δ. These convex subspaces are called *quads* if $\delta = 2$ and *hexes* if $\delta = 3$. The fundamental property of dense near polygons regarding the existence of convex subspaces allows to build a typical valuation theory for dense near polygons.

Let $\mathcal{S} = (\mathcal{P}, \mathcal{L}, \mathrm{I})$ be a dense near polygon. A map $f : \mathcal{P} \to \mathbb{N}$ is called a *valuation* if the following three properties are satisfied.

(V1) There exists at least one point x with $f(x) = 0$.
(V2) f is a semi-valuation.
(V3) Every point x of $\mathcal{S}$ is contained in a convex subspace F_x that satisfies the following properties:

- $f(y) \leq f(x)$ for every point y of F_x;
- every point z of $\mathcal{S}$ which is collinear with a point y of F_x and which satisfies $f(z) = f(y) - 1$ also belongs to F_x.

Valuations of dense near polygons were introduced in [18]. It can be shown, see [18, Proposition 2.5], that the convex subspace F_x mentioned in Property (V3) is unique. For several classes of dense near polygons, it can be shown that Property (V3) is a consequence of Property (V2), see [10, Corollary 1.4].

That the above definition of valuation can serve as a suitable basis for a "valuation theory of dense near polygons" follows from the following two results. The first result was proved in [18, Proposition 2.6].

Proposition 3.1. *Let $\mathcal{S}' = (\mathcal{P}', \mathcal{L}', \mathrm{I}')$ be a dense near 2d-gon and let $\mathcal{S} = (\mathcal{P}, \mathcal{L}, \mathrm{I})$ be a dense near polygon which is isometrically embedded as a full subgeometry in $\mathcal{S}'$. (So, $d_{\mathcal{S}'}(x_1, x_2) = d_{\mathcal{S}}(x_1, x_2)$ for any two points x_1 and x_2 of $\mathcal{S}$.) Let x be a point of $\mathcal{S}'$ and put $m_x := d(x, \mathcal{P})$. For every point $y \in \mathcal{P}$, we define $f_x(y) := d(x, y) - m_x$. Then f_x is a valuation of $\mathcal{S}$, the so-called valuation of $\mathcal{S}$ induced by x.*

Proposition 3.2. *Let $\mathcal{S}' = (\mathcal{P}', \mathcal{L}', \mathrm{I}')$ be a dense near 2d-gon and let $\mathcal{S} = (\mathcal{P}, \mathcal{L}, \mathrm{I})$ be a dense near 2d-gon which is fully and isometrically embedded into $\mathcal{S}'$. Then for every line L' of $\mathcal{S}'$, there exists a line L in $\mathcal{S}$ at distance $d - 1$ from L'. As a consequence the lines of $\mathcal{S}'$ will induce L-sets of valuations of $\mathcal{S}$.*

Proof. We give a sketch of the proof. We implicitly make use of some known properties of dense near polygons.

Let y be a point of $\mathcal{S}$ at maximal distance from L' and let x denote the point of L' nearest to y. Then L' and y are contained in a unique convex subspace F_1 of diameter δ of $\mathcal{S}'$ where $\delta = \mathrm{d}(x, y) + 1$. The points x and y are contained in a unique convex subspace F_2 of diameter $\delta - 1$ of $\mathcal{S}'$.

We show that $\delta = d$. Suppose to the contrary that $\delta < d$. Since $F_1 \cap \mathcal{P}$ is a convex subspace of $\mathcal{S}$ of diameter at most $\delta < d$, there exists a line K of $\mathcal{S}$ through y not contained in F_1. Any point of $K \setminus \{y\}$ would lie at distance $\mathrm{d}(y, L') + 1$ from L' which is impossible. Hence, $\delta = d$ and $\mathrm{d}(x, y) = d - 1$. Since $F_2 \cap \mathcal{P}$ is a convex subspace of $\mathcal{S}$ of diameter at most $d - 1$, there exists a line L of $\mathcal{S}$ through y not contained in F_2. This line L lies at distance $d - 1$ from L'.

By Proposition 2.4, it now follows that the lines of $\mathcal{S}'$ will induce L-sets of valuations of $\mathcal{S}$.

The first claim in the following proposition is precisely Lemma 4.2 of [9].

Proposition 3.3. *Suppose f is a valuation of a dense near 2d-gon $\mathcal{S}$. Then every point with non-maximal value is collinear with a point whose value is one more. As a consequence, the valuation f can be reconstructed from its associated hyperplane H_f in the sense of Proposition 2.2.*

3.2 The Known Valuations

The aim of this subsection is to describe all known classes of valuations of dense near polygons. Nine classes have been described in the literature. With exception of one class, all known classes were already mentioned in the paper [18] where the notion of valuation was introduced. The remaining class of the so-called semi-diagonal valuations was introduced and studied in [7]. The SDPS-valuations have been studied in [19] (finite case) and [6, Chap. 5] (general case). Throughout this subsection, we suppose that $\mathcal{S} = (\mathcal{P}, \mathcal{L}, \mathrm{I})$ is a dense near $2d$-gon with $d \geq 2$.

(1) **The classical valuations.** Let x be a fixed point of $\mathcal{S}$ and put $f_x(y) := \mathrm{d}(x, y)$ for every point y of $\mathcal{S}$. Then f_x is a so-called *classical valuation* of $\mathcal{S}$.

(2) **The ovoidal valuations.** Let O be an *ovoid* of $\mathcal{S}$, that is, a set of points having a unique point on each line of $\mathcal{S}$. If we put $f_O(x) := 0$ for every $x \in O$ and $f_O(x) := 1$ for every $x \in \mathcal{P} \setminus O$, then f_O is a so-called *ovoidal valuation* of $\mathcal{S}$.

(3) **The hybrid valuations.** Suppose x is a point of $\mathcal{S}$ and $\delta \in \{2, 3, \ldots, d\}$. Then let $\mathcal{S}'$ be the subgeometry of $\mathcal{S}$ whose points consist of all the points of $\mathcal{S}$ at distance at least δ from x and whose lines consist of all lines of $\mathcal{S}$ at distance at least $\delta - 1$ from x. For every ovoid O' of $\mathcal{S}'$, let $f_{x,O'} : \mathcal{P} \rightarrow \mathbb{N}$ be the map which maps a point $y \in \mathcal{P}$ to $\mathrm{d}(x, y)$ if $\mathrm{d}(x, y) \leq \delta - 1$, to $\delta - 2$ if $\mathrm{d}(x, y) \geq \delta$ and $y \in O'$ and to $\delta - 1$ if $\mathrm{d}(x, y) \geq \delta$ and $y \notin O'$. Then $f_{x,O'}$ is a so-called *hybrid valuation of $\mathcal{S}$ with index δ*. A hybrid valuation of index 2 is just an ovoidal valuation. Hybrid valuations of index d are also called *semi-classical valuations*. We can regard classical valuations as "hybrid valuations of index $d + 1$".

(4) **The distance-j-ovoidal valuations.** Let $j \in \{2, 3, \ldots, d\}$. A *distance-$j$-ovoid* of $\mathcal{S}$ is a set X of points satisfying the following properties:

- $|X| \geq 2$ and the minimal distance between two distinct points of X is equal to j;
- for every point a of $\mathcal{S}$, there exists a point $x \in X$ such that $\mathrm{d}(a, x) \leq \frac{j}{2}$;
- for every line L of $\mathcal{S}$, there exists a point $x \in X$ such that $\mathrm{d}(x, L) \leq \frac{j-1}{2}$.

A distance-2-ovoid is just an ovoid. Distance-j-ovoids (of generalized polygons) were introduced in [30]. If X is a distance-j-ovoid of $\mathcal{S}$ with $j \in \{2, 3, \ldots, d\}$ even, then the map $f : \mathcal{P} \rightarrow \mathbb{N}; x \mapsto \mathrm{d}(x, X)$ is a valuation of $\mathcal{S}$. Any valuation of $\mathcal{S}$ which can be obtained in this way is called a *distance-j-ovoidal valuation*. The distance-2-ovoidal valuations are precisely the ovoidal valuations.

(5) **The extended valuations.** Suppose $\mathcal{S}' = (\mathcal{P}', \mathcal{L}', \mathrm{I}')$ is a convex subspace of $\mathcal{S}$ such that for every point x of $\mathcal{S}$ there exists a (unique) point x' of $\mathcal{S}'$ such that $\mathrm{d}(x, y) = \mathrm{d}(x, x') + \mathrm{d}(x', y)$ for every point y of $\mathcal{S}'$. Suppose f' is a valuation of $\mathcal{S}'$. Then the map $f : \mathcal{P} \rightarrow \mathbb{N}$ defined by $x \mapsto \mathrm{d}(x, x') + f'(x')$ is a valuation of $\mathcal{S}$, the so-called *extension of f'*.

(6) **The SDPS-valuations.** Suppose $\mathcal{S}$ is a dual polar space of rank $2n$ where $n \in \mathbb{N}$. A nonempty set X of points of $\mathcal{S}$ is called an *SDPS-set* if it satisfies the following properties.

- No two distinct points of X are collinear.
- If a quad Q contains two points of X then it intersects X in an ovoid.
- The partial linear space Δ whose points are the elements of X and whose lines are the quad intersections of size at least 2 (natural incidence) is a dual polar space of rank n.

- If $x_1, x_2 \in X$, then $\mathrm{d}_{\mathcal{S}}(x_1, x_2) = 2 \cdot \mathrm{d}_{\Delta}(x_1, x_2)$.
- Every line of $\mathcal{S}$ meeting X is contained in a quad which intersects X in at least two points.

An SDPS-set of the near 0-gon consists of the unique point of that geometry. An SDPS-set of a generalized quadrangle is just an ovoid of that geometry. The word SDPS is an abbreviation of *Sub Dual Polar Space* and refers to the fact that Δ can be regarded as a sub dual polar space of $\mathcal{S}$. If X is an SDPS-set of $\mathcal{S}$, then the map $f_X : \mathcal{P} \rightarrow \mathbb{N}; x \mapsto \mathrm{d}(x, X)$ is a valuation of $\mathcal{S}$. Any valuation of the dual polar space $\mathcal{S}$ which can be obtained in this way is called an *SDPS-valuation*.

(7) **The product valuations.** Suppose $\mathcal{S}$ is the direct product $\mathcal{S}_1 \times \mathcal{S}_2$ of two other dense near polygons $\mathcal{S}_1 = (\mathcal{P}_1, \mathcal{L}_1, \mathrm{I}_1)$ and $\mathcal{S}_2 = (\mathcal{P}_2, \mathcal{L}_2, \mathrm{I}_2)$ whose diameters are at least 1. Then the point set $\mathcal{P}$ of $\mathcal{S}$ is equal to the cartesian product $\mathcal{P}_1 \times \mathcal{P}_2$, see [6, Sect. 1.6] for more details. Suppose $f_i, i \in \{1, 2\}$, is a valuation of $\mathcal{S}_i$. For every point $x = (x_1, x_2)$ of $\mathcal{S}$, we define $f(x) = f_1(x_1) + f_2(x_2)$. Then f is a valuation of $\mathcal{S}$. Any valuation of $\mathcal{S}$ which can be obtained in this way is called a *product valuation*.

(8) **The diagonal valuations.** Suppose $\mathcal{S} = \mathcal{S}' \times \mathcal{S}'$ where $\mathcal{S}' = (\mathcal{P}', \mathcal{L}', \mathrm{I}')$ is some dense near polygon. Define $X := \{(x, x) \,|\, x \in \mathcal{P}'\}$. Then the map from $\mathcal{P} = \mathcal{P}' \times \mathcal{P}'$ to $\mathbb{N}$ mapping a point x of $\mathcal{S}$ to $\mathrm{d}(x, X)$ is a valuation of $\mathcal{S}$. Any valuation of $\mathcal{S}$ which can be obtained in this way is called a *diagonal valuation*.

(9) **The semi-diagonal valuations.** Suppose $\mathcal{Q}_1$ and $\mathcal{Q}_2$ are two generalized quadrangles and $\mathcal{S}$ is a glued near hexagon of type $\mathcal{Q}_1 \otimes \mathcal{Q}_2$ as discussed in [6, Chap. 4]. Then there exists a partition T_1 of the point set $\mathcal{P}$ in quads isomorphic to $\mathcal{Q}_1$ and a partition T_2 of $\mathcal{P}$ in quads isomorphic to $\mathcal{Q}_2$ such that the following properties are satisfied: (a) every point of $\mathcal{S}$ is contained in a unique quad of T_1 and a unique quad of T_2; (b) every quad of T_1 intersects every quad of T_2 in a line. Suppose X is a set of points of $\mathcal{S}$ satisfying the following properties:

(1) every element of $T_1 \cup T_2$ has a unique point in common with X;
(2) any two distinct points of X lie at distance 2 from each other.

Then the map $f : \mathcal{P} \rightarrow \mathbb{N}; x \mapsto \mathrm{d}(x, X)$ is a valuation of $\mathcal{S}$. Any valuation of $\mathcal{S}$ which can be obtained in this way is called a *semi-diagonal valuation* of $\mathcal{S}$.

3.3 Some Properties of Valuations

The following characterization of classical and ovoidal valuations was proved in [18, Proposition 2.10].

Proposition 3.4. *Suppose f is a valuation of a dense near $2d$-gon $\mathcal{S}$ and let M_f denote the maximal value attained by f. Then f is an ovoidal valuation if and only if $M_f = 1$, and f is a classical valuation if and only if $M_f = d$.*

The following is a consequence of Proposition 3.4.

Corollary 3.5 *Every valuation of a generalized quadrangle with at least three points on each line is either classical or ovoidal.*

We now give two results regarding valuations of dense near hexagons. The first proposition was proved in [18, Propositions 4.1 and 4.2].

Proposition 3.6. *Let f be a valuation of a dense near hexagon $\mathcal{S}$, and let O_f denote the set of points with f-value 0. If f is not classical, semi-classical nor ovoidal, then $|O_f| \geq 2$ and every two distinct points of O_f lie at distance 2 from each other.*

The next proposition is taken from [19, Proposition 3].

Proposition 3.7. *Every valuation of a thick dual polar space of rank 3 is either classical, semi-classical, ovoidal or the extension of an ovoidal valuation of a quad.*

Suppose f is a valuation of a dense near polygon $\mathcal{S}$ and X is a subspace of $\mathcal{S}$ such that the point-line geometry $\widetilde{X}$ induced on X is a dense near polygon that is isometrically embedded into $\mathcal{S}$. If m is the minimal value that f attains on the set X, then the map $f' : X \to \mathbb{N}; x \mapsto f(x) - m$ is a valuation of $\widetilde{X}$ by [18, Proposition 2.8]. This valuation is called the *valuation of $\widetilde{X}$ induced by f*. The following proposition was proved in [19, Theorems 5 and 6] (for finite dual polar spaces), but its proof easily extends to the infinite case, see [6, Theorems 5.30 and 5.32].

Proposition 3.8. *Let f be a valuation of a thick dual polar space. Then the following are equivalent:*

- *every induced hex valuation is either classical or the extension of an ovoidal valuation of a quad;*
- *there exists a convex subspace X of even diameter in Δ and an SDPS-valuation f' of $\widetilde{X}$ such that f is the extension of f'.*

3.4 Applications of Valuations

We give a number of instances were valuations have been used to obtain new classification results regarding dense near polygons.

(1) All dense near octagons with three points per line have been classified in [23]. By [3, Theorem 4], we know that every such dense near octagon must contain convex subspaces of diameter 3, the hexes. Each such hex is a dense near hexagon with three points per line. By [1], we know that there are eleven such near hexagons. The valuations of the eleven dense near hexagons with three points per line have been classified in the papers [20, 21]. This classification of the valuations was subsequently used in [23] to obtain a complete classification of all dense near octagons with three points per line.

(2) A classification of the dense near octagons with four points per line was obtained in [13]. This classification is not complete in the sense that besides the 28

known examples a few open cases still remain. Every hex of a dense near octagon with four points per line must be a dense near hexagon with four points per line. A partial classification of such dense near hexagons was obtained in [5]. In [5], ten examples were described along with four open cases. In [9], the valuations of the dense near hexagons with four points per line were studied, and these results were subsequently used in [13] to obtain the above-mentioned classification for the octagons. Very recently, in [27, Proposition 4.7], one of the four open cases in the classification of the near hexagons was killed. In view of this, there remain now three open cases in the classification of the hexagons and also three open cases in the classification of the octagons.

Valuations have also found applications in the study of hyperplanes of near polygons, in particular, of dual polar spaces. It was the study of certain valuations in dual polar spaces which led to a new interesting class of hyperplanes, the so-called *SDPS-hyperplanes* associated with SDPS-valuations. In the case of rank 4 these hyperplanes were independently discovered by Pralle and Shpectorov in [31]. In [8], the (extensions of) SDPS-hyperplanes were characterized. Valuations played a crucial role in the proof of that characterization result.

Valuations are (potentially) useful for studying isometric embeddings between dense near polygons. They provide information on how such embeddings (can) look like.

Further classification results regarding valuations of dense near polygons were obtained in [11, 12, 15, 22, 24, 25]. We hope that these results will be useful for obtaining new classification results regarding dense near polygons.

4 Valuations of Generalized Polygons

4.1 Basic Definitions

Suppose $\mathcal{S} = (\mathcal{P}, \mathcal{L}, \mathrm{I})$ is a generalized $2d$-gon with $d \in \mathbb{N} \setminus \{0, 1\}$. A map $f : \mathcal{P} \to \mathbb{N}$ is called a *polygonal valuation* if the following three conditions are satisfied.

(PV1) There exists a point x whose f-value $f(x)$ equals 0.
(PV2) f is a semi-valuation.
(PV3) If x is a point of $\mathcal{S}$ with non-maximal f-value, then there is at most one line through x containing a (necessary unique) point with f-value $f(x) - 1$.

The notion of a polygonal valuation was introduced in [14]. Polygonal valuations seem to be suitable objects to build a valuation theory for generalized polygons.

If f is a polygonal valuation of $\mathcal{S}$, then we denote by $\mathcal{O}_f$ the set of points with f-value 0 and by $\mathcal{M}_f$ the set of all points x of $\mathcal{S}$ that are not collinear with a point having f-value $f(x) - 1$. We denote by M_f the maximal value attained by f. Clearly, $\mathcal{O}_f \subseteq \mathcal{M}_f$ and $M_f \in \{1, 2, \ldots, d\}$.

We now describe three classes of polygonal valuations which were introduced in [14].

(1) Let x be a fixed point of $\mathcal{S}$ and put $f_x(y) := \mathrm{d}(x, y)$ for every $y \in \mathcal{P}$. Then f_x is a so-called *classical polygonal valuation* of $\mathcal{S}$.

(2) Suppose $x \in \mathcal{P}$ and $O \subseteq \Gamma_d(x)$ such that every line of $\mathcal{S}$ at distance $d-1$ from x has a unique point in common with O. If y is a point of $\mathcal{S}$ at distance at most $d-1$ from x, then we define $f_{x,O}(y) := \mathrm{d}(x, y)$. If y is a point of $\mathcal{S}$ at distance d from x, then $f_{x,O}(y) := d-2$ if $y \in O$ and $f_{x,O}(y) := d-1$ otherwise. Then $f_{x,O}$ is a so-called *semi-classical polygonal valuation* of $\mathcal{S}$.

(3) Let O be a set of points of $\mathcal{S}$ having a unique point in common with every line. We put $f_O(x) = 0$ if $x \in O$ and $f_O(x) = 1$ if $x \in \mathcal{P} \setminus O$. Then f_x is a so-called *ovoidal polygonal valuation* of $\mathcal{S}$.

(4) Let $j \in \{2, 3, \ldots, d\}$ and let X be a *distance-j-ovoid* of $\mathcal{S}$ with j even (as defined by the three properties mentioned in Sect. 3.2). The map $\mathcal{P} \to \mathbb{N}; x \mapsto \mathrm{d}(x, X)$ is a so-called *distance-j-ovoidal polygonal valuation* of $\mathcal{S}$. The distance-2-ovoidal polygonal valuations are precisely the *ovoidal polygonal valuations*.

4.2 Properties of Polygonal Valuations

In this subsection, we suppose that $\mathcal{S} = (\mathcal{P}, \mathcal{L}, \mathrm{I})$ is a generalized $2d$-gon with $d \in \mathbb{N} \setminus \{0, 1\}$. The next proposition says that polygonal valuations satisfy a property that is similar to the one mentioned in Proposition 3.3.

Proposition 4.1. *If f is a polygonal valuation of $\mathcal{S}$, then every point with non-maximal value is collinear with a point whose value is one more. As a consequence, the valuation f can be reconstructed from its associated hyperplane H_f in the sense of Proposition 2.2.*

Proof. The first claim is an immediate consequence of Property (PV3) in the definition of polygonal valuation.

The following proposition was proved in Sect. 3 of [14].

Proposition 4.2. *Suppose f is a polygonal valuation of $\mathcal{S}$. Then $M_f = 1$ if and only if f is ovoidal, $M_f = d-1$ if and only if f is semi-classical and $M_f = d$ if and only if f is classical. Also, $\mathcal{O}_f = \mathcal{M}_f$ if and only if f is either classical or distance-j-ovoidal for some even j.*

The following is a consequence of Proposition 4.2.

Corollary 4.3 (1) *Every polygonal valuation of a generalized quadrangle is either classical or ovoidal.*

(2) *Every polygonal valuation of a generalized hexagon is either classical, semi-classical or ovoidal.*

Let $f_i, i \in I$, be a collection of mutually distinct polygonal valuations of $\mathcal{S}$, where I is some index set of size at least two. Put $\mathcal{F} = \{f_i \mid i \in I\}$. If $\mathcal{F}$ is an L-set, then by [14, Proposition 4.8], there exists a line of $\mathcal{S}$ containing precisely $|\mathcal{F}| = |I|$ points. The set $\mathcal{F}$ is called *admissible* if the following holds for all $i_1, i_2 \in I$ with $i_1 \neq i_2$, for every $x \in \mathcal{M}_{f_{i_1}}$ and every $y \in \mathcal{M}_{f_{i_2}}$:

- if f_{i_1} and f_{i_2} are classical, then $\mathrm{d}(x, y) = 1$;
- if $x = y$, then $(f_{i_1}(x) - M_{f_{i_1}}) - (f_{i_2}(x) - M_{f_{i_2}}) \in \{-1, 0, 1\}$;
- if $x \neq y$ and at least one of f_{i_1}, f_{i_2} is not classical, then $\mathrm{d}(x, y) + f_{i_1}(x) + f_{i_2}(y) - M_{f_{i_1}} - M_{f_{i_2}} + 1 \geq 0$.

The *valuation geometry* $\mathcal{V}_{\mathcal{S}}$ of $\mathcal{S}$ is the point-line geometry whose points are the polygonal valuations of $\mathcal{S}$ and whose lines are the admissible L-sets of polygonal valuations of $\mathcal{S}$, with incidence being containment. The properties mentioned in the following proposition were proved in Sect. 6 of [14].

Proposition 4.4. *Suppose $\mathcal{S}' = (\mathcal{P}', \mathcal{L}', \mathrm{I}')$ is a generalized $2d$-gon containing $\mathcal{S}$ as a full subgeometry. Let x be a point of $\mathcal{S}'$ and put $m_x := d(x, \mathcal{P})$. For every point $y \in \mathcal{P}$, we define $f_x(y) := d(x, y) - m_x$. Then:*

(1) f_x is a polygonal valuation of $\mathcal{S}$, the so-called polygonal valuation of $\mathcal{S}$ induced by x. Moreover, $M_{f_x} = d - m_x$.

(2) If x_1 and x_2 are two distinct collinear points of $\mathcal{S}'$, then the polygonal valuations f_{x_1} and f_{x_2} are distinct.

(3) The map $\theta : x \mapsto f_x$ between the point-sets of $\mathcal{S}'$ and $\mathcal{V}_{\mathcal{S}}$ maps every line of $\mathcal{S}'$ to a full line of $\mathcal{V}_{\mathcal{S}}$.

The properties mentioned in the following proposition were proved in Sect. 5 of [14].

Proposition 4.5. *Let $\mathcal{S}$ be a generalized $2d$-gon having three points on each line. Then:*

(1) If $\{f_1, f_2, f_3\}$ is an L-set of polygonal valuations of $\mathcal{S}$, then H_{f_3} equals the complement $\overline{H_{f_1} \Delta H_{f_2}}$ of the symmetric difference $H_{f_1} \Delta H_{f_2}$ of H_{f_1} and H_{f_2}.

(2) Let f_1, f_2 and f_3 be three polygonal valuations of $\mathcal{S}$ and let H_i, $i \in \{1, 2, 3\}$, be the hyperplane associated with f_i. If $H_3 = \overline{H_1 \Delta H_2}$ and $\{f_1, f_2, f_3\}$ is admissible, then $\{f_1, f_2, f_3\}$ is an admissible L-set of $\mathcal{S}$.

(3) The valuation geometry of $\mathcal{S}$ is a partial linear space.

4.3 Applications of Polygonal Valuations

Polygonal valuations can be used to obtain classification results for generalized polygons containing a given generalized polygon as a full subgeometry. We give three instances where success has been achieved.

(1) In [28], it was shown that every generalized hexagon of order $(3, 3)$ which contains a subhexagon of order $(3, 1)$ must be isomorphic to the split Cayley hexagon $H(3)$. This result was proved without the aid of a computer. The same result can also be easily proved with the aid of polygonal valuations, but by relying on some computer computations using the computer algebra system GAP [29], see [17].

(2) The existence of semi-finite thick generalized polygons is one of the most important open problems in the theory of generalized polygons (see [35, Appendix E, Problem 5]). A generalized polygon is said to be *semi-finite* if it has a finite number of points on each line, but an infinite number of lines through each point. By relying on polygonal valuations and computer computations (with GAP), it will be shown in [17] that there do not exist semi-finite generalized hexagons of order $(2, t)$ containing a subhexagon of order $(2, 2)$. Note that by [4], there are up to isomorphism two generalized hexagons of order $(2, 2)$, the split Cayley hexagon $H(2)$ and its point-line dual $H^D(2)$.

(3) Up to isomorphism, a unique generalized octagon of order $(2, 4)$ is known to exist. It belongs to the class of the so-called *Ree-Tits generalized octagons*. This family of generalized octagons was first constructed by Tits [34] using a family of simple groups discovered by Ree [32]. In [16], it was shown that the Ree-Tits octagon of order $(2, 4)$ is the up to isomorphism unique generalized octagon of order $(2, 4)$ containing a suboctagon of order $(2, 1)$. The idea of the proof consists of considering the valuation geometry $\mathcal{V}$ of the unique generalized octagon of order $(2, 1)$. By studying $\mathcal{V}$, it was shown in [16] that $\mathcal{S}$ can be reconstructed in a unique way from its suboctagon of order $(2, 1)$. This then implies that $\mathcal{S}$ must be isomorphic to the Ree-Tits generalized octagon of order $(2, 4)$.

Remarks. (1) In [4] all finite generalized hexagons of order $(2, t)$ have been classified.

(2) Let $s \in \mathbb{N} \setminus \{0, 1\}$. In [26, Corollary 3.20] it was shown that there exist no semi-finite generalized hexagons of order (s, t) that contain a subhexagon of order (s, s^3) as a full subgeometry.

References

1. Brouwer, A.E., Cohen, A.M., Hall, J.I., Wilbrink, H.A.: Near polygons and Fischer spaces. Geom. Dedicata **49**, 349–368 (1994)
2. Brouwer, A.E., Cohen, A.M., Neumaier, A : Distance-regular graphs. In: Results in Mathematics and Related Areas (3), vol. 18. Springer, Berlin (1989)
3. Brouwer, A.E., Wilbrink, H.A.: The structure of near polygons with quads. Geom. Dedicata **14**, 145–176 (1983)
4. Cohen, A.M., Tits, J.: On generalized hexagons and a near octagon whose lines have three points. Eur. J. Comb. **6**, 13–27 (1985)
5. De Bruyn, B.: Near hexagons with four points on a line. Adv. Geom. **1**, 211–228 (2001)
6. De Bruyn, B.: Near polygons. In: Frontiers in Mathematics. Birkhäuser Verlag, Basel (2006)
7. De Bruyn, B.: Valuations of glued near hexagons. J. Comb. Des. **15**, 35–48 (2007)

8. De Bruyn, B.: A characterization of the SDPS-hyperplanes of dual polar spaces. Eur. J. Comb. **28**, 705–714 (2007)
9. De Bruyn, B.: Dense near octagons with four points on each line, II. Adv. Geom. **7**, 191–206 (2007)
10. De Bruyn, B.: An alternative definition of the notion valuation in the theory of near polygons. Electron. J. Comb. **16**, 14 (2009) (Research paper 16)
11. De Bruyn, B.: The valuations of the near polygon $\mathbb{G}_n$. Electron. J. Comb. **16**, 29 (2009) (Research paper 137)
12. De Bruyn, B.: The valuations of the near $2n$-gon $\mathbb{I}_n$. Ars Comb. **98**, 321–336 (2011)
13. De Bruyn, B.: Dense near octagons with four points on each line, III. Ann. Comb. **15**, 19–35 (2011)
14. De Bruyn, B.: Polygonal valuations. Discrete Math. **313**, 84–93 (2013)
15. De Bruyn, B.: On the valuations of the near polygon $\mathbb{H}_n$. Preprint 2012
16. De Bruyn, B.: The uniqueness of a certain generalized octagon of order $(2, 4)$. Preprint 2012
17. De Bruyn, B.: On semi-finite generalized hexagons of order $(2, t)$ and $(3, t)$ containing subhexagons. In preparation
18. De Bruyn, B., Vandecasteele, P.: Valuations of near polygons. Glasg. Math. J. **47**, 347–361 (2005)
19. De Bruyn, B., Vandecasteele, P.: Valuations and hyperplanes of dual polar spaces. J. Com. Theory, Ser. A **112**, 194–211 (2005)
20. De Bruyn, B., Vandecasteele, P.: The valuations of the near hexagons related to the Witt designs $S(5, 6, 12)$ and $S(5, 8, 24)$. J. Comb. Des. **14**, 214–228 (2006)
21. De Bruyn, B., Vandecasteele, P.: The distance-2-sets of the slim dense near hexagons. Ann. Comb. **10**, 193–210 (2006)
22. De Bruyn, B., Vandecasteele, P.: The valuations of the near octagon $\mathbb{I}_4$. Electron. J. Comb. **13**, 23 (2006) (Research paper 76)
23. De Bruyn, B., Vandecasteele, P.: The classification of the slim dense near octagons. Eur. J. Comb. **28**, 410–428 (2007)
24. De Bruyn, B., Vandecasteele, P.: The valuations of the near octagon $\mathbb{H}_4$. Graphs and Combinatorics **23**, 601–623 (2007)
25. De Bruyn, B., Vandecasteele, P.: The valuations of the near octagon $\mathbb{G}_4$. Discrete Math. **310**, 755–766 (2010)
26. De Bruyn, B., Vanhove, F.: Inequalities for regular near polygons, with applications to m-ovoids. Eur. J. Comb. **34**, 522–538 (2013)
27. De Bruyn, B., Vanhove, F.: On Q-polynomial regular near $2d$-gons. Combinatorica (to appear)
28. De Medts, T., Van Maldeghem, H.: The uniqueness of a generalized hexagon of order 3 containing a subhexagon of order $(1, 3)$. Discrete Math. **309**, 714–720 (2009)
29. The GAP Group, GAP—Groups, Algorithms, and Programming, Version 4.4.12. http://www.gap-system.org (2008)
30. Offer, A., Van Maldeghem, H.: Distance-j ovoids and related structures in generalized polygons. Discrete Math. **294**, 147–160 (2005)
31. Pralle, H., Shpectorov, S.: The ovoidal hyperplanes of a dual polar space of rank 4. Adv. Geom. **7**, 1–17 (2007)
32. Ree, R.: A family of simple groups associated with the simple Lie algebra of type (F_4). Am. J. Math. **83**, 401–420 (1961)
33. Shult, E.E., Yanushka, A.: Near n-gons and line systems. Geom. Dedicata **9**, 1–72 (1980)
34. Tits, J.: Les groupes simples de Suzuki et de Ree. Séminaire Bourbaki 13 (1960/61), No. 210, pp. 18 (1961)
35. Van Maldeghem, H.: Generalized polygons. Birkhäuser, Basel (1998)

Chapter 3
An Outline of Polar Spaces: Basics and Advances

Ilaria Cardinali

Abstract This paper is an extended version of a series of lectures on polar spaces given during the workshop and conference ‘Groups and Geometries’ held at the Indian Statistical Institute in Bangalore in December 2012. We firstly give a concise exposition of the theory of polar spaces, ending up with the classification of polar spaces of rank at least 3. Then we present a few related research topics, as polar spaces of infinite rank, non-linear embeddings of polar spaces, projective embeddings of dual polar spaces and polar grassmannians.

Keywords Polar spaces · Sesquilinear forms · Quadratic forms · Embeddings · Dual polar spaces · Polar grassmanians

1 Introduction

The content of this paper has been the subject of three lectures on polar spaces given during the workshop and conference ‘Groups and Geometries’ held at the Indian Statistical Institute in Bangalore in December 2012.

The theory of polar spaces is essentially due to Tits, who gave it a complete setting in his celebrated monograph [72]. However, many of the basic ideas of this theory can be found already in Dieudonné [45], Tits [71] and an earliest layout was carried out by Veldkamp [76].

In the first part of this paper (Sect. 2) we give a concise exposition of the theory of polar spaces starting from classical polar spaces, associated to sesquilinear and pseudoquadratic forms, next widening our perspective to abstract polar spaces and

I. Cardinali (✉)
Department of Information Engineering and Mathematics, University of Siena,
Via Roma 56, 53100 Siena, Italy
e-mail: ilaria.cardinali@unisi.it

N. S. N. Sastry (ed.), *Groups of Exceptional Type, Coxeter Groups and Related Geometries*, Springer Proceedings in Mathematics & Statistics 82,
DOI: 10.1007/978-81-322-1814-2_3, © Springer India 2014

ending up with the Classification Theorem of ordinary non-degenerate polar spaces of rank at least 3.

In the rest of this paper (Sects. 3–6) we present a few research topics. In view of the constraint of keeping this exposition within a reasonable length, I have decided to choose only a few subjects, namely polar spaces of infinite rank, non-linear embeddings of polar spaces and projective embeddings of dual polar spaces and polar grassmannians.

In particular, in Sect. 3 we discuss polar spaces of infinite rank, focusing on the differences between them and the finite rank case. Section 4 is devoted to non-linear embeddings of classical polar spaces. In Sect. 5 we survey a number of results on projective embeddings of dual polar spaces. Section 6 is about projective embeddings of polar grassmannians.

Many more research subjects exist which are linked to polar spaces but are not discussed in this survey. Some of them, as the theory of generalized quadrangles (i.e. polar spaces of rank 2) are well established research fields of their own. The investigation of remarkable combinatorial structures contained in classical finite polar spaces such as ovoids, spreads, partial spreads, m-systems, tight-sets and the study of codes arising from polar spaces, such as grassmann and affine grassmann codes, are other lively research subjects.

2 Basic Results

Several monographs and lecture notes have been written on polar spaces, e.g.: Tits [72, Chapters 7–9], Cameron [19], Shult [64, Chapter 7], Ueberberg [74, Chapters 4 and 5], Buekenhout and Cohen [15, Chapters 7-10]. In my exposition I will mainly follow Tits [72] and Buekenhout and Cohen [15]. I will skip all proofs. The reader is referred to [72] and [15] for them (also Bourbaki [11] and Dieudonné [46] for Sect. 2.1).

We start by introducing sesquilinear and pseudoquadratic forms (Sects. 2.1 and 2.2). In Sect. 2.3 we define abstract polar spaces. The main theorems are reported in Sect. 2.4.

2.1 Sesquilinear Forms

Definition 2.1. Let $\mathbb{K}$ be a division ring, σ is an anti-automorphism of $\mathbb{K}$ and $\varepsilon \in \mathbb{K}^*$ such that $\varepsilon^\sigma = \varepsilon^{-1}$ and $t^{\sigma^2} = \varepsilon t \varepsilon^{-1}$ for any $t \in \mathbb{K}$.

Let $\mathbb{V}$ be a right vector space over $\mathbb{K}$. A function $\phi\colon \mathbb{V} \times \mathbb{V} \to \mathbb{K}$ is a *reflexive* (σ, ε)*-sesquilinear form* (also (σ, ε)*-sesquilinear form* for short) if

(i) $\phi(x, y\alpha + z\beta) = \phi(x, y)\alpha + \phi(x, z)\beta, \quad \forall \alpha, \beta \in \mathbb{K}, \forall x, y \in \mathbb{V};$
(ii) $\phi(y, x) = \phi(x, y)^\sigma \varepsilon, \quad \forall x, y \in \mathbb{V}.$

The following are straightforward consequences of (i) and (ii).

(iii) $\phi(\sum_i x_i\alpha_i, \sum_j y_j\beta_j) = \sum_{i,j} \alpha_i^\sigma \phi(x_i, y_j)\beta_j, \quad \forall\alpha_i, \beta_j \in \mathbb{K}, \ \forall x_i, y_j \in \mathbb{V};$
(iv) $\phi(x, x) = \phi(x, x)^\sigma \varepsilon, \quad \forall x \in \mathbb{V}.$

We adopt the following terminology and notation. If $a, b \in \mathbb{V}$ we write $a \perp b$ when $\phi(a, b) = 0$. We say that a is *isotropic* if $a \perp a$ and we put $a^\perp := \{x \in \mathbb{V} : x \perp a\}$. Note that $a^\perp$ is a linear subspace of $\mathbb{V}$, as it follows from (i) and (ii) of Definition 2.1.

If $X \subseteq \mathbb{V}$, then we say that X is *totally isotropic* if $\phi(x, y) = 0$ for every $x, y \in X$ and we define $X^\perp := \cap_{x \in X} x^\perp$. Clearly, $X^\perp = \langle X \rangle^\perp$ and $X^\perp$ is a linear subspace of $\mathbb{V}$. Obviously, X is totally isotropic if and only if $X \subseteq X^\perp$.

The following proposition is straightforward.

Proposition 2.2. *Let $a, b, c \in \mathbb{V} \setminus \{0\}$.*
If $\langle b \rangle \neq \langle c \rangle$ and $\langle b, c \rangle \not\subseteq a^\perp$ then $\dim(a^\perp \cap \langle b, c \rangle) = 1$*. Consequently, if $a^\perp \neq \mathbb{V}$ then $a^\perp$ is a hyperplane of $\mathbb{V}$.*

By Zorn's Lemma, every totally isotropic subspace is contained in a maximal totally isotropic subspace.

Definition 2.3. We say that a (σ, ε)-sesquilinear form ϕ admits a Witt index if all maximal totally isotropic subspaces have the same dimension. If that is the case, the common dimension of the maximal totally isotropic subspaces is called the Witt index of ϕ.

We warn the reader that a few authors (e.g. Buekenhout and Cohen [15]) define the Witt index in a different way, as the least upper bound of the dimensions of the totally singular subspaces. According to their definition, every form admits a Witt index, which is not the case according to our definition, as we will see in Sect. 3.

Theorem 2.4. *Let ϕ be a (σ, ε)-sesquilinear form. If there exists a maximal isotropic subspace of finite dimension then ϕ admits a Witt index.*

(See Tits [72, 8.1.3], also Theorem 3.1 cf Sect. 3 of the present paper.)

Definition 2.5. The subspace $\mathbb{V}^\perp = \{a : a^\perp = \mathbb{V}\}$ is called the radical of ϕ and denoted by $Rad(\phi)$. The form ϕ is degenerate (respectively, non-degenerate) if $Rad(\phi) \neq \{0\}$ (respectively, $Rad(\phi) = \{0\}$).

Note that $Rad(\phi) \subseteq S$ for every maximal totally isotropic subspace S of $\mathbb{V}$.

The following straightforward proposition shows that we can always replace a degenerate sesquilinear form with a non-degenerate one.

Proposition 2.6. *Let ϕ be a degenerate (σ, ε)-sesquilinear form and W a complement of $Rad(\phi)$ in $\mathbb{V}$. Then the form $\bar{\phi} := \phi|_{W \times W}$ induced by ϕ on $W \times W$ is non-degenerate and the function mapping every totally isotropic subspace S of $\mathbb{V}$ containing $Rad(\phi)$ to $S \cap W$ is an isomorphism from the partially ordered set of totally isotropic subspaces of $\mathbb{V}$ containing $Rad(\phi)$ to the partially ordered set of totally isotropic subspaces contained in W.*

Recall that the *codimension* $\mathrm{codim}(X)$ of a subspace X of $\mathbb{V}$ is defined as the (common) dimension of all vector complements of X in $\mathbb{V}$. In particular, if $\dim(\mathbb{V})$ is finite then $\mathrm{codim}(X) = \dim(\mathbb{V}) - \dim(X)$.

Proposition 2.7. *Let ϕ be a non-degenerate (σ, ε)-sesquilinear form and let X be a totally isotropic subspace. If $\dim(X) < \infty$ then $\mathrm{codim}(X^{\perp}) = \dim(X)$ and $X^{\perp\perp} = X$ (hence $\dim(X^{\perp}) = \mathrm{codim}(X)$).*

The first claim of this proposition is straightforward. The second one can be proved by the same argument used for scalar products in real vector spaces in any course of linear algebra. The equalities $X^{\perp\perp} = X$ and $\mathrm{codim}(X^{\perp}) = \dim(X)$ might fail to hold when $\dim(X)$ is infinite. A counterexample will be given at the end of Sect. 3.1.

Consider the following subgroups of the additive group of $\mathbb{K}$:

$$\mathbb{K}_1 := \{t : t = t^{\sigma}\varepsilon\} \text{ and } \mathbb{K}_2 := \{t + t^{\sigma}\varepsilon\}_{t \in \mathbb{K}}. \tag{1}$$

Clearly, $\mathbb{K}_2 \subseteq \mathbb{K}_1$. Moreover, $\phi(x, x) \in \mathbb{K}_1$ for every $x \in \mathbb{V}$, by (ii) of Definition 2.1.

Definition 2.8. A (σ, ε)-sesquilinear form ϕ is said to be trace-valued if $\phi(x, x) \in \mathbb{K}_2$ for every $x \in \mathbb{V}$.

It immediately follows from the remark preceding Definition 2.8 that for a given pair (σ, ε), we have $\mathbb{K}_1 = \mathbb{K}_2$ if and only if every (σ, ε)-sesquilinear form is trace-valued.

Theorem 2.9. [see [72, 8.1.6]] *Let ϕ be a (σ, ε)-sesquilinear form different from the null form such that not all isotropic vectors belong to $Rad(\phi)$. Then $\mathbb{V}$ is spanned by the isotropic vectors of ϕ if and only if ϕ is trace-valued.*

Let $\mathrm{char}(\mathbb{K})$ be the characteristic of $\mathbb{K}$ and $Z(\mathbb{K})$ the center of $\mathbb{K}$.

Proposition 2.10. [see [72, 8.1.5]] *If (i) $\mathrm{char}(\mathbb{K}) \neq 2$ or (ii) $\mathrm{char}(\mathbb{K}) = 2$ and $\sigma|_{Z(\mathbb{K})} \neq id$ then $\mathbb{K}_2 = \mathbb{K}_1$, so ϕ is trace-valued.*

By Proposition 2.10, if $\mathbb{K}$ is a field (namely, it is commutative), non-trace-valued (σ, ε)-sesquilinear forms only exist for $\mathrm{char}(\mathbb{K}) = 2$ and $\sigma|_{Z(\mathbb{K})} = id$. Suppose now $\mathrm{char}(\mathbb{K}) = 2$ and $\sigma|_{Z(\mathbb{K})} = id$. Suppose also that the form ϕ is not trace-valued and that non-zero isotropic vectors exist. Note that ϕ is not the null form since the null form is always trace-valued. Let $\mathbb{V}_0$ be the subspace of $\mathbb{V}$ spanned by the isotropic vectors. By Theorem 2.9, $\mathbb{V}_0$ is a proper subspace of $\mathbb{V}$. Let $\phi_0 := \phi|_{\mathbb{V}_0 \times \mathbb{V}_0}$ be the restriction of ϕ to $\mathbb{V}_0 \times \mathbb{V}_0$. By Theorem 2.9, if ϕ_0 is not the null form, then ϕ_0 is trace-valued. Moreover, every totally isotropic subspace of $\mathbb{V}$ is contained in $\mathbb{V}_0$.

So, given a non-trace-valued sesquilinear form ϕ, we can always consider the associated non-degenerate trace-valued sesquilinear form ϕ_0, defined as above.

For the rest of this section we will only deal with (σ, ε)-sesquilinear forms admitting a finite Witt index $n \in \mathbb{N}$.

Theorem 2.11. [see [15, *Proposition 7.5.3*]] *Let ϕ be a sesquilinear form of $\mathbb{V}$ of finite Witt index. Then, for every maximal totally isotropic subspace A there exists a maximal totally isotropic subspace B such that $A \cap B = Rad(\phi)$.*

Corollary 2.12. *Let ϕ be a non-degenerate trace-valued sesquilinear form of $\mathbb{V}$ of finite Witt index n. Then $2n \leq \dim(\mathbb{V})$.*

Assume that ϕ is non-degenerate and trace-valued. We say that ϕ has *maximal Witt index* if $2n + 1 \geq \dim(\mathbb{V}) \geq 2n$.

If the dimension of $\mathbb{V}$ is finite then we can associate a matrix M to ϕ, called the *Gram matrix* of ϕ. We recall that if $(e_i)_{i\in I}$ is a basis of $\mathbb{V}$, then the entries M_{ij} of M are defined as $M_{ij} := \phi(e_i, e_j)$ for every $i, j \in I$. Clearly, every (σ, ε)-sesquilinear form is completely determined by its Gram matrix together with σ.

Theorem 2.11 can be used to construct a basis $\mathbf{B}$ of $\mathbb{V}$ so that the Gram matrix of ϕ with respect to $\mathbf{B}$ has a canonical form. We briefly explain how to proceed.

Suppose the form ϕ is non-degenerate and trace-valued with finite Witt index n. By Theorem 2.11, there exist two maximal totally isotropic subspaces A and B of $\mathbb{V}$ such that $A \cap B = 0$. Let $\{a_1, \ldots, a_n\}$ be a basis of A. With the help of Proposition 2.2 and some elementary linear algebra one can prove the following.

Proposition 2.13. *All the following hold.*
(i) For every $k = 1, \ldots, n$, $B \cap (\cap_{i\neq k} a_i^{\perp})$ is a point $\langle b_k \rangle$ of PG($\mathbb{V}$) *not in $a_k^{\perp}$.*
(ii) The vectors $b_1, \ldots, b_n$ form a basis of B.
(iii) $(A + B)^{\perp}$ does not contain any isotropic point.

Since $\mathbb{V} = (A + B) \oplus (A + B)^{\perp}$, we can construct the basis $\mathbf{B}$ by taking the vectors $a_1, \ldots, a_n, b_1, \ldots, b_n$ and then completing this set with a basis $c_1, c_2 \ldots$ of $(A + B)^{\perp} \neq 0$. We can always choose $b_1, \ldots, b_n$ in such a way that $\phi(a_i, b_i) = 1$ for $i = 1, \ldots, n$. With this choice of $\mathbf{B}$ the Gram matrix of ϕ is

$$M = \begin{pmatrix} 0 & I_n & 0 \\ \varepsilon I_n & 0 & 0 \\ 0 & 0 & M_0 \end{pmatrix} \qquad (2)$$

where 0 stands for null matrices, I_n is the $(n \times n)$-identity matrix and M_0 is the Gram matrix of the restriction of ϕ to the space $(A + B)^{\perp}$. Moreover, as $\mathbb{V}$ has finite dimension by assumption, we can always choose the vectors $c_1, c_2, \ldots$ in such a way that $\phi(c_i, c_j) = 0$ for $i \neq j$, so that M_0 is a diagonal matrix.

Bilinear and hermitian forms. If σ is the identity then ϕ is called a *bilinear form*, $\mathbb{K}$ must be a field and $\varepsilon = \pm 1$. The *alternating* and *symmetric bilinear forms*, described below, are the only possible examples of (id, ε)-sesquilinear forms. A bilinear form is *alternating* if $\phi(x, x) = 0$ for every $x \in \mathbb{V}$. If ϕ is not the null form this condition forces $\varepsilon = -1$. Conversely, if char($\mathbb{K}$) $\neq 2$ and $\varepsilon = -1$ then ϕ is alternating. Clearly, all alternating bilinear forms are trace-valued. Non-degenerate alternating forms of

Witt index n exist only in vector spaces of dimension $2n$. The canonical Gram matrix of a non-degenerate alternating form is $\begin{pmatrix} 0 & I_n \\ -I_n & 0 \end{pmatrix}$.

A bilinear form ϕ is *symmetric* if $\varepsilon = 1$. If $\mathrm{char}(\mathbb{K}) = 2$ then all non-degenerate trace-valued symmetric bilinear forms are alternating.

A (σ, ε)-sesquilinear form ϕ is *hermitian* (respectively *anti-hermitian*) if $\sigma \neq id$ and $\varepsilon = 1$ (respectively, $\varepsilon = -1$). Note that $\varepsilon = \pm 1$ forces $\sigma^2 = id$. Moreover, in view of Proposition 2.10, if $\mathbb{K}$ is a field then ϕ is trace-valued.

Recall that two sesquilinear forms ϕ and ψ are *proportional* if there exists a scalar $\lambda \in \mathbb{K}^*$ such that $\phi = \lambda\psi$. The following holds.

Theorem 2.14. [see [72, 8.1.2]] *Every reflexive trace-valued (σ, ε)-sesquilinear form is either symmetric, alternating or proportional to a hermitian form.*

Two special cases. (1) Let $\mathbb{K}$ be a field of characteristic $\mathrm{char}(\mathbb{K}) \neq 2$. Suppose moreover that the subgroup of the square elements of $\mathbb{K}$ has index 2 in the multiplicative group $\mathbb{K}^*$ of $\mathbb{K}$. Let ϕ be a non-degenerate symmetric bilinear form and let M be its canonical Gram matrix as in Equation (2) (with $\varepsilon = 1$, of course). Then we have the following possibilities.

(a) $\dim(\mathbb{V}) = 2n$, hence M_0 is the empty matrix, namely $M = \begin{pmatrix} 0 & I_n \\ I_n & 0 \end{pmatrix}$.
In this case the maximal totally isotropic subspaces fall into two families $\mathcal{M}^+$ and $\mathcal{M}^-$ and every totally isotropic subspace of dimension $n-1$ is contained in exactly one maximal subspace of each family $\mathcal{M}^+$ and $\mathcal{M}^-$.

(b) $\dim(\mathbb{V}) = 2n+1$, hence $M_0 = (\lambda)$ (a matrix of order 1) with $\lambda \neq 0$. Up to rescaling the form if necessary, one can always assume that $\lambda = 1$.

(c) $\dim(\mathbb{V}) = 2n+2$, whence M_0 is diagonal of order 2. Up to rescaling the form, one can assume that $M_0 = \begin{pmatrix} 1 & 0 \\ 0 & \eta \end{pmatrix}$ where η is a non-square of $\mathbb{K}$.

(d) $\dim(\mathbb{V}) > 2n+2$. Up to rescaling the form we can assume that M_0 is the identity matrix of order $m = \dim(\mathbb{V}) - 2n$.
This case can occur only if -1 is neither a square nor the sum of two squares.

If $\mathbb{K}$ is a finite field of odd characteristic then -1 is either a square or the sum of two squares. In this case only cases (a), (b), (c) occur.

(2) Let $\mathbb{K}$ be a finite field $\mathbb{K} = \mathbb{F}_q$. Let $\sigma \neq id$. In this case $q = q_0^2$ and $t^\sigma = t^{q_0}$ for every $t \in \mathbb{K}$. In view of Theorem 2.14 we can assume that $\varepsilon = 1$. If ϕ is a non-degenerate hermitian form with finite Witt index n then $2n \leq \dim(\mathbb{V}) \leq 2n+1$ (i.e. ϕ has maximal Witt index). Moreover, when $\dim(\mathbb{V}) = 2n+1$, up to rescaling ϕ, we can assume $M_0 = (1)$.

2.2 Pseudoquadratic Forms

A motivation to introduce pseudoquadratic forms is the existence of an important class of classical polar spaces which cannot be defined by means of trace-valued sesquilinear forms.

As in Sect. 2.1, denote by $\mathbb{K}$, σ respectively ε a division ring, an anti-automorphism of $\mathbb{K}$ and an element of $\mathbb{K}^*$ as in Definition 2.1. By Proposition 2.10, if $\mathbb{K}$ is a division ring of characteristic different from 2 or $\mathrm{char}(\mathbb{K}) = 2$ and $\sigma|_{Z(\mathbb{K})} \neq id$ then all (σ, ε)-sesquilinear forms are trace-valued. Put $\mathbb{K}_{\sigma,\varepsilon} := \{t - t^{\sigma}\varepsilon\}_{t\in\mathbb{K}}$. Then $\mathbb{K}_{\sigma,\varepsilon}$ is a subgroup of the additive group of $\mathbb{K}$.

The following is straightforward.

Proposition 2.15. $\mathbb{K}_{\sigma,\varepsilon} = \mathbb{K}$ *if and only if* $\mathrm{char}(\mathbb{K}) \neq 2$ *and* $(\sigma, \varepsilon) = (id, -1)$. *On the other hand,* $\mathbb{K}_{\sigma,\varepsilon} = 0$ *if and only if* $(\sigma, \varepsilon) = (id, 1)$.

For every $\lambda \in \mathbb{K}$ and every $t \in \mathbb{K}_{\sigma,\varepsilon}$ we have $\lambda^{\sigma}t\lambda \in \mathbb{K}_{\sigma,\varepsilon}$. Hence $\lambda^{\sigma}\tau\lambda$ is a well defined element of $\mathbb{K}/\mathbb{K}_{\sigma,\varepsilon}$, for every $\tau \in \mathbb{K}/\mathbb{K}_{\sigma,\varepsilon}$.

Definition 2.16. A function $f\colon \mathbb{V} \to \mathbb{K}/\mathbb{K}_{\sigma,\varepsilon}$ is a (σ, ε)-pseudoquadratic form (quadratic form, when $(\sigma, \varepsilon) = (id, 1)$) if

(i) $f(xt) = t^{\sigma} f(x)t$ for every $x \in \mathbb{V}$ and every $t \in \mathbb{K}$;
(ii) $f(x + y) = f(x) + f(y) + (\phi(x, y) + \mathbb{K}_{\sigma,\varepsilon})$ for every $x, y \in \mathbb{V}$ and a suitable (σ, ε)-sesquilinear form ϕ.

It is not so difficult to see that $f(0) = 0$, $f(-x) = f(x)$ for every $x \in \mathbb{V}$ and the form ϕ mentioned in (ii) is trace-valued. Moreover (see [72, 8.2.1]):

Proposition 2.17. *If* $\mathbb{K}_{\sigma,\varepsilon} \neq \mathbb{K}$ *then* ϕ *is uniquely determined by* f.

If $\mathbb{K}_{\sigma,\varepsilon} = \mathbb{K}$ then the factor group $\mathbb{K}/\mathbb{K}_{\sigma,\varepsilon}$ is trivial. In this case, pseudoquadratic forms are just null functions.

From here on we assume $\mathbb{K}_{\sigma,\varepsilon} \neq \mathbb{K}$. By Proposition 2.17 the sesquilinear form ϕ is uniquely determined by f. The form ϕ is called the *sesquilinearization of* f (the *bilinearization of* f when $(\sigma, \varepsilon) = (id, 1)$). The following proposition yields a partial converse of Proposition 2.17.

Proposition 2.18. *Let* ϕ *be a trace-valued* (σ, ε)*-sesquilinear form.*
(a) Let $\mathrm{char}(\mathbb{K}) \neq 2$. *Then the function* $f\colon \mathbb{V} \to \mathbb{K}/\mathbb{K}_{\sigma,\varepsilon}$ *defined by the clause* $f(x) := \phi(x, x)/2 + \mathbb{K}_{\sigma,\varepsilon}$ *for every* $x \in \mathbb{V}$ *is a* (σ, ε)*-pseudoquadratic form and* ϕ *is its sesquilinearization.*
(b) If $\mathrm{char}(\mathbb{K}) = 2$ *and* $\sigma|_{Z(\mathbb{K})} \neq id$, *let* $t \in Z(\mathbb{K})$ *with* $t^{\sigma} \neq t$. *Put*

$$f(x) = \phi(x, x)/(1 + (t^{\sigma}/t)^2) + \mathbb{K}_{\sigma,\varepsilon} \quad \forall x \in \mathbb{V}.$$

Then the function f *defined in this way is a* (σ, ε)*-pseudoquadratic form and* ϕ *is its sesquilinearization.*

Keeping the hypothesis that $\mathbb{K}_{\sigma,\varepsilon} \neq \mathbb{K}$, let f be a (σ, ε)-pseudoquadratic form with ϕ as its sesquilinearization.

Definition 2.19. A vector $x \in \mathbb{V}$ is singular if $f(x) = 0$. A subspace S of $\mathbb{V}$ is *totally singular* if $f(x) = 0$ for every $x \in S$.

Proposition 2.20. [see [72, 8.2.6]] *A subspace $S \subseteq \mathbb{V}$ is totally singular for f if and only if it is totally isotropic for ϕ and it is spanned by a set of vectors singular for f. In particular, all points of $\mathbb{V}$ singular for f are isotropic for ϕ.*

By Proposition 2.18, if $\mathrm{char}(\mathbb{K}) \neq 2$ or $\mathrm{char}(\mathbb{K}) = 2$ and $\sigma|_{Z(\mathbb{K})} \neq id$ then every isotropic vector for ϕ is also singular for f. Consequently, totally isotropic subspaces and totally singular subspaces are the same.

For the rest of this subsection we assume $\mathrm{char}(\mathbb{K}) = 2$ and $\sigma|_{Z(\mathbb{K})} = id$.

Let $S(f)$ be the set of vectors of $\mathbb{V}$ singular for f. Then for every totally isotropic subspace S of $\mathbb{V}$ the set $S \cap S(f)$ is a (possibly trivial) subspace of S. In particular, $S(f) \cap Rad(\phi)$ is a subspace of $Rad(\phi)$.

Definition 2.21. The defect of f is the codimension of $S(f) \cap Rad(\phi)$ as a subspace of $Rad(\phi)$. If $S(f) \cap Rad(\phi) = 0$ (respectively, $S(f) \cap Rad(\phi) \neq 0$) then f is said to be *non-singular* (respectively, *singular*).

Many results stated in Sect. 2.1 for (non-degenerate) sesquilinear forms (such as Proposition 2.6 and Theorems 2.9 and 2.11) have an analogue in the present setting. For instance if $S(f) \not\subseteq Rad(\phi)$ then $\langle S(f) \rangle = \mathbb{V}$ (see [72, 8.2.7]).

We say that f *admits a Witt index* if all maximal singular subspaces have the same dimension, the common dimension of these subspaces being the *Witt index* of f. As for sesquilinear forms (Theorem 2.4), if f admits a maximal singular subspace of finite dimension then f admits a Witt index.

Suppose that both f and its sesquilinearization ϕ admit a Witt index. Let n_f and n_ϕ be the Witt indices of f and ϕ respectively. Then

Proposition 2.22. $\dim(Rad(\phi)) + n_f \leq n_\phi + \dim(Rad(\phi) \cap S(f))$.

Let S be totally isotropic for ϕ and $\langle S(f) \cap S \rangle = S$. Then S is totally singular for f. However, it can happen that $\langle S(f) \cap S \rangle \subset S$. For instance, this is the case when f is of elliptic type (see below). If $\mathbb{F}$ is non-perfect, it can even happen that the radical of ϕ is infinite dimensional while f has Witt index (finite or infinite) smaller than $\dim(Rad(\phi))$.

To conclude this subsection we give some additional information on quadratic forms. Recall that we have assumed that $\mathrm{char}(\mathbb{F}) = 2$. The following is straightforward.

Proposition 2.23. *Let $\mathbb{K}$ be perfect. Let f be a quadratic form and ϕ its bilinearization. Let X be a subspace of $\mathbb{V}$ totally isotropic for ϕ. Then either $X \subseteq S(f)$ or $X \cap S(f)$ is a hyperplane of X.*

Corollary 2.24. *With $\mathbb{K}$, f and ϕ as in Proposition 2.23, suppose moreover that f is non-singular. Then* $\dim(Rad(\phi)) \leq 1$. *If* $\dim(Rad(\phi)) = 1$ *(so, $Rad(\phi)$ is a projective point of* $PG(\mathbb{V})$*) then every line of* $PG(\mathbb{V})$ *through the point $Rad(\phi)$ contains only one singular point of* $PG(\mathbb{V})$.

In the sequel we still assume that $\mathbb{K}$ is perfect of characteristic 2, f is a non-singular quadratic form and ϕ is its sesquilinearization, as above. Moreover, we assume that f has finite Witt index $n_f = n$. The form ϕ, being bilinear and trace-valued, is alternating (since $\mathrm{char}(\mathbb{K}) = 2$). By Proposition 2.23, ϕ has Witt index $n_\phi = n + 1$ or $n_\phi = n$. By Corollary 2.24, $2n \leq \dim(\mathbb{V}) \leq 2n + 2$. The following two cases can occur.

Case 1. $\boxed{Rad(\phi) = 0.}$ Then $n \leq n_\phi \leq n + 1$. Two subcases are possible.

Case 1.1. *Hyperbolic case*: $n_\phi = n_f = n$. Hence $\dim(\mathbb{V}) = 2n$. In this case we can choose a suitable basis $\mathbf{B} = \{a_1, \ldots, a_n, b_1, \ldots b_n\}$ of $\mathbb{V}$ so that f has the following expression, where $x = \sum_{i=1}^n x_i a_i + \sum_{i=1}^n y_i b_i$ is an arbitrary vector of $\mathbb{V}$: $f(x) = \sum_{i=1}^n x_i y_i$.

Case 1.2. *Elliptic case*: $n_\phi = n_f + 1 = n + 1$. Then $\dim(\mathbb{V}) = 2n + 2$ and we can choose a suitable basis $\mathbf{B} = \{a_1, \ldots, a_n, b_1, \ldots b_n, c_1, c_2\}$ of $\mathbb{V}$ so that f has the following expression: $f(x) = \sum_{i=1}^n x_i y_i + u^2 + \eta uv + v^2$, where $\eta \in \mathbb{K}$ is such that the equation $t^2 + \eta t + 1 = 0$ has no solution in $\mathbb{K}$ and $x = \sum_{i=1}^n x_i a_i + \sum_{i=1}^n y_i b_i + uc_1 + vc_2$ is an arbitrary vector of $\mathbb{V}$.

Case 2. $\boxed{\dim(Rad(\phi)) = 1.}$ Then $n_\phi = n_f + 1 = n + 1$ and $\dim(\mathbb{V}) = 2n + 1$. The maximal isotropic subspaces of $\mathbb{V}$ for ϕ have the form $Rad(\phi) \oplus S$ where S is a maximal totally singular subspace for f. So, $n = n_\phi = n_f + 1$. The projective point $Rad(\phi)$ is said to be the *nucleus* of f and it is characterized as the unique non-singular point of $PG(\mathbb{V})$ such that every projective line through it meets $S(f)$ in exactly one point.

It is possible to choose a suitable basis $\mathbf{B} = \{a_1, \ldots, a_n, b_1, \ldots b_n, c\}$ of $\mathbb{V}$ in such a way that f has the following expression, where $x = \sum_{i=1}^n x_i a_i + \sum_{i=1}^n y_i b_i + zc$ is an arbitrary vector of $\mathbb{V}$: $f(x) = \sum_{i=1}^n x_i y_i + z^2$.

Above we have assumed that $\mathbb{K}$ is perfect. If $\mathbb{K}$ is not perfect then Proposition 2.23 fails to hold and no upper bound can be established for $\dim(Rad(\phi))$ or $n_\phi - n_f$ in general (see the last paragraph of Sect. 2.5.2).

2.3 Abstract Polar Spaces

We recall that a *partial linear space* is a connected point-line geometry $\mathcal{P} = (P, L)$ where P is the point-set, L is the line-set, any line is incident with at least two points and any pair of distinct points is incident with at most one line. As usual, we assume that lines are subsets of P. Thus, when saying that a point p is incident with a line l we just mean that $p \in l$.

Let $\mathcal{P} = (P, L)$ be a partial linear space. We adopt the following notation. If $a, b \in P$ then $a \perp b$ means that a and b are collinear. Put $a^{\perp} := \{p \in P : p \perp a\} \cup \{a\}$ for $a \in P$ and $X^{\perp} := \cap_{x \in X} x^{\perp}$ for $X \subseteq P$. We recall that a subset $S \subseteq P$ is a *subspace* of $\mathcal{P}$ if it contains all lines $l \in L$ such that $|l \cap S| > 1$. Intersection of subspaces is a subspace. Thus, for every subset $X \subset P$, we can define the *subspace* $\langle X \rangle$ *spanned by* X as the smallest subspace containing X, namely the intersection of all subspaces containing X. A subspace S is said to be *singular* if $S \subseteq S^{\perp}$.

Definition 2.25. (Buekenhout-Shult [17]**).**
A partial linear space $\mathcal{P} = (P, L)$ is a polar space if

(i) (1-all axiom) For every point $p \in P$ and every line $l \in L$, either $l \subseteq p^{\perp}$ or $|l \cap p^{\perp}| = 1$.
A polar space $\mathcal{P}$ is said to be non-degenerate if
(ii) $P^{\perp} = \emptyset$.
A polar space $\mathcal{P}$ is said to be ordinary if
(iii) each line has at least three points.

We have chosen partial linear spaces as our starting point, but more lax settings can be chosen (see Buekenhout [13]). In this paper we will only consider ordinary polar spaces, but polar spaces with lines of size 2 have also been studied (Buekenhout and Sprague [18]). As for degenerate polar spaces, we can always replace a degenerate polar space $\mathcal{P} = (P, L)$ with a non-degenerate one by replacing P with the set of singular subspaces $\langle P^{\perp} \cup \{x\} \rangle$ for $x \in P \setminus P^{\perp}$ and L with the set of singular subspaces $\langle P^{\perp} \cup l \rangle$ for $l \in L$ such that $l \cap P^{\perp} = \emptyset$ (see [15, 7.4.9]). In view of this fact, we feel free to keep degenerate polar spaces out of the scope of this survey. This choice is closer to Tits [72, Chapter 7] than Buekenhout and Cohen [15]. It will save us some little troubles when stating certain definitions and results.

Thus, from here on, $\mathcal{P}$ is a non-degenerate ordinary polar space.

Theorem 2.26. (Buekenhout and Shult). *Every singular subspace of $\mathcal{P}$ is a projective space.*

The *dimension* of a singular subspace is its dimension as a projective space. Its *rank* is its dimension, augmented by 1. A *plane* is a singular subspace of dimension 2. By Zorn's Lemma, every singular subspace is contained in a maximal singular subspace.

Definition 2.27. The rank rank($\mathcal{P}$) of $\mathcal{P}$ is the least upper bound of the ranks of the singular subspaces of $\mathcal{P}$. We say that $\mathcal{P}$ *admits a Witt index* if all maximal singular subspaces have the same rank. If this is the case, then the common rank of the maximal singular subspaces of $\mathcal{P}$ is taken as the Witt index of $\mathcal{P}$. The *Witt index* of $\mathcal{P}$, if it exists, coincides with the rank of $\mathcal{P}$.

Theorem 2.28. (Buekenhout and Shult [17]**).** *Suppose that $\mathcal{P}$ has finite rank. Then all the following hold.*

(1) For every maximal singular subspace M and every subspace $X \subseteq M$ (possibly $X = \emptyset$) there exists a maximal singular subspace M' such that $M \cap M' = X$.
(2) For every maximal singular subspace M and every point $p \notin M$, $p^{\perp} \cap M$ is a hyperplane of M and $\langle p, p^{\perp} \cap M \rangle$ is a maximal singular subspace.
(3) $\mathcal{P}$ admits a Witt index.

Properties (1)–(3) of Theorem 2.28 together with Theorem 2.26 are just rephrasing of the axioms chosen by Tits [72, Chapter 7] to define polar spaces. Hence non-degenerate ordinary polar spaces of finite rank as defined in [17] are the same as polar spaces in the meaning of [72].

The following theorem describes the so-called classical polar spaces.

Theorem 2.29. (see [72, Chapter 8])

(i) Let ϕ be a (non-degenerate) trace-valued sesquilinear form on a vector space $\mathbb{V}$. Then the points and lines of PG($\mathbb{V}$) *totally isotropic for ϕ form a (non-degenerate) ordinary polar space $\mathcal{P}_{\phi}$. If ϕ admits a Witt index n_{ϕ} then n_{ϕ} is the Witt index of $\mathcal{P}_{\phi}$.*
(ii) Let f be a (non-singular) pseudoquadratic form on a vector space $\mathbb{V}$. Then the points and lines of PG($\mathbb{V}$) *totally singular for f form a (non-degenerate) ordinary polar space $\mathcal{P}_f$. If f admits a Witt index n_f then n_f is the Witt index of $\mathcal{P}_f$.*

Definition 2.30. A non-degenerate ordinary polar space obtained from a non-degenerate trace-valued sesquilinear form or a non-singular pseudoquadratic form as explained in Theorem 2.29 is called a classical polar space.

We warn that this definition slightly differs from the one given in [15]. According to [15, Definition 10.3.12], any line-grassmannian (see Sect. 6) of a thick projective space of dimension three deserves to be called a classical polar space while, according to our Definition 2.30, only embeddable line-grassmannians can be regarded as classical polar spaces.

We say that a classical polar space is *symplectic* if it can be obtained from an alternating bilinear form; it is *orthogonal* if it can be obtained from a quadratic form (equivalently, a symmetric bilinear form when the underlying field has characteristic different from 2); it is *hermitian* if it arises from a hermitian (or anti-hermitian) form.

We will use the following notation, where $\mathbb{F}$ stands for a given field. All polar spaces considered in the following list are meant to be non-degenerate, in accordance with Definition 2.30. The number n is the rank of the space.

When $\mathbb{F}$ is finite, say $\mathbb{F} = \mathbb{F}_q$, we adopt the following usual shortenings: $W(2n-1, q)$ for $W(2n-1, \mathbb{F}_q)$, $Q^+(2n-1, q)$ for $Q^+(2n-1, \mathbb{F}_q)$, and so on.

$W(2n-1,\mathbb{F})$	symplectic in $PG(2n-1,\mathbb{F})$
$Q^+(2n-1,\mathbb{F})$	orthogonal of hyperbolic type in $PG(2n-1,\mathbb{F})$
$Q^-(2n+1,\mathbb{F})$	orthogonal of elliptic type in $PG(2n+1,\mathbb{F})$
$Q(2n,\mathbb{F})$	orthogonal in $PG(2n,\mathbb{F})$
$H(m,\mathbb{F})$	hermitian in $PG(m,\mathbb{F})$, $m=2n$ or $m=2n+1$

2.4 Fundamental Theorems

Definition 2.31. A polar space $\mathcal{P}=(P,L)$ is said to be embeddable if there exists a vector space $\mathbb{V}$ and an injective map $\varepsilon\colon P\to PG(\mathbb{V})$ from P to the set of points of $PG(\mathbb{V})$ such that

(E1) $\varepsilon(l):=\{\varepsilon(p)\}_{p\in l}$ is a line of $PG(\mathbb{V})$ for every line $l\in L$;
(E2) $\langle\varepsilon(P)\rangle=PG(\mathbb{V})$.

All classical polar spaces are embeddable, by definition. An embeddable polar space is necessarily ordinary and all of its planes are Desarguesian.

Theorem 2.30. (Tits [72], **Buekenhout and Lefèvre** [16]) *Every non-degenerate ordinary embeddable polar space of rank $n\geq 2$ is classical.*

Definition 2.33. Let $\mathcal{P}$ be a non-degenerate ordinary polar space of finite rank n. Then $\mathcal{P}$ is said to be thick if every non-maximal singular subspace is contained in at least three maximal singular subspaces. We say that $\mathcal{P}$ is top-thin if every singular subspace of dimension $n-2$ is contained in exactly two maximal singular subspaces.

For instance, the grassmannian of lines of $PG(3,\mathbb{K})$ is a top-thin polar space of rank 3. The lines of $PG(3,\mathbb{K})$ are taken as points, a line is a pencil of lines in a plane while the planes of the polar space are either the planes of $PG(3,\mathbb{K})$ or the stars of the points in $PG(3,\mathbb{K})$. When $\mathbb{K}$ is commutative this polar space is isomorphic to $Q^+(5,\mathbb{K})$.

Theorem 2.34. (Tits [72]) *Let $\mathcal{P}$ be a non-degenerate ordinary polar space with $rank(\mathcal{P})=n\geq 3$ (allowing n to be infinite).*

(1) Let $n\geq 4$. Then $\mathcal{P}$ is embeddable.
(2) Let $n=3$. Then $\mathcal{P}$ is embeddable if and only if all of its planes are Desarguesian.
(3) Let $n=3$ and let $\mathcal{P}$ be top-thin. Then $\mathcal{P}$ is isomorphic to the grassmannian of lines of $PG(3,\mathbb{K})$ *for a suitable division ring $\mathbb{K}$. It is embeddable if and only if $\mathbb{K}$ is commutative.*
(4) There exists a unique family of non-embeddable non-degenerate, ordinary thick polar spaces of rank 3. The planes of these polar spaces are Moufang (but non-Desarguesian).

A proof of (1)-(3) of this theorem different from that given in [72] can be found in [15]. The argument used in [15] for (2) and (3) is inspired by Veldkamp [76].

Another proof of (1) is given in Cuypers, Johnson and Pasini [34]. The reader is referred to Johnson [50] for a thoroughful discussion of the case where n is infinite.

A description of non-embeddable polar spaces of rank 3 as in (4) of Theorem 2.34 is given in [72, Chapter 9]. Other descriptions have been given by Mühlherr [54] and De Bruyn and Van Maldeghem [43].

By Artin-Zorn's Theorem, every Moufang finite projective plane is Desarguesian. So,

Corollary 2.35. *Every non-degenerate ordinary finite polar space of rank at least 3 is classical.*

Generalized quadrangles. Ordinary non-degenerate polar spaces of rank 2 are called *generalized quadrangles.* We have kept generalized quadrangles out of the scope of our exposition. A reason for this decision is that generalized quadrangles form a research topic of their own.

A classification of generalized quadrangles is perhaps hopeless due to the existence of too many families of non-classical generalized quadrangles.

We refer the interested reader to the following monographs for more insight on this branch of research: Payne and Thas [60], Tits and Weiss [73], Van Maldeghem [75], Thas, Thas and Van Maldeghem [70], Cardinali and Payne [25], Thas [68].

2.5 More on Embeddings

2.5.1 Generalities on Projective Embeddings of Point-Line Geometries

Let $\mathcal{P} = (P, L)$ be a point-line geometry. A *projective embedding* of $\mathcal{P}$ in a projective space $\Sigma = \mathrm{PG}(\mathbb{V})$ is an injective mapping ε from the point-set P of $\mathcal{P}$ to the point-set of Σ satisfying properties (E1) and (E2) of Definition 2.31.

We say that ε is a $\mathbb{K}$-*embedding* if $\mathbb{K}$ is the underlying division ring of the vector space $\mathbb{V}$. The dimension $\dim(\mathbb{V})$ of $\mathbb{V}$ is called the *dimension of* ε and denoted by $\dim(\varepsilon)$.

Given two projective $\mathbb{K}$-embeddings $\varepsilon_1 : \Gamma \to \mathrm{PG}(\mathbb{V}_1)$ and $\varepsilon_2 : \Gamma \to \mathrm{PG}(\mathbb{V}_2)$, a *morphism* from ε_1 to ε_2 is a semilinear mapping $f : \mathbb{V}_1 \to \mathbb{V}_2$ such that $\varepsilon_2 = f \cdot \varepsilon_1$. If a morphism $f : \varepsilon_1 \to \varepsilon_2$ exists from ε_1 to ε_2 then we say that ε_2 is a *morphic image* of ε_1 (also a *quotient* of ε_1) and we write $\varepsilon_1 \geq \varepsilon_2$. Note that (E2) of Definition 2.31 combined with the clause $\varepsilon_2 = f \cdot \varepsilon_1$ forces f to be surjective. If f is bijective then we say that ε_1 and ε_2 are *isomorphic* and we write $\varepsilon_1 \cong \varepsilon_2$. If $\varepsilon_1 \geq \varepsilon_2$ but $\varepsilon_1 \ncong \varepsilon_2$ then we say that ε_2 is a *proper quotient* of ε_1.

Let $f : \mathbb{V}_1 \to \mathbb{V}_2$ be a morphism from ε_1 to ε_2. Let π_f be the canonical projection of $\mathbb{V}_1$ onto $\mathbb{V}_1/ker(f)$. Then $\pi_f \cdot \varepsilon_1$ is a $\mathbb{K}$-embedding and $\varepsilon_2 \cong \pi_f \cdot \varepsilon_1$. This fact gives a motivation for the use of the word 'quotient' as a synonym of 'morphic image'. It also suggests the following definition: a subspace U of $\mathbb{V}$ defines a *quotient* of an embedding $\varepsilon : \Gamma \to \mathrm{PG}(\mathbb{V})$ if the composition $\varepsilon_U := \pi_U \cdot \varepsilon$ of ε with the projection $\pi_U : \mathbb{V} \to \mathbb{V}/U$ is an embedding. This happens precisely when $U \cap \varepsilon(p) = 0$ for every point $p \in P$ and $\langle U, \varepsilon(p_1)\rangle \neq \langle U, \varepsilon(p_2)\rangle$ for any two distinct points p_1, $p_2 \in P$. We call ε_U the *quotient of ε by U*.

A projective embedding $\varepsilon\colon \mathcal{P} \to \Sigma$ is said to be *relatively universal* (also *dominant*, following Tits [72, 8.6]) if it is not a proper quotient of any other projective embedding of $\mathcal{P}$. For every embedding $\varepsilon\colon \mathcal{P} \to \Sigma$ there exists a relatively universal embedding $\tilde{\varepsilon}\colon \mathcal{P} \to \tilde{\Sigma}$ such that $\tilde{\varepsilon} \geq \varepsilon$ (Ronan [62]; see also Sect. 4 of this paper). The embedding $\tilde{\varepsilon}$ is called the *(linear) hull* of ε and it is uniquely determined up to isomorphisms. Moreover, if ε' is any other embedding of $\mathcal{P}$ such that $\varepsilon' \geq \varepsilon$, then $\tilde{\varepsilon} \geq \varepsilon'$. Clearly, a projective embedding is relatively universal if and only if it is its own hull.

A projective $\mathbb{K}$-embedding is said to be *absolutely universal* if it is the hull of all projective $\mathbb{K}$-embeddings of $\mathcal{P}$. Clearly, the absolutely universal projective embedding, if it exists, is uniquely determined up to isomorphisms. Sufficient conditions for a point-line geometry to admit the absolutely universal projective $\mathbb{K}$-embedding have been obtained by Kasikova and Shult [51].

2.5.2 Projective Embeddings of Polar Spaces

Let now $\mathcal{P}$ be an embeddable non-degenerate polar space of Witt index $n \geq 2$. If either $n \geq 3$ or $\mathcal{P}$ is thick then all projective embeddings of $\mathcal{P}$ are defined over the same division ring $\mathbb{K}$ (Tits [72, 8.6]). We say that $\mathcal{P}$ is *defined over* $\mathbb{K}$.

Theorem 2.36. [see [72, 8.6]] *Let $\mathbb{K}_1$ and $\mathbb{K}_2$ be as in formula* (1). *Let $\mathcal{P}$ be a polar space defined by a non-singular (σ, ε)-pseudoquadratic form or by a non-degenerate (σ, ε)-sesquilinear form. Then*
(1) The polar space $\mathcal{P}$ admits an absolutely universal projective embedding.
(2) Let $\mathbb{K}_1 = \mathbb{K}_2$. Then, up to isomorphism, $\mathcal{P}$ admits a unique projective embedding (necessarily universal).
(3) Let $\mathbb{K}_1 \subset \mathbb{K}_2$. Then, up to isomorphism, there is a unique way to embed $\mathcal{P}$ as the space of totally singular subspaces for a non-singular (σ, ε)-pseudoquadratic form f. This embedding is absolutely universal.

Note that not all projective embeddings of a polar space arise from a pseudoquadratic form or from a sesquilinear form. Indeed, in case (3) of Theorem 2.36, let φ be the (universal) embedding defined by f and let ϕ be the sesquilinearization of f. It is possible that $\dim(Rad(\phi)) > 1$. Let U be a subspace $0 < U < Rad(\phi)$ and consider the quotient φ_U of φ by U. The embedding φ_U cannot be realized neither by a pseudoquadratic form nor by a sesquilinear form.

For instance, let $\mathcal{P} \cong W(2n-1, \mathbb{F})$ for a field $\mathbb{F}$ with $\mathrm{char}(\mathbb{F}) = 2$ and let $\bar{\varphi}$ embed $\mathcal{P}$ in $\mathrm{PG}(2n-1, \mathbb{F})$ as a symplectic polar space. Then the absolutely universal embedding $\tilde{\varphi}$ of $\mathcal{P}$ embeds $\mathcal{P}$ as a quadric in $\mathrm{PG}(\widetilde{\mathbb{V}})$, the embedding $\bar{\varphi}$ is a proper quotient of $\tilde{\varphi}$ and $\dim(\widetilde{\mathbb{V}}) = 2n + \delta$ where δ is the degree of $\mathbb{F}$ over its subfield of square elements (see De Bruyn and Pasini [42]). If $\mathbb{F}$ is perfect then $\delta = 1$ while, when $\mathbb{F}$ is non-perfect, δ is larger than 1 (and it can be even infinite). When $\delta > 1$ the situation discussed in the previous paragraph occurs: $\mathcal{P}$ admits embeddings. These embeddings cannot be obtained by a pseudoquadratic form or by a sesquilinear form.

3 Polar Spaces of Infinite Rank

The material of this section is mainly taken from Johnson [50] and Pasini [58]. We take the following theorem as a starting point for our discussion.

Theorem 3.1. *Let $\mathcal{P}$ be a non-degenerate ordinary polar space. If at least one of the maximal singular subspaces of $\mathcal{P}$ has finite dimension then all maximal singular subspaces of $\mathcal{P}$ have the same dimension.*

(Compare Theorem 2.4 and Claim (3) of Theorem 2.28 of Sect. 2.) It is worth giving a sketch of the proof of Theorem 3.1, in order to show where the existence of a finite dimensional maximal singular subspace enters the game. Claim (2) of Theorem 2.28 is the main tool in this proof. It holds for any polar space, even of infinite rank (see [15, 7.4.13]). Let now M and M' be two maximal singular subspaces of $\mathcal{P}$. Put $X := M \cap M'$. By claim (2) of Theorem 2.28, the mapping $f\colon M \to M'$ sending $x \in M$ to $x^{\perp} \cap M'$ induces an injective morphism of projective spaces from M/X to the dual $(M'/X)^*$ of M'/X (recall that M and M' are projective spaces). We have $X = \bigcap_{x \in M} f(x)$. Assume now that the dimension of M is finite. Then X is the intersection of a finite number of hyperplanes of M', corresponding to a basis of M/X. Therefore the codimension of X in M' equals the codimension of X in M. Hence $\dim(M) = \dim(M')$.

If we remove the finiteness hypothesis in Theorem 3.1, it is no more true that all maximal singular subspaces must have the same dimension. Indeed, it is not difficult to construct polar spaces which do not admit a Witt index, as we will show in Sect. 3.1.

It is worth noting that the main achievement of the theory of polar spaces, namely the classification Theorem 2.34, holds for polar spaces of rank at least 3 regardless of the finiteness or infiniteness of their rank. All differences between polar spaces of finite or infinite rank only depend on the fact that sesquilinear or pseudoquadratic forms of infinite rank can behave rather differently from those of finite rank.

3.1 An Example of a Polar Space of Infinite Rank

The following construction is due to Buekenhout and Cohen [15, 7.5.6]. The reader can see Pasini [58] for a more detailed investigation of this example. Let $\mathbb{K}$ be a division ring, σ an anti-automorphism of $\mathbb{K}$ and $\varepsilon \in \mathbb{K} \setminus \{0\}$ as in Definition 2.1. Let $\mathbb{V}$ be a right $\mathbb{K}$-vector space of infinite dimension and let $\mathbb{V}^*$ be its dual, but still regarded as a right vector space over $\mathbb{K}$ according to the rule: $\xi \cdot t := t^{\sigma}\xi$ for every $\xi \in \mathbb{V}^*$ and every $t \in \mathbb{K}$. For $\xi \in \mathbb{V}^*$ and $x \in \mathbb{V}$ we put $\langle \xi, x\rangle := \xi(x)$ and we call $\langle .,.\rangle$ the *natural scalar product* of the pair $(\mathbb{V}^*, \mathbb{V})$.

Put $\overline{\mathbb{V}} := \mathbb{V} \oplus \mathbb{V}^*$ and define a non-degenerate (σ, ε)-sesquilinear form ϕ on $\overline{\mathbb{V}}$ as follows:

$$\phi(a \oplus \alpha, b \oplus \beta) := \langle \alpha, b \rangle + \langle \beta, a \rangle^\sigma \varepsilon. \tag{3}$$

Define a polar space $\mathcal{P}$ where the points and the lines are the isotropic points and the totally isotropic lines of $\mathrm{PG}(\overline{\mathbb{V}})$ with respect to ϕ.

We immediately see that $\mathcal{P}$ does not admit any Witt index. Indeed, $\mathbb{V}$ and $\mathbb{V}^*$ are maximal totally isotropic subspaces. However $\dim(\mathbb{V}^*) \geq 2^{\dim(\mathbb{V})} > \dim(\mathbb{V})$, as $\dim(\mathbb{V})$ is infinite. So, ϕ (hence $\mathcal{P}$) admits no Witt index.

Let $\perp$ be the orthogonality relation associated to ϕ. For a subset A of $\mathbb{V}$ we have $A^\perp = \mathbb{V} \oplus (A^\perp \cap \mathbb{V}^*)$ and $A^\perp \cap \mathbb{V}^*$ is the subspace of $\mathbb{V}^*$ formed by all linear functionals $\xi \in \mathbb{V}^*$ such that $\xi(x) = 0$ for all $x \in A$. Similarly, for $B \subseteq \mathbb{V}^*$, $B^\perp = (B^\perp \cap \mathbb{V}) \oplus \mathbb{V}^*$ and $B^\perp \cap \mathbb{V}$ is formed by all vectors $x \in \mathbb{V}$ such that $\xi(x) = 0$ for all $\xi \in B$. Let now A and B be subspaces of $\mathbb{V}$ and $\mathbb{V}^*$ respectively such that $A_0 := B^\perp \cap \mathbb{V} \subset A$ and $B_0 := A^\perp \cap \mathbb{V}^* \subset B$. Then $A_0 = B^\perp \cap A$ and $B_0 = A^\perp \cap B$. Hence $A_0 \oplus B_0$ is singular. Moreover $\mathbb{V}^*/B_0$ is isomorphic to the dual A^* of A, the quotient $\mathbb{V}^*/B \cong (\mathbb{V}^*/B_0)/(B/B_0)$ is isomorphic to A_0^* and B_0 is isomorphic to the dual $(\mathbb{V}/A)^*$ of $\mathbb{V}/A$. As a consequence, B/B_0 is isomorphic to a subspace of $(A/A_0)^*$ and A/A_0 is isomorphic to a subspace of $(B/B_0)^*$. The natural scalar product $\langle .,. \rangle$ of $(\mathbb{V}^*, \mathbb{V})$ induces a scalar product $\langle .,. \rangle_{B,A}$ for the pair $(B/B_0, A/A_0)$, defined as follows:

$$\langle \Xi, X \rangle_{B,A} = \xi(x) \text{ where } \xi \in \Xi \text{ and } x \in X, \text{ for } \Xi \in B/B_0 \text{ and } X \in A/A_0.$$

As $A_0 \subseteq B^\perp$ and $B_0 \subseteq A^\perp$, $\xi(x)$ does not depend on the choice of $\xi \in \Xi$ and $x \in X$. So, $\langle .,. \rangle_{B,A}$ is well defined. Suppose that $A/A_0 \cong B/B_0$ and let $f : A/A_0 \to B/B_0$ be an isomorphism satisfying

$$\langle f(X), Y \rangle_{B,A} + \langle f(Y), X \rangle^\sigma_{B,A} \varepsilon = 0, \quad \forall X, Y \in A/A_0.$$

Put $M_{A,B,f} = \{x \oplus \xi | x \in A, \xi \in f(x + A_0)\}$. It can be proved that $M_{A,B,f}$ is a maximal totally isotropic subspace of $\overline{\mathbb{V}}$ and every maximal totally isotropic subspace can be obtained as above, for a suitable choice of A, B and f. Moreover, $\dim(\mathbb{V}) \leq \dim(M_{A,B,f}) \leq \dim(\mathbb{V}^*)$.

Note that $A = A^{\perp\perp}$ for every subspace A of $\mathbb{V}$. Nevertheless, many totally isotropic subspaces S of $\overline{\mathbb{V}}$ exist such that S is properly contained in $S^{\perp\perp}$. For instance, chosen a basis $(e_i)_{i\in I}$ of $\mathbb{V}$, let $\lambda_i \in \mathbb{V}^*$ be defined by the clause $\langle \lambda_i, e_j \rangle = \delta_{ij}$ (Kronecker symbol). Put $S := \langle \lambda_i \rangle_{i\in I}$. Then $S \subset S^\perp = S^{\perp\perp} = \mathbb{V}^*$. Moreover, $\dim(S) = \dim(\mathbb{V}) < \dim(\mathbb{V}^*)$.

3.2 Open Problems

As said in Sect. 2 (Theorem 2.28 Claim (1)) the following holds in all non-degenerate ordinary polar spaces of finite rank.

(SS) (Strong Separation Property) For every maximal singular subspace M, there exists a maximal singular subspace M' such that $M \cap M' = \emptyset$.

We can also consider the following weaker version of (SS).

(WS) (Weak Separation Property) There exists at least one pair of mutually disjoint maximal singular subspaces.

Properties (SS) and (WS) also hold in non-degenerate ordinary polar spaces of infinite rank provided there exists at least one maximal singular subspace of countable dimension (see Pasini [58]). The following question is quite natural.

(Q1) Does (SS) (or at least (WS)) hold in any ordinary polar space of non-countable rank? So far no example is known where (SS) fails to hold.

Here is another problem.

(Q2) Let $\mathcal{P}$ be a polar space of infinite rank, admitting no Witt index. Is there any function f from the class of cardinal numbers to itself such that, if d is the minimal dimension of a maximal singular subspace of $\mathcal{P}$, then every maximal singular subspace of $\mathcal{P}$ has dimension at most $f(d)$?

The following is another question that can be asked.

(Q3) Suppose that f is a non-singular (σ, ε)-pseudoquadratic form defined on a division ring $\mathbb{K}$ of characteristic 2, with $\sigma|_{Z(\mathbb{K})} = id$. Let ϕ be the sesquilinearization of f and suppose that the maximal singular subspaces of ϕ have infinite dimension. Suppose moreover that f admits a Witt index. Does this force ϕ to admit a Witt index as well?

4 Non-Linear Embeddings of Polar Spaces

In this section we address the problem of embeddability of polar spaces in groups. In Sect. 4.1 we give the basic ingredients of a general theory of embeddings for poset geometries allowing the subgroup lattice of any group to be a feasible codomain. This abstract approach will provide a framework where both projective embeddings as defined in Sect. 2.5 and representation groups in the sense of Ivanov and Shpectorov [49] (see also [48]) can be placed quite naturally. In Sect. 4.3 we focus on embeddings of polar spaces in groups and we will report on the main achievements obtained in this respect. The source for the material of this section is Pasini [57].

4.1 Basics on Geometries

We firstly recall some standard facts and set some terminology from diagram geometry. Our reference is Pasini [56].

A geometry of rank n is recursively defined as follows. A geometry of rank 1 is a set with at least two elements. A geometry of rank $n \geq 2$ is a connected graph equipped with an n-partition Θ such that, for every vertex x of Γ, the set of vertices adjacent to x equipped with the partition induced by Θ is a geometry of rank $n-1$. Moreover, a set I and a surjective mapping $\tau\colon \Gamma \to I$ are given such that the classes of Θ are just the fibers of τ. The function τ and the set I are called the *type function* and the *set of types* of Γ. This emphasis on types is useful in practice but we warn the reader not to be mislead by it. Indeed the partition Θ is uniquely determined by the graph Γ (see [56, Section 1.7]).

The vertices of Γ are called the *elements* of the geometry. Two elements are said to be *incident* when they are adjacent in Γ. The cliques of Γ are the *flags* of the geometry and the *rank* (*corank*) of a flag F is its size $|F|$ (respectively, $n - |F|$). Clearly, every flag picks at most one element out of each type. Flags of corank 1 are called *panels*. Flags of rank n are called *chambers*. Given a flag F, the subset $\tau(F) \subseteq I$ (respectively $I \setminus \tau(F)$) is called the *type* (*cotype*) of F.

The residue $Res_\Gamma(F)$ of F (also $Res(F)$ for short) is the set of elements of Γ incident with (all elements of) F. The inductive nature of our definition implies that Γ is residually connected ($Res(F)$ is connected for every flag F of corank at least 2) and firm (every flag of rank less than n is contained in at least two chambers).

Example 4.1. A geometry of rank 2 is just a connected bipartite graph were every vertex has valency at least 2. For instance, projective planes, affine planes, generalized quadrangles are geometries of rank 2. The proper non-empty subspaces of a n-dimensional projective or affine space form a geometry of rank n. Also, the singular subspaces of a non-degenerate polar space of rank n form a geometry of rank n where the types are naturally identified with the dimension of the singular subspaces. Projective, affine and polar spaces yield in fact examples of poset-geometries as defined below.

Poset-geometries. A geometry Γ is a *poset-geometry* when its set of types is equipped with a total ordering $\leq$ such that, for any three elements X, Y, Z of Γ, if $\tau(X) \leq \tau(Y) \leq \tau(Z)$ and Y is incident with both X and Z, then X is incident with Z. If X and Y are two elements of Γ, we write $X \leq Y$ when X and Y are incident and $\tau(X) \leq \tau(Y)$.

Assuming that $\{0, 1, \ldots, n-1\}$ is the set of types of Γ, ordered in the natural way, we denote by $P(\Gamma)$ and $L(\Gamma)$ the set of elements of Γ of type 0 and respectively 1. The elements of $P(\Gamma)$ (respectively $L(\Gamma)$) are called *points* (*lines*). If X is an element of Γ then $P(X)$ denotes the set of points $p \leq X$ and $L(X)$ the set of lines incident with X. Moreover, we assume the following weaker version of the Intersection Property (see [56, Chapter 6]): $X \leq Y$ if and only if $P(X) \subseteq P(Y)$.

Coverings. As geometries are graphs, a *morphism* of geometries is just a morphism of graphs. Given two geometries Γ' and Γ of rank $n \geq 2$ over the same set of types I and with type functions τ' and τ respectively, a morphism $\varphi\colon \Gamma' \to \Gamma$ is *type preserving* if $\tau'(x) = \tau(\varphi(x))$ for every $x \in \Gamma'$. A type preserving morphism is an *isomorphism* if it is an isomorphism of graphs.

A *covering* from Γ' to Γ is a type-preserving morphism $\varphi\colon \Gamma' \to \Gamma$ such that, for every element x of Γ', the restriction of φ to $Res_{\Gamma'}(x)$ is an isomorphism to $Res_{\Gamma}(\varphi(x))$. If $\varphi\colon \Gamma' \to \Gamma$ is a covering, we say that Γ' is a *cover* of Γ and Γ is a *quotient* of Γ'. We say that a covering $\bar{\varphi}\colon \bar{\Gamma} \to \Gamma$ is *universal* if for every covering $\varphi'\colon \Gamma' \to \Gamma$ there exists a covering $\psi\colon \bar{\Gamma} \to \Gamma'$ such that $\bar{\varphi} = \varphi'\psi$. If a covering $\bar{\varphi}\colon \bar{\Gamma} \to \Gamma$ exists and is universal, then $\bar{\Gamma}$ is called the *universal cover* of Γ. The universal cover always exists and it is uniquely determined up to isomorphisms (see e.g. [56, Chapter 12]).

4.2 Embeddings of Poset-Geometries

Definition 4.2. Let Γ be a poset-geometry and G a group. An *embedding* of Γ in G is an injective mapping $\varepsilon\colon \Gamma \to G$ from the set of elements of Γ to the set of proper non-trivial subgroups of G such that

(E1) for $X, Y \in \Gamma$, we have $\varepsilon(X) \leq \varepsilon(Y)$ if and only if $X \leq Y$;
(E2) $\varepsilon(X) = \langle \varepsilon(p) \rangle_{p \in P(X)}$ for every element X in Γ;
(E3) $G = \langle \varepsilon(p) \rangle_{p \in P(\Gamma)}$.

Definition 4.3. Given an embedding $\varepsilon\colon \Gamma \to G$ of a poset-geometry Γ of rank n over the set of types $\{0, 1, \ldots, n-1\}$ in a group G, we define a poset-geometry $Exp(\varepsilon)$ of rank $n+1$ as follows. The points of $Exp(\varepsilon)$ are the elements of G and, for $i = 1, 2, \ldots, n$ the elements of $Exp(\varepsilon)$ of type i are the cosets $g \cdot \varepsilon(X)$, for $g \in G$ and $X \in \Gamma$ with $\tau(X) = i - 1$. The incidence relation is the inclusion between cosets and between elements and cosets. We call $Exp(\varepsilon)$ the *expansion* of Γ to G via ε.

Note that G, in its action by left multiplication on itself, is turned into a subgroup of $Aut(Exp(\varepsilon))$.

Example 4.4. Take a geometry S of rank 1, namely just a set of size at least 2. Let G be a group and $\varepsilon\colon S \to G$ an embedding of S. So, ε bijectively maps the elements of S onto the members of a collection $\mathcal{P}$ of proper non-trivial subgroups of G such that no member of $\mathcal{P}$ contains any other member of $\mathcal{P}$. If $\mathcal{P}$ is a partition of G then $Exp(\varepsilon)$ is a translation Andrè structure. If moreover any two distinct members of $\mathcal{P}$ commute and generate G, then $Exp(\varepsilon)$ is an affine translation plane.

Example 4.5. Let $\mathbb{V}$ be a vector space of finite dimension n, let $\Gamma = \mathrm{PG}(\mathbb{V})$ and let G be the additive group of $\mathbb{V}$. Let $\varepsilon\colon \Gamma \to G$ be the identity embedding sending every proper non-trivial subspace of $\mathbb{V}$ to itself. Then $Exp(\varepsilon)$ is an affine space of dimension $n + 1$.

4.2.1 A Few Special Cases

Let $\varepsilon\colon \Gamma \to G$ be an embedding of a poset-geometry Γ into a group G. When G is commutative then ε is said to be *abelian*. In particular, let G be the additive group of a vector space $\mathbb{V}$ defined over a given division ring $\mathbb{K}$. If $\varepsilon(X)$ is a linear subspace of $\mathbb{V}$ for every $X \in \Gamma$, then we say that ε is a *$\mathbb{K}$-linear embedding* of Γ in $\mathbb{V}$.

If ε is a $\mathbb{K}$-linear embedding of a poset-geometry of rank at least 2 such that $\dim(\varepsilon(p)) = 1$ for all points $p \in P(\Gamma)$, $\dim(\varepsilon(L)) = 2$ and $\varepsilon(L) = \cup_{p\in P(L)}\varepsilon(p)$ for every line L of Γ, then ε is a *projective embedding* as defined in Sect. 2.5.

Note that if ε is a projective embedding of Γ in PG($\mathbb{V}$) then the elements of $Exp(\varepsilon)$ are points and subspaces of the affine space $AG(\mathbb{V})$ (i.e. vectors and cosets of subspaces of $\mathbb{V}$).

Let $\varepsilon\colon \Gamma \to G$ be an embedding of Γ in a group G. Suppose that two families $\{V(p)\}_{p\in P(\Gamma)}$ and $\{V(L)\}_{L\in L(\Gamma)}$ of $\mathbb{K}$-vector spaces are given such that the following hold:

(LP1) For every point p, $\dim(V(p)) = 1$ and $\varepsilon(p)$ is the additive group of $V(p)$.
(LP2) For every line L, $\dim(V(L)) = 2$ and $\varepsilon(L)$ is the additive group of $V(L)$. Moreover $\{V(p)\}_{p\in P(L)}$ is the family of all 1-dimensional linear subspaces of $V(L)$.

Then we say that ε is a *locally $\mathbb{K}$-projective embedding*.

4.2.2 Morphisms and Abstract Hulls

Given two embeddings $\varepsilon\colon \Gamma \to G$ and $\eta\colon \Gamma \to F$, a *morphism* from ε to η is a group homomorphism $f\colon G \to F$ such that, for every $X \in \Gamma$, the restriction of f to $\varepsilon(X)$ is an isomorphism to $\eta(X)$. If furthermore $f\colon G \to F$ is an isomorphism, then we say that ε and η are *isomorphic* and we write $\varepsilon \cong \eta$. If a morphism exists from ε to η, then we say that η is a *morphic image* of ε and that ε *dominates* η.

Given a morphism f from ε to η, for every $X \in \Gamma$ the homomorphism f maps the coset $g\cdot\varepsilon(X)$ of G onto the coset $f(g)\cdot\eta(X)$ of F. Accordingly, f defines a morphism $Exp(f)\colon Exp(\varepsilon) \to Exp(\eta)$ which is indeed a covering, i.e. the residues of the elements of $Exp(\varepsilon)$ are isomorphically mapped onto residues of elements of $Exp(\eta)$.

Let $\varepsilon\colon \Gamma \to G$ be an embedding. We will define a further embedding $\tilde{\varepsilon}\colon \Gamma \to U(\varepsilon)$, which we call the *abstract hull* of ε. The group $U(\varepsilon)$ is defined as follows.

Consider the amalgam $\mathcal{A}_\varepsilon = (\{\varepsilon(X)\}_{X\in\Gamma}, \{\iota_{X,Y}\}_{X<Y})$ where $\iota_{X,Y}$ is the inclusion map between $\varepsilon(X)$ and $\varepsilon(Y)$, for $X < Y$. Denote by $\widetilde{X}$ the disjoint union of the subgroups $\varepsilon(X)$'s where $\varepsilon(X)$ is regarded as an abstract group (not as a subgroup of G) and let $F(\widetilde{X})$ be the free group of $\widetilde{X}$. Let $\mathcal{R} = (\cup_{X\in\Gamma}\mathcal{R}_X)\cup(\cup_{X<Y}\mathcal{R}_{X,Y})$ where $\mathcal{R}_X$ is the set of all words $\phi = \phi(x_1,\dots,x_n)$ for $x_1,\dots,x_n \in \varepsilon(X)$ such that $\phi = 1$ in $\varepsilon(X)$ and $\mathcal{R}_{X,Y}$ is the set of all words $\iota_{X,Y}(x)x^{-1}$ for $x \in \varepsilon(X)$ and $X < Y$. Then

$U(\varepsilon) := F(\widetilde{X})/N_{\mathcal{R}}$ where $N_{\mathcal{R}}$ is the normal subgroup of $F(\widetilde{X})$ generated by $\mathcal{R}$. In short, $U(\varepsilon)$ is the universal completion of the amalgam $\mathcal{A}_\varepsilon$.

For every $X \in \Gamma$, let ϕ_X be the natural embedding of the group $\varepsilon(X)$ in $U(\varepsilon)$. Put $\tilde{\varepsilon}(X) := \phi_X(\varepsilon(X))$ for every $X \in \Gamma$. Then $\tilde{\varepsilon}\colon \Gamma \to U(\varepsilon)$ embeds Γ in $U(\varepsilon)$. The canonical projection of $U(\varepsilon)$ onto G induces a morphism $\pi_\varepsilon\colon \tilde{\varepsilon} \to \varepsilon$, which we call the *canonical projection* of $\tilde{\varepsilon}$ onto ε. It can be proved that for every embedding η of Γ and every morphism $f : \eta \to \varepsilon$ there exists a unique morphism $g\colon \tilde{\varepsilon} \to \eta$ such that $f \circ g = \pi_\varepsilon$. The following is proved in Pasini [57].

Theorem 4.6. *The geometry $Exp(\tilde{\varepsilon})$ is the universal cover of $Exp(\varepsilon)$.*

Definition 4.7. If $\varepsilon \cong \tilde{\varepsilon}$ then we say that ε is *abstractly dominant.*

If we restrict ourselves to consider only special classes of embeddings, like projective embeddings or embeddings of geometries in abelian groups, then we can define the concept of abelian hull and linear hull of an embedding in a similar way as we have defined the concept of abstract hull. More precisely, if G is an abelian group and $\varepsilon\colon \Gamma \to G$ is an embedding of Γ, then the *abelian hull* of ε is the embedding $\tilde{\varepsilon}_{ab}\colon \Gamma \to U(\varepsilon)_{ab}$ where $U(\varepsilon)_{ab}$ is the quotient of $U(\varepsilon)$ over its commutator subgroup.

Analogously, if G is the additive group of a vector space and $\varepsilon\colon \Gamma \to G$ is a linear embedding, we consider the largest (possibly improper) quotient $U(\varepsilon)_{lin}$ of $U(\varepsilon)_{ab}$ that carries the structure of a vector space compatible with the vector space structure of $\varepsilon(X)$ for every $X \in \Gamma$. Explicitly, $U(\varepsilon)_{lin}$ is obtained by factorizing $U(\varepsilon)_{ab}$ over the subgroup generated by the elements $\sum_{i=1}^m ta_i$ for $t \in \mathbb{K}$ and $a_1, a_2, \dots a_m \in \cup_{X\in\Gamma}\varepsilon_{ab}(X)$ such that $\sum_{i=1}^m a_i = 0$ in $U(\varepsilon)_{ab}$. If $\pi\colon U(\varepsilon)_{ab} \to U(\varepsilon)_{lin}$ is the canonical projection of $U(\varepsilon)_{ab}$ onto $U(\varepsilon)_{lin}$ then the composition $\tilde{\varepsilon}_{lin} = \pi \cdot \tilde{\varepsilon}_{ab}$ is an embedding of Γ in $U(\varepsilon)_{lin}$, called the *linear hull* of ε. In particular, if Γ has rank 2 and ε is a projective embedding of Γ then $\tilde{\varepsilon}_{lin}$ is the linear hull of ε as defined in Sect. 2.5.1.

A projective embedding which is its own linear hull is seldom abstractly dominant. On the other hand the abstract hull of a projective embedding is always locally projective.

In the next section we will deal with the abstract hull of a projective embedding of a polar space. Another example is briefly discussed below.

Example 4.8. Let Γ = DQ(6, p) be the dual polar space of orthogonal type and rank 3 defined over the finite field $\mathbb{F}_p$ where p is an odd prime and let ε be the spin embedding of DQ(6, p) into (the additive group of) $V(8, p)$ (see Buekenhout and Cameron [14], also Sect. 5 of the present paper). The abstract hull of ε embeds DQ(6, p) into the unipotent radical of the stabilizer of an element x of a building of type $F_4(p)$ where x is chosen in such a way that $Res(x) \cong$ DQ(6, p). In other words, $U(\varepsilon)$ is extraspecial of type p_+^{1+14} (see [57, Section 9]).

4.3 Hulls of Embeddings of Polar Spaces

Before focusing on polar space we need a more general definition. Let Γ be a thick building of connected spherical type and rank at least 2. Given a flag $F \neq \emptyset$ and a chamber C of Γ, there is a unique chamber $C_F \supseteq F$ at minimal distance from C, where distance is computed in the graph having as vertices the chambers and with two chambers adjacent when they intersect in a panel. We denote the distance between C and C_F by $d(C, C_F)$. For every non-empty flag X, the distance $d(X, F)$ from X to F is the minimal distance $d(C, F)$ from F to a chamber C containing X.

We say that *a flag X is far from F* if $d(X, F)$ is maximal, compatibly with the types of F and X.

Definition 4.9. The geometry $Far_\Gamma(F)$ is the substructure of Γ formed by the elements far from F. Two elements $X, Y \in Far_\Gamma(F)$ are incident in $Far_\Gamma(F)$ if and only if they are incident in Γ and the flag $\{X, Y\}$ is far from F.

The structure $Far_\Gamma(F)$ is indeed a geometry as defined in Sect. 4.1 (in particular, it is residually connected, except for a few small cases defined over $\mathbb{F}_2$ (see Blok and Brouwer [4]), which however do not fall into the scope of what we are going to say.

From here on Γ denotes a classical polar space of rank $n \geq 2$, regarded as a geometry of rank n as explained in Example 4.1, and $\varepsilon\colon \Gamma \to \mathrm{PG}(\mathbb{V})$ is a projective embedding of Γ. In Sect. 2.5 we have defined projective embeddings only for the point-line geometry $(P(\Gamma), L(\Gamma))$ of Γ, but such an embedding ε can obviously be extended to the whole of Γ by setting $\varepsilon(X) := \cup_{p \in P(X)} \varepsilon(p)$ for any $X \in \Gamma$.

Let $\Delta := Exp(\varepsilon)$ be the expansion of ε and $\widetilde{\Delta}$ the universal cover of Δ. The following is proved in Cuypers and Pasini [33].

Proposition 4.10. *We have $\widetilde{\Delta} \cong Far_\Pi(p_0)$ for a polar space Π of rank $n+1$ and a point p_0 of Π, such that $\Gamma \cong Res_\Pi(p_0)$.*

Let $\tilde{\varepsilon}\colon \Gamma \to U(\varepsilon)$ be the abstract hull of ε. According to Theorem 4.6,

Proposition 4.11. $\widetilde{\Delta} \cong Exp(\tilde{\varepsilon})$.

The next theorem follows from Propositions 4.10 and 4.11. It completely characterizes the abstract hull $\tilde{\varepsilon}$ of a projective embedding ε of a classical polar space of finite rank at least 2. A special case of this theorem also appears in Sahoo and Sastry [63], for symplectic polar spaces defined over $\mathbb{F}_q$ for q an odd prime power.

Theorem 4.12. *The codomain $U(\varepsilon)$ of $\tilde{\varepsilon}$ is isomorphic to the unipotent radical of the stabilizer of p_0 in $Aut(\Pi)$, with Π and p_0 as in Proposition 4.10.*

It is not so difficult to see that $\widetilde{\Delta} = \Delta$ if and only if $\varepsilon(\Gamma)$ is a quadric. Consequently,

Corollary 4.13. *The embedding ε is abstractly dominant if and only if $\varepsilon(\Gamma)$ is a quadric.*

As an example, let $\Gamma = W(2n-1, p)$ for an odd prime p and let $\varepsilon\colon \Gamma \to \mathrm{PG}(2n-1, p)$ be its natural embedding in $\mathrm{PG}(2n-1, p)$. Then $\tilde{\varepsilon}$ is the unipotent radical of the stabilizer of a point of $W(2n+1, p)$. It is well known that this group is isomorphic to the extra special group p_+^{1+2n}.

4.4 Problems

The following two problems concern the investigation of the universal completion of remarkable geometries of rank $n \geq 2$. More open problems can be found in [57].

(1) Describe the abstract hull of the half-spin embedding of a building of type D_n (see Buekenhout and Cameron [14]).
(2) Describe the abstract hull of a projective embedding of a dual polar space (see Sect. 5).

5 Dual Polar Spaces

In this section we focus on projective embeddings of dual polar spaces of rank $n \geq 3$. We do not consider the case $n = 2$ since dual polar spaces of rank 2 are generalized quadrangles, hence polar spaces of rank 2, which we have decided not to discuss in this paper, as already said at the end of Sect. 2.4. We firstly expose some basics about projective embeddings of a dual polar space. After that, we provide a survey of known results concerning the universal and the minimal polarized projective embeddings of a classical dual polar space. The material of this section is taken from [7, 9, 26, 32, 35, 40, 41, 65, 66], and [51].

5.1 Definition and General Results

Let $\mathcal{P}$ be a thick polar space of rank $n \geq 3$ and let Δ be its associated dual polar space, i.e., Δ is the point-line geometry whose points, respectively lines, are the maximal, respectively next-to-maximal, singular subspaces of $\mathcal{P}$ with inclusion as the incidence relation. We keep the notation introduced in Sect. 2.3, adding the following to it. We will denote a dual polar space by putting a "D" in front of the name of the corresponding polar space. So, $\mathrm{DQ}(2n, \mathbb{F})$ is the dual of $Q(2n, \mathbb{F})$, $\mathrm{DW}(2n-1, \mathbb{F})$ is the dual of $W(2n-1, \mathbb{F})$ and so on.

If x and y are two points of Δ, then (x, y) denotes the distance between x and y in the collinearity graph of Δ. The maximal distance between two points of Δ is the *diameter* of Δ. It is equal to the rank n of $\mathcal{P}$. We also call n the *rank* of Δ.

A subspace X of Δ is called *convex* if every point on a shortest path (in the collinearity graph) between two points of X is also contained in X. In particular, convex subspaces of diameter 2 are called *quads*. They are generalized quadrangles and correspond to the singular subspaces of $\mathcal{P}$ of (projective) dimension $n-3$.

A proper subspace of Δ meeting every line of Δ non-trivially is called a *hyperplane.* For every point p of Δ, let H_p be the set of points of Δ at non-maximal distance from p. Then H_p is a hyperplane (see Shult and Yanushka [67]). We call it the *singular hyperplane* of Δ with p as its *deepest point*. It is well-known that H_p is a maximal subspace of Δ, see e.g. Brouwer and Wilbrink [12].

Let $\varepsilon\colon \Delta \to \Sigma = \mathrm{PG}(\mathbb{V})$ be a projective embedding of Δ (defined as in Sect. 2.5) and p a point of Δ. Note firstly that the underlying division ring of ε is uniquely determined by Δ, as it follows from [72, 8.6] applied to the embedding ε_Q induced by ε on a quad Q of Δ. This allows us to talk about projective embeddings of thick dual polar spaces, without mentioning the underlying division rings.

As noticed above, the singular hyperplane H_p is a maximal subspace of Δ. Consequently, its image $\varepsilon(H_p)$ spans either a hyperplane of Σ or the whole of Σ. Following Thas and Van Maldeghem [69], we state the following.

Definition 5.1. A projective embedding $\varepsilon\colon \Delta \to \Sigma = \mathrm{PG}(\mathbb{V})$ is said to be *polarized* if $\langle\varepsilon(H_p)\rangle$ is a hyperplane of Σ for every point p of Δ.

If ε is polarized, then $\langle\varepsilon(H_p)\rangle \cap \varepsilon(P) = \varepsilon(H_p)$. As noticed in [41, Remarks 2 and 3], if the rank n of Δ is sufficiently large, then Δ admits non-polarized embeddings. It is not difficult to see that if ε admits a polarized quotient, then ε is polarized as well.

Definition 5.2. Let $\varepsilon\colon \Delta \to \Sigma = \mathrm{PG}(\mathbb{V})$ be a projective embedding. An automorphism g of Δ *lifts to* Σ *through* ε if there exists an automorphism $\varepsilon(g)$ of Σ such that $\varepsilon(g)\cdot\varepsilon = \varepsilon\cdot g$. Clearly $\varepsilon(g)$, if it exists, is uniquely determined by g. We call it the *lifting* of g to Σ *through* ε. Let G be a subgroup of the automorphism group $\mathrm{Aut}(\Delta)$ of Δ. If all elements of G lift to Σ through ε then we say that G *lifts* to Σ, we put $\varepsilon(G) := \{\varepsilon(g)\}_{g\in G}$ and we call $\varepsilon(G)$ the *lifting* of G to Σ (also, the lifting of G to $\mathrm{Aut}(\Sigma)$). Clearly, $\varepsilon(G)$ is a group. If G lifts to Σ through ε then we also say that ε is *G-homogeneous*.

Let ε be G-homogeneous. A subspace U of $\mathbb{V}$ is said to be *stabilized* by G (also, *G-invariant*), if it is stabilized by $\varepsilon(G)$. Clearly, if U defines a quotient ε_U of ε and U is G-invariant, then ε_U is G-homogeneous.

We denote by $\mathrm{Aut}(\Delta)_0$ the normal subgroup of $\mathrm{Aut}(\Delta)$ generated by the root groups of the building associated to Δ. We say that a projective embedding $\varepsilon\colon \Delta \to \Sigma$ is *$\mathrm{Aut}(\Delta)_0$-homogeneous* if $\mathrm{Aut}(\Delta)_0$ lifts through ε to a subgroup of the full automorphism group $Aut(\Sigma)$ of Σ.

In the rest of this subsection we summarize the major achievements obtained for homogeneous, polarized and absolute universal embeddings of dual polar spaces. In Sect. 5.2 we will investigate in detail the situation in certain special cases. The following is proved in Blok, Cardinali, De Bruyn and Pasini [9].

Theorem 5.3. *If ε is* $\mathrm{Aut}(\Delta)_0$*-homogeneous, then it is polarized.*

The next theorem follows from Kasikova and Shult [51, 4.6].

Theorem 5.4. *The dual polar space Δ admits the absolutely universal projective embedding, provided that it admits at least one projective embedding.*

Let Δ admit a polarized embedding. The first one of the next two theorems is proved in [26] (but see also [9]). The second one is due to De Bruyn and Pasini [41].

Theorem 5.5. *Up to isomorphisms, there exists a unique polarized embedding $\bar{\varepsilon}$ such that every polarized embedding ε of Δ has a quotient isomorphic to $\bar{\varepsilon}$.*

The embedding $\bar{\varepsilon}$ is called the *minimal polarized embedding* of Δ.

Theorem 5.6. $\dim(\bar{\varepsilon}) \geq 2^n$.

For the rest of this section when we say that a projective embedding is *universal* we mean it is absolutely universal in the sense of Sect. 2.5.1.

Let $\tilde{\varepsilon} : \Delta \to \mathrm{PG}(\widetilde{V})$ be the universal projective embedding of Δ (which exists by Theorem 5.4) and let $\bar{\varepsilon}$ be the minimal polarized embedding (which exists by Theorem 5.6). So, with every polarized embedding ε of Δ there are associated two projections: $\tilde{\varepsilon} \to \varepsilon \to \bar{\varepsilon}$.

The universal projective embedding is homogeneous, as it readily follows from the definition of universality stated in Sect. 2.5.1. Hence every subgroup of the full automorphism group $\mathrm{Aut}(\Delta)$ of Δ lifts through $\tilde{\varepsilon}$ to a subgroup of $\mathrm{P}\Gamma\mathrm{L}(\widetilde{\mathbb{V}})$. In particular, $\tilde{\varepsilon}$ is also $\mathrm{Aut}(\Delta)_0$-homogeneous. Hence it is polarized by Theorem 5.3 (see also Cardinali, De Bruyn and Pasini [27, Corollary 1.8] for a different, more straightforward proof of this claim).

Put $R = \bigcap_{x \in P} \langle \tilde{\varepsilon}(H_x) \rangle$ where P stands for the point-set of Δ. We call R the *nucleus* of $\tilde{\varepsilon}$. By [26], R defines a quotient $\tilde{\varepsilon}_R$ of $\tilde{\varepsilon}$, which is polarized. Moreover, a projective embedding of Δ is polarized if and only if it admits a (possibly improper) quotient isomorphic to $\tilde{\varepsilon}_R$. For this reason, $\tilde{\varepsilon}_R$ is the minimal polarized projective embedding of Δ. Note that R might be trivial. If that is the case then $\tilde{\varepsilon}$ is the unique polarized embedding of Δ.

Clearly R is an $\mathrm{Aut}(\Delta)$-invariant subspace of $\mathrm{PG}(\widetilde{\mathbb{V}})$. In particular, R is $\mathrm{Aut}(\Delta)_0$-invariant. Hence $\tilde{\varepsilon}_R$ is $\mathrm{Aut}(\Delta)_0$-homogeneous. As proved in [9],

Theorem 5.7. *All* $\mathrm{Aut}(\Delta)_0$*-invariant proper subspaces of* $\mathrm{PG}(\widetilde{\mathbb{V}})$ *are contained in R.*

In other words R, regarded as a subspace of $\widetilde{\mathbb{V}}$, is the largest proper $Z{\cdot}\tilde{\varepsilon}(G)$-submodule of $\widetilde{\mathbb{V}}$, where $G = \mathrm{Aut}(\Gamma)_0$, Z stands for the center of $\mathrm{GL}(\widetilde{\mathbb{V}})$ and $Z{\cdot}\tilde{\varepsilon}(G)$ is the preimage of $\tilde{\varepsilon}(G)$ by the projection of $\mathrm{GL}(\widetilde{\mathbb{V}})$ onto $\mathrm{PGL}(\widetilde{\mathbb{V}})$.

5.2 Embeddings of Certain Dual Polar Spaces

In the sequel $\mathbb{F}$ stands for a given field. In a few occasions we shall mention generating sets. A subset X of the point-set P of Δ is said to *generate* Δ if $\langle X \rangle = P$. Clearly, if X generates Δ then $\dim(\varepsilon) \leq |X|$ for every projective embedding ε of Δ. Consequently, if $\dim(\varepsilon) = |X|$ for a generating set X, then ε is universal.

5.2.1 Embeddings of DW$(2n-1, \mathbb{F})$

Let $\Delta = \mathrm{DW}(2n-1, \mathbb{F})$. Then $\mathrm{Aut}(\Delta)_0 = \mathrm{PSp}(2n, \mathbb{F})$. The dual polar space Δ admits the so-called *Grassmann embedding* $\varepsilon^{\mathrm{gr}} : \Delta \rightarrow \mathrm{PG}(W_n)$, where $W_n = V(N, \mathbb{F}) \subset \bigwedge^n V$, $V = V(2n, \mathbb{F})$, $N = \binom{2n}{n} - \binom{2n}{n-2}$. It maps every subspace $\langle v_1, v_2, \ldots, v_n \rangle$ of V totally isotropic for the alternating form associated to $W(2n-1, \mathbb{F})$ to the point $\langle v_1 \wedge v_2 \wedge \cdots \wedge v_n \rangle$ of $\mathrm{PG}(W_n)$.

The grassmann embedding is polarized (Cardinali, De Bruyn and Pasini [26]) and it is proved to be universal when $\mathbb{F} \neq \mathbb{F}_2$ (see Cooperstein [28] for the finite case and De Bruyn and Pasini [40] for the infinite case). When $\mathbb{F} = \mathbb{F}_2$, the universal embedding of Δ has dimension equal to $\frac{(2^n+1)(2^{n-1}+1)}{3}$ (Li [52], Blokhuis and Brouwer [10]).

As far as the minimal polarized embedding is concerned, if $\mathrm{char}(\mathbb{F}) = 0$ then the Grassmann embedding is the minimal polarized embedding of Δ. Indeed, let R be the nucleus of $\varepsilon^{\mathrm{gr}}$, regarded as a subspace of $\mathbb{V} = V(N, \mathbb{F})$. If $\mathrm{char}(\mathbb{F}) = 0$ then $R = 0$ (see e.g. De Bruyn [38]). As the Grassmann embedding is also universal, it is the unique polarized embedding of Δ.

When $\mathrm{char}(\mathbb{F}) \neq 0$ then in general $R \neq 0$.

Suppose $\mathrm{char}(\mathbb{F}) = 2$. By Blok, Cardinali and De Bruyn [8] (see also Cardinali and Lunardon [20] for the rank 3 case), if $\mathrm{char}(\mathbb{F}) = 2$ then R has vector dimension $\dim(R) = N - 2^n$. In this case, the minimal polarized embedding $\varepsilon^{\mathrm{gr}}/R$ has dimension 2^n. In particular, if $\mathbb{F}$ is perfect then $\Delta \cong \mathrm{DQ}(2n, \mathbb{F})$ and $\varepsilon^{\mathrm{gr}}/R$ is just the spin embedding of $\mathrm{DQ}(2n, \mathbb{F})$.

If $\mathrm{char}(\mathbb{F}) > 2$ the dimension of the minimal polarized embedding can be computed by a recursive formula due to Premet and Suprunenko [61] (see also Adamovich [1]) which indeed describes the dimensions of the modules in the composition series of R, where R is regarded as a module for $\mathrm{PSp}(2n, \mathbb{F})$. A complete geometric insight of that composition series is still missing. Some progress aimed to explain the situation in more geometric terms have been made in Cardinali and Pasini [21].

5.2.2 Embeddings of DQ$(2n, \mathbb{F})$

Let $\Delta = \mathrm{DQ}(2n, \mathbb{F})$. Then $\mathrm{Aut}(\Delta)_0 = \mathrm{P}\Omega(2n+1, \mathbb{F})$. The dual polar space Δ admits a polarized projective embedding $\varepsilon_{\mathrm{spin}} : \Delta \rightarrow \mathrm{PG}(2^n - 1, \mathbb{F})$, called the *spin embedding* of Δ (Buekenhout and Cameron [14, Section 7]; see also De Bruyn [37]).

When char($\mathbb{F}$) $\neq 2$ the embedding $\varepsilon_{\text{spin}}$ is universal (Wells [78]) and minimal polarized. Hence it is the unique polarized projective embedding of Δ. The universality of the spin embedding also follows from the fact that Δ admits a generating set of size 2^n (Blok and Brouwer [5], Cooperstein and Shult [32]).

Let char($\mathbb{F}$) $= 2$. Suppose furthermore that $\mathbb{F}$ is perfect. Then $Q(2n, \mathbb{F}) \cong W(2n-1, \mathbb{F})$. If N_0 denotes the radical of the bilinear form associated to the quadratic form defining $Q(2n, \mathbb{F})$, then the projection from $V := V(2n+1, \mathbb{F})$ to $V/N_0 \cong V(2n, \mathbb{F})$ induces an isomorphism from $Q(2n, \mathbb{F})$ to $W(2n-1, \mathbb{F})$, which can be regarded as an isomorphism from $\mathrm{DQ}(2n, \mathbb{F})$ to $\mathrm{DW}(2n-1, \mathbb{F})$. Thus, we are lead back to Sect. 5.2.1.

As for the minimal polarized embedding, if $\mathbb{F}$ is perfect then the spin embedding is the minimal polarized embedding. If $\mathbb{F}$ is non-perfect the minimal polarized embedding still has dimension 2^n, as when $\mathbb{F}$ is perfect (see De Bruyn [39]).

5.2.3 Embeddings of $\mathrm{DQ}^-(2n+1, \mathbb{F})$

We recall that $\mathrm{Q}^-(2n+1, \mathbb{F})$ arises from a non-singular quadratic form of $V(2n+2, \mathbb{F})$ of Witt index n. That form becomes a quadratic form of Witt index $n+1$ when $\mathbb{F}$ is replaced by a quadratic Galois extension $\mathbb{F}_1$ of $\mathbb{F}$.

Let $\Delta = \mathrm{DQ}^-(2n+1, \mathbb{F})$. Then $\mathrm{Aut}(\Delta)_0 = \mathrm{P}\Omega^-(2n+2, \mathbb{F})$. The dual polar space Δ admits a polarized projective embedding in $\mathrm{PG}(2^n-1, \mathbb{F}_1)$, called the *spin embedding* of Δ. It arises from the half-spin embedding of the half-spin geometry of $\mathrm{Q}^+(2n+1, \mathbb{F}_1)$ (see Sect. 6.2). We refer to Cooperstein and Shult [32] and De Bruyn [39] for details. We shall denote this embedding by the symbol $\varepsilon^-_{\text{spin}}$. The embedding $\varepsilon^-_{\text{spin}}$ is universal and minimal polarized. Indeed Δ admits a generating set of size 2^n (see [32] for the finite case and [39] for the general case).

5.2.4 Embeddings of $\mathrm{DH}(2n-1, \mathbb{F})$

Let $\Delta = \mathrm{DH}(2n-1, \mathbb{F})$. Then $\mathrm{Aut}(\Delta)_0 = \mathrm{PSU}(2n, \mathbb{F})$. The dual polar space Δ admits a polarized projective embedding in $\mathrm{PG}(N-1, \mathbb{F}_0)$, where $N = \binom{2n}{n}$ and $\mathbb{F}_0$ is the subfield of $\mathbb{F}$ fixed by the automorphism σ of $\mathbb{F}$ used to define $H(2n-1, \mathbb{F})$. See Cooperstein [29]; also [26], De Bruyn [36]. We recall that $|\mathbb{F} : \mathbb{F}_0| = 2$. We call this embedding the *Grassmann embedding* of Δ and we denote it by $\varepsilon^{\text{gr}}_H$. The attribute 'Grassmann' is motivated by the fact that $\varepsilon^{\text{gr}}_H$ arises from the usual embedding of the Grassmannian of n-subspaces of $V(2n, \mathbb{F})$ in $\mathrm{PG}(N-1, \mathbb{F})$, via the choice of a suitable Baer subgeometry $\mathrm{PG}(N-1, \mathbb{F}_0)$ of $\mathrm{PG}(N-1, \mathbb{F})$. If $|\mathbb{F}_0| > 2$ then $\varepsilon^{\text{gr}}_H$ is universal. Indeed, in this case Δ admits a generating set of size $\binom{2n}{n}$ (see Cooperstein [29] for the finite case and De Bruyn and Pasini [40] for the general case). If $\mathbb{F}_0 = \mathbb{F}_2$ then $\varepsilon^{\text{gr}}_H$ is not universal. In this case the universal embedding has dimension $\frac{(4^n+2)}{3}$ (Li [53]).

In any case, the grassmann embedding $\varepsilon_H^{\mathrm{gr}}$ is the minimal polarized embedding of Δ (see [26]).

We add a few words on $\mathrm{DH}(2n,\mathbb{F})$. The generalized quadrangle $\mathrm{DH}(4,\mathbb{F})$ is not classical. Hence it is not embeddable (Theorem 2.32). Therefore $\mathrm{DH}(2n,\mathbb{F})$ does not admit any projective embedding. Nevertheless, a grassmann embedding $\varepsilon_H^{\mathrm{gr}}$ can be defined for $\mathrm{DH}(2n,\mathbb{F})$ too, just as for $\mathrm{DH}(2n-1,\mathbb{F})$. The embedding $\varepsilon_H^{\mathrm{gr}}$ is linear in the sense of Sect. 4, but is not projective. In fact, it maps lines onto hermitian curves (see Cardinali and Pasini [24]).

5.2.5 The Dual of a Non-Embeddable Polar Space.

Let $\mathcal{P}$ be a non-embeddable thick polar space of rank 3 as in (4) of Theorem 2.34 and let Δ be its dual. Then Δ admits a 56-dimensional projective embedding, which is $Aut(\Delta)_0$-homogeneous (hence polarized, by Theorem 5.3.) and universal (De Bruyn and Van Maldeghem [44]). It is not known if that embedding admits any proper quotient.

6 Polar Grassmannians

Let $\mathrm{PG}(\mathbb{V})$ be a projective space. For $1 \leq k \leq \dim(\mathbb{V})$, denote by $\mathcal{G}_k$ the k–Grassmannian of $\mathrm{PG}(\mathbb{V})$, i.e. the point–line geometry whose points are the k–dimensional subspaces of $\mathbb{V}$ and whose lines are sets of the form $l_{X,Y} := \{Z \mid X \subset Z \subset Y,\ \dim(Z) = k\}$, where X and Y are any two subspaces of $\mathbb{V}$ with $\dim(X) = k-1$, $\dim(Y) = k+1$ and $X \subset Y$.

Definition 6.1. Let $\mathcal{P}$ be a classical non-degenerate ordinary polar space of finite rank n naturally embedded in a projective space $\mathrm{PG}(\mathbb{V})$. For $1 \leq k \leq n$, the *k-polar grassmannian* Δ_k is the proper subgeometry of $\mathcal{G}_k$ whose points are the (singular) subspaces of $\mathcal{P}$ of vector dimension k. If $k \neq n$ then the lines of Δ_k are the sets $l_{X,Y} := \{Z \mid X \subset Z \subset Y,\ \dim(Z) = k\}$ where X and Y are singular subspaces of $\mathcal{P}$ with vector dimension $\dim(X) = k-1$ and $\dim(Y) = k+1$. If $k = n$ then the lines of Δ_n are the sets $l_X := \{Z \mid X \subset Z,\ \dim(Z) = n\}$ where X is a singular subspaces of $\mathcal{P}$ with vector dimension $\dim(X) = n-1$.

Note that a 1-polar grassmannian is just a polar space and an n-polar grassmannian is a dual polar space. Polar grassmannians of symplectic, orthogonal or hermitian polar spaces are usually called *symplectic grassmannians*, *orthogonal grassmannians* or *hermitian grassmannians*, respectively.

There are three main research fields related to polar grassmannians: (1) properties and classification of hyperplanes; (2) generating sets; (3) projective embeddings.

To my knowledge, not so much is known on topic (1), except in the extreme cases $k = 1$ and $k = n$. For $k = 1$, it follows from Ronan's theory of [62] that, if the embedding of the polar space $\mathcal{P}$ in PG($\mathbb{V}$) is the absolutely universal one, all hyperplanes of $\mathcal{P}$ arise as intersection of the point-set of $\mathcal{P}$ (regarded as a subset of PG($\mathbb{V}$)) with hyperplanes of PG($\mathbb{V}$). Consequently, all hyperplanes of $\mathcal{P}$ are (possibly degenerate) polar spaces. On the contrary, the situation looks rather wild for $k = n$ (dual polar spaces). A lot of work has been done on hyperplanes of dual polar spaces during the last ten years (by Pralle and, especially, De Bruyn), but an accurate report of that work would take too long. As for (2), this field is covered by Cooperstein [30]. Thus, I feel free to skip (2), focusing only on the last aforementioned topic. I will try to give an updated report of what is currently known about projective embeddings of polar grassmannians. Since the cases $k = 1$ and $k = n$ (respectively polar and dual polar spaces) have been already discussed in previous sections of this paper, I will only consider k-polar grassmannians with $1 < k < n$. We shall say a few more on the case $k = n$ at the end of this section (Sect. 6.2).

First of all, it follows from Kasikova and Shult [51] that every embeddable k-polar grassmannian admits the absolutely universal projective embedding (henceforth called *universal* for short, as we have done in Sect. 5).

As noticed in Sect. 5.2, there is a tight link between generating sets and universal embeddings. The smallest cardinality of a generating set is called the *generating rank* of Δ_k and it is denoted by $gr(\Delta_k)$. As said for dual polar spaces in Sect. 5.2, $\dim(\varepsilon) \leq gr(\Delta_k)$ for every projective embedding of Δ_k. If $\dim(\varepsilon) = gr(\Delta_k)$ then ε is universal. As $1 < k < n$ by assumption, Δ_k contains subgeometries isomorphic to the line-grassmannian of PG(3, $\mathbb{K}$), where $\mathbb{K}$ is the underlying division ring of $\mathbb{V}$. The latter is embeddable if and only if $\mathbb{K}$ is commutative. Therefore Δ_k admits a projective embedding only if $\mathbb{K}$ is commutative. The converse also holds true, as we will see in a few lines.

Grassmann embedding. Let $\mathbb{K}$ be commutative. Let $W_k := \wedge^k\mathbb{V}$ be the k-th exterior power of $\mathbb{V}$ and $\varepsilon_k^{\mathrm{gr}}$ the mapping from the set of points of Δ_k to the set of points of PG(W_k) sending every point $X = \langle v_1, ..., v_k\rangle$ of Δ_k to the 1-dimensional subspace $\langle v_1 \wedge ... \wedge v_k\rangle$ of W_k.

Then $\varepsilon_k^{\mathrm{gr}}$ is a projective embedding of Δ_k in the subspace $W_k^{\mathrm{gr}} := \langle\varepsilon_k^{\mathrm{gr}}(\Delta_k)\rangle$ of W_k spanned by $\varepsilon_k^{\mathrm{gr}}(\Delta_k)$. We call $\varepsilon_k^{\mathrm{gr}}$ the *Grassmann embedding* of Δ_k (compare Sects. 5.2.1 and 5.2.4).

6.1 Special Cases

Symplectic Grassmannians. Let Δ_k be the k-grassmannian of $W(2n-1, \mathbb{F})$, for $1 < k < n$. The following theorem is due to Blok and Cooperstein, see [2, 3, 6, 29].

Theorem 6.2. *Let* $\mathrm{char}(\mathbb{F}) \neq 2$. *Then the grassmann embedding* $\varepsilon_k^{\mathrm{gr}} : \Delta_k \to$ PG(W_k^{gr}) *is universal. It has dimension* $\binom{2n}{k} - \binom{2n}{k-2}$.

Let char$(\mathbb{K}) = 2$. If $\mathbb{F}$ is perfect then $W(2n-1,\mathbb{F}) \cong Q(2n,\mathbb{F})$. In this case Δ_k can also be regarded as the k-grassmannian of $Q(2n,\mathbb{F})$. Let $\bar{\varepsilon}_k$ be the grassmann embedding of the k-grassmannian of $Q(2n,\mathbb{F})$. It is proved in Cardinali and Pasini [29] that $\varepsilon_k^{\mathrm{gr}}$ is a proper quotient of $\bar{\varepsilon}_k$. Consequently:

Proposition 6.3. *Let $\mathbb{F}$ be a perfect field of characteristic 2 and let $k < n$. Then $\varepsilon_k^{\mathrm{gr}}$ is not universal.*

Most likely the same holds when $\mathbb{F}$ is non-perfect. (Compare the last paragraph of Sect. 2.5). We warn the reader that, on the contrary, $\varepsilon_n^{\mathrm{gr}}$ is universal provided that $\mathbb{F} \neq \mathbb{F}_2$ (see Sect. 5.2.1).

Hermitian Grassmannians.
Let Δ_k be the k-grassmannian of $H(2n-1,\mathbb{F})$, for $1 < k < n$. Let $\varepsilon_k^{\mathrm{gr}}$ be its grassmann embedding. (Compare Sect. 5.2.4.) By Blok and Cooperstein [7] we have

Theorem 6.4. *Let $\mathbb{F} \neq \mathbb{F}_4$. Then the embedding $\varepsilon_k^{\mathrm{gr}}$ is universal. It has dimension $\binom{2n}{k}$ (whence $W_k^{gr} = W_k$).*

To my knowledge not so much is known on the grassmann embedding of the k-grassmannians of $H(2n,\mathbb{F})$.

Orthogonal Grassmannians.
For an up-to-date survey of orthogonal grassmannian I refer to Pasini [59] where not only projective embeddings are discussed. Not so much is known on universal embeddings of orthogonal grassmannians. Theorem 6.6 (to be stated below) contains all we presently know on this problem.

Before going on, we need to define a further embedding of Δ_k which relies on the theory of Weyl modules. We refer to Humphreys [47] for basics about Lie algebra.

Weyl embeddings. Let $\lambda_1, \lambda_2, \ldots, \lambda_n$ be the fundamental dominant weights for the root system of type B_n.

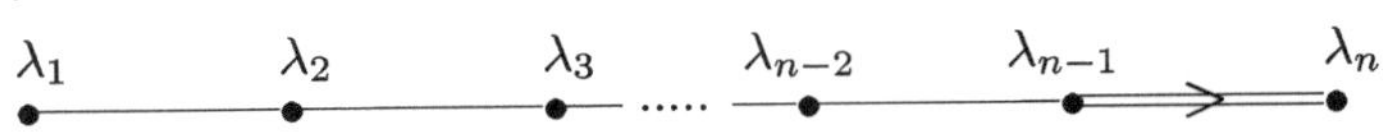

For $\lambda = \lambda_1, \ldots, \lambda_n$ or $\lambda = 2 \cdot \lambda_n$, let $V(\lambda)$ be the Weyl module with λ as the highest weight. An embedding ε_λ of Δ_k (for $\lambda = \lambda_k$ or $k = n$ and $\lambda = 2\lambda_n$) can be created in $V(\lambda)$ as follows. Let v_0 be a highest weight vector of $V(\lambda)$. Let G be the universal Chevalley group of type B_n. Then the G-orbit of $\langle v_0 \rangle$ corresponds to the set of points of Δ_k and, if P_k is the minimal fundamental parabolic subgroup of G of type k and L_0 is the P_k-orbit of $\langle v_0 \rangle$, then the G-orbit of L_0 corresponds to the set of lines of Δ_k. If X_0 is the point of Δ_k corresponding to $\langle v_0 \rangle$ then ε_λ maps $g(X_0)$ to $g(\langle v_0 \rangle)$, for every $g \in G$. If $\lambda = \lambda_k$ then ε_λ is projective.

On the other hand, $\varepsilon_{2\lambda_n}$ is not projective. Indeed lines of Δ_n are mapped by $\varepsilon_{2\lambda_n}$ onto non-singular conics, as one can see by computing L_0 explicitly. We denote the embedding $\varepsilon_{2\lambda_n}$ by the symbol $\varepsilon_n^{\mathrm{W}}$ and we call it the *Weyl quadratic embedding* of

Δ_n. We extend this notation to the case of $k < n$: when $k < n$ we put $\varepsilon_k^{\mathrm{W}} = \varepsilon_{\lambda_k}$ and we call $\varepsilon_k^{\mathrm{W}}$ the *Weyl embedding* of Δ_k.

We have $\dim(V(2\lambda_n)) = \binom{2n+1}{n}$ and $\dim(V(\lambda_k)) = \binom{2n+1}{k}$ when $k < n$, as one can check by using the Weyl dimension formula.

Theorem 6.5. [Cardinali and Pasini [23]]

(i) Let $\mathrm{char}(\mathbb{F}) \neq 2$. *Then* $\varepsilon_k^{\mathrm{W}} \cong \varepsilon_k^{\mathrm{gr}}$.
(ii) Let $\mathrm{char}(\mathbb{F}) = 2$. *Then* $\dim(\varepsilon_k^{\mathrm{gr}}) = \binom{2n+1}{k} - \binom{2n+1}{k-2} < \binom{2n+1}{k} = \dim(\varepsilon_k^{W})$ *and* $\varepsilon_k^{\mathrm{gr}}$ *is a proper quotient of* ε_k^{W}.

Theorem 6.6. [Cardinali and Pasini [23]] *Let* $\mathbb{F}$ *be a perfect field of positive characteristic or a number field.*
(1) If $n > 2$ *then* $\varepsilon_2^{\mathrm{W}}$ *is universal.*
(2) Let $n > 3$ *and* $\mathbb{F} \neq \mathbb{F}_2$. *Then* $\varepsilon_3^{\mathrm{W}}$ *is universal.*

We conjecture that the Weyl embedding $\varepsilon_k^{\mathrm{W}}$ is universal for every $k = 2, 3, \ldots, n-1$, with no restriction on $\mathbb{F}$.

6.2 More on Polar Grassmannians

Half-spin geometries and their embeddings have been occasionally mentioned in the previous pages as well as linear but non-projective embeddings, where lines are mapped onto conics, for instance. We cannot discuss these two topics properly here. Nevertheless, we cannot avoid to say a very few words on them.

Half Spin Geometries. A half spin geometry is a rank two geometry having as point-set one of the two families $\mathcal{M}^+$ or $\mathcal{M}^-$ of maximal singular subspaces for a non-singular quadratic form of Witt index n defined in a $2n$-dimensional vector space. To fix ideas, suppose to have chosen the members of $\mathcal{M}^+$ as points. Then a line is the set of maximal singular subspaces of $\mathcal{M}^+$ containing a given $(n-2)$-dimensional subspace. The unique polarized (hence universal) projective embedding of the half spin-geometry is the so-called *half-spin embedding*, of dimension 2^{n-1}. We refer the reader to [14] and [31] for details.

Linear non-projective embeddings of polar grassmannians. Projective embeddings are a particular case of linear embeddings as defined in Sect. 4.2.1. Quadratic embeddings (called Veronesean in [22]) are particularly interesting. They map lines onto non-degenerate conics. For instance, the Weyl embedding ε_n^W (Sect. 6.1) and the Grassmann embedding of $\mathrm{DQ}(2n, \mathbb{F})$ are quadratic. The Grassmann embedding of $\mathrm{DH}(2n, \mathbb{F})$ is also linear but non-projective (see Sect. 5.2.4). A lot of work is still waiting to be done for linear non-projective embeddings. For more on this topic, we refer the interested reader to Cardinali and Pasini [24].

More on grassmannians. Polar grassmannians as defined in this paper are a very special case of shadow geometries of buildings, defined and discussed by Tits [72, Chapter 12]. The reader is referred to Pankov [55] for a thorough investigation of grassmannians of buildings of spherical type. A generalization of Tits' ideas from [72, Chapter 12] to arbitrary geometries can be found in Pasini [56, Chapter 5].

References

1. Adamovich, A.M.: The submodule lattice of Weyl modules for symplectic groups with fundamental highest weights. Mosc. Univ. Math. Bull. **41**, 6–9 (1986)
2. Blok, R.J.: The generating rank of the symplectic line-Grassmannian. Beiträge Algebra Geom. **44**, 575–580 (2003)
3. Blok, R.J.: The generating rank of the symplectic Grassmannians: hyperbolic and isotropic geometry. European J. Combin. **28**, 1368–1394 (2007)
4. Blok, R.J., Brouwer, A.E.: The geometry far from a residue.In: Groups and Geometries (Siena: (1996), pp. 29–38. Trends Math , Birkhäuser, Basel, (1998)
5. Blok, R.J., Brouwer, A.E.: Spanning point-line geometries in buildings of spherical type. J. Geom. **62**, 26–35 (1998)
6. Blok, R.J., Cooperstein, B.N.: Projective subgrassmannians of polar Grassmannian. Bull. Belg. Math. Soc. Simon Stevin **17**(4), 675–691 (2010)
7. Blok, R.J., Cooperstein, B.N.: The generating rank of the unitary and symplectic grassmannians. J. Combin. Theory Ser. **A119**(1), 1–13 (2012)
8. Blok, R.J., Cardinali, I., De Bruyn, B.: On the nucleus of the Grassmann embedding of the symplectic dual polar space $DSp(2n, \mathbb{F})$, char$(\mathbb{F}) = 2$. European J. Combin. **30**, 468–472 (2009)
9. Blok, R.J., Cardinali, I., De Bruyn, B., Pasini, A.: Polarized and homogeneous embeddings of dual polar spaces. J. Alg. Combin. **30**, 381–399 (2009)
10. Blokhuis, A., Brouwer, A.E.: The universal embedding dimension of the binary symplectic dual polar space. Discrete Math. **264**, 3–11 (2003)
11. Bourbaki, N., Algèbre, Chp. 9: Formes sesquilinéares et formes quadratiques. Actu. Sci. Ind. n 1272, Hermann, Paris, (1959)
12. Brouwer, A.E., Wilbrink, H.A.: The structure of near polygons with quads. Geom. Dedicata **14**, 145–176 (1983)
13. Buekenhout, F.: On the foundations of polar geometry II. Geom. Dedicata **33**, 21–26 (1990)
14. Buekenhout, F., Cameron, P.J.: Projective and affine geometry over division rings. (Chapter 2) of the In: F. Buekenhout, F.(ed.) "Handbook of Incidence Geometry" Elsevier, Amsterdam (1995)
15. Buekenhout, F., Cohen, A.: Diagram geometry: related to classical groups and buildings. Springer, Berlin (2013)
16. Buekenhout, F., Lefèvre-Percsy, C.: Generalized quadrangles in projective spaces. Arch. Math. (Basel) **25**, 540–552 (1974)
17. F. Buekenhout and E. E. Shult. On the foundations of polar geometry. Geom. Dedicata **3**, 155–170 (1974)
18. Buekenhout, F., Sprague, A.: Polar spaces having some line of cardinality two. J. Combin. Theory Ser. **A33**(2), 223–228 (1982)
19. Cameron, P.J.: Projective and polar spaces. QMW Maths Notes, 13. Queen Mary and Westfield College, School of Mathematical Sciences, London, viii + 147pp (1990)
20. Cardinali, I., Lunardon, G.: A geometric description of the spin-embedding of symplectic dual polar spaces of rank 3. J. Combin. Theory Ser. A **115**, 1056–1064 (2008)
21. Cardinali, I., Pasini, A.: On Weyl modules for the symplectic group. Innov. Incidence Geom. **12**, 85–110 (2011)
22. Cardinali, I., Pasini, A.: Veronesean embeddings of dual polar spaces of orthogonal type. J. Combin. Th. A **120**, 1328–1350 (2013)

23. Cardinali, I., Pasini, A.: Grassmann and Weyl embeddings of orthogonal grassmannians. J. Algebraic Combin. **38**,863–888 (2013)
24. Cardinali, I., Pasini, A.: Embeddings of Line-grassmannians of Polar Spaces in Grassmann Varieties. (this volume)
25. Cardinali, I., Payne, S.E.: q-Clan Geometries in Characteristic 2. Birkhäuser Verlag, Basel (2007)
26. Cardinali, I., De Bruyn, B., Pasini, A.: Minimal full polarized embeddings of dual polar spaces. J. Algebraic Combin. **25**, 7–23 (2007)
27. Cardinali, I., De Bruyn, B., Pasini, A.: On the simple connectedness of hyperplane complements in dual polar spaces. Discrete Math. **309**, 294–303 (2009)
28. Cooperstein, B.N.: On the generation of dual polar spaces of unitary type over finite fields. European J. Combin. **18**, 849–856 (1997)
29. Cooperstein, B.N.: On the generation of dual polar spaces of symplectic type over finite fields. J. Combin. Th. **A83**, 221–232 (1998)
30. Cooperstein, B.N.: Generation of Lie Incidence Geometries: a Survey (this volume)
31. Cooperstein, B.N., Shult, E.E.: Frames and bases of Lie incidence geometries. J. Geometry **60**, 17–46 (1997)
32. Cooperstein, B.N., Shult, E.E.: A note on embedding and generating dual polar spaces. Adv. Geom. **1**, 37–48 (2001)
33. Cuypers, H., Pasini, A.: Locally polar geometries with affine planes. European J. Combin. **13**, 39–57 (1992)
34. Cuypers, H., Johnson, P.M., Pasini, A.: On the embeddability of polar spaces. Geom. Dedicata **44**, 349–358 (1992)
35. De Bruyn, B.: A decomposition of the natural embedding spaces for the symplectic dual polar spaces. Linear Algebra Appl. **426**, 462–477 (2007)
36. De Bruyn, B.: On the Grassmann-embeddings of the hermitian dual polar spaces. Linear Multilinear Algebra **56**, 665–677 (2008)
37. De Bruyn, B.: The structure of the spin-embeddings of dual polar spaces and related geometries. European J. Combin. **29**, 1242–1256 (2008)
38. De Bruyn, B.: On the Grassmann modules for the symplectic groups. J. Algebra **324**(2), 218–230 (2010)
39. De Bruyn, B.: A note on the spin-embedding of the dual polar space $DQ^-(2n+1, \mathbb{K})$. Ars Combin. **99**, 365–375 (2011)
40. De Bruyn, B., Pasini, A.: Generating symplectic and Hermitian dual polar spaces over arbitrary fields nonisomorphic to $\mathbb{F}_2$. Electron. J. Combin. **14** R54, pp.1–17 (2007)
41. De Bruyn, B. Pasini, A.: Minimal scattered sets and polarized embeddings of dual polar spaces. European J. Combin. **28**(7):1890–1909 (2007)
42. De Bruyn, B., Pasini, A.: On symplectic polar spaces over non-perfect fields of characteristic 2. Linear and Multilinear Algebra **57**, 567–575 (2009)
43. De Bruyn, B., Van Maldeghem, H.: Non-embeddable polar spaces. Münster J. Math. (To appear)
44. De Bruyn, B., Van Maldeghem, H.: Universal and homogeneous embeddings of dual polar spaces of rank 3 defined over quadratic alternative division algebras. J. Reine Angew. (To appear)
45. Dieudonné, J.: Sur les groupes classiques. Actualits Sci. Ind., no. 1040 = Publ. Inst. Math. Univ. Strasbourg (N.S.) 1, (1945)
46. J. Dieudonné. La géométrie des groupes classiques. Ergebnisse der Math. N.F. Bd.5, 2de éd, Springer, Berlin (1963)
47. Humphreys, J.E.: Introduction to Lie Algebras and Representation Theory. Springer-Verlag, New York (1972)
48. Ivanov, A.A.: Non-abelian representation of geometries. Groups and combinatorics-In: memory of Michio Suzuki, Adv. Stud. Pure Math., 32, Math. Soc. 301–314, Japan, Tokyo (2001)
49. Ivanov, A.A., Shpectorov, S.V.: Geometry of sporadic groups. II. Representations and amalgams. Encyclopedia of Mathematics and its Applications, 91, Cambridge University Press, Cambridge (2002)

50. Johnson, P.M.: Polar spaces of arbitrary rank. Geom. Dedicata **35**, 229–250 (1990)
51. Kasikova, A., Shult, E.E.: Absolute embeddings of point-line geometries. J. Algebra **238**, 265–291 (2001)
52. Li, P.: On the universal embedding of the $Sp_{2n}(2)$ dual polar space. J. Combin. Theory Ser. **A94**, 100–117 (2001)
53. Li, P.: On the universal embedding of the $U_{2n}(2)$ dual polar space. J. Combin. Theory Ser. **A98**, 235–252 (2002)
54. Mühlherr, B.: A geometric approach to non-embeddable polar spaces of rank 3. Algebra, groups and geometry. Bull. Soc. Math. Belg. Sr.**A42**(3):577–594 (1990)
55. Pankov, M.: Grassmannians of classical buildings. Algebra and Discrete Mathematics, 2. World Scientific Publishing Co., Pte. Ltd., Hackensack, NJ (2010)
56. Pasini, A.: Diagram Geometries. Oxford University Press, Oxford (1994)
57. Pasini, A.: Embeddings and expansions. Bull. Belg. Math. Soc. Simon Stevin **10**, 585–626 (2003)
58. Pasini, A.: On polar spaces of infinite rank. J. Geom. **91**, 1–2, 84–118 (2009)
59. Pasini, A.: Embeddings of Orthogonal Grassmannians. Innov. Incidence Geom. **13**, 107–133 (2013)
60. Payne, S.E., Thas, J.A.: Finite Generalized quadrangles, 2nd edn. EMS Series of Lectures in Mathematics, Zürich (2009)
61. Premet, A.A., Suprunenko, I.D.: The Weyl modules and the irreducible representations of the symplectic group with the fundamental highest weights. Comm. Algebra **11**, 1309–1342 (1983)
62. Ronan, M.A.: Embeddings and hyperplanes of discrete geometries. European J. Combin. **8**, 179–185 (1987)
63. Sahoo, B.K., Sastry, N.S.N.: A characterization of finite symplectic polar spaces of odd prime order. J. Combin. Theory Ser. A **114**, 52–64 (2007)
64. Shult, E.E.: Points and lines. Characterizing the classical geometries. Universitext. Springer, Heidelberg (2011)
65. Shult, E.E.: Embeddings and hyperplanes of Lie incidence geometries. In: Di Martino, L., Kantor, W. (eds.) Groups of Lie Type and their Geometries, pp. 215–232 Cambridge Univ. Press, Cambridge (1995)
66. Shult, E.E.: On Veldkamp lines. Bull. Belg. Math. Soc. Simon Stevin **4**, 299–316 (1997)
67. Shult, E.E., Yanushka, A.: Near n-gons and line systems. Geom. Dedicata **9**, 1–72 (1980)
68. Thas, K.: A course on elation quadrangles. E.M.S. Publishing House, Zürich (2012)
69. Thas, J.A. Van Maldeghem, H.: Embeddings of small generalized polygons. Finite Fields Appl. **12**, 4, 565–594 (2006)
70. Thas, J.A., Thas, K., Van Maldeghem, H.: Translation generalized quadrangles. Ser. Pure Math. 26, World Scientific, Singapore (2006)
71. Tits, J.: Les groupes de Lie exceptionnels et leur interprétation géométrique. Bull. Soc. Math. Belg. **8**, 48–81 (1956)
72. Tits. J.: Buildings of Spherical type and Finite BN-pairs. In: Lecture Notes in Mathematics, vol. 386 Springer, Berlin (1974)
73. Tits, J., Weiss, R.M.: Moufang Polygons. Springer, Berlin (2002)
74. Ueberberg, J.: Foundations of incidence geometry. Projective and polar spaces. Springer Monographs in Mathematics. Springer, Heidelberg (2011)
75. Van Maldeghem, H.: Generalized polygons. In: Monographs in Mathematics, vol. 93. Birkhäuser, Basel (1998)
76. Veldkamp. F.D.: Polar geometry. I, II, III, IV, V. Nederl. Akad. Wetensch. In: Proc. Ser. A 62; 63 = Indag. Math. 21, (1959)
77. Weiss, R.M.: The Structure of Spherical Buildings. Princeton Univ. Press, Princeton (2003)
78. Wells, A.: Universal projective embeddings of the Grassmannian, half spinor, and dual orthogonal geometries. Quart. J. Math. Oxford **34**, 375–386 (1983)

Chapter 4
Embeddings of Line-Grassmannians of Polar Spaces in Grassmann Varieties

Ilaria Cardinali and A. Pasini

Abstract An embedding of a point-line geometry Γ is usually defined as an injective mapping ε from the point-set of Γ to the set of points of a projective space such that $\varepsilon(l)$ is a projective line for every line l of Γ. However, different situations are considered in the literature, where $\varepsilon(l)$ is allowed to be a subline of a projective line or a curve. In this paper we propose a more general definition of embedding which includes all the above situations and we focus on a class of embeddings, which we call Grassmann embeddings, where the points of Γ are firstly associated to lines of a projective geometry $\mathrm{PG}(V)$, next they are mapped onto points of $\mathrm{PG}(V \wedge V)$ via the usual projective embedding of the line-grassmannian of $\mathrm{PG}(V)$ in $\mathrm{PG}(V \wedge V)$. In the central part of our paper we study sets of points of $\mathrm{PG}(V \wedge V)$ corresponding to lines of $\mathrm{PG}(V)$ totally singular for a given alternating, hermitian or quadratic form of V. Finally, we apply the results obtained in that part to the investigation of Grassmann embeddings of several generalized quadrangles.

Keywords Embeddings · Polar spaces · Generalized quadrangles · Exterior products · Tensor products · Grassmann varieties · Veronese varieties

---

MSC 2000: 51A45, 51A50, 51E12, 15A75, 15A69, 14A10.

I. Cardinali (✉) · A. Pasini
Department of Information Engineering and Mathematics,
University of Siena, Via Roma 56, 53100 Siena, Italy
e-mail: ilaria.cardinali@unisi.it

A. Pasini
e-mail: antonio.pasini@unisi.it

N. S. N. Sastry (ed.), *Groups of Exceptional Type, Coxeter Groups and Related Geometries*, Springer Proceedings in Mathematics & Statistics 82,
DOI: 10.1007/978-81-322-1814-2_4,

1 Introduction

According to a well established definition, an embedding of a point-line geometry Γ is an injective mapping ε from the point-set $\mathcal{P}$ of Γ to the set of points of the projective geometry PG(V) of subspaces of V such that $\varepsilon(\mathcal{P})$ spans PG(V) and $\varepsilon(l)$ is a projective line, for every line l of Γ. However, different interesting situations are also considered in the literature, where $\varepsilon(l)$ is allowed to be a subline of a projective line or a curve (a conic or a unital, for instance) or even a surface. In Sect. 2 we propose a more general definition of embedding which includes all the above situations and we sketch a bit of theory for it. More general definitions can be considered, where geometries are embedded in groups instead of projective spaces (see Ivanov [16], Pasini [19]), but in this paper we are not willing to go so far. At the end of Sect. 2 we consider a class of embeddings, which we call Grassmann embeddings, where the points of Γ are firstly associated to lines of PG(V) for a vector space V of finite dimension n over a commutative division ring $\mathbb{F}$, next they are mapped onto points of PG($V \wedge V$) via the usual projective embedding of the line-grassmannian of PG(V) in PG($V \wedge V$).

In Sect. 3 we study polar line-grassmannians, namely sets of points of PG($V \wedge V$) corresponding to lines of PG(V) totally isotropic or totally singular for a given sesquilinear or quadratic form of V, describing them as solutions of suitable sets of equations. Recall that the exterior power $V \wedge V$ is isomorphic to the vector space $S_n(\mathbb{F})$ of skew-symmetric matrices of order n with entries in $\mathbb{F}$. A matrix $(x_{i,j})_{i,j=1}^n \in S_n(\mathbb{F})$ represents a line of PG(V) if and only if it satisfies the following set of equations (Hirschfeld and Thas [15]):

$$x_{i,j}x_{k,h} - x_{i,k}x_{j,h} + x_{i,h}x_{j,k} = 0, \qquad (1 \le i < j < k < h \le n). \tag{1}$$

The next lemma, to be proved in Sect. 3.2.2, is our main tool in the investigation of polar line-grassmannians:

Lemma 1.1. *Let x and y be two non-proportional vectors of V and θ and ζ two non-proportional linear functionals of V. Let $A_{x \wedge y}$ and $A^*_{\theta \wedge \zeta}$ be the skew-symmetric matrices representing the vectors $x \wedge y \in V \wedge V$ and $\theta \wedge \zeta \in V^* \wedge V^*$, where V^* stands for the dual of V. Then we have $\theta(x) = \theta(y) = \zeta(x) = \zeta(y) = 0$ if and only if $A^*_{\theta \wedge \zeta} A_{x \wedge y} = O$, where O stands for the null matrix.*

This lemma allows us to describe polar line-grassmannians by very simple matrix equations, where a number of scalar equations are packed and by which equations for tangent spaces can be deduced quite easily. For instance, let φ be a non-degenerate reflexive bilinear form of $V = V(n, \mathbb{F})$ and Φ the matrix representing φ with respect to the natural basis of V, namely $\varphi(x, y) = x^T \Phi y$ for any two vectors $x, y \in V$. Let $\perp$ be the orthogonality relation associated to Φ. It is not so difficult to see that, for a 2-subspace $\langle x, y \rangle$ of V, if $X \in S_n(\mathbb{F})$ represents $x \wedge y$ then $\Phi^T X \Phi$ represents a vector of $V^* \wedge V^*$ corresponding to $\langle x, y \rangle^\perp$. By Lemma 1.1, we have $\langle x, y \rangle \subseteq \langle x, y \rangle^\perp$ if and only if $\Phi^T X \Phi X = O$, namely $X \Phi X = O$, since Φ is non-singular. Therefore:

Theorem 1.2. *Let $\mathcal{G}_\varphi$ be the set of points of* $\mathrm{PG}(V \wedge V)$ *that correspond to lines of* $\mathrm{PG}(V)$ *totally isotropic for* φ. *Then* $\mathcal{G}_\varphi$ *is the projective variety described by the scalar equations* (1) *together with the following matrix equation:*

$$X\Phi X = O, \quad (\text{where } X \in S_n(\mathbb{F})). \tag{2}$$

Equation (2) involves $\binom{n+1}{2}$ quadratic scalar equations in $\binom{n}{2}$ unknowns. Let $A \in S_n(\mathbb{F})$ represent a point of $\mathcal{G}_\varphi$. By differentiating (2) at A we obtain the following matrix equation, where $\binom{n+1}{2}$ linear scalar equations are packed:

$$A\Phi X + X\Phi A = O.$$

This equation, combined with the scalar equations obtained by differentiating (1), describes the tangent space of $\mathcal{G}_\varphi$ at the point represented by A.

We have only considered bilinear forms in the statement of Theorem 1.2 in order to avoid a few complications that arise with hermitian forms (see for instance Sect. 3.4.3), but similar results are obtained in Sect. 3 for hermitian forms too, as well as quadratic forms in even characteristic. In fact Theorem 1.2 is a special case of a more general result, namely Theorem 3.2 of Sect. 3, where hermitian forms and degenerate forms are also considered.

The reader might now wonder why we stick to polar line-grassmannians instead of considering polar k-grassmannians for any $k \geq 2$. A reason is that, in order to deal with the case $k > 2$, we should preliminarily generalize Lemma 1.1 to k-subspaces of V and skew-symmetric tensors of degree k. (Recall that the grassmannian of k-subspaces of V is naturally embedded in $\mathrm{PG}(\wedge^k V)$ and $\wedge^k V$ is isomorphic to the space of skew-symmetric tensors of degree k.) Regretfully, we presently have no idea of what such generalized statement should look like.

Turning back to the contents of this paper, in the last section (Sect. 4) we apply the results of Sect. 3 to Grassmann embeddings of a number of generalized quadrangles. We also compare those embeddings with other embeddings, which we call quadratic veronesean embeddings, obtained by composing a projective embedding ε with the usual quadratic veronesean mapping of the projective space hosting ε.

In Sect. 3.4 we will freely use a few elementary notions from algebraic geometry. We might refer the reader to Shafarevich [24] or the first paper of Hartshorne's celebrated book [12] for them, but even a quick introduction to affine and projective varieties as offered in Chap. 1 of Humphreys [14] can be sufficient.

As for generalized quadrangles, to be discussed in Sect. 4, we refer to Payne and Thas [21].

To finish this introduction we fix some conventions to be used throughout this paper.

Notation. We will often go back and forth from a vector space V to its projective geometry $\mathrm{PG}(V)$. In order to avoid any confusion, we keep the usual symbol $\langle . \rangle$ for spans in V while we denote spans in $\mathrm{PG}(V)$ by the symbol $\langle . \rangle^{\mathrm{pr}}$. Given a non-zero vector $v \in V$ we denote by $[v]$ the point of $\mathrm{PG}(V)$ represented by v. Given a set X

of vectors of V we put $[X] = \{[v]\}_{v\in X, v\neq 0}$. In particular, if X is a subspace of V then $[X]$ is a subspace of PG(V) (and $\dim(X) = \dim([X]) + 1$). On the other hand, let X be a set of points of PG(V). Then X is a set of 1-dimensional subspaces of V. By a harmless abuse we can switch from that set of subspaces to its union $\cup_{p\in X} p$, thus regarding X as a subset of V. Accordingly, we write $\langle X\rangle$ for $\langle \cup_{p\in X} p\rangle$.

2 A Generalized Notion of Embedding

Throughout this section $\Gamma = (\mathcal{P}, \mathcal{L})$ is a point-line geometry, $\mathcal{P}$ is its set of points and $\mathcal{L}$ its set of lines. The members of $\mathcal{L}$ are regarded as subsets of $\mathcal{P}$, as usual. We assume that Γ is connected and no two distinct lines of Γ meet in more than one point.

2.1 Definitions

Given a point-line geometry $\Gamma = (\mathcal{P}, \mathcal{L})$, a vector space V and a positive integer d, a *locally d-dimensional embedding* (*d-embedding* for short) of Γ in PG(V) is an injective mapping ε from $\mathcal{P}$ to the set of points of PG(V) satisfying the following:

$(E1)$ for every line $l \in \mathcal{L}$ the image $\varepsilon(l) := \{\varepsilon(p)\}_{p\in l}$ of l spans a d-dimensional subspace $\langle\varepsilon(l)\rangle^{\mathrm{pr}}$ of PG(V).

$(E2)$ $\langle\varepsilon(l)\rangle^{\mathrm{pr}} \cap \varepsilon(\mathcal{P}) = \varepsilon(l)$ for every line $l \in \mathcal{L}$;

$(E3)$ $\langle\varepsilon(\mathcal{P})\rangle^{\mathrm{pr}} = \mathrm{PG}(V)$.

Let l and m be distinct lines of Γ. The injectivity of ε implies that $|\varepsilon(l) \cap \varepsilon(m)| \leq 1$ while property $(E2)$ implies that $\langle\varepsilon(l)\rangle^{\mathrm{pr}} \neq \langle\varepsilon(m)\rangle^{\mathrm{pr}}$. Note also that, by $(E1)$, if Γ admits a d-embedding then every line of Γ has at least $d + 1$ points.

We write $\varepsilon : \Gamma \stackrel{d}{\to} \mathrm{PG}(V)$ to mean that ε is a d-embedding of Γ in PG(V). We call d the *local dimension* of ε. The *global dimension* $\dim(\varepsilon)$ of ε (also *dimension* of ε for short) is the dimension of PG(V), but we warn that many authors, including ourselves in earlier papers, when dealing with particular classes of d-embeddings as projective or quadratic embeddings (to be defined below), take $\dim(V)$ instead of $\dim(\mathrm{PG}(V))$ as the dimension of ε. If $\mathbb{F}$ is the underlying division ring of V we say that ε is *defined over* $\mathbb{F}$, also that ε is an *$\mathbb{F}$-embedding of local dimension d*, or an *$\mathbb{F}$-d-embedding*, for short.

Let ε be a 1-embedding. Then ε maps the lines of Γ into lines of PG(V). We say that ε is *full* if $\varepsilon(l) = \langle\varepsilon(l)\rangle^{\mathrm{pr}}$ for every line $l \in \mathcal{L}$. If ε is not full then we say that it is *lax*.

Full 1-embeddings are often called *projective embeddings* in the literature. We shall follow this habit in this paper, but we warn the reader that many authors use a different terminology, calling 1-embeddings 'lax projective embeddings' (also just lax embeddings, as in Van Maldeghem [29]) and projective embeddings in our sense 'full projective embeddings'.

We also slightly change our notation as follows: if ε is a projective embedding then we simply write $\varepsilon : \Gamma \to \mathrm{PG}(V)$, without keeping a record of the local dimension $d = 1$ in our notation.

A *quadratic embedding* is a 2-embedding ε defined over a field (namely a commutative division ring) and such that $\varepsilon(l)$ is a non-singular conic of the projective plane $\langle\varepsilon(l)\rangle^{\mathrm{pr}}$, for every line l of Γ. For instance, the quadratic veronesean embedding of the projective space $\mathrm{PG}(n, \mathbb{F})$ in $\mathrm{PG}(\binom{n+1}{2} - 1, \mathbb{F})$ (see Sect. 2.2.1) is a quadratic embedding. Quadratic embeddings are called veronesean embeddings in [6] and [7], but we are not going to use that terminology in this paper. We shall define veronesean embedding in Sect. 2.2.1, but as a special case of quadratic embeddings.

Not all 2-embeddings are quadratic. For instance, a projective space admits many non-quadratic 2-embeddings (see [26] and [1], where these embeddings are called generalized veronesean and lax generalized veronesean embeddings). The 2-embeddings of $\mathrm{PG}(n, \mathbb{F})$ in $\mathrm{PG}(m, \mathbb{K})$ for a division ring $\mathbb{K}$ possibly different from $\mathbb{F}$, with $|\mathbb{F}| > 2$ and $m \geq \binom{n+1}{2} - 1$ are classified in [26] and [1]. In particular, it is proved that $\mathbb{F} \subseteq \mathbb{K}$ and $m = \binom{n+1}{2} - 1$ in any case.

We have mentioned 2-embeddings of $\mathrm{PG}(n, \mathbb{F})$, but $\mathrm{PG}(n, \mathbb{F})$ also admits d-embeddings with $d > 2$. For instance, a veronesean embedding of $\mathrm{PG}(n, \mathbb{F})$ of degree $d > 2$ (Herzer [13], Timmermann [27], Gmainer and Havlicek [11]) is a d-embedding, provided that $|\mathbb{F}| \geq d$. Hermitian veronesean embeddings of projective spaces (Lunardon [18], Cooperstein, Thas and Van Maldeghem [9], also Pasini [19, 6.2]) are 3-embeddings where lines are mapped onto elliptic quadrics. Dual hyperovals of dimension d (see e.g. Yoshiara [30]) yield full d-embeddings of dual circular spaces. It is also worth noticing that the well known André-Bose-Bruck representation of a translation affine plane in $\mathrm{PG}(2d, \mathbb{F})$ (André [2], Bose and Bruck [5]) is an $\mathbb{F}$-d-embedding.

2.1.1 Morphisms and Quotients

Morphisms, quotients and hulls can be defined for arbitrary d-embeddings just in the same way as for projective embeddings. We consider morphisms and quotients in this subsection. In the next one we shall turn to hulls.

Given two $\mathbb{F}$-embeddings $\varepsilon_1 : \Gamma \stackrel{d_1}{\to} \mathrm{PG}(V_1)$ and $\varepsilon_2 : \Gamma \stackrel{d_2}{\to} \mathrm{PG}(V_2)$, a *morphism* $f : \varepsilon_1 \to \varepsilon_2$ from ε_1 to ε_2 is a semi-linear mapping $f : V_1 \to V_2$ such that $\varepsilon_2 = f \cdot \varepsilon_1$. To be precise, in this definition we should refer to the mapping $\mathrm{PG}(f)$ from the attenuated space $\mathrm{PG}(V_1) \setminus [\mathrm{Ker}(f)]$ to $\mathrm{PG}(V_2)$ induced by f rather than to f itself (compare Kasikova and Shult [17]), but we prefer to take f in place of $\mathrm{PG}(f)$. This is an abuse, but it is harmless. It makes a few statements slightly clumsy (see for instance the description of $\mathrm{Aut}(\varepsilon)$, a few lines below), but it saves us the trouble of introducing attenuated spaces.

Note that, since $\langle\varepsilon_2(\mathcal{P})\rangle^{\mathrm{pr}} = \mathrm{PG}(V_2)$, the equality $\varepsilon_2 = f\varepsilon_1$ forces the mapping $f : V_1 \to V_2$ to be surjective. If f is bijective then we say that f is an *isomorphism* from ε_1 to ε_2. If ε_1 and ε_2 are isomorphic in this sense, we write $\varepsilon_1 \cong \varepsilon_2$. If a morphism exists from ε_1 to ε_2 then we write $\varepsilon_1 \geq \varepsilon_2$ and we say that ε_2 is a *morphic image* of ε_1. If moreover $\varepsilon_1 \ncong \varepsilon_2$ then we say that ε_2 is *proper morphic image* of ε_1 and we write $\varepsilon_1 > \varepsilon_2$.

The automorphisms of a d-embedding $\varepsilon : \Gamma \stackrel{d}{\to} \mathrm{PG}(V)$ form a group $\mathrm{Aut}(\varepsilon)$, equal to the point-wise stabilizer of $\varepsilon(\mathcal{P})$ in the group $\Gamma\mathrm{L}(V)$ of all invertible semi-linear transformations of V. When ε is projective or quadratic then $\mathrm{Aut}(\varepsilon)$ is equal to the center $Z(\mathrm{GL}(V))$ of $\mathrm{GL}(V)$ (see [20, Proposition 9] for a proof of this claim when ε is projective), but in general $\mathrm{Aut}(\varepsilon) \neq Z(\mathrm{GL}(V))$. Given two d-embeddings $\varepsilon_1 : \Gamma \stackrel{d}{\to} \mathrm{PG}(V_1)$ and $\varepsilon_2 : \Gamma \stackrel{d}{\to} \mathrm{PG}(V_2)$ such that $\varepsilon_1 \geq \varepsilon_2$, let $f : \varepsilon_1 \to \varepsilon_2$ be a morphism. Then $\mathrm{Aut}(\varepsilon_2) \cdot f$ is the family of all morphisms from ε_1 to ε_2. In particular, when ε_1 and ε_2 are projective or quadratic and $\varepsilon_1 \geq \varepsilon_2$, the morphism $f : \varepsilon_1 \to \varepsilon_2$ is unique modulo scalars.

Given a d-embedding $\varepsilon : \Gamma \stackrel{d}{\to} \mathrm{PG}(V)$, let K be a subspace of V satisfying the following:

($Q1$) If $p, q \in \mathcal{P}$ (possibly $p = q$) then $\langle\varepsilon(p) \cup \varepsilon(q)\rangle \cap K = 0$. In particular $K \cap \varepsilon(p) = 0$ for every point $p \in \mathcal{P}$.
($Q2$) For $p \in \mathcal{P}$ and $l \in \mathcal{L}$, if $p \notin l$ then $\varepsilon(p) \cap \langle\varepsilon(l) \cup K\rangle = 0$.
($Q3$) There exists a nonnegative integer $k < d$ such that $\dim(K \cap \langle\varepsilon(l)\rangle) = k$ for every line $l \in \mathcal{L}$.

Then the function ε/K mapping $p \in \mathcal{P}$ to $[\langle\varepsilon(p) \cup K\rangle/K]$ is a $(d-k)$-embedding of Γ in $\mathrm{PG}(V/K)$ and the canonical projection of V onto V/K is a morphism from ε to ε/K. Indeed ($Q1$) forces ε/K to be injective, ($Q2$) implies ($E2$) for ε/K and ($Q3$) implies ($E1$) with d replaced by $d - k$ (> 0 as $k < d$; note also that the condition $k < d$ could be removed from ($Q3$), since it follows from ($Q1$) and ($Q2$)).

We call ε/K a *k-quotient* of ε (also a *quotient* of ε, omitting to mention k), and we say that K *defines a k-quotient* of ε (a *quotient* of ε, for short).

If ε is a projective embedding then ($Q1$) implies both ($Q2$) and ($Q3$). If ε is a 1-embedding then ($Q2$) does not follow from ($Q1$) but ($Q1$) still implies ($Q3$). Clearly, all quotients of a 1-embedding are 0-quotients.

If ε is a quadratic embedding defined over a field $\mathbb{F}$ with $\mathrm{char}(\mathbb{F}) \neq 2$, then ($Q1$) implies ($Q3$) with $k = 0$. In this case all quotients of ε are quadratic (see [6, 7]). On the other hand, let $\mathrm{char}(\mathbb{F}) = 2$ and let K be a subspace of V satisfying ($Q1$) and ($Q2$). Then $\dim(K \cap \langle\varepsilon(l)\rangle) \leq 1$ for every line $l \in \mathcal{L}$ (see [6, 7]). Thus, assuming ($Q3$) is equivalent to assume that $k := \dim(K \cap \langle\varepsilon(l)\rangle)$ does not depend on the choice of l. If $k = 0$ then ε/K is quadratic. If $k = 1$ then $[K] \cap \langle\varepsilon(l)\rangle^{\mathrm{pr}}$ is the nucleus of the conic $\varepsilon(l)$. In this case ε/K is a 1-embedding. It is a projective embedding if $\mathbb{F}$ is perfect, otherwise it is lax (see [6, 7]).

2.1.2 Hulls and Universality

A d-embedding of Γ is said to be *relatively universal* (dominant in Tits [28] and in [19]) when it is not a proper morphic image of any other d-embedding of Γ. As we will show in a few lines, every d-embedding ε of Γ admits a *hull* $\tilde{\varepsilon}$, uniquely determined up to isomorphisms by the following properties: $\tilde{\varepsilon}$ is a d-embedding of Γ and $\tilde{\varepsilon} \geq \varepsilon'$ for every d-embedding ε' of Γ such that $\varepsilon' \geq \varepsilon$. In particular, $\tilde{\varepsilon} \geq \varepsilon$. Clearly, $\tilde{\varepsilon}$ is relatively universal. A d-embedding is relatively universal if and only if it is its own hull.

The hull $\tilde{\varepsilon}$ of a d-embedding $\varepsilon : \Gamma \xrightarrow{d} \mathrm{PG}(V)$ can be constructed as follows, by a word for word rephrasing of the construction given by Ronan [23] for hulls of projective embeddings. Let $\mathcal{F}$ be the set of point-line flags of Γ. Consider the following presheaf:

$$\mathbf{F}(\varepsilon) = (\{V_x\}_{x\in\mathcal{P}\cup\mathcal{L}}, \{\iota_{p,l}\}_{(p,l)\in\mathcal{F}})$$

where if $x \in \mathcal{P}$ then $V_x = \varepsilon(x)$ (regarded as a 1-dimensional vector space), if $x \in \mathcal{L}$ then $V_x = \langle\varepsilon(x)\rangle = \langle V_p\rangle_{p\in x}$ (a $(d+1)$-dimensional vector space) and $\iota_{p,l}$ is the inclusion embedding of V_p in V_l, for every flag $(p,l) \in \mathcal{F}$.

The vector spaces V_p and V_l for $p \in \mathcal{P}$ and $l \in \mathcal{L}$ are taken from V, but when forming the presheaf $\mathbf{F}(\varepsilon)$ we regard them as abstract 1- and $(d+1)$-dimensional vector spaces. So, we can consider their formal direct sum $O(\varepsilon) := \bigoplus_{x\in\mathcal{P}\cup\mathcal{L}} V_x$. Let J be the subspace of $O(\varepsilon)$ spanned by the vectors $v - \iota_{p,l}(v)$ for every flag $(p,l) \in \mathcal{F}$ and every vector $v \in \varepsilon(p)$.

Put $V(\varepsilon) = O(\varepsilon)/J$ and define the mapping $\tilde{\varepsilon}$ from $\mathcal{P}$ to the set of 1-dimensional subspaces of $V(\varepsilon)$ as follows: $\tilde{\varepsilon}(p) = (V_p + J)/J$ for every point $p \in \mathcal{P}$. Then $\tilde{\varepsilon}$ is a d-embedding of Γ in $\mathrm{PG}(V(\varepsilon))$ and the natural projection of $V(\varepsilon)$ onto V is a morphism from $\tilde{\varepsilon}$ to ε.

Proposition 2.1. *The embedding $\tilde{\varepsilon}$ defined as above is the hull of ε.*

The proof is straightforward. We leave it to the reader.

The previous proposition answers a theoretical existence question but it is not of great help when we want to check if a given embedding is relatively universal or compute the dimension of its hull. When dealing with projective embeddings, questions like these can be answered, sometimes easily, sometimes with some hard work. They are more difficult for 1-embeddings in general and become extremely difficult when we turn to 2-embeddings. Nevertheless, something can be said in this latter case too. For instance, it is proved in [26] and [1] that every 2-embedding of a projective space of finite dimension n with at least four points on each line is relatively universal (and has dimension equal to $\binom{n+1}{2} - 1$) provided that its dimension is at least $\binom{n+1}{2} - 1$. The same conclusion, but only for quadratic embeddings, is obtained in [6] by an easier argument (see also [7]). The 2-embedding of a hermitian unital of $\mathrm{PG}(2,\mathbb{F})$ induced by the hermitian veronesean embedding of $\mathrm{PG}(3,\mathbb{F})$ is relatively universal (Pepe and Van Maldeghem [22]). One more result in this trend, taken from [7], will be mentioned in Sect. 4, Proposition 4.4.

So far for hulls and relative universality. Turning to absolute universality, let **C** be a nonempty class of d-embeddings of Γ defined over $\mathbb{F}$, for a given positive integer d and a given division ring $\mathbb{F}$. Assume that **C** is closed under isomorphism and under taking 0-quotients and hulls and that if $\varepsilon \in \mathbf{C}$ then $\varepsilon \cdot g \in \mathbf{C}$ for every automorphism g of Γ. A member ε of **C** is said to be *absolutely universal* (in **C**) if it is the hull of all members of **C**. Clearly, the absolutely universal member of **C**, if it exists, is uniquely determined up to isomorphisms.

The class **C** must be not too large, otherwise there is no chance for it to admit an absolutely universal member. For instance, **C** can be chosen as the class of all projective embeddings of Γ defined over a given division ring.

Absolutely universal projective embeddings have been thoroughly studied. We refer the reader to Kasikova and Shult [17] for a very far-reaching sufficient condition for the existence of the absolutely universal projective embedding. A bit more on this topic can be found in Blok and Pasini [4].

2.1.3 Homogeneity

Given a d-embedding $\varepsilon : \Gamma \xrightarrow{d} \mathrm{PG}(V)$ and an automorphism g of Γ, a *lifting* of g to $\mathrm{PG}(V)$ through ε is an invertible semilinear mapping $\hat{g} \in \Gamma\mathrm{L}(V)$ such that $\hat{g}\varepsilon = \varepsilon g$. Clearly, if $\hat{g}$ is a lifting of g then the coset $\hat{g} \cdot \mathrm{Aut}(\varepsilon)$ is the family of all liftings of g. In particular, if $\mathrm{Aut}(\varepsilon) = Z(\mathrm{GL}(V))$ (as when ε is projective or quadratic) then the lifting $\hat{g}$ is unique modulo scalars.

The set of all elements of $\mathrm{Aut}(\Gamma)$ that lift to $\mathrm{PG}(V)$ through ε is a subgroup $\mathrm{Aut}_\varepsilon(\Gamma)$ of $\mathrm{Aut}(\Gamma)$. Given a subgroup G of $\mathrm{Aut}(\Gamma)$, if $G \leq \mathrm{Aut}_\varepsilon(\Gamma)$ we say that ε is G-*homogeneous*. Let ε be G-homogeneous. The *lifting* $\widehat{G}$ of G is the subgroup of $\Gamma\mathrm{L}(V)$ formed by the liftings of the elements of G. It contains $\mathrm{Aut}(\varepsilon)$ as a normal subgroup and $\widehat{G}/\mathrm{Aut}(\varepsilon) \cong G$.

If ε is $\mathrm{Aut}(\Gamma)$-homogenous then we say it is *fully homogeneous*, also just *homogeneous*, for short.

2.2 Two Ways to Construct d-Embeddings

2.2.1 Veronesean Embeddings

Given a point-line geometry $\Gamma = (\mathcal{P}, \mathcal{L})$ let $\varepsilon : \Gamma \to \mathrm{PG}(V)$ be a projective embedding. If $\eta : \mathrm{PG}(V) \xrightarrow{d} \mathrm{PG}(V')$ is a d-embedding of $\mathrm{PG}(V)$ then the composition $\eta \cdot \varepsilon$ is a d-embedding of Γ in the subspace of $\mathrm{PG}(V')$ spanned by $\eta(\varepsilon(\mathcal{P}))$. We shall discuss only a particular case of this setting, with $d = 2$.

Let V be a vector space of finite dimension n over a field $\mathbb{F}$. The tensor squares $v \otimes v$ and the sums $v \otimes w + w \otimes v$ for $v, w \in V$ are called *pure symmetric tensor of degree* 2. A *symmetric tensors of degree* 2 is a sum of pure symmetric tensors of degree 2. The symmetric tensors of degree 2 form a vector subspace of $V \otimes V$, called the *space of symmetric tensors of degree* 2, which we shall denote by the symbol $V^{\otimes}$. Note that $v \otimes w + w \otimes v = (v - w) \otimes (v + w) - v \otimes v - w \otimes w$. Hence $V^{\otimes} = \langle v \otimes v \rangle_{v \in V}$.

Given a basis $E = \{e_i\}_{i=1}^n$ of V put $\epsilon_{i,j} = e_i \otimes e_j + e_j \otimes e_i$ for $i < j$ and $e_{i,i} = e_i \otimes e_i$. Clearly $E^{\otimes} := \{e_{i,j}\}_{1 \leq i \leq j \leq n}$ is a basis of $V^{\otimes}$. The *quadratic veronesean map* from V to $V^{\otimes}$ is the function η^{ver} mapping a vector $v = \sum_{i=1}^n e_i x_i$ of V onto the vector $\eta^{\mathrm{ver}}(v) = v \otimes v = \sum_{i \leq j} e_{i,j} x_i x_j$ of $V^{\otimes}$.

The mapping η^{ver} naturally defines a quadratic embedding of $\mathrm{PG}(V)$ in $\mathrm{PG}(V^{\otimes})$, which we still denote by the symbol η^{ver}. The composition $\eta^{\mathrm{ver}}\varepsilon$ is a quadratic embedding of Γ in a subspace of $\mathrm{PG}(V^{\otimes})$. We call it the *quadratic veronesean embedding* of Γ *induced* by ε and we denote it by the symbol $\varepsilon^{\mathrm{ver}}$.

The image $\mathcal{V} = \eta^{\mathrm{ver}}(\mathrm{PG}(V))$ of the set of points of $\mathrm{PG}(V)$ by η^{ver} is a well known projective variety of dimension $n - 1$, called the *veronesean variety* (*of degree* 2), described by the following set of equations in the unknowns $x_{i,j}$ for $1 \leq i \leq j \leq n$, where we put $x_{j,i} := x_{i,j}$ when $j > i$:

$$\left.\begin{array}{ll} x_{i,i}x_{j,j} = x_{i,j}^2 & \text{for } i < j, \\ x_{i,j}x_{i,k} = x_{i,i}x_{j,k} & \text{for } j < k \text{ and } i \neq j, k, \\ x_{i,j}x_{k,h} = x_{i,k}x_{j,h} & \text{for } i < j < h, k \text{ and } h \neq k. \end{array}\right\} \tag{3}$$

(See Hirschfeld and Thas [15, Chap. 25]) Clearly $\mathcal{V} \supseteq \varepsilon^{\mathrm{ver}}(\mathcal{P})$. In many interesting cases $\varepsilon^{\mathrm{ver}}(\mathcal{P})$ is a projective variety, whence a subvariety of $\mathcal{V}$. For instance, if $\varepsilon(\mathcal{P})$ is a quadric in $\mathrm{PG}(V)$ then $\varepsilon^{\mathrm{ver}}(\mathcal{P})$ is a hyperplane section of $\mathcal{V}$.

The embedding η^{ver} is homogeneous. Therefore, if ε is G-homogeneous for a subgroup G of $\mathrm{Aut}(\Gamma)$ then $\varepsilon^{\mathrm{ver}}$ is G-homogeneous.

More generally, let $\varepsilon_1 : \Gamma \to \mathrm{PG}(V_1)$ and $\varepsilon_2 : \Gamma \to \mathrm{PG}(V_2)$ be two projective embeddings of Γ and $f : V_1 \to V_2$ a morphism from ε_1 to ε_2. A unique semilinear mapping $f^{\otimes} : V_1^{\otimes} \to V_2^{\otimes}$ exists such that $f^{\otimes}\eta_1^{\mathrm{ver}} = \eta_2^{\mathrm{ver}} f$, where η_1^{ver} and η_2^{ver} are the quadratic veronesean mappings of V_1 in $V_1^{\otimes}$ and V_2 in $V_2^{\otimes}$ respectively. Clearly, $f^{\otimes}$ is a morphism from $\varepsilon_1^{\mathrm{ver}}$ to $\varepsilon_2^{\mathrm{ver}}$.

We have focused on quadratic veronesean embeddings since in the sequel of this paper we will have no occasion to consider veronesean embeddings of degree $d > 2$, but these can be defined too, starting from the veronesean mapping of degree d from $V(n, \mathbb{F})$ to the space of symmetric tensors of degree d. The precise definition of that mapping is rather technical. We prefer to skip it, referring the interested reader to Herzer [13] or Gmainer and Havlicek [11, Sect. 2]. We only mention that a veronesean embedding of degree $d \leq |\mathbb{F}|$ is a d-embedding.

2.2.2 Grassmann Embeddings

Given a point-line geometry $\Gamma = (\mathcal{P}, \mathcal{L})$, for every point $p \in \mathcal{L}$ let $L(p)$ be the set of lines of Γ through p. Put $\mathcal{P}^* = \mathcal{L}$, $\mathcal{L}^* = \{L(p)\}_{p \in \mathcal{P}}$ and $\Gamma^* = (\mathcal{P}^*, \mathcal{L}^*)$. Then Γ^* is a point-line geometry, isomorphic to the dual $(\mathcal{L}, \mathcal{P})$ of Γ.

Given a field $\mathbb{F}$ and a vector space V over $\mathbb{F}$ of finite dimension at least 4, suppose that a projective embedding $\varepsilon : \Gamma^* \to \mathrm{PG}(V)$ is given such that

(E_d^*) $\dim(\langle \varepsilon(L(p)) \rangle_{p \in l}^{pr}) = d + 1$ for every point l of Γ^* (line of Γ).

We recall that the *line-grassmannian* $\mathrm{Gr}_2(\mathrm{PG}(V))$ of $\mathrm{PG}(V)$ is the point-line geometry where the lines of $\mathrm{PG}(V)$ are taken as points and the lines are the sets of lines of $\mathrm{PG}(V)$ incident to a point-plane flag of $\mathrm{PG}(V)$. Put $V^\wedge := V \wedge V$ and let ι_2 be the function mapping every line $\langle [x], [y] \rangle^{\mathrm{pr}}$ of $\mathrm{PG}(V)$ to the point $[x \wedge y]$ of $\mathrm{PG}(V^\wedge)$. It is well known that ι_2 is a projective embedding of $\mathrm{Gr}_2(\mathrm{PG}(V))$ in $\mathrm{PG}(V^\wedge)$. We call it the *natural embedding* of $\mathrm{Gr}_2(\mathrm{PG}(V))$.

For every point $p \in \mathcal{P}$ of Γ put $\varepsilon^{\mathrm{gr}}(p) := \iota_2(\varepsilon(L(p)))$.

Proposition 2.2. *The mapping $\varepsilon^{\mathrm{gr}}$ defined as above is a d-embedding of Γ in the subspace of* $\mathrm{PG}(V^\wedge)$ *spanned by* $\iota_2(\mathcal{L}^*)$.

Proof. The mapping $\varepsilon^{\mathrm{gr}}$ is injective, since both ε and ι_2 are injective and distinct points of Γ are incident with distinct sets of lines of Γ. Property $(E3)$ trivially holds, since we have chosen $\langle \iota_2(\mathcal{L}^*) \rangle^{\mathrm{pr}}$ as the codomain of $\varepsilon^{\mathrm{gr}}$.

Every line $l \in \mathcal{L}$ of Γ is mapped by $\varepsilon^{\mathrm{gr}}$ onto the set $\{\iota_2(\varepsilon(L(p)))\}_{p \in l}$. By (E_d^*) the set $X = \{\varepsilon(L(p))\}_{p \in l}$ spans a $d+1$-dimensional subspace $\langle X \rangle^{\mathrm{pr}}$ of $\mathrm{PG}(V)$ containing the point $\varepsilon(l)$. On the other hand, it is well known that, for every point a of $\mathrm{PG}(V)$, if $\mathrm{St}(a)$ is the set of lines of $\mathrm{PG}(V)$ through a, then the set $\iota_2(\mathrm{St}(a))$ is a subspace of $\mathrm{PG}(V^\wedge)$ and ι_2 induces an isomorphism from the residue of a in $\mathrm{PG}(V)$ to the subspace $\iota_2(\mathrm{St}(a))$. It follows that $\langle \varepsilon^{\mathrm{gr}}(l) \rangle^{\mathrm{pr}} = \langle \iota_2(\varepsilon(L(p))) \rangle_{p \in l}^{\mathrm{pr}}$ is a d-dimensional subspace of $\mathrm{PG}(V^\wedge)$. So, $\varepsilon^{\mathrm{gr}}$ satisfies $(E1)$.

Finally, $\langle \varepsilon^{\mathrm{gr}}(l) \rangle^{\mathrm{pr}} \cap \iota_2(\mathrm{PG}(V)) = \iota_2(\mathrm{St}(\varepsilon(l)))$. It follows that $\langle \varepsilon^{\mathrm{gr}}(l) \rangle^{\mathrm{pr}} \cap \varepsilon^{\mathrm{gr}}(\mathcal{P}) = \varepsilon^{\mathrm{gr}}(l)$ (recall that ε maps lines of Γ^* surjectively onto lines of $\mathrm{PG}(V)$, since it is a projective embedding). So, $\varepsilon^{\mathrm{gr}}$ also satisfies $(E2)$. □

We call $\varepsilon^{\mathrm{gr}}$ the *Grassmann embedding* of Γ *induced* by ε, also a *Grassmann embedding* for short. The ι_2-image $\mathcal{G} := \iota_2(\mathrm{Gr}_2(\mathrm{PG}(V)))$ of the set of points of $\mathrm{Gr}_2(\mathrm{PG}(V))$ (lines of $\mathrm{PG}(V)$) is a well known projective variety, called the *line-Grassmann variety* of $\mathrm{PG}(V)$. Clearly, $\varepsilon^{\mathrm{gr}}(\mathcal{P}) \subseteq \mathcal{G}$.

The embedding ι_2 is homogeneous. Moreover $\mathrm{Aut}(\Gamma)$ and $\mathrm{Aut}(\Gamma^*)$ are canonically isomorphic. Therefore, if ε is G^*-homogeneous for a subgroup G^* of $\mathrm{Aut}(\Gamma^*)$ and G is the subgroup of $\mathrm{Aut}(\Gamma)$ corresponding to G^* then $\varepsilon^{\mathrm{gr}}$ is G-homogeneous.

More generally, let $\varepsilon_1 : \Gamma^* \to \mathrm{PG}(V_1)$ and $\varepsilon_2 : \Gamma^* \to \mathrm{PG}(V_2)$ be two projective embeddings of Γ^* satisfying condition (E_d^*), for possibly different choices of d. Let $f : V_1 \to V_2$ be a morphism from ε_1 to ε_2. There exists a semilinear mapping $f^\wedge : V_1^\wedge \to V_2^\wedge$, unique modulo scalars, which commutes with the natural embeddings of $\mathrm{Gr}_2(\mathrm{PG}(V_1))$ and $\mathrm{Gr}(\mathrm{PG}(V_2))$. Clearly, $f^\wedge$ is morphism from $\varepsilon_1^{\mathrm{gr}}$ to $\varepsilon_2^{\mathrm{gr}}$.

2.2.3 A More General Setting

The previous definition of $\varepsilon^{\mathrm{gr}}$ is a special case of a more general setting, quite familiar to everyone working on embeddings of dual structures, as dual polar spaces for instance. We recall it below, even if we will make no use of it in this paper.

With $\Gamma = (\mathcal{P}, \mathcal{L})$ as in the previous subsection, let Γ^* be a point-line geometry equipped with two families of subspaces $\mathcal{A}$ and $\mathcal{B}$ such that the pair $(\mathcal{A}, \mathcal{B})$ equipped with inclusion as the incidence relation is a rank 2 geometry isomorphic to the dual $(\mathcal{L}, \mathcal{P})$ of Γ. Given an isomorphism from $(\mathcal{L}, \mathcal{P})$ to $(\mathcal{A}, \mathcal{B})$ let λ be the bijection from $\mathcal{P}$ to $\mathcal{B}$ induced by it. Let $\varepsilon : \Gamma^* \to \mathrm{PG}(V)$ be a projective embedding such that for a given integer $k \geq 2$ and any choice of $X \in \mathcal{A}$ and $Y \in \mathcal{B}$ we have $\dim(\langle\varepsilon(X)\rangle^{\mathrm{pr}}) = k-2$, $\dim(\langle\varepsilon(Y)\rangle^{\mathrm{pr}}) = k-1$ and $\langle\varepsilon(X)\rangle^{\mathrm{pr}} \subseteq \langle\varepsilon(Y)\rangle^{\mathrm{pr}}$ if and only if $X \subseteq Y$. Then, if ι_k is the usual projective embedding of the k-grassmannian of $\mathrm{PG}(V)$ in $\mathrm{PG}(\wedge^k V)$ mapping a $(k-1)$-subspace $\langle[x_1], \ldots, [x_k]\rangle^{\mathrm{pr}}$ of $\mathrm{PG}(V)$ onto the point $[\wedge_{i=1}^k x_i]$, the composition $\varepsilon^{\mathrm{gr}} := \iota_k\varepsilon\lambda$, which maps $p \in \mathcal{P}$ onto $\iota_k(\varepsilon(\lambda(p)))$, is an injective mapping from the point-set $\mathcal{P}$ of Γ to the set of points of $\mathrm{PG}(\wedge^k V)$. Suppose moreover that for every $X \in \mathcal{A}$ the set $\{\varepsilon(Y)\}_{X\subset Y\in\mathcal{B}}$ spans a $(k-1+d)$-dimensional subspace of $\mathrm{PG}(V)$. Then $\varepsilon^{\mathrm{gr}}$ is a d-embedding of Γ in a subspace of $\mathrm{PG}(\wedge^k V)$.

3 Line Grassmannians of Polar Spaces

Throughout this section V is a vector space of finite dimension $n \geq 4$ over a field $\mathbb{F}$, $V^\wedge := V \wedge V$ and ι_2 is the natural embedding of $\mathrm{Gr}_2(\mathrm{PG}(V))$ in $\mathrm{PG}(V^\wedge)$, as in Sect. 2.2.2. As in that subsection, $\mathcal{G} := \iota_2^{\mathrm{gr}}(\mathrm{Gr}_2(\mathrm{PG}(V)))$, the line-Grassmann variety in $\mathrm{PG}(V^\wedge)$.

Given an ordered basis $E = (e_1, \ldots, e_n)$ of V, the vectors $e_{i,j} := e_i \wedge e_j$ for $1 \leq i < j \leq n$ form a basis $E^\wedge$ of $V^\wedge$. We call $E^\wedge$ the basis of $V^\wedge$ *canonically associated* to E. The embedding ι_2 maps the line $\langle[\sum_{i=1}^n e_i x_i], [\sum_{i=1}^n e_i y_i]\rangle^{\mathrm{pr}}$ of $\mathrm{PG}(V)$ onto the point $[\sum_{i<j} e_{i,j}x_{i,j}]$ of $\mathrm{PG}(V^\wedge)$, where $x_{i,j} = x_i y_j - x_j y_i$ for any choice of $i < j$.

3.1 *Properties of $\mathcal{G}$*

In this subsection we recall a few well known properties of $\mathcal{G}$. This variety is described by the set of equations (1) of Introduction. We recall them here:

$$x_{i,j}x_{k,h} - x_{i,k}x_{j,h} + x_{i,h}x_{j,k} = 0, \qquad (1 \leq i < j < k < h \leq n).$$

Given any two non-proportional vectors $a = \sum_{i=1}^n e_i a_i$ and $b = \sum_{i=1}^n e_i b_i$ of V let $c = \sum_{i<j} e_{i,j}c_{i,j} = a \wedge b$, namely $c_{i,j} = a_i b_j - a_j b_i$ for $1 \leq i < j \leq n$. The tangent space $\mathrm{Tan}(\mathcal{G})_{[c]}$ of $\mathcal{G}$ at the point $[c]$ is the subspace of $V^\wedge$ described by the following set of linear equations, obtained by differentiating (1):

$$c_{k,h}x_{i,j} + c_{i,j}x_{k,h} - c_{j,h}x_{i,k} - c_{i,k}x_{j,h} + c_{j,k}x_{i,h} + c_{i,h}x_{j,k} = 0, \qquad (4)$$

where $1 \leq i < j < k < h \leq n$, as in (1). This linear system contains $\binom{n}{4}$ equations but it has rank equal to $\binom{n-2}{2}$, independently of the choice of the point $[c] \in \mathcal{G}$ (Hirschfeld and Thas [15]). Accordingly, $\dim(\mathrm{Tan}(\mathcal{G}_{[c]})) = 2n - 3 = \binom{n}{2} - \binom{n-1}{2}$. Hence $\mathcal{G}$ has dimension $\dim(\mathcal{G}) = 2n - 4$.

The tangent space $\mathrm{Tan}(\mathcal{G})_{[c]}$ also admits another description, as we shall see in a few lines.

Given a non-zero vector $a = \sum_{i=1}^{n} e_i a_i$ of V, let $\mathrm{St}([a])$ be the set of lines of $\mathrm{PG}(V)$ containing the point $[a]$, as in Sect. 2.2.2. Then $S_{[a]} := \iota_2(\mathrm{St}([a]))$ is a subspace of $\mathrm{PG}(V^\wedge)$ and it is described by the following $\binom{n}{3}$ linear equations:

$$a_i x_{j,k} - a_j x_{i,k} + a_k x_{i,j} = 0, \qquad (1 \leq i < j < k \leq n). \tag{5}$$

This linear system has rank equal to $\binom{n-1}{2}$. Indeed, suppose that $a_1 \neq 0$, to fix ideas. Then the Eq. (5) with $i = 1$ form a maximal independent subset of the whole set (5). Accordingly $S_{[a]}$, regarded as a subspace of $V^\wedge$, has vector dimension $\dim(S_{[a]}) = n - 1$.

Given another point $[b] \neq [a]$ in $\mathrm{PG}(V)$, let $c = a \wedge b$. Then $S_{[a]} \cap S_{[b]} = [c]$. Hence, regarded $S_{[a]}$ and $S_{[b]}$ as subspaces of $V^\wedge$, the sum $S_{[a]} + S_{[b]}$ has dimension equal to $2(n-1) - 1 = 2n - 3 = \dim(\mathrm{Tan}(\mathcal{G})_{[c]})$. It is not difficult to check that both $S_{[a]}$ and $S_{[a]}$ are contained in $\mathrm{Tan}(\mathcal{G})_{[c]}$. Therefore $\mathrm{Tan}(\mathcal{G})_{[c]} = S_{[a]} + S_{[b]}$.

3.2 A Matrix Notation

It is well known that $V^\wedge$ is isomorphic to the space $S_n(\mathbb{F})$ of skew-symmetric $n \times n$ matrices over $\mathbb{F}$. Explicitly, for every vector $v = \sum_{i<j} e_{i,j} v_{i,j}$ of $V^\wedge$ let

$$A_v = (v_{i,j})_{i,j=1}^{n} \tag{6}$$

where we put $v_{i,i} = 0$ for every i and $v_{i,j} = -v_{j,i}$ for $i > j$. So, $A_v \in S_n(\mathbb{F})$ and the function α mapping $v \in V^\wedge$ to $\alpha(v) := A_v$ is an isomorphism from $V^\wedge$ to $S_n(\mathbb{F})$. In this way the points of $\mathcal{G}$ are represented by matrices of the form $xy^T - yx^T$ for $x, y \in V$ with $\dim(\langle x, y\rangle) = 2$. (Needless to say, when writing $xy^T - yx^T$ we regard the vectors of V as $(n \times 1)$-matrices with the E-coordinates as the entries.)

We can dualize the above as follows. The linear functionals of V form the dual V^* of V. Let $E^* = (e_1^*, ..., e_n^*)$ be the basis of V^* dual of E, namely $e_i^*(e_j) = \delta_{i,j}$ (Kronecker symbol) for every choice of $i, j = 1, 2, ..., n$. The vectors $e_{i,j}^* := e_i^* \wedge e_j^*$ for $i < j$ form the basis $E^{*\wedge}$ of $V^{*\wedge} := V^* \wedge V^*$ canonically associated to E^*. Given a vector $\xi = \sum_{i<j} e_{i,j}^* \xi_{i,j}$ of $V^{*\wedge}$ we put $A_\xi^* = (\xi_{i,j})_{i,j=1}^{n}$, with $\xi_{i,i} = 0$ and $\xi_{i,j} = -\xi_{j,i}$ if $i > j$, as in (6). Thus $A_\xi^* \in S_n(\mathbb{F})$ and the clause $\alpha^*(\xi) = A_\xi^*$ defines an isomorphism α^* from $V^{*\wedge}$ to $S_n(\mathbb{F})$. In particular, if θ and ζ are non-proportional vectors of V^* then $\alpha^*(\theta \wedge \zeta) = \theta\zeta^T - \zeta\theta^T$, where θ and ζ are regarded as $n \times 1$-matrices with the E^*-coordinates as the entries.

Equation (1) can be rephrased as properties of skew-symmetric matrices: a vector $v \in V^\wedge$ satisfies (1) if and only if all principal (4×4)-submatrices of A_v are singular. Admittedly, this rephrasing is not a great improvement, but skew-symmetric matrices are useful in other contexts, such as the description of the lifting of a linear mapping of V to $V^\wedge$ and the characterization of the inclusion of a line of $\mathrm{PG}(V)$ in a subspace of codimension 2.

3.2.1 Lifting Linear Mappings from V to $V^\wedge$

The embedding ι_2 is homogeneous. In particular, let M be the matrix representing $f \in \mathrm{GL}(V)$ with respect to E by a matrix M, namely $f(x) = Mx$ for every $x \in V$. Then

$$\begin{aligned} A_{f(x)\wedge f(y)} &= Mx(My)^T - My(Mx)^T \\ &= Mxy^TM^T - Myx^TM^T = MA_{x\wedge y}M^T \end{aligned}$$

for any two non-proportional vectors $x, y \in V$. So, if $\mathrm{PG}(f)$ is the automorphism of $\mathrm{PG}(V)$ represented by f, then the linear mapping $f^\wedge$ of $V^\wedge$ defined by the clause

$$f^\wedge(v) = \alpha^{-1}(MA_vM^T) \quad (\text{for all } v \in V^\wedge) \tag{7}$$

is a representative of the lifting of $\mathrm{PG}(f)$ to $\mathrm{PG}(V^\wedge)$. We call $f^\wedge$ the *lifting* of f to $V^\wedge$.

The mapping f induces on V^* the linear mapping f^* acting as follows: $f^*(\theta) = M^{-T}\theta$ for every $\theta \in V^*$. We have

$$\begin{aligned} A^*_{f^*(\theta)\wedge f^*(\zeta)} &= M^{-T}\theta(M^{-T}\zeta)^T - M^{-T}\zeta(M^{-T}\theta)^T \\ &= M^{-T}\theta\zeta^TM^{-1} - M^{-T}\zeta\theta^TM^{-1} = M^{-T}A^*_{\theta\wedge\zeta}M^{-1} \end{aligned}$$

for any two non-proportional vectors $\theta, \zeta \in V^*$. Thus, the linear mapping $f^{*\wedge}$ of $V^{*\wedge}$ defined by the clause

$$f^{*\wedge}(\xi) := \alpha^{*-1}(M^{-T}A^*_\xi M^{-1}) \quad (\text{for all } \xi \in V^{*\wedge}) \tag{8}$$

is the lifting of f^* to $V^{*\wedge}$.

3.2.2 Inclusion of a Line in a Dual Line

The subspaces of $\mathrm{PG}(V)$ of codimension 2, also called *dual lines* of $\mathrm{PG}(V)$, naturally correspond to the lines of $\mathrm{PG}(V^*)$. Explicitly, recalling that the vectors of V^* are the linear functionals of V, a line $\langle[\theta], [\zeta]\rangle^{\mathrm{pr}}$ of $\mathrm{PG}(V^*)$ corresponds to the subspace $[\mathrm{Ker}(\theta) \cap \mathrm{Ker}(\zeta)]$ of $\mathrm{PG}(V)$.

We shall now prove Lemma 1.1 of Sect. 1. We repeat the statement, for the commodity of the reader.

Lemma 3.1. *Let $x, y \in V$ and $\theta, \zeta \in V^*$ be such that $\dim(\langle x, y\rangle) = \dim(\langle \theta, \zeta\rangle) = 2$. Then we have $\langle x, y\rangle \subseteq \mathrm{Ker}(\theta) \cap \mathrm{Ker}(\zeta)$ if and only if*

$$A^*_{\theta\wedge\zeta} A_{x\wedge y} = O \tag{9}$$

where O stands for the null square matrix of order n.

Proof. Put $X = A_{x\wedge y}$ and $\Theta = A^*_{\theta\wedge\zeta}$, for short. Then

$$\begin{aligned}\Theta X &= (\theta\zeta^T - \zeta\theta^T)(xy^T - yx^T)\\ &= \theta(\zeta^T x)y^T - \zeta(\theta^T x)y^T - \theta(\zeta^T y)x^T + \zeta(\theta^T y)x^T\\ &= (\zeta^T x)\theta y^T - (\theta^T x)\zeta y^T - (\zeta^T y)\theta x^T + (\theta^T y)\zeta x^T.\end{aligned}$$

(Recall that $\zeta^T x$, $\theta^T x$, $\zeta^T y$ and $\theta^T y$ are scalars.) Suppose that $x, y \in \mathrm{Ker}(\theta) \cap \mathrm{Ker}(\zeta)$. In matrix notation, $\theta^T x = \zeta^T x = \theta^T y = \zeta^T y = 0$. However $\Theta X = (\zeta^T x)\theta y^T - (\theta^T x)\zeta y^T - (\zeta^T y)\theta x^T + (\theta^T y)\zeta x^T$, as shown above. Hence $\Theta X = O$.

Conversely, let $\Theta X = O$. Recall that, if $x = \sum_{i=1}^n e_i x_i$, $y = \sum_{i=1}^n e_i y_i$, $\theta = \sum_{i=1}^n e_i^* \theta_i$ and $\zeta = \sum_{i=1}^n e_i^* \zeta_i$, then $x_{i,j} = x_i y_j - x_j y_i$ is the (i, j)-entry of X and $\zeta_{i,j} = \theta_i\zeta_j - \theta_j\zeta_i$ is the (i, j)-entry of Θ.

By (7) and (8), if we replace x and y by $f(x)$ and $f(y)$ for a linear mapping $f \in \mathrm{GL}(V)$ then Θ is replaced by $\Theta' := M^{-T}\Theta M^{-1}$ and X by $X' := MXM^T$, where M is the representative matrix of f. Accordingly, the product ΘX is replaced by $\Theta' X' = M^{-T}\Theta X M^T$. As M is non-singular, we have $\Theta X = O$ if and only if $\Theta' X' = O$. Moreover, $\langle x, y\rangle \subseteq \mathrm{Ker}(\theta) \cap \mathrm{Ker}(\zeta)$ if and only if $\langle f(x), f(y)\rangle \subseteq \mathrm{Ker}(f^*(\theta)) \cap \mathrm{Ker}(f^*(\zeta))$. We can always choose f in such a way that $f(x) = e_1$ and $f(y) = e_2$.

By the above, we can assume without loss that $x = e_1$ and $y = e_2$. With this choice of x and y we have $x_{1,2} = -x_{2,1} = 1$ and $x_{i,j} = 0$ for $\{i, j\} \neq \{1, 2\}$. Thus the equation $\Theta X = O$ forces $\zeta_{i,1} = \zeta_{i,2} = 0$ for every $i = 1, 2, ..., n$, namely $\theta_i\zeta_1 - \theta_1\zeta_i = \theta_i\zeta_2 - \theta_2\zeta_i = 0$ for every $i = 1, 2, ..., n$. Suppose that at least one of $\theta_1, \zeta_1, \theta_2$ or ζ_2 is non-zero, say $\theta_1 \neq 0$. Then $\zeta_i = \frac{\zeta_1}{\theta_1}\theta_i$ for every i, namely $\zeta = \frac{\zeta_1}{\theta_1}\theta$, contrary to the hypotheses of the lemma. It follows that $\theta_1 = \theta_2 = \zeta_1 = \zeta_2 = 0$, namely $\theta^T e_1 = \theta^T e_2 = \zeta^T e_1 = \zeta^T e_2 = 0$. So, $e_1, e_2 \in \mathrm{Ker}(\theta) \cap \mathrm{Ker}(\zeta)$, namely $x, y \in \mathrm{Ker}(\theta) \cap \mathrm{Ker}(\zeta)$. □

3.3 Lines Totally Isotropic for a Sesquilinear Form

Let $\sigma \in \mathrm{Aut}(\mathbb{F})$ and $\varepsilon \in \mathbb{F} \setminus \{0\}$ be such that $\sigma^2 = \mathrm{id}_{\mathbb{F}}$ and $\varepsilon^\sigma\varepsilon = 1$. Let $\varphi : V \times V \to \mathbb{F}$ be a reflexive (σ, ε)-sequilinear form on V, represented as follows in matrix notation with respect to the basis E of V:

$$\varphi(x, y) = (x^\sigma)^T \Phi y \tag{10}$$

for an $n \times n$ matrix $\Phi = (\varphi_{i,j})_{i,j=1}^n$ such that $\Phi^T = \varepsilon\Phi^\sigma$, where $\Phi^\sigma := (\varphi_{i,j}^\sigma)_{i,j=1}^n$. We can rewrite (10) as follows. Consider the semi-linear mapping $\lambda_\varphi : V \to V^*$ mapping x to the linear functional $\lambda_\varphi(x)$ defined by the clause $\lambda_\varphi(x)(y) := (x^\sigma)^T \Phi y$ for every $y \in V$, namely $\lambda_\varphi(x) = \Phi^T x^\sigma = \varepsilon\Phi^\sigma x^\sigma$ with respect to the basis E^* of V^*. Then (10) is equivalent to $\lambda_\varphi(x)(y) = 0$.

We recall that the *radical* $\mathrm{Rad}(\varphi)$ of φ is the subspace $V^\perp$ of V, where $\perp$ denotes the orthogonality relation with respect to φ. The form φ is said to be *degenerate* if $\mathrm{Rad}(\varphi) \neq 0$. Clearly, $\mathrm{Rad}(\varphi) = \mathrm{Ker}(\lambda_\varphi) = \mathrm{Ker}(\Phi^\sigma) = (\mathrm{Ker}(\Phi))^\sigma$. So, λ_φ is an isomorphism if and only if φ is non-degenerate if and only if Φ is non-singular.

Let $S = \langle x, y\rangle$ be a 2-dimensional subspace of $\mathrm{PG}(V)$. Then $\lambda_\varphi(S)$ is a subspace of V^* of dimension at most 2. Suppose firstly that $\dim(\lambda_\varphi(S)) = 2$ (which is always the case when φ is non-degenerate). Then

$$\begin{aligned} A^*_{\lambda_\varphi(x)\wedge\lambda_\varphi(y)} &= \varepsilon^2(\Phi^\sigma x^\sigma(\Phi^\sigma y^\sigma)^T - \Phi^\sigma y^\sigma(\Phi^\sigma x^\sigma)^T) \\ &= \varepsilon(\Phi^\sigma(xy^T)^\sigma \Phi - \Phi^\sigma(yx^T)^\sigma \Phi) = \varepsilon\Phi^\sigma A^\sigma_{x\wedge y}\Phi. \end{aligned}$$

Namely,

$$A^*_{\lambda_\varphi(x)\wedge\lambda_\varphi(y)} = \varepsilon\Phi^\sigma A^\sigma_{x\wedge y}\Phi. \tag{11}$$

Suppose now that $\dim(\lambda_\varphi(S)) \leq 1$. In this case φ is degenerate and S either intersects the radical $\mathrm{Rad}(\varphi)$ of φ in a 1-dimensional subspace or is entirely contained in $\mathrm{Rad}(\varphi)$. In either case $A^*_{\lambda_\varphi(x)\wedge\lambda_\varphi(y)} = O$. On the other hand $\mathrm{Rad}(\varphi) = \mathrm{Ker}(\Phi)$. Hence either $\Phi x = 0$ or $\Phi y = t\Phi x$ for a scalar t (possibly $t = 0$). In each of these two cases $\Phi xy^T\Phi^T - \Phi yx^T\Phi^T = O$, namely $\Phi A_{x\wedge y}\Phi^\sigma = O$ (recall that $\Phi^T = \varepsilon\Phi^\sigma$). However $\Phi A_{x\wedge y}\Phi^\sigma = (\Phi^\sigma A^\sigma_{x\wedge y}\Phi)^\sigma$ (recall that $\sigma^2 = \mathrm{id}_{\mathbb{F}}$). So, equation (11) remains valid in this case too.

Theorem 3.2. *Let $S = \langle x, y\rangle$ and $S' = \langle u, v\rangle$ be 2-dimensional linear subspaces of V. Suppose that $S \cap \mathrm{Rad}(\varphi) = 0$. Put $X = A_{x\wedge y}$ and $Y = A_{u\wedge v}$. Then $S' \subseteq S^\perp$ if and only if*

$$\Phi^\sigma X^\sigma \Phi Y = O. \tag{12}$$

In particular, S is totally isotropic (namely $S \subseteq S^\perp$) if and only if

$$\Phi^\sigma X^\sigma \Phi X = O. \tag{13}$$

Proof. The main claim of the theorem immediately follows from equations (9) and (11) while the second claim is just a special case of the first one. The hypothesis $S \cap \mathrm{Rad}(\varphi) = 0$ is needed in order to apply (9). Indeed in the present context $\lambda_\varphi(x)$ and $\lambda_\varphi(y)$ play the role of θ and ζ of Lemma 3.1, but in that lemma we assume that $\dim(\langle\theta, \zeta\rangle) = 2$. □

Corollary 3.3. *Suppose that φ is non-degenerate. Let $S = \langle x, y\rangle$ and $S' = \langle u, v\rangle$ be 2-dimensional linear subspaces of V. Put $X = A_{x\wedge y}$ and $Y = A_{u\wedge v}$, as in Theorem 3.2. Then $S' \subseteq S^{\perp}$ if and only if*

$$X^{\sigma}\Phi Y = O. \tag{14}$$

In particular, S is totally isotropic if and only if

$$X^{\sigma}\Phi X = O. \tag{15}$$

Proof. Trivial from Theorem 3.2, recalling that Φ is non-singular since φ is non-degenerate. □

Corollary 3.4. *Let $S = \langle x, y\rangle$ be a 2-dimensional subspace of V and put $X = A_{x\wedge y}$. We have $S \cap \mathrm{Rad}(\varphi) \neq 0$ if and only if*

$$\Phi X \Phi^{\sigma} = O. \tag{16}$$

Proof. The 'only if' part of this claim is implicit in the comments after formula (11). Conversely, suppose that (16) holds. Then $\Phi^{\sigma}X^{\sigma}\Phi = O$, hence $\Phi^{\sigma}X^{\sigma}\Phi Y = O$ for every $n \times n$ matrix Y. By way of contradiction, suppose that $S \cap \mathrm{Rad}(\varphi) = 0$. Since $\Phi^{\sigma}X^{\sigma}\Phi Y = O$ for any Y, Theorem 3.2 implies that $S' \subseteq S^{\perp}$ for every 2-space S' of V. But this is impossible. Indeed $\dim(S^{\perp}) = n - 2$ since $\dim(S) = 2$ and $S \cap \mathrm{Rad}(\varphi) = 0$. Therefore $S \cap \mathrm{Rad}(\varphi) \neq 0$. □

Corollary 3.5. *Let $S = \langle x, y\rangle$ be a 2-dimensional subspace of V. The matrix $X := A_{x\wedge y}$ satisfies equation* (13) *of Theorem 3.2 if and only if either S is totally isotropic or $S \cap \mathrm{Rad}(\varphi) \neq 0$.*

Proof. Since $\Phi^{\sigma}X^{\sigma}\Phi = (\Phi X \Phi^{\sigma})^{\sigma}$ Eq. (16) implies (13). The claim immediately follows from this remark, Theorem 3.2 and Corollary 3.4. □

Each of the matrix Eqs. (12), (13), (14), (15) and (16) packs in a unique formal container a number of scalar equations. For instance, (14) is equivalent to the following set of equations:

$$\sum_{i,j=1}^{n} x^{\sigma}_{k,i}\varphi_{i,j}x_{j,h} = 0 \tag{17}$$

for $k, h = 1, 2, \ldots, n$, with the implicit convention that $x_{j,i} = -x_{i,j}$ and $x_{i,i} = 0$. Needless to say, in each particular case some or even many of the n^2 Eq. (17) become trivial or follow from other equations of (17), possibly combined with those of (1), which define the Grassmann variety $\mathcal{G}$. It can also happen that some or even all of the Eq. (1) follow from (17).

3.4 The Variety $\mathcal{G}_\varphi$

Let φ be a (σ, ε)-sesquilinear form on V, as in the previous subsection, but now we assume that φ has Witt index $m \geq 2$ and it is not the null form, namely some but not all of the lines of $\mathrm{PG}(V)$ are totally isotropic for φ. Let $\mathcal{L}_\varphi$ be the set of lines of $\mathrm{PG}(V)$ totally isotropic for φ. Put $\mathcal{G}_\varphi := \iota_2(\mathcal{L}_\varphi) \subseteq \mathcal{G}$. By the assumptions made on φ, we have $\emptyset \neq \mathcal{G}_\varphi \subset \mathcal{G}$. We call $\mathcal{G}_\varphi$ the φ-*subset* of $\mathcal{G}$.

When φ is non-degenerate the set $\mathcal{G}_\varphi$ is described by the equations packed in (15) of Corollary 3.3 combined with those of (1).

Let φ be degenerate. Let $\mathrm{St}(\mathrm{Rad}(\varphi)) = \cup_{[x]\in[\mathrm{Rad}(\varphi)]}\mathrm{St}([x])$ be the set of lines of $\mathrm{PG}(V)$ meeting $[\mathrm{Rad}(\varphi)]$ non-trivially and put $\mathcal{R}_\varphi := \iota_2(\mathrm{St}(\mathrm{Rad}(\varphi)))$. As remarked in Sect. 3.1, for every point $[x]$ of $\mathrm{PG}(V)$ the set $\iota_2(\mathrm{St}([x]))$ is an $(n-2)$-dimensional subspace of $\mathrm{PG}(V)$. Thus, $\mathcal{R}_\varphi$ is a union of subspaces of $\mathrm{PG}(V)$. Given a basis $U = \{u_1, ..., u_n\}$ of V such that $\langle u_1, ..., u_r\rangle = \mathrm{Rad}(\varphi)$, the subspace $\langle \mathcal{R}_\varphi\rangle$ of $V^\wedge$ is spanned by the vectors $u_i \wedge u_j$ with $i < j$ and $i \leq r$. Hence $\dim(\langle \mathcal{R}_\varphi\rangle) = (n-r)r+\binom{r}{2}$. The equations packed in (16) decribe $\langle \mathcal{R}_\varphi\rangle$ while the set $\mathcal{R}_\varphi$ is described by those equations together with (1). By Corollary 3.5 the equations packed in (13) combined with those of (1) describe $\mathcal{R}_\varphi \cup \mathcal{G}_\varphi$. We have $\mathcal{R}_\varphi \cup \mathcal{G}_\varphi = \mathcal{G}_\varphi$ if and only if all points of $\mathrm{PG}(V)$ are isotropic, namely φ is alternating. In this case (13) and (1) characterize $\mathcal{G}_\varphi$.

The subgroup of $\mathrm{GL}(V)$ preserving φ acts transitively on the set of totally φ-isotropic lines of $\mathrm{PG}(V)$ that meet $\mathrm{Rad}(\varphi)$ trivially. Moreover, the embedding of $\mathrm{Gr}_2(\mathrm{PG}(V))$ in $\mathrm{PG}(V^\wedge)$ is homogeneous. Therefore the setwise stabilizer of $\mathcal{G}_\varphi$ in $\mathrm{PGL}(V^\wedge)$ acts transitively on $\mathcal{G}_\varphi \setminus \mathcal{R}_\varphi$.

3.4.1 The Algebraic Variety $\widetilde{\mathcal{G}}_\varphi$

Let φ be non-degenerate or alternating and suppose that either $\sigma = \mathrm{id}_{\mathbb{F}}$ or $\mathbb{F} = \mathbb{F}_{q^2}$ for a prime power q and $t^\sigma = t^q$. Then all equations (17) are algebraic. In this case $\mathcal{G}_\varphi$ is the set of $\mathbb{F}$-rational points of the algebraic variety $\widetilde{\mathcal{G}}_\varphi$ defined over the algebraic closure $\overline{\mathbb{F}}$ of $\mathbb{F}$ by the equations (17) and (1) (if φ is non-degenerate) or those packed in (13) together with (1) (if φ is alternating but degenerate). We warn that the variety $\widetilde{\mathcal{G}}_\varphi$ might be reducible, even disconnected, and it might contain some singularities. We will see two examples of this kind in Sect. 4. In Sect. 4.4 the variety $\widetilde{\mathcal{G}}_\varphi$ is just a finite set of points while in Sect. 4.6 it consists of two disjoint conics. In the first case $\mathbb{F} = \mathbb{F}_{q^2}$ and $\sigma(t) = t^q$ while $\sigma = \mathrm{id}_{\mathbb{F}}$ in the second case.

Put $\overline{V} := \overline{\mathbb{F}} \otimes_{\mathbb{F}} V$ and $\overline{V}^\wedge := \overline{V} \wedge \overline{V} \cong \overline{\mathbb{F}} \otimes_{\mathbb{F}} V^\wedge$. Let $\overline{\mathcal{G}} \subset \mathrm{PG}(\overline{V}^\wedge)$ be the Grassmann variety defined over $\overline{\mathbb{F}}$. We might wonder if $\widetilde{\mathcal{G}}_\varphi$ is the $\bar{\varphi}$-subset $\overline{\mathcal{G}}_{\bar{\varphi}}$ of $\overline{\mathcal{G}}$ for a suitable sesquilinear form $\bar{\varphi}$ of $\overline{V}$. When $\sigma = \mathrm{id}_{\mathbb{F}}$ the answer is affirmative. Indeed in this case a unique bilinear form $\bar{\varphi}$ exists on $\overline{V}$ inducing φ on V. Clearly, $\widetilde{\mathcal{G}}_\varphi = \overline{\mathcal{G}}_{\bar{\varphi}}$.

On the other hand, when $\mathbb{F} = \mathbb{F}_{q^2}$ and $t^\sigma = t^q$ the answer is negative. In this case, let $\tilde{\sigma}$ be the natural extension of σ to $\overline{\mathbb{F}}$, defined by the clause $t^{\tilde{\sigma}} := t^q$ for every $t \in \overline{\mathbb{F}}$. Then $\tilde{\sigma}$ is an automorphism of $\overline{\mathbb{F}}$ but $\tilde{\sigma}^2 \neq \mathrm{id}_{\overline{\mathbb{F}}}$. Indeed the elements of $\overline{\mathbb{F}}$ fixed by $\tilde{\sigma}^2$ are just those of $\mathbb{F}$. Consequently there is no sesquilinear form $\bar{\varphi}$ on $\overline{V}$ such that $\widetilde{\mathcal{G}}_\varphi$ is the $\bar{\varphi}$-subset of $\overline{\mathcal{G}}$. Nevertheless we can extend σ to an involutory automorphism $\hat{\sigma}$ of $\overline{\mathbb{F}}$, and we can always do it in infinitely many ways. Let $\bar{\varphi}$ be the unique $(\hat{\sigma}, \varepsilon)$-sesquilinear form on $\overline{V}$ inducing φ on V. Then $\mathcal{G}_\varphi$ is the set of $\mathbb{F}$-rational points of the $\bar{\varphi}$-subset $\overline{\mathcal{G}}_{\bar{\varphi}}$ of $\overline{\mathcal{G}}$. However $\widetilde{\mathcal{G}}_\varphi \neq \overline{\mathcal{G}}_{\bar{\varphi}}$ for any choice of $\hat{\sigma}$. Moreover, since no non-trivial involutory automorphism of an infinite field can be expressed as a polynomial function, no polynomial $p(t) \in \overline{\mathbb{F}}[t]$ exists such that $t^{\hat{\sigma}} = p(t)$ for every $t \in \overline{\mathbb{F}}$. Thus, apparently, the set $\overline{\mathcal{G}}_{\bar{\varphi}}$ is not an algebraic variety.

In order to define $\widetilde{\mathcal{G}}_\varphi$ we have assumed that either $\sigma = \mathrm{id}_{\mathbb{F}}$ or $\mathbb{F}$ is finite. It remains to consider the case where $\mathbb{F}$ is infinite and $\sigma \neq \mathrm{id}_{\mathbb{F}}$ but we are not going to discuss it here. We only say that in this case an obvious trick to use is to switch from $\mathbb{F}$ to its subfield formed by the elements fixed by σ.

3.4.2 Tangent Spaces and Dimension

Assume that φ is either non-degenerate or alternating and either $\sigma = \mathrm{id}_{\mathbb{F}}$ or $\mathbb{F}$ is finite. Then we can define the tangent space $\mathrm{Tan}(\mathcal{G}_\varphi)_{[a\wedge b]}$ of $\mathcal{G}_\varphi$ at a point $[a \wedge b]$ of $\mathcal{G}_\varphi$ as the subspace of $\mathrm{PG}(V^\wedge)$ formed by the $\mathbb{F}$-rational points of the tangent space $\mathrm{Tan}(\widetilde{\mathcal{G}}_\varphi)_{[a\wedge b]}$ of the algebraic variety $\widetilde{\mathcal{G}}_\varphi$ at the point $[a \wedge b]$. Thus $\mathrm{Tan}(\mathcal{G}_\varphi)_{[a\wedge b]}$ is the subspace of $V^\wedge$ described by the linear equations (4) together with those that can be obtained by differentiating (15) (when φ is non-degenerate) or (13) (if φ is degenerate and alternating). Explicitly, from (15) with $\sigma = \mathrm{id}_{\mathbb{F}}$ we obtain

$$X\Phi A_{a\wedge b} + A_{a\wedge b}\Phi X = O. \tag{18}$$

From (15) with $\mathbb{F} = \mathbb{F}_{q^2}$ and $t^\sigma = t^q$ we get

$$A^q_{a\wedge b}\Phi X = O \tag{19}$$

(where if $A_{a\wedge b} = (a_{i,j})^n_{i,j=1}$ we put $A^q_{a\wedge b} = (a^q_{i,j})^n_{i,j=1}$). Finally, from (13) with $\sigma = \mathrm{id}_{\mathbb{F}}$ we get

$$\Phi X\Phi A_{a\wedge b} + \Phi A_{a\wedge b}\Phi X = O. \tag{20}$$

By Corollary 3.4, when $[a\wedge b] \in \mathcal{R}_\varphi$ the previous equation reduces to $\Phi X\Phi A_{a\wedge b} = O$, which in its turn is implied by (16). So, if $[a\wedge b] \in \mathcal{R}_\varphi$ then $\langle \mathcal{R}_\varphi \rangle \subseteq \mathrm{Tan}(\mathcal{G}_\varphi)_{[a\wedge b]}$.

If all tangent spaces of $\mathcal{G}_\varphi$ have the same dimension, as when φ is non-degenerate, we define the *dimension* of $\mathcal{G}_\varphi$ as the projective dimension of its tangent spaces.

3.4.3 A Warning

In the Eqs. (12)–(16) of Sect. 3.3 the unknown X stands for an arbitrary skew-symmetric matrix. Hence $X^T = -X$. Moreover $\Phi^T = \varepsilon\Phi^\sigma$ and $\sigma^2 = \mathrm{id}_{\mathbb{F}}$. It follows that $(X^\sigma \Phi X)^T = \varepsilon(X^\sigma \Phi X)^\sigma$. Consequently, $X^\sigma \Phi X = O$ if and only if $(X^\sigma \Phi X)^T = O$. So, we might conclude that there is no loss in taking $k \leq h$ in (17). This conclusion is correct if $\sigma = \mathrm{id}_{\mathbb{F}}$ but it is wrong when $\mathbb{F} = \mathbb{F}_{q^2}$ and $t^\sigma = t^q$. Indeed in the latter case if we differentiate the equations (17) to compute a system of linear equations for $\mathrm{Tan}(\mathcal{G}_\varphi)_{[a\wedge b]}$, but we only take those with $k \leq h$, then we miss some of the equations that we obtain if we differentiate all of (17).

Here is an explanation of this puzzle. Let $\mathbb{F} = \mathbb{F}_{q^2}$ with $t^\sigma = t^q$ and let $\tilde{\sigma}$ be the natural extension of σ to $\overline{\mathbb{F}}$, as in Sect. 3.4.1. Then $\tilde{\sigma}^2 \neq \mathrm{id}_{\overline{\mathbb{F}}}$. Therefore, in general $(X^{\tilde{\sigma}} \Phi X)^T \neq \varepsilon(X^{\tilde{\sigma}} \Phi X)^{\tilde{\sigma}}$. Hence the equations $X^{\tilde{\sigma}} \Phi X = O$ and $(X^{\tilde{\sigma}} \Phi X)^T = O$ are not equivalent over $\overline{\mathbb{F}}$. Consequently, the subset of (17) formed by the equations with $k \leq h$ describes an algebraic variety $\widehat{\mathcal{G}}_\varphi$ larger than $\widetilde{\mathcal{G}}_\varphi$. Accordingly, if we only differentiate those equations then we obtain a description of the space $\mathrm{Tan}(\widehat{\mathcal{G}}_\varphi)_{[a\wedge b]}$, which properly contains $\mathrm{Tan}(\widetilde{\mathcal{G}}_\varphi)_{[a\wedge b]}$.

Of course, in principle there is nothing wrong in choosing $\widehat{\mathcal{G}}_\varphi$ instead of $\widetilde{\mathcal{G}}_\varphi$. Indeed $\mathcal{G}_\varphi$ is also the set of $\mathbb{F}$-rational points of $\widehat{\mathcal{G}}_\varphi$. But we should not forget that when $\sigma \neq \mathrm{id}_{\mathbb{F}}$ the tangent spaces of $\mathcal{G}_\varphi$ (whence the dimension of $\mathcal{G}_\varphi$) depend on which of $\widetilde{\mathcal{G}}_\varphi$ or $\widehat{\mathcal{G}}_\varphi$ we choose. In Sect. 4, where we will apply the theory developed so far to a number of special cases, $\widetilde{\mathcal{G}}_\varphi$ is our choice.

3.4.4 The Span of $\mathcal{G}_\varphi$

Keeping on φ the hypotheses assumed at the beginning of Sect. 3.4, now we also assume that φ is trace-valued, namely V is spanned by the set of vectors isotropic for φ. We recall that φ is trace-valued if and only if $\varphi(x, x) \in \{t + \varepsilon t^\sigma\}_{t\in\mathbb{F}}$ for every $x \in V$ (Tits [28, Chap. 8]). The latter condition is satisfied whenever either $\mathrm{char}(\mathbb{F}) \neq 2$ or $\sigma \neq \mathrm{id}_{\mathbb{F}}$ (recall that $\mathbb{F}$ is a field, namely it is commutative).

Theorem 3.6. *Under the previous hypotheses, let φ be non-degenerate or alternating.*

If φ is neither degenerate nor alternating then $\langle\mathcal{G}_\varphi\rangle = V^\wedge$.

If φ is alternating (possibly degenerate) then $\langle\mathcal{G}_\varphi\rangle$ is a hyperplane of $V^\wedge$.

Proof. Let φ be non-alternating, whence non-degenerate by the hypotheses of the theorem. Modulo multiplying φ by a suitable scalar when $\sigma \neq \mathrm{id}_{\mathbb{F}}$ we may assume that $\varepsilon = 1$. We argue by induction on n, $n = 4$ being the smallest case to consider.

Suppose firstly that $n > 4$. We may assume to have chosen the basis E of V in such a way that

$$\varphi((x_i)_{i=1}^n, (y_i)_{i=1}^n) = \sum_{i=1}^{m}(x_{2i-1}^\sigma y_{2i} + x_{2i}^\sigma y_{2i-1}) + \sum_{i=2m+1}^{n} \lambda_i x_i^\sigma y_i$$

for an anisotropic reflexive σ-hermitian form $\sum_{i=2m+1}^{n} \lambda_i x_i^\sigma y_i$ in the unknowns $x_{2m+1}, ..., x_n, y_{2m+1}, ..., y_n$ when $n > 2m$ and with $\sum_{i=2m+1}^{n} \lambda_i x_i^\sigma y_i := 0$ when $n = 2m$. Suppose firstly that $n > 2m$. Let $H_1 = \mathrm{Ker}(e_n^*)$ and $H_2 = \mathrm{Ker}(e_1^* + e_n^*)$ and let φ_1 and φ_2 be the forms induced by φ on H_1 and H_2 respectively. Then φ_1 and φ_2 are non-degenerate, non-alternating and trace-valued and have Witt index equal to m. By the inductive hypothesis, $\langle \mathcal{G}_{\varphi_1} \rangle = H_1 \wedge H_1$ and $\langle \mathcal{G}_{\varphi_2} \rangle = H_2 \wedge H_2$. We have $(H_1 \wedge H_1) \cap (H_2 \wedge H_2) = (H_1 \cap H_2) \wedge (H_1 \cap H_2)$. It follows that

$$\dim((H_1 \wedge H_1) + (H_2 \wedge H_2)) = 2\binom{n-1}{2} - \binom{n-1}{2} = \binom{n}{2} - 1$$

namely $H^\wedge := (H_1 \wedge H_1) + (H_2 \wedge H_2) = \langle \mathcal{G}_{\varphi_1} \cup \mathcal{G}_{\varphi_2} \rangle$ is a hyperplane of $V^\wedge$. However $\mathcal{G}_\varphi \supseteq \mathcal{G}_{\varphi_1} \cup \mathcal{G}_{\varphi_2}$. Hence $\langle \mathcal{G}_\varphi \rangle \supseteq H^\wedge$. In order to prove that $\langle \mathcal{G}_\varphi \rangle = V^\wedge$ it remains to show that $\mathcal{G}_\varphi$ also contains a vector of $V^\wedge \setminus H^\wedge$.

The spaces $H_1 \wedge H_1$ and $H_2 \wedge H_2$ are represented by the equations $x_{i,n} = 0$ for $i = 1, 2, ..., n-1$ and respectively $x_{1,i} + x_{i,n} = 0$ for $i = 1, 2, ..., n-1$. Note that, since $x_{n,n} = x_{1,1} = 0$, the equations $x_{1,1} + x_{1,n} = 0$ and $x_{1,n} + x_{n,n} = 0$ are both equivalent to $x_{1,n} = 0$. So, $H^\wedge$ is described by the equation $x_{1,n} = 0$. Put now $u = e_1$ and $v = e_{2m-1} + e_{2m}s + e_n$ for a scalar s such that $s + s^\sigma + \lambda_n = 0$. A scalar s with this property exists because $\lambda_n = \varphi(e_n, e_n) \in \{t + t^\sigma\}_{t \in \mathbb{F}}$ (recall that φ is trace-valued and $\varepsilon = 1$). If $t \in \mathbb{F}$ is such that $\lambda_n = t + \varepsilon t^\sigma$ then take $s = -t$.

With this choice of u and v the 2-space $\langle u, v \rangle$ is totally isotropic. Hence $[u \wedge v] \in \mathcal{G}_\varphi$. On the other hand, $u \wedge v = e_{1,2m-1} + e_{1,2m}s + e_{1,n}$, which does not satisfy the equation $x_{1,n} = 0$ of $H^\wedge$. Hence $u \wedge v \notin H^\wedge$, as we wanted.

The case where $n = 2m$ can be dealt with in a similar way, considering the hyperplanes $H_1 = \mathrm{Ker}(e_1^* + e_2^*)$ and $H_2 = \mathrm{Ker}(e_{2m-1}^* + e_m^*)$. The forms φ_1 and φ_2 induced on H_1 and H_2 have Witt index $m-1$, but now $m > 2$ since $2m = n > 4$. Hence $m - 1 \geq 2$ and the induction hypothesis can be applied. The hyperplane $H^\wedge = (H_1 \wedge H_1) + (H_2 \wedge H_2)$ is described by the equation $x_{1,n-1} + x_{1,n} + x_{2,n-1} + x_{2,n} = 0$. If we take $u = e_1$ and $v = e_n$ then $\langle u, v \rangle$ is totally isotropic but $u \wedge v$ does not satisfy the equation of $H^\wedge$.

Finally, let $n = 2m = 4$. In this case we must find six totally isotropic 2-subspaces $\langle u_i, v_i \rangle$ for $i = 1, 2, ..., 6$ such that the vectors $u_i \wedge v_i$ are linearly independent. The following choice does the job:

$$\begin{array}{ll} u_1 = e_1,\ v_1 = e_3, & u_2 = e_2,\ v_2 = e_4, \\ u_3 = e_1,\ v_3 = e_4, & u_4 = e_2,\ v_4 = e_3, \\ u_5 = e_1 - e_3, & v_5 = e_2 + e_4, \\ u_6 = -e_1 t + e_2 + e_3 t + e_4, & v_6 = -e_1 t + e_3(t - t^\sigma) + e_4 \end{array}$$

where $t + t^\sigma = 1$. Note that in the present case if $\mathrm{char}(\mathbb{F}) = 2$ then $\sigma \neq \mathrm{id}_\mathbb{F}$ otherwise, since $\varepsilon = 1$ and $n = 2m = 4$, the form φ is alternating, contrary to our assumptions. So, the equation $t + t^\sigma = 1$ admits a solution.

Assume now that φ is alternating. If φ is non-degenerate then it is well known that $\mathcal{G}_\varphi$ spans a hyperplane of $V^\wedge$ (see Blok and Cooperstein [3], for instance). We have assumed that $m \geq 2$ but the previous claim holds true for $m = 1$ too. Indeed in this case $\mathcal{G}_\varphi = \emptyset$ but $V^\wedge$ is 1-dimensional. Suppose now that φ is degenerate. Then

$$\varphi((x_i)_{i=1}^n, (y_i)_{i=1}^n) = \sum_{i=1}^k (x_{2i-1}y_{2i} - x_{2i}y_{2i-1})$$

for a positive integer k such that $2k < n$ and $\mathrm{Rad}(\varphi) = \langle e_{2k+1}, ..., e_n \rangle$. Let $S = \langle e_1, ..., e_{2k} \rangle$. Then φ induces a non-degenerate alternating form φ_S on S. By the above, $\mathcal{G}_{\varphi_S}$ spans a hyperplane of $S^\wedge := S \wedge S$. On the other hand, $\mathcal{G}_\varphi$ contains $\mathcal{R}_\varphi$, which is a subspace of $\mathrm{PG}(V^\wedge)$. Regarded $\mathcal{R}_\varphi$ as a subspace of $V^\wedge$, as we may, we have $V^\wedge = S^\wedge \oplus \mathcal{R}_\varphi$. So, $\langle \mathcal{G}_\varphi \rangle = \langle \mathcal{G}_{\varphi_S} \rangle \oplus \mathcal{R}_\varphi$ is a hyperplane of $V^\wedge$. □

3.4.5 The Set $\iota_2(\mathrm{St}([x])) \cap \mathcal{G}_\varphi$ for a Point $[x] \in \mathcal{S}_\varphi$

In this subsection we recall a few well known facts, for further reference. Let $\mathcal{S}_\varphi$ be the set of points of $\mathrm{PG}(V)$ isotropic for φ and $\overline{\mathcal{S}}_\varphi := \mathcal{S}_\varphi \setminus [\mathrm{Rad}(\varphi)]$ ($= \mathcal{S}_\varphi$ when φ is non-degenerate). The group of linear transformations of V that preserve φ acts transitively on $\overline{\mathcal{S}}_\varphi$ and on $[\mathrm{Rad}(\varphi)]$ (when φ is degenerate). So, for a point $[a] \in \overline{\mathcal{S}}_\varphi$ and a point $[b] \in [\mathrm{Rad}(\varphi)]$, the set $\iota_2(\mathrm{St}([x])) \cap \mathcal{G}$ is isomorphic to either $\iota_2(\mathrm{St}([a])) \cap \mathcal{G}$ or $\iota_2(\mathrm{St}([b])) \cap \mathcal{G}$, according to whether $[x] \in \overline{\mathcal{S}}_\varphi$ or $[x] \in [\mathrm{Rad}(\varphi)]$.

We may assume to have chosen the basis $E = (e_1, ..., e_n)$ of V in such a way that $[e_n] \in \mathcal{S}_\varphi$. If $e_n \in \mathrm{Rad}(\varphi)$ then we can take $b = e_n$, otherwise $a = e_n$. With this choice of E the matrix Φ can be described as follows

$$\Phi = \begin{bmatrix} \Phi_0 & \upsilon \\ \varepsilon(\upsilon^\sigma)^T & 0 \end{bmatrix}$$

where Φ_0 is an $(n-1) \times (n-1)$ matrix such that $\Phi_0^T = \varepsilon \Phi_0^\sigma$ and υ is an $(n-1) \times 1$ matrix. We have $\upsilon = 0$ if and only if $e_n \in \mathrm{Rad}(\varphi)$. Given a non-zero vector $x = \sum_{i=1}^n e_i x_i$, we have

$$A_{x \wedge e_n} = \begin{bmatrix} O & \hat{x} \\ -\hat{x}^T & 0 \end{bmatrix}$$

where O stands for the $(n-1) \times (n-1)$ null matrix and $\hat{x} = (x_i)_{i=1}^{n-1}$, written as an $(n-1) \times 1$ matrix. If $e_n \notin \mathrm{Rad}(\varphi)$ (whence $\upsilon \neq 0$) by equation (13) with $X = A_{x \wedge e_n}$ we obtain that $\langle x, e_n \rangle$ is totally isotropic if and only if

$$(\upsilon^\sigma)^T \hat{x} = 0 \quad \text{and} \quad (\hat{x}^\sigma)^T \Phi_0 \hat{x} = 0. \tag{21}$$

The set $\iota_2(\mathrm{St}([e_n]))\cap\mathcal{G}_\varphi$ is described by the Eq. (21) together with the linear equations (5) describing the subspace $S_{[e_n]} = \iota_2(\mathrm{St}([e_n]))$ of $\mathrm{PG}(V^\wedge)$. If φ is trace-valued then $\iota_2(\mathrm{St}([e_n]) \cap \mathcal{G}_\varphi$ spans the hyperplane $S^{\upsilon}_{[e_n]}$ of $S_{[e_n]}$ described by the first equation of (21). When φ is alternating the second equation of (21) is trivial. In this case $\iota_2(\mathrm{St}([e_n])) \cap \mathcal{G}_\varphi = S^{\upsilon}_{[e_n]}$.

Let $e_n \in \mathrm{Rad}(\varphi)$. Then (13) trivially holds for any $X = A_{x\wedge e_n}$, hence it gives no conditions on $\hat{x}$. However in this case $\langle x, e_n\rangle$ is totally isotropic if and only if $x - e_n x_n$ is isotropic. As the entries of $\hat{x}$ are just the first $n-1$ coordinates of $x - e_n x_n$, the vector $x - e_n x_n$ is isotropic if and only if the second equation of (21) holds. The first one is trivial, since now $\upsilon = 0$.

If φ is alternating and $e_n \in \mathrm{Rad}(\varphi)$ then both equations (21) are trivial.

3.5 Lines Totally Singular for a Quadratic Form

Let χ be a pseudoquadratic form on V and φ its sesquilinearization (Tits [28, Chap. 8]). Since $\mathbb{F}$ is a field, if either $\mathrm{char}(\mathbb{F}) \neq 2$ or $\sigma \neq \mathrm{id}_{\mathbb{F}}$ then we have $\chi(x) = 0$ if and only if $\varphi(x, x) = 0$. In this case there is no need to consider χ in addition to φ. Thus, in this subsection we assume that $\mathrm{char}(\mathbb{F}) = 2$ and $\sigma = \mathrm{id}_{\mathbb{F}}$, whence $\varepsilon = 1$. So χ is a quadratic form. We also assume that χ is non-singular and has Witt index $m \geq 2$, namely at least one line of $\mathrm{PG}(V)$ is totally singular for χ. The bilinearization φ of χ is a possibly degenerate alternating form and $\chi(x) \neq 0$ for every non-zero vector $x \in \mathrm{Rad}(\varphi)$, if any. Let Φ be the representative matrix of φ.

Theorem 3.7. *For two non-proportional vectors $x = \sum_{i=1}^n x_i$ and $y = \sum_{i=1}^n e_i y_i$ of V, let $X = (x_{i,j})_{i,j=1}^n := A_{x\wedge y}$. The 2-subspace $\langle x, y\rangle$ of V is totally singular for χ if and only $\Phi X \Phi X = O$ and*

$$\chi((x_{i,k})_{i=1}^n) = 0 \quad \textit{for } k = 1, 2, \ldots, n. \tag{22}$$

Proof. Recalling that $x_{i,j} = x_i y_j + x_j y_i$, it is easy to see that

$$\chi((x_{i,k})_{i=1}^n) = y_k^2 \chi(x) + x_k^2 \chi(y) + x_k y_k \varphi(x, y). \tag{23}$$

Put $S = \langle x, y\rangle$. By Corollary 3.5 and since $\mathcal{R}_\varphi \subseteq \mathcal{G}_\varphi$ because φ is alternating, the space S is totally isotropic for φ if and only if $\Phi X \Phi X = O$. It is totally singular for χ if and only if it is totally isotropic and $\chi(x) = \chi(y) = 0$, if and only if $\chi(x) = \chi(y) = \varphi(x, y) = 0$. So, if S is totally singular it follows from (23) that $\chi((x_{i,k})_{i=1}^n) = 0$ for every $k = 1, 2, \ldots, n$. Conversely, assuming that S is totally isotropic for φ, suppose that all Eq. (22) hold for S. Modulo replacing x and y with other two non-proportional vectors of S if necessary, we may assume that two indices $r < s$ exist such that $x_r = 1, x_s = 0, y_r = 0$ and $y_s = 1$. Then $x_{i,r} = y_i$ and $x_{i,s} = x_i$ for $i = 1, 2, \ldots, n$. By (22) and (23) with $k = r$ and $k = s$ we obtain $\chi(y) = 0$ and $\chi(x) = 0$, respectively. So, S is totally singular. □

Note that in general the equation $\Phi X \Phi X = O$ is not a consequence of (22) and (1), as we can see by the following example.

Example 3.8. Let $n = 4$ and $\chi(x_1, x_2, x_3, x_4) = x_1x_2 + x_3x_4$. Then the equations $\Phi X \Phi X = O$, (22) and (1) can be written as follows:

$$\left.\begin{array}{l}(A.1)\ x_{1,2}^2 + x_{1,4}x_{2,3} + x_{1,3}x_{2,4} = 0,\\ (A.2)\ x_{1,2} = x_{3,4},\end{array}\right\}\quad (\text{from } \Phi X \Phi X = O)$$

$$\left.\begin{array}{l}(B.1)\ x_{1,3}x_{1,4} = 0,\\ (B.2)\ x_{2,3}x_{2,4} = 0,\\ (B.3)\ x_{1,3}x_{2,3} = 0,\\ (B.4)\ x_{1,4}x_{2,4} = 0,\end{array}\right\}\quad (\text{from } (3.22))$$

$$(C)\ x_{1,2}x_{3,4} + x_{1,3}x_{2,4} + x_{1,4}x_{2,3} = 0.\quad (\text{from } (1.1))$$

Modulo $(A.2)$, the equations (C) and $(A.1)$ are equivalent, but there is no way to deduce $(A.2)$ from $(B.1)$, $(B.2)$, $(B.3)$, $(B.4)$ and (C).

3.5.1 The Variety $\mathcal{G}_\chi$

Let $\mathcal{L}_\chi$ be the set of lines of PG(V) totally singular for χ. Put $\mathcal{G}_\chi := \iota_2(\mathcal{L}_\chi) \subseteq \mathcal{G}_\varphi$. By Theorem 3.7 the set $\mathcal{G}_\chi$ is a projective variety described by the equations (22) together with those packed in $\Phi X \Phi X = O$ and the Eq. (1) describing $\mathcal{G}$. Its tangent space at a point $[a \wedge b]$ is described by the linear equations (4) together with those packed in $\Phi A_{a\wedge b} \Phi X + \Phi X \Phi A_{a\wedge b} = O$ (compare (20)) and the following ones, obtained by differentiating (22):

$$\sum_{j=1}^{n} \left[\frac{\partial \chi((x_{i,k})_{i=1}^n)}{\partial x_{j,k}}\right]_{a\wedge b} x_{j,k} = 0 \quad \text{for } k = 1, 2, ..., n. \tag{24}$$

Clearly, $\langle \mathcal{G}_\chi \rangle$ is contained in the hyperplane $\langle \mathcal{G}_\varphi \rangle$ of $V^\wedge$ (compare Theorem 3.6).

Theorem 3.9. *We have* $\langle \mathcal{G}_\chi \rangle = \langle \mathcal{G}_\varphi \rangle$.

Proof. The proof is quite similar to that of Theorem 3.6. We will give only a sketch of it, leaving details for the reader. We may assume to have chosen the basis E of V in such a way that

$$\chi((x_i)_{i=1}^n) = \sum_{i=1}^{m} x_{2i-1}x_{2i} + \chi_0(x_{2m+1}, ..., x_n)$$

where if $n > 2m$ then χ_0 is a totally anisotropic quadratic form as follows

$$\chi_0(x_{2m+1}, ..., x_n) = \sum_{i=m+1}^{r} (\lambda_{2i-1}x_{2i-1}^2 + x_{2i-1}x_{2i} + \lambda_{2i}x_{2i}^2) + \sum_{i=2r+1}^{n} \lambda_i x_i^2$$

while $\chi_0 := 0$ if $n = 2m$. Note that if $\mathbb{F}$ is perfect then $n - 2m \leq 2$, but in the general case n can be larger than $m + 2$. We allow $r = m$ and $n = 2r$.

We argue by induction on n. Suppose firstly that $n > 4$. Let $n = 2m$. Put $H_1 = \mathrm{Ker}(e_1^* + e_2^*)$, $H_2 = \mathrm{Ker}(e_{n-1}^* + e_n^*)$ and $H^\wedge = (H_1 \wedge H_1) + (H_2 \wedge H_2)$. By the inductive hypothesis on the forms induced by χ on H_1 and H_2, the intersection $H^\wedge \cap \langle \mathcal{G}_\varphi \rangle$ is a hyperplane of $\langle \mathcal{G}_\varphi \rangle$. In order to prove that $\mathcal{G}_\chi$ spans $\langle \mathcal{G}_\varphi \rangle$ we only must find a point of $\mathcal{G}_\chi$ outside $H^\wedge$. The point $[e_1 \wedge e_n]$ has the required properties.

Suppose now that $n > 2r \geq 2m$. In this case we choose $H_1 = \mathrm{Ker}(e_n^*)$ and $H_2 = \mathrm{Ker}(e_1^* + e_n^*)$. With $H^\wedge$ defined as above, let $v = e_{2m-1} + e_{2m}\lambda_n + e_n$. Then $[e_1 \wedge v]$ is a point of $\mathcal{G}_\chi$ outside $H^\wedge$.

Finally, let $n = 2r > 2m$. Put $H_1 = \mathrm{Ker}(e_{n-1}^*)$, $H_2 = \mathrm{Ker}(e_n^*)$ and $v = e_{2m-1} + e_{2m}(\lambda_{n-1} + \lambda_n + 1) + e_{n-1} + e_n$. The point $[e_1 \wedge v]$ belongs to $\mathcal{G}_\chi$ but not to $H^\wedge$.

The case $n = 4$ remains to be examined. In this case $\mathcal{G}_\chi$ contains the points represented by the vectors $e_{1,3}$, $e_{1,4}$, $e_{2,3}$, $e_{2,4}$ and $e_{1,2} + e_{1,3} + e_{2,4} + e_{3,4} = (e_2 + e_3) \wedge (e_1 + e_4)$ (compare the equations describing $\mathcal{G}_\chi$ in Example 3.8). These five vectors are linearly independent. So, they span $\langle \mathcal{G}_\varphi \rangle$, which is 5-dimensional. □

3.5.2 The Set $\iota_2(\mathrm{St}([a])) \cap \mathcal{G}_\chi$ for a Point $[a] \in \mathcal{G}_\chi$

The group of linear transformations of V preserving χ acts transitively on $\mathcal{G}_\chi$. Therefore, for a point $[a] \in \mathcal{G}_\chi$, we have $\iota_2(\mathrm{St}([x])) \cap \mathcal{G}_\chi \cong \iota_2(\mathrm{St}([a])) \cap \mathcal{G}_\chi$ for every point $[x] \in \mathcal{G}_\chi$. Assuming to have chosen the basis E of V in such a way that $[e_n] \in \mathcal{G}_\chi$, take $a = e_n$. As in Sect. 3.4.5,

$$\Phi = \begin{bmatrix} \Phi_0 & v \\ \varepsilon(v^\sigma)^T & 0 \end{bmatrix}$$

for an $(n-1) \times 1$ matrix v. Note that $v \neq 0$ since $[e_n] \in \mathcal{G}_\chi$ and $\mathcal{G}_\chi \cap [\mathrm{Rad}(\varphi)] = \emptyset$. Given a vector $x = \sum_{i=1} e_i x_i$ of V non-proportional to e_n let $\hat{x} := (x_i)_{i=1}^{n-1}$ and put $\chi_0(\hat{x}) := \chi(x_1, \ldots, x_{n-1}, 0)$. The 2-subspace $\langle x, e_n \rangle$ is totally singular if and only if

$$(v^\sigma)^T \hat{x} = 0 \quad \text{and} \quad \chi_0(\hat{x}) = 0. \tag{25}$$

4 Grassmann Embeddings of Quadrangles

Throughout this section $\Gamma = (\mathcal{P}, \mathcal{L})$ is a generalized quadrangle, $\Gamma^* = (\mathcal{P}^*, \mathcal{L}^*)$ is its dual (see Sect. 2.2.2), ε is a projective embedding of Γ^* in $\mathrm{PG}(V)$ for a vector space V of dimension $n \geq 4$ over a given field $\mathbb{F}$ and $\varepsilon^{\mathrm{gr}}$ is the Grassmann embedding of Γ induced by ε (see Sect. 2.2.2). We keep the notation of Sect. 3. In particular, $V^\wedge := V \wedge V$ and, given an ordered basis $E = (e_1, \ldots, e_n)$ of V, the vectors $e_{i,j} =$

$e_i \wedge e_j$ for $i < j$ form the basis $E^\wedge$ of $V^\wedge$ canonically associated to E. Moreover, ι_2 is the natural embedding of $\mathrm{Gr}_2(\mathrm{PG}(V))$ in $\mathrm{PG}(V^\wedge)$ and $\mathcal{G} = \iota_2(\mathrm{Gr}_2(\mathrm{PG}(V)))$.

Given a trace-valued non-degenerate (σ, ε)-sesquilinear form φ on V of Witt index 2, we denote by $Q(\varphi)$ the generalized quadrangle associated to φ, namely $Q(\varphi) = (\mathcal{P}_\varphi, \mathcal{L}_\varphi)$ where $\mathcal{P}_\varphi$ is the set of points of $\mathrm{PG}(V)$ isotropic for φ and $\mathcal{L}_\varphi$ is the set of lines of $\mathrm{PG}(V)$ totally isotropic for φ. Similarly, given a non-singular quadratic form χ of Witt index 2, $Q(\chi) = (\mathcal{P}_\chi, \mathcal{L}_\chi)$ is the generalized quadrangle formed by the points and the lines of $\mathrm{PG}(V)$ that are singular and respectively totally singular for χ. Also, $\mathcal{G}_\varphi := \iota_2(\mathcal{L}_\varphi)$ and $\mathcal{G}_\chi := \iota_2(\mathcal{L}_\chi)$, as in Sect. 3. So, if $\varepsilon(\Gamma^*) = Q(\varphi)$ then $\varepsilon^{\mathrm{gr}}$ embeds Γ in $\langle \mathcal{G}_\varphi \rangle^{\mathrm{pr}}$ and we have $\varepsilon^{\mathrm{gr}}(\mathcal{P}) = \mathcal{G}_\varphi$. We recall that $\langle \mathcal{G}_\varphi \rangle^{\mathrm{pr}}$ is a hyperplane of $\mathrm{PG}(V^\wedge)$ when φ is alternating while $\langle \mathcal{G}_\varphi \rangle^{\mathrm{pr}} = \mathrm{PG}(V^\wedge)$ in all other cases (Theorem 3.6).

Similarly, if $\mathrm{char}(\mathbb{F}) = 2$ and $\varepsilon(\Gamma^*) = Q(\chi)$ for a quadratic form χ then $\varepsilon^{\mathrm{gr}}$ embeds Γ in $\langle \mathcal{G}_\chi \rangle^{\mathrm{pr}}$ and we have $\varepsilon^{\mathrm{gr}}(\mathcal{P}) = \mathcal{G}_\chi$, but now $\langle \mathcal{G}_\chi \rangle^{\mathrm{pr}}$ is a hyperplane of $\mathrm{PG}(V^\wedge)$ (Theorem 3.6).

We shall examine several special cases of the situation described above. In each of the cases we will consider but the last two Γ also admits a projective embedding ε_0. We will also consider the quadratic veronesean embedding $\varepsilon_0^{\mathrm{ver}}$ induced by ε_0 (Sect. 2.2.1), discussing its relations with $\varepsilon^{\mathrm{gr}}$, if any.

4.1 Quadrangles of Symplectic Type

Let $\Gamma \cong W(3, \mathbb{F})$, the generalized quadrangle associated to a non-degenerate alternating form of $V(4, \mathbb{F})$. Then $\Gamma^* \cong Q(4, \mathbb{F})$. Accordingly, V has dimension $n = 5$ and $\varepsilon(\Gamma^*) = Q(\chi)$ with χ a non-singular quadratic form of V of Witt index 2. Let φ be the bilinearization of χ. If $\mathrm{char}(\mathbb{F}) \neq 2$ then $Q(\chi) = Q(\varphi)$. By (21) of Sect. 3.4.5 when $\mathrm{char}(\mathbb{F}) \neq 2$ and (25) of Sect. 3.5.2 when $\mathrm{char}(\mathbb{F}) = 2$ it follows that $\varepsilon^{\mathrm{gr}}(l)$ is a non-singular conic for every line l of Γ. Thus, the embedding $\varepsilon^{\mathrm{gr}}$ is quadratic.

4.1.1 The Case char($\mathbb{F}$) $\neq$ 2

Let $\mathrm{char}(\mathbb{F}) \neq 2$. Then $Q(\chi) = Q(\varphi)$. The form φ is symmetric. By Theorem 3.6, the set $\varepsilon^{\mathrm{gr}}(\mathcal{P}) = \mathcal{G}_\varphi$ $(= \mathcal{G}_\chi)$ spans $V^\wedge \cong V(10, \mathbb{F})$.

Modulo multiplying φ by a scalar, we can assume to have chosen the basis E of V in such a way that φ is represented by the following matrix with respect to E:

$$\Phi = \begin{bmatrix} 0 & 1 & 0 & 0 & 0 \\ 1 & 0 & 0 & 0 & 0 \\ 0 & 0 & 0 & 1 & 0 \\ 0 & 0 & 1 & 0 & 0 \\ 0 & 0 & 0 & 0 & 1 \end{bmatrix}.$$

With the convention that $x_{i,i} = 0$ and $x_{j,i} = -x_{i,j}$, the variety $\mathcal{G}_\varphi$ is described by the equations (1) together with the following ones, packed in $X\Phi X = O$ (Corollary 3.3, (15)):

$$x_{k,1}x_{2,h} + x_{k,2}x_{1,h} + x_{k,3}x_{4,h} + x_{k,4}x_{3,h} + x_{k,5}x_{5,h} = 0, \quad (1 \le k \le h \le 5).$$

The equation $A_{a\wedge b}\Phi X + X\Phi A_{a\wedge b} = O$, together with (4), describes the tangent space $\mathrm{Tan}(\mathcal{G}_\varphi)_{[a\wedge b]}$ of $\mathcal{G}_\varphi$ at a point $[a \wedge b] \in \mathcal{G}_\varphi$. Clearly, $\varepsilon^{\mathrm{gr}}$ is homogeneous. So, we are free to choose the point $[a \wedge b]$ as we like. We choose $a = e_1$ and $b = e_3$, so that $a \wedge b = e_{1,3}$. With this choice of $[a \wedge b]$ we get the following linear system of rank 6 for $\mathrm{Tan}(\mathcal{G}_\varphi)_{[a\wedge b]}$:

$$x_{1,2} - x_{3,4} = x_{1,4} = x_{2,3} = x_{2,4} = x_{2,5} = x_{4,5} = 0.$$

As $\dim(V^\wedge) = 10$ we obtain that $\dim(\mathrm{Tan}(\mathcal{G}_\varphi)_{[a\wedge b]}) = 4$ (vector dimension). Consequently, $\mathcal{G}_\varphi$ is 3-dimensional.

4.1.2 The Case Char($\mathbb{F}$) = 2

Let $\mathrm{char}(\mathbb{F}) = 2$. By Theorems 3.6 and 3.9, the set $\varepsilon^{\mathrm{gr}}(\mathcal{P}) = \mathcal{G}_\chi$ spans a hyperplane of $V^\wedge$.

Modulo multiplying χ by a scalar we can assume to have chosen the basis E of V in such a way that $\chi(x_1, \ldots, x_5) = x_1x_2 + x_3x_4 + x_5^2$. Accordingly, the bilinearization φ of χ is represented by the following matrix:

$$\Phi = \begin{bmatrix} 0&1&0&0&0 \\ 1&0&0&0&0 \\ 0&0&0&1&0 \\ 0&0&1&0&0 \\ 0&0&0&0&0 \end{bmatrix}.$$

The variety $\mathcal{G}_\chi$ is described by the following equations, obtained from $\Phi X \Phi X = O$ (Corollary 3.5) and (22):

$$\begin{array}{ll} x_{k,1}x_{2,h} + x_{k,2}x_{1,h} + x_{k,3}x_{4,h} + x_{k,4}x_{3,h} = 0, & (1 \le k < h \le 5); \\ x_{1,k}x_{2,k} + x_{3,k}x_{4,k} + x_{5,k}^2 = 0, & (k = 1, \ldots, 5). \end{array}$$

It is not difficult to check that these equations imply (1). Moreover, they are equivalent to the following set of equations:

$$(A) \quad x_{1,2} + x_{3,4} = 0,$$

$$(B)\ \begin{cases} x_{1,2}^2 + x_{1,3}x_{2,4} + x_{1,4}x_{2,3} = 0, \\ x_{1,5}^2 + x_{1,3}x_{1,4} = 0, \\ x_{2,5}^2 + x_{2,3}x_{2,4} = 0, \\ x_{3,5}^2 + x_{1,3}x_{2,3} = 0, \\ x_{4,5}^2 + x_{1,4}x_{2,4} = 0, \end{cases}$$

$$(C)\ \begin{cases} x_{1,5}x_{2,5} + x_{3,5}x_{4,5} = 0, \\ x_{1,2}x_{1,5} + x_{1,3}x_{4,5} + x_{1,4}x_{3,5} = 0, \\ x_{1,2}x_{2,5} + x_{2,3}x_{4,5} + x_{2,4}x_{3,5} = 0, \\ x_{1,3}x_{2,5} + x_{2,3}x_{1,5} + x_{1,2}x_{3,5} = 0, \\ x_{1,4}x_{2,5} + x_{2,4}x_{1,5} + x_{1,2}x_{4,5} = 0. \end{cases}$$

Equation (A) describes the hyperplane of $V^{\wedge}$ spanned by $\mathcal{G}_{\chi}$. Note that the squares of the equations listed in (C) follow from the equations (B), but this fact does not allow us to drop (C) (compare Hilbert's Nullstellensatz). In order to compute $\mathrm{Tan}(\mathcal{G}_{\chi})_{[a\wedge b]}$ we choose $a = e_1$ and $b = e_3$, as when $\mathrm{char}(\mathbb{F}) \neq 2$. We get just the same linear system as in that case, namely:

$$x_{1,2} - x_{3,4} = x_{1,4} = x_{2,3} = x_{2,4} = x_{2,5} = x_{4,5} = 0.$$

(Note that if we drop the set of equations (C) then we miss the last two equations of this linear system.) So, $\mathcal{G}_{\chi}$ has dimension 3 in this case too.

4.1.3 The Veronesean Embedding $\varepsilon_0^{\mathrm{ver}}$

As $\Gamma \cong W(3, \mathbb{F})$, a projective embedding $\varepsilon_0 : \Gamma \to \mathrm{PG}(3, \mathbb{F})$ also exists such that $\varepsilon_0(\Gamma) = Q(\alpha)$ for a non-degenerate alternating form α of $V(4, \mathbb{F})$. Since $\varepsilon_0(\mathcal{P})$ is equal to the set of points of $\mathrm{PG}(3, \mathbb{F})$, the image $\varepsilon_0^{\mathrm{ver}}(\mathcal{P})$ of $\mathcal{P}$ by the quadratic veronesean embedding $\varepsilon_0^{\mathrm{ver}}$ induced by ε_0 is equal to the Veronese variety $\mathcal{V} = \eta^{\mathrm{ver}}(\mathrm{PG}(3, \mathbb{F}))$.

As proved in [6], we have $\varepsilon_0^{\mathrm{ver}} \geq \varepsilon^{\mathrm{gr}}$, namely there exists a linear mapping π from $\langle\mathcal{V}\rangle = V(10, \mathbb{F})$ to $\langle\varepsilon^{\mathrm{gr}}(\mathcal{P})\rangle$ inducing a bijection from $\mathcal{V}$ to $\varepsilon^{\mathrm{gr}}(\mathcal{P})$ and mapping every conic $\varepsilon_0^{\mathrm{ver}}(l)$ for $l \in \mathcal{L}$ onto the corresponding conic $\varepsilon^{\mathrm{gr}}(l)$. We call π the *projection* of $\varepsilon_0^{\mathrm{ver}}$ onto $\varepsilon^{\mathrm{gr}}$.

Proposition 4.1. *If* $\mathrm{char}(\mathbb{F}) \neq 2$ *then* π *is an isomorphism. If* $\mathrm{char}(\mathbb{F}) = 2$ *then* $\dim(\mathrm{Ker}(\pi)) = 1$.

This result is proved in [6] in a more general context, but it can also be obtained quite easily from Theorems 3.6 and 3.9 of the present paper, by comparing the dimensions of $\langle\mathcal{V}\rangle$ and $\langle\mathcal{G}_{\varphi}\rangle$. The next corollary immediately follows from Proposition 4.1.

Corollary 4.2. *If* $\mathrm{char}(\mathbb{F}) \neq 2$ *then* π *induces an isomorphism from the Veronese variety* $\mathcal{V}$ *of* $\mathrm{PG}(3, \mathbb{F})$ *to the subvariety* $\mathcal{G}_{\varphi}$ *of* $\mathcal{G}$.

In any case, since π is a morphism of embeddings from $\varepsilon_0^{\mathrm{ver}}$ to $\varepsilon^{\mathrm{gr}}$ and $\mathcal{V} = \varepsilon_0^{\mathrm{ver}}(\mathcal{P})$, it induces a bijective morphism of varieties $\pi_{|\mathcal{V}}$ from $\mathcal{V}$ to $\mathcal{G}_\chi$. We do not know if $\pi_{|\mathcal{V}}$ is also an isomorphism of algebraic varieties when $\mathrm{char}(\mathbb{F}) = 2$.

Let $\mathrm{char}(\mathbb{F}) = 2$. As proved in [7], the nuclei of the conics $\varepsilon_0^{\mathrm{ver}}(l)$ for $l \in \mathcal{L}$ form a copy $\mathcal{N}$ of the quadric $Q(4, \mathbb{F})$ ($\cong \Gamma^*$), naturally embedded in a 4-subspace of $\mathrm{PG}(9, \mathbb{F})$. Let $n(\mathcal{N})$ be the nucleus of $\mathcal{N}$. The following is proved in [7]:

Proposition 4.3. *Let* $\mathrm{char}(\mathbb{F}) = 2$. *Then* $\mathrm{Ker}(\pi) = n(\mathcal{N})$. *Whence the image* $\pi(\mathcal{N})$ *of* $\mathcal{N}$ *by* π *is isomorphic to* $W(3, \mathbb{F})$. *It consists of the nuclei of the conics* $\varepsilon^{\mathrm{gr}}(l)$ *for* $l \in \mathcal{L}$. *The vector space* $K := \langle \pi(\mathcal{N}) \rangle$ *is 4-dimensional and defines a quotient of* $\varepsilon^{\mathrm{gr}}$. *The quotient* $\varepsilon^{\mathrm{gr}}/K$ *is a 1-embedding of* Γ *in* $\mathrm{PG}(4, \mathbb{F})$. *It is projective if and only if* $\mathbb{F}$ *is perfect.*

The image of Γ by $\varepsilon^{\mathrm{gr}}/K$ is a subgeometry of a copy of Γ^* in $\mathrm{PG}(4, \mathbb{F})$ (De Bruyn and Pasini [10, Theorem 3.11]). We recall that if $\mathbb{F}$ is perfect then $\Gamma \cong \Gamma^*$, whence ε is also a projective embedding of Γ. In this case $\varepsilon^{\mathrm{gr}}/K \cong \varepsilon$ and the restriction of π to $\langle \mathcal{N} \rangle$ is the natural projection of ε onto ε_0.

Turning to universality, we mention the following result, proved in [7]:

Proposition 4.4. *Let* $\mathbb{F} = \mathbb{F}_q$ *for a prime power* $q > 3$. *Then* $\varepsilon_0^{\mathrm{ver}}$ ($\cong \varepsilon^{\mathrm{gr}}$ *by Proposition 4.1) is relatively universal.*

We guess that this statement holds true for any field of characteristic different from 2. On the other hand, if $\mathrm{char}(\mathbb{F}) = 2$ then $\varepsilon_0^{\mathrm{ver}}$ is not relatively universal. Indeed in this case Γ also admits a projective embedding ε_δ in $\mathrm{PG}(4 + \delta, \mathbb{F})$ for any $\delta \leq [\mathbb{F} : \mathbb{F}^2]$, where $\mathbb{F}^2$ stands for the subfield of square elements of $\mathbb{F}$ (De Bruyn and Pasini [10]). Clearly, if $\mathbb{F}$ is perfect then $\delta = 1$, otherwise δ can be larger than 1, even infinite. To make things easier, when $[\mathbb{F} : \mathbb{F}^2]$ is infinite assume to have chosen a finite value for δ. The image of $\mathcal{P}$ by ε_δ is a non-singular quadric Q_δ and $\varepsilon_0 \cong \varepsilon_\delta/N_\delta$, where N_δ is the radical of the bilinearization of Q_δ, namely $[N_\delta]$ is the nucleus of Q_δ. We can always consider the veronesean embedding $\varepsilon_\delta^{\mathrm{ver}}$ induced by ε_δ. The projection from ε_δ to ε_0 lifts to a morphism from $\varepsilon_\delta^{\mathrm{ver}}$ to $\varepsilon_0^{\mathrm{ver}}$, but $\dim(\varepsilon_\delta^{\mathrm{ver}}) > \dim(\varepsilon_0^{\mathrm{ver}})$. Hence $\varepsilon_0^{\mathrm{ver}}$ cannot be relatively universal.

In particular, if $[\mathbb{F} : \mathbb{F}^2]$ is infinite then the hull of $\varepsilon_0^{\mathrm{ver}}$ is infinite-dimensional. This is not surprising, since in this case the hull of ε_0 is infinite-dimensional as well.

4.2 *Quadrics in* PG(4, $\mathbb{F}$)

In this subsection we consider a setting dual to that of Sect. 4.1. Now $\Gamma \cong Q(4, \mathbb{F})$, the generalized quadrangle associated to a non-singular quadratic form of $V(5, \mathbb{F})$ of Witt index 2. Accordingly, $\Gamma^* \cong W(3, \mathbb{F})$ and ε is a projective embedding of $W(3, \mathbb{F})$.

We recall that when $\mathrm{char}(\mathbb{F}) \neq 2$ the natural embedding of $W(3, \mathbb{F})$ in $\mathrm{PG}(3, \mathbb{F})$ (uniquely determined up to isomorphisms) is the unique projective embedding of

$W(3, \mathbb{F})$. As noticed at the end of the previous subsection, this is no more true when $\mathrm{char}(\mathbb{F}) = 2$. In this case we must say which is the embedding ε of Γ^* that we consider. We assume that ε is the natural one in this case too. So, in either case V is 4-dimensional and $\varepsilon(\Gamma^*) = Q(\varphi)$ for a non-degenerate alternating form φ.

4.2.1 The Embedding $\varepsilon^{\mathrm{gr}}$

It is well known that $\varepsilon^{\mathrm{gr}}$ is isomorphic to the natural projective embedding ε_0 of Γ as $Q(4, \mathbb{F})$ in $\mathrm{PG}(4, \mathbb{F})$. This conclusion can be obtained from the results of Sect. 3 of this paper, as we are going to show.

The space $V^\wedge$ is 6-dimensional and $\mathcal{G}_\varphi$ spans a hyperplane of $V^\wedge$, by Theorem 3.6. So, $\dim(\langle \mathcal{G}_\varphi \rangle) = 5$ (vector dimension). The equation $\hat{x}^T \Phi_0 \hat{x} = 0$ of (21) of Sect. 3.4.5 is trivial, since Φ_0 is skew-symmetric. Consequently, $\varepsilon^{\mathrm{gr}}(l)$ is a projective line. Thus, $\varepsilon^{\mathrm{gr}}$ is projective. Choose the representative matrix Φ of φ as follows,

$$\Phi = \begin{bmatrix} 0 & 0 & 1 & 0 \\ 0 & 0 & 0 & 1 \\ -1 & 0 & 0 & 0 \\ 0 & -1 & 0 & 0 \end{bmatrix}.$$

The matrix equation $X \Phi X = O$ of Corollary 3.3 is equivalent to the following pair of scalar equations:

$$x_{1,3} + x_{2,4} = 0 \quad \text{(which describes } \langle \mathcal{G}_\varphi \rangle\text{)},$$
$$x_{1,4} x_{2,3} + x_{1,2} x_{2,4} + x_{1,3}^2 = 0.$$

Hence $\varepsilon^{\mathrm{gr}} \cong \varepsilon_0$. Note that all Eq. (1) follow from the previous two equations.

4.2.2 The Veronesean Embedding $\varepsilon_0^{\mathrm{ver}}$

The quadratic veronesean embedding η^{ver} embeds $\mathrm{PG}(4, \mathbb{F})$ in $\mathrm{PG}(14, \mathbb{F})$ and $\varepsilon_0^{\mathrm{ver}}(\mathcal{P})$ spans a hyperplane H of $\mathrm{PG}(14, \mathbb{F})$. Explicitly, let $\varepsilon_0(\mathcal{P})$ be the quadric described by the equation $x_1 x_2 + x_3 x_4 + x_5^2 = 0$, as we can always assume. Then H is described by the equation $x_{1,2} + x_{3,4} - x_{5,5} = 0$.

When $\mathrm{char}(\mathbb{F}) \neq 2$ no morphism can exist between $\varepsilon_0^{\mathrm{ver}}$ and ε_0 ($\cong \varepsilon^{\mathrm{gr}}$). Let $\mathrm{char}(\mathbb{F}) = 2$ and let N be the subspace of $\mathrm{PG}(14, \mathbb{F})$ spanned by the nuclei of the conics $\eta^{\mathrm{ver}}(l)$ for l a line of $\mathrm{PG}(4, \mathbb{F})$. Let N_Γ be the subspace of N spanned by the nuclei of the conics $\varepsilon_0^{\mathrm{ver}}(l) = \eta^{\mathrm{ver}}(\varepsilon_0(l))$ for l a line of Γ. Clearly, $N_\Gamma \subseteq H \cap N$.

It is known that N has codimension 5 in $\mathrm{PG}(14, \mathbb{F})$, it defines a quotient of η^{ver} and the quotient η^{ver}/N is a 1-embedding of $\mathrm{PG}(4, \mathbb{F})$ in itself (Thas and Van Maldeghem [25]). In particular, if $\mathbb{F}$ is perfect then η^{ver}/N is isomorphic to the identity embedding of $\mathrm{PG}(4, \mathbb{F})$ in itself. Moreover, the mapping ν that maps every line of $\mathrm{PG}(4, \mathbb{F})$ onto its nucleus is isomorphic to the projective embedding ι_2 of $\mathrm{Gr}_2(\mathrm{PG}(4, \mathbb{F}))$ in $\mathrm{PG}(9, \mathbb{F})$ ($\cong N$).

According to the above, the subspace $N_H := H \cap N$ defines a quotient of $\varepsilon_0^{\mathrm{ver}}$. Clearly, $\varepsilon_0^{\mathrm{ver}}/N_H$ is a 1-embedding of Γ.

Lemma 4.5. *The subspace N_H is a hyperplane of N.*

Proof. We must prove that $N \not\subseteq H$, namely there exists a line l of $\mathrm{PG}(4, \mathbb{F})$ such that $\nu(l) \notin H$. Let l be the line described by the linear system $x_3 = x_4 = x_5 = 0$. Then $\nu(l)$ is the point $[x_{i,j}]_{1 \leq i \leq j \leq 5}$ with $x_{i,j} = 0$ for every $(i, j) \neq (1, 2)$. This point does not satisfy the equation of H. □.

By Lemma 4.5, the subspace $N_H \subset H$ has codimension 5 in H. The claims gathered in the next proposition follow from this fact, the remarks before Lemma 4.5 and those after Corollary 4.3 (modulo permuting the roles of ε and ε_0). We leave the details of the proof to the reader.

Proposition 4.6. *We have $N_\Gamma = N_H$, $\dim(\varepsilon_0^{\mathrm{ver}}/N_H) = 4$ and the composition $\nu \cdot \varepsilon_0^{\mathrm{ver}}$ is isomorphic to the Grassmann embedding $\varepsilon_0^{\mathrm{gr}}$ of $\Gamma^* \cong W(3, \mathbb{F})$ induced by ε_0 (recall that Γ is the dual of Γ^*). Moreover, if $\mathbb{F}$ is perfect then $\varepsilon_0^{\mathrm{ver}}/N_H \cong \varepsilon_0$.*

Let $\mathbb{F}$ be non-perfect. Then the 1-embedding $\varepsilon_0^{\mathrm{ver}}/N_H$ is lax and the image of Γ by $\varepsilon_0^{\mathrm{ver}}/N_H$ is a proper subgeometry of $\varepsilon_0(\Gamma)$. Hence Γ is isomorphic to a proper subgeometry of itself. Clearly, the same holds for Γ^*. Moreover, each of the geometries Γ and Γ^* is isomorphic to a proper subgeometry of the other one (De Bruyn and Pasini [10]).

4.3 Hermitian Varieties in PG(3, $\mathbb{K}$)

Given a field $\mathbb{F}$, let $\mathbb{K}$ be an extension of $\mathbb{F}$ of degree $[\mathbb{K} : \mathbb{F}] = 2$ and σ the unique involutory automorphism of $\mathbb{K}$ fixing $\mathbb{F}$ elementwise. Let Γ be isomorphic to the generalized quadrangle associated to a non-degenerate σ-hermitian form of $V(4, \mathbb{K})$ of Witt index 2. It is well known that $\Gamma^* \cong Q^-(5, \mathbb{F})$. So, V is a 6-dimensional vector space over $\mathbb{F}$ and $\varepsilon(\Gamma^*) = Q(\chi)$ for a non-singular quadratic form χ of Witt index 2 with non-degenerate bilinearization φ. Recall that if $\mathrm{char}(\mathbb{F}) \neq 2$ then $\mathcal{G}_\chi = \mathcal{G}_\varphi$ while $\mathcal{G}_\chi \subset \mathcal{G}_\varphi$ when $\mathrm{char}(\mathbb{F}) = 2$.

The space $V^\wedge$ is 15-dimensional. By Theorems 3.6 and 3.9, if $\mathrm{char}(\mathbb{F}) \neq 2$ then $\mathcal{G}_\varphi$ spans $V^\wedge$ while $\langle \mathcal{G}_\chi \rangle = \langle \mathcal{G}_\varphi \rangle$ is a hyperplane of $V^\wedge$ when $\mathrm{char}(\mathbb{F}) = 2$.

By (21) of Sect. 3.4.5 when $\mathrm{char}(\mathbb{F}) \neq 2$ and (25) of Sect. 3.5.2 when $\mathrm{char}(\mathbb{F}) = 2$ it follows that $\dim(\langle \varepsilon^{\mathrm{gr}}(l) \rangle^{\mathrm{pr}}) = 3$ for every line l of Γ and $\varepsilon^{\mathrm{gr}}(l)$ is an elliptic quadric of the 3-space $\langle \varepsilon^{\mathrm{gr}}(l) \rangle^{\mathrm{pr}}$. Thus $\varepsilon^{\mathrm{gr}}$ is a 3-embedding.

Let $\mathrm{char}(\mathbb{F}) \neq 2$. We can assume that φ is represented by the following matrix, for a suitable non-square element η of $\mathbb{F}$:

$$\Phi = \begin{bmatrix} 0&1&0&0&0&0 \\ 1&0&0&0&0&0 \\ 0&0&0&1&0&0 \\ 0&0&1&0&0&0 \\ 0&0&0&0&1&0 \\ 0&0&0&0&0&-\eta \end{bmatrix}.$$

The variety $\mathcal{G}_\varphi$ is described by the equations (1) together with the following ones, packed in $X\Phi X = O$ (Corollary 3.3, (15)):

$$x_{k,1}x_{2,h} + x_{k,2}x_{1,h} + x_{k,3}x_{4,h} + x_{k,4}x_{3,h} + x_{k,5}x_{5,h} - \eta x_{k,6}x_{6,h} = 0$$

for $1 \leq k \leq h \leq 6$. The equation $A_{a\wedge b}\Phi X + X\Phi A_{a\wedge b} = O$, together with (4), describes the tangent space $\mathrm{Tan}(\mathcal{G}_\varphi)_{[a\wedge b]}$. We choose $a = e_1$ and $b = e_3$. With this choice of $[a \wedge b]$ we get the following linear system of rank 9 for $\mathrm{Tan}(\mathcal{G}_\varphi)_{[a\wedge b]}$:

$$x_{1,2} - x_{3,4} = x_{1,4} = x_{2,3} = x_{2,4} = x_{2,5} = x_{2,6} = x_{4,5} = x_{4,6} = x_{5,6} = 0.$$

As $\dim(V^\wedge) = 15$ we obtain that $\dim(\mathrm{Tan}(\mathcal{G}_\varphi)_{[a\wedge b]}) = 6$ (vector dimension). Consequently, $\mathcal{G}_\varphi$ is 5-dimensional.

Let $\mathrm{char}(\mathbb{F}) = 2$. We can assume that

$$\chi(x_1, \ldots, x_6) = x_1x_2 + x_3x_4 + x_5x_6 + x_5^2 + \lambda x_6^2$$

for a scalar $\lambda \in \mathbb{F}$ such that the polynomial $t^2+t+\lambda$ is irreducible over $\mathbb{F}$. Accordingly, φ is represented by the following matrix

$$\Phi = \begin{bmatrix} 0&1&0&0&0&0 \\ 1&0&0&0&0&0 \\ 0&0&0&1&0&0 \\ 0&0&1&0&0&0 \\ 0&0&0&0&0&1 \\ 0&0&0&0&1&0 \end{bmatrix}.$$

The variety $\mathcal{G}_\chi$ is described by the equations (1) together with the following equations, obtained from $X\Phi X = O$ (Corollary 3.3) and (22):

$$x_{k,1}x_{2,h} + x_{k,2}x_{1,h} + x_{k,3}x_{4,h} + x_{k,4}x_{3,h} + x_{k,5}x_{6,h} + x_{k,6}x_{5,h} = 0$$

for $1 \leq k < h \leq 6$ and

$$x_{1,k}x_{2,k} + x_{3,k}x_{4,k} + x_{5,k}x_{6,k} + x_{5,k}^2 + \lambda x_{5,k}^2 = 0$$

for $k = 1, \ldots, 6$. The tangent space $\mathrm{Tan}(\mathcal{G}_\chi)_{[e_{1,3}]}$ is described by the same system of linear equations as when $\mathrm{char}(\mathbb{F}) \neq 2$. Hence $\mathcal{G}_\chi$ has dimension 5.

No morphism exists between the embeddings $\varepsilon_0^{\mathrm{ver}}$ and $\varepsilon^{\mathrm{gr}}$, since they are defined over different fields, namely $\mathbb{K}$ and $\mathbb{F}$ respectively. We do not know if any interesting direct relation exists between the variety $\mathcal{G}_\chi$ and the image $\varepsilon_0^{\mathrm{ver}}(\mathcal{P})$ of the point-set $\mathcal{P}$ of Γ under $\varepsilon_0^{\mathrm{ver}}$.

On the other hand, let η^{hv} the the hermitian veronesean embedding of $\mathrm{PG}(3,\mathbb{K})$ in $\mathrm{PG}(15,\mathbb{F})$ (Lunardon [18], Cooperstein, Thas and Van Maldeghem [9]; also Pasini [19, 6.2], where embeddings like this are called *twisted tensor embeddings*). Put $\varepsilon_0^{\mathrm{hv}} := \eta^{\mathrm{hv}} \cdot \varepsilon_0$. It is not difficult to see that $\varepsilon^{\mathrm{hv}}(\mathcal{P})$ spans a hyperplane of $\mathrm{PG}(15,\mathbb{F})$. So, $\varepsilon_0^{\mathrm{hv}}$ has global dimension $\dim(\varepsilon_0^{\mathrm{hv}}) = 14 = \mathrm{PG}(V^\wedge)$. Moreover, if $l \in \mathcal{L}$ is a line of Γ then both $\varepsilon_0^{\mathrm{hv}}(l)$ and $\varepsilon^{\mathrm{gr}}(l)$ are elliptic quadrics. These facts, combined with Theorems 3.6 and 3.9, suggest the following conjecture: if $\mathrm{char}(\mathbb{K}) \neq 2$ then $\varepsilon_0^{\mathrm{hv}} \cong \varepsilon^{\mathrm{gr}}$ while if $\mathrm{char}(\mathbb{K}) = 2$ then $\varepsilon^{\mathrm{gr}}$ is a 1-quotient of $\varepsilon_0^{\mathrm{hv}}$.

4.4 Elliptic Quadrics in PG(5, *q*)

Let $\Gamma \cong Q^-(5,q)$, the generalized quadrangle associated to a non-singular quadratic form of $V(6,q)$ of Witt index 2, for a prime power q. Then $\Gamma^* \cong H(3,q^2)$, namely V is a 4-dimensional vector space over $\mathbb{F}_{q^2}$ and $\varepsilon(\Gamma^*) = Q(\varphi)$ for a non-degenerate hermitian form φ of V of Witt index 2. It is well known that $\varepsilon^{\mathrm{gr}}$ is a lax 1-embedding. It embeds Γ as $Q^-(5,q)$ in a Baer subgeometry of $\mathrm{PG}(5,q^2)$ (see Cooperstein and Shult [8], for instance). Thus, $\varepsilon^{\mathrm{gr}}$ is obtained from the natural embedding $\varepsilon_0 : \Gamma \to \mathrm{PG}(5,q)$ by extending the field $\mathbb{F}_q$ to $\mathbb{F}_{q^2}$.

Some of the previous statements can be obtained from the results of Sect. 3 of our paper. By Theorem 3.6 the set $\mathcal{G}_\varphi$ spans $V^\wedge$, which is 6-dimensional. Given a point $[a] \in \mathcal{G}_\varphi$, the space $\mathrm{St}([a])$ is a projective plane. By the first equation of (21) of Sect. 3.4.5, $\langle \varepsilon^{\mathrm{gr}}(l)\rangle^{\mathrm{pr}}$ is a line. By the second equation of (21), $\varepsilon^{\mathrm{gr}}(l)$ is a Baer subline of $\langle \varepsilon^{\mathrm{gr}}(l)\rangle^{\mathrm{pr}}$. Thus, $\varepsilon^{\mathrm{gr}}$ is a lax 1-embedding.

We may assume to have chosen the basis E of V in such a way that φ is represented by the identity matrix. In this way $\mathcal{G}_\varphi$ is represented by the matrix equation $X^q X = O$ (Corollary3.3), combined with equations (1). Explicitly, $X^q X = O$ can be written as follows:

$$x_{k,1}^q x_{1,h} + x_{k,2}^q x_{2,h} + x_{k,3}^q x_{3,h} + x_{k,4}^q x_{4,h} = 0, \quad (1 \le k, h \le 4).$$

The tangent space $\mathrm{Tan}(\mathcal{G}_\varphi)_{[a\wedge b]}$ is described by (4) together with the equation $A_{a\wedge b}^q X = O$ (see (19)). Choose $a = e_1 + e_2 t$ and $b = e_3 + e_4 t$ for a given t such that $t^{q+1} + 1 = 0$, and $t \neq 1$ if q is even. So, $a \wedge b = e_{1,2} + (e_{1,4} + q_{2,3})t + e_{2,4}t^2$. With this choice of $a \wedge b$ we get the following linear system of rank 5 for the tangent space $\mathrm{Tan}(\mathcal{G}_\varphi)_{[a\wedge b]}$:

$$x_{2,3} + t^q x_{2,4} = x_{1,2} = x_{1,3} = x_{1,4} = x_{3,4} = 0.$$

As $\dim(V^\wedge) = 6$ we obtain that $\dim(\mathrm{Tan}(\mathcal{G}_\varphi)_{[a\wedge b]}) = 1$ (vector dimension). Consequently, $\mathcal{G}_\varphi$ is 0-dimensional.

In fact, let $\overline{\mathbb{F}}$ be the algebraic closure of $\mathbb{F}_{q^2}$ and put $\overline{V}^\wedge = \overline{\mathbb{F}} \otimes V^\wedge$. It is straightforward to check that the solutions of the equation $X^q X = O$ in $\mathrm{PG}(\overline{V}^\wedge)$ are just the same as in $\mathrm{PG}(V^\wedge)$. In other words, $\widetilde{\mathcal{G}}_\varphi = \mathcal{G}_\varphi$ (notation as in Sect. 3.4.1). Thus, $\widetilde{\mathcal{G}}_\varphi$ is a finite set of points.

4.5 The Dual of $H(4, q^2)$

Let $\Gamma^* \cong H(4, q^2)$, namely V is a 5-dimensional vector space over $\mathbb{F}_{q^2}$ and $\varepsilon(\Gamma^*) = Q(\varphi)$ for a non-degenerate hermitian form φ of Witt index 2. The space $V^\wedge$ is 10-dimensional and $\mathcal{G}_\varphi$ spans $V^\wedge$, by Theorem 3.6. Given a point $[a] \in \mathcal{G}_\varphi$, $\mathrm{St}([a])$ is a 3-dimensional subspace of $\mathrm{PG}(V^\wedge)$. By (21) of Sect. 3.4.5, the set $\varepsilon^{\mathrm{gr}}(l)$ is a unital, spanning a plane of $\mathrm{St}([a])$. So, $\varepsilon^{\mathrm{gr}}$ is a 2-embedding, but it is not quadratic.

As in the previous subsection, we may assume to have chosen the basis E of V in such a way that φ is represented by the identity matrix. In this way $\mathcal{G}_\varphi$ is represented by the matrix equation $X^q X = O$ combined with (1). Explicitly, $X^q X = O$ can be written as follows:

$$x_{k,1}^q x_{1,h} + x_{k,2}^q x_{2,h} + x_{k,3}^q x_{2,h} + x_{k,4}^q x_{4,h} + x_{k,5}^q x_{5,h} = 0, \quad (1 \le k, h \le 5).$$

The tangent space $\mathrm{Tan}(\mathcal{G}_\varphi)_{[a\wedge b]}$ is described by the matrix equation $A_{a\wedge b}^q X = O$ together with (4). Choose $a = e_1 + e_2 t$ and $b = e_3 + e_4 t$ for a given t such that $t^{q+1} + 1 = 0$, and $t \neq 1$ if q is even. So, $a \wedge b = e_{1,2} + (e_{1,4} + q_{2,3})t + e_{2,4}t^2$. With this choice of $a \wedge b$ we get the following linear system of rank 7 for $\mathrm{Tan}(\mathcal{G}_\varphi)_{[a\wedge b]}$:

$$\begin{aligned} &x_{1,2} = x_{1,3} = x_{1,4} = x_{3,4} = 0, \\ &x_{2,3} + t^q x_{2,4} = x_{1,5} + t^q x_{2,5} = x_{3,5} + t^q x_{4,5} = 0. \end{aligned}$$

As $\dim(V^\wedge) = 10$, it follows that $\dim(\mathrm{Tan}(\mathcal{G}_\varphi)_{[a\wedge b]}) = 3$. Consequently, $\mathcal{G}_\varphi$ is 2-dimensional.

It is well known that the dual of $H(4, q^2)$ admits no projective embedding. So, no projective embedding exists for Γ.

4.6 Dual Grids

Let $\Gamma^* \cong Q^+(3, \mathbb{F})$ and let $\varepsilon : \Gamma^* \to \mathrm{PG}(3, \mathbb{F})$ be a projective embedding of Γ^*. Thus, $\varepsilon(\Gamma^*) = Q(\chi)$ for a non-singular quadratic form χ of Witt index 2 in $V = V(4, \mathbb{F})$. The space $V^\wedge$ is 6-dimensional.

We may assume that $\chi(x_1, x_2, x_3, x_4) = x_1x_2 + x_3x_4$. We have already given the equations of $\mathcal{G}_\chi$ when char($\mathbb{F}$) $= 2$ in Example 3.8. When char($\mathbb{F}$) $\neq 2$, we still obtain the equations $(B.1) - (B.4)$ of Example 3.8 but $(A.1)$ and $(A.2)$ are replaced by $x_{1,2}^2 = x_{1,3}x_{2,4} + x_{1,4}x_{2,3}$ and $x_{1,2}^2 = x_{2,4}^2$ respectively. In addition to these equations we also obtain the following ones:

$$x_{1,3}(x_{1,2} - x_{3,4}) = 0, \quad x_{2,4}(x_{1,2} - x_{3,4}) = 0,$$
$$x_{1,4}(x_{1,2} + x_{3,4}) = 0, \quad x_{2,3}(x_{1,2} + x_{3,4}) = 0.$$

Moreover, $x_{1,2}x_{3,4} - x_{1,3}x_{2,4} + x_{1,4}x_{3,4} = 0$ by (1).

It follows that $\mathcal{G}_\chi$ is the union of two disjoint conics, say C_1 and C_2. When char($\mathbb{F}$) $\neq 2$ the planes $\langle C_1\rangle^{\mathrm{pr}}$ and $\langle C_2\rangle^{\mathrm{pr}}$ are disjoint. In this case $C_1 \cup C_2$ spans PG($V^\wedge$). If char($\mathbb{F}$) $= 2$ then C_1 and C_2 have the same nucleus, say n, and $\langle C_1\rangle^{\mathrm{pr}} \cap \langle C_2\rangle^{\mathrm{pr}} = n$. In this case $C_1 \cup C_2$ spans a hyperplane of PG($V^\wedge$) (compare Theorems 3.6 and 3.9).

References

1. Akça, Z., Bayar, A., Ekmekçi, S., Kaya, R., Thas, J.A., Van Maldeghem, H.: Generalized Veronesean embeddings of projective spaces, Part II. The lax case. Ars Combin. **103**, 65–80 (2012)
2. André, J.: Über nicht-Desarguessche Ebenen mit transitiver Translationsgruppe. Math. Z. **60**, 156–186 (1954)
3. Blok, R.J., Cooperstein, B.N.: The generating rank of the unitary and symplectic Grassmannians. J. Combin. Th. A **119**, 1–13 (2012)
4. Blok, R.J., Pasini, A.: On absolutely universal embeddings. Discrete Math. **267**, 45–62 (2003)
5. Bruck, R.H., Bose, R.C.: The construction of translation planes from projective spaces. J. Algebra **1**, 85–102 (1964)
6. Cardinali, I., Pasini, A.: Grassmann and Weyl embeddings of orthogonal grassmannians. J. Alg. Combin **38**(4), 863–888 (2013)
7. Cardinali, I., Pasini, A.: Veronesean embeddings of dual polar spaces of orthogonal type. J. Combin. Th. A **120**, 1328–1350 (2013)
8. Cooperstein, B.N., Shult, E.E.: A note on embedding and generating dual polar spaces. Adv. Geometry **1**, 37–48 (2001)
9. Cooperstein, B.N., Thas, J.A., Van Maldeghem, H.: Hermitian Veronesean over finite fields. Forum Math. **16**, 365–381 (2004)
10. De Bruyn, B., Pasini, A.: On symplectic polar spaces over non-perfect fields of characteristic 2. Linear Multilinear Algebra **57**, 567–575 (2009)
11. Gmainer, J., Havlicek, H.: A dimension formula for the nucleus of a Veronese variety. Linear Algebra Appl. **305**, 191–201 (2000)
12. Hartshorne, R.: Algebraic Geometry. Springer, Berlin (1977)
13. Herzer, A.: Die Schmieghyperebenen and die Veronese-Manningfaltigkeit bei beliebiger Charakteristik. J. Geometry **18**, 140–154 (1982)
14. Humphreys, J.E.: Linear Algebraic Groups. Springer, Berlin (1981)
15. Hirschfeld, J.W.P., Thas, J.A.: General Galois Geometries. Oxford University Press, Oxford (1991)
16. Ivanov, A.A.: Non-abelian representation of geometries. Adv. Stud. Pure Math. **32**, 301–314 (2001)

17. Kasikova, A., Shult, E.E.: Absolute embeddings of point-line geometries. J. Algebra **238**, 265–291 (2001)
18. Lunardon, G.: Normal spreads. Geom. Dedicata **75**, 245–261 (1999)
19. Pasini, A.: Embeddings and expansions. Bull. Belg. Math. Soc. Simon Stevin **10**, 585–626 (2003)
20. Pasini, A., Van Maldeghem, H.: Some constructions and embeddings of the tilde geometry. Note di Matematica **21** (2002/2003), 1–33
21. Payne, S.E., Thas, J.A.: Finite Generalized Quadrangles. Pitman, Boston (1984)
22. Pepe, V., Van Maldeghem, H.: Lax embeddings of the Hermitian unital. Des. Codes Crypt. **68**, 325–347 (2013)
23. Ronan, M.A.: Embeddings and hyperplanes of discrete geometries. Eur. J. Combin. **8**, 179–185 (1987)
24. Shafarevich, I.: Basic Algebraic Geometry 1: Varieties in Projective Spaces. Springer, Berlin (1994)
25. Thas, J.A., Van Maldeghem, H.: Characterizations of the finite quadric veroneseans ${\mathcal{V}_n}^{2^n}$. Q. J. Math. **55**(1), 99–113 (2004)
26. Thas, J.A., Van Maldeghem, H.: Generalized veronesean embeddings of projective spaces. Combinatorica **31**, 615–629 (2011)
27. Timmermann, H.: Zur Geometrie der Veronesemannigfaltigkeit bei endlicher Characteristik. University of Hamburg, Hamburg, Habilitationsschrift (1978)
28. Tits, J.: Buildings of Spherical Type and Finite BN-pairs. Lecture Notes in Mathematical Sciences, vol. 386, Springer, Berlin (1974)
29. Van Maldeghem, H.: Generalized Polygons. Birkhäuser, Basel (1998)
30. Yoshiara, S.: Ambient spaces of dimensional dual arcs. J. Alg. Combin. **19**, 5–23 (2004)

Chapter 5
Generation of Lie Incidence Geometries: A Survey

Bruce N. Cooperstein

Abstract A subspace of a point-line geometry $(\mathcal{P}, \mathcal{L})$ is a subset S of the point set $\mathcal{P}$ with the property that whenever a line L meets S in at least two points then L is contained in S. For an arbitrary subset X of $\mathcal{P}$ the subspace generated by X, denoted by $\langle X \rangle_\Gamma$, is the intersection of all subspaces which contain X. A subset X is said to generate Γ if $\langle X \rangle_\Gamma = \mathcal{P}$. The generating rank of Γ is the minimal cardinality of a generating set. In this paper we survey what is currently know about the generating rank of the Lie incidence geometries arising as the shadow of a spherical building.

Keywords Point line geometry · Lie incidence geometry · Generating rank of a geometry · Embedding of a geometry

1 Preliminaries

We begin with several definitions:

A *point-line geometry* is a pair $\Gamma = (\mathcal{P}, \mathcal{L})$ where $\mathcal{P}$ is a set whose elements are called 'points' and $\mathcal{L}$ is a collection of subsets of $\mathcal{P}$ called 'lines' with the property that any two points belong to at most one line. If $\mathcal{P}$ and $\mathcal{L}$ are not mentioned explicitly, the sets of points and lines of a point-line geometry Γ are denoted $\mathcal{P}(\Gamma)$ and $\mathcal{L}(\Gamma)$.

A *subspace* of Γ is a subset $X \subseteq \mathcal{P}$ such that any line containing at least two points of X entirely belongs to X. A *hyperplane* of Γ is a proper subspace that meets every line.

Projective Embeddings and Generating Sets The *span* of a set $\mathcal{S} \subseteq \mathcal{P}$ is the smallest subspace containing $\mathcal{S}$; it is the intersection of all subspaces containing $\mathcal{S}$ and is denoted by $\langle \mathcal{S} \rangle_\Gamma$. We say that $\mathcal{S}$ is a *generating set* (or *spanning set*) for Γ if

B. N. Cooperstein (✉)
Department of Mathematics, University of California, Santa Cruz, USA
e-mail: coop@ucsc.edu

N. S. N. Sastry (ed.), *Groups of Exceptional Type, Coxeter Groups and Related Geometries*, Springer Proceedings in Mathematics & Statistics 82,
DOI: 10.1007/978-81-322-1814-2_5,

$\langle \mathcal{S}\rangle_\Gamma = \mathcal{P}$. The *generating rank* of Γ, denote by $\mathrm{gr}(\Gamma)$ is the smallest cardinality of a generating set of Γ.

For a vector space V over some field $\mathbb{F}$, the *projective geometry* associated to V is the point-line geometry $\mathbb{P}(V) = (\mathcal{P}(V), \mathcal{L}(V))$ whose points are the 1-spaces of V and whose lines are the sets of 1-spaces contained in some 2-space of V.

A *projective embedding* of a point-line geometry $\Gamma = (\mathcal{P}, \mathcal{L})$ is a pair (ϵ, V), where ϵ is an injective map $\mathcal{P} \xrightarrow{\epsilon} \mathcal{P}(V)$ that sends every line of $\mathcal{L}$ onto a line of $\mathcal{L}(V)$, and with the property that

$$\langle \epsilon(\mathcal{P})\rangle_{\mathbb{P}(V)} = \mathcal{P}(V).$$

In the literature, this is often referred to as a *full projective embedding*. We will often abuse notation and refer to the map ϵ as an embedding when the vector space V is understood. The *dimension* of the embedding is the dimension of the vector space V. A geometry Γ is *embeddable* if some embedding (ϵ, V) exists for Γ. With only one or two exceptions all the geometries Γ we will consider in this survey are embeddable and there will exist an embedding (ϵ, V) with finite dimension.

Morphisms of Embeddings Assume (ϵ, V) and (ϵ', V') are embeddings of $\Gamma = (\mathcal{P}, \mathcal{L})$. A *morphism* from (ϵ, V) to (ϵ', V') is a mapping $\Psi\colon \mathrm{PG}(V) \to \mathrm{PG}(V')$ induced by a semi-linear transformation $f\colon V \to V'$ such that $\Psi \circ \epsilon = \epsilon'$.

Given an embedding (ϵ, V), an embedding $(\hat{\epsilon}, \hat{V})$ is said to be *universal relative* to (ϵ, V) if there is a morphism $\hat{\Psi}\colon \hat{\epsilon} \to \epsilon$ such that for any morphism Ψ' from an embedding ϵ' of Γ to ϵ, $\hat{\Psi}$ factors through Φ', that is, there is a morphism $\gamma\colon \hat{\epsilon} \to \epsilon'$ such that $\hat{\Psi} = \Psi' \circ \gamma$. An embedding ϵ of a geometry Γ is *relatively universal* if it is universal relative to itself, while an embedding is *absolutely universal* if it is universal relative to every embedding of Γ.

Bases of Embeddable Geometries It is an immediate consequence of the definitions that if (ϵ, V) is an embedding of $\Gamma = (\mathcal{P}, \mathcal{L})$ and S is a generating set then $dim(V) \leq |S|$ and therefore, in particular, $dim(V) \leq gr(\Gamma)$. Moreover, if $dim(V) = gr(\Gamma)$ then (ϵ, V) is relatively universal. In this situation, a generating set S with $|S| = gr(\Gamma) = dim(V)$ will be referred to as a *basis* of Γ.

2 The Lie Geometries

In this section we give a brief description of the geometries which concern us.

2.1 Ordinary Grassmannians

Let $\mathbb{F}$ be a field and V a vector space of dimension n over $\mathbb{F}$. For $1 \leq l \leq n-1$ let $L_l(V)$ be the collection of all l-dimensional subspaces of V. Fix $l, 2 \leq l \leq n-2$

and set $\mathcal{P} = L_l(V)$. For pairs (C, A) of incident subspaces $(A \subset C)$ of V with $dim(A) = a$, $dim(C) = c$, satisfying $a \leq l \leq c$, we denote by $S_l(C, A)$ the set of all $l-$subspaces B of V such that $A \subset B \subset C$. Finally, let $\mathcal{L}$ consists of all the sets $S_l(C, A)$ where $dim\ A = l - 1, dim\ C = l + 1$, and $A \subset C$. Denote by $\mathcal{G}_l(V)$ the rank two incidence geometry $(\mathcal{P}, \mathcal{L})$. We use the notation $\mathcal{G}_{n,l}(\mathbb{F})$ to denote the isomorphism type of this geometry.

2.2 Symplectic and Unitary Grassmannians

Assume (V, f) is either a non-degenerate symplectic space or unitary space with $dim\ V = 2n$ and Witt index n. For $l \leq n$ denote by $L_l(V)$ the totally isotropic subspaces of V of dimension l. If A is a totally isotropic subspace of V, $dim\ A < l$ and $A \subset C \subset A^{\perp}$ then denote by $S_l(C, A) = \{B \in L_l(V) | A \subset B \subset C\}$. Now fix $l < n$ and set $\mathcal{P} = L_l(V)$ and set $\mathcal{L} = \{S_l(C, A) | A \in L_{l-1}(V), C \in L_{l+1}(V), A \subset C\}$. Then $(\mathcal{P}, \mathcal{L})$ is a symplectic or unitary Grassmannian. The geometry of symplectic Grassmannians is denote by $C_{n,l}(\mathbb{F})$ and the geometry of unitary Grassmannians is denoted by ${}^2A_{m,l}(\mathbb{F})$.

2.3 Symplectic and Unitary Dual Polar Spaces

We continue with the notation of Sect. (2.2). If $\mathcal{P} = L_n(V)$ and $\mathcal{L} = \{S(A^{\perp}, A) | A \in L_{n-1}(V)\}$ then $\Gamma = (\mathcal{P}, \mathcal{L})$ is a symplectic or unitary dual polar space. The dual symplectic geometry is denote by $DS(2n, \mathbb{F})$ or $C_{n,n}(\mathbb{F})$. The dual unitary polar space is denoted by $DU(2n, \mathbb{F})$ or ${}^2A_{2n-1,n}(\mathbb{F})$.

2.4 Orthogonal Grassmannians

Assume that (V, Q) is a non-degenerate orthogonal space with Witt index n and $dim(V) = d \in \{2n, 2n+1, 2n+2\}$. For $l \leq n$ denote by $L_l(V)$ the collection of totally singular subspaces of V of dimension l. If $A \in L_{l-1}(V), l < n$ and C is a subspace of V such that $A \subset C \subset A^{\perp}$ then denote by $S_l(C, A)$ the collection of all $B \in L_l(V)$ such that $A \subset B \subset C$. Assume $l < n - 1$ and set $\mathcal{P} = L_l(V), \mathcal{L} = \{S_l(C, A) | A \in L_{l-1}(V), C \in L_{l+1}(V). A \subset C\}$. Then $(\mathcal{P}, \mathcal{L})$ is an orthogonal Grassmannian. When $dim(V) = 2n$ we denote this geometry by $D_{n,l}(\mathbb{F})$. When $dim(V) = 2n + 1$ we denote the geometry by $B_{n,l}(\mathbb{F})$. Finally, when $dim(V) = 2n + 2$ we denote this geometry by ${}^2D_{n+1,l}(\mathbb{F})$.

2.5 Orthogonal Dual Polar Spaces

Assume (V, Q) is a non-degenerate orthogonal space with Witt index n and dimensional $d \in \{2n+1, 2n+2\}$. Now set $\mathcal{P} = L_n(V)$, $\mathcal{L} = \{S_n(A^\perp, A) | A \in L_{n-1}(V)\}$. The geometry $(\mathcal{P}, \mathcal{L})$ is an orthogonal dual polar space. When $dim(V) = 2n+1$ we denote this dual polar space by either $B_{n,n}(\mathbb{F})$ or $DO(2n+1, \mathbb{F})$. When $dim(V) = 2n+2$ the dual polar space is denoted by $DO^-(2n+2, \mathbb{F})$.

2.6 Half-Spin Geometries

Let (V, Q) be a non-degenerate orthogonal space with Witt index n and dimension $2n$. Define a relation $\sim$ on $L_n(V)$ by $M \sim N$ if $dim(M) - dim(M \cap N)$ is even. Then $\sim$ is an equivalence class. Let $\mathcal{P}$ denote one of the classes. For $A \in L_{n-2}(V)$ let $S_{\mathcal{P}}(A) = \{M \in \mathcal{P} | A \subset M\}$ and set $\mathcal{L} = \{S_{\mathcal{P}}(A) | A \in L_{n-2}(V)\}$. The geometry $(\mathcal{P}, \mathcal{L})$ is the half-spin geometry. This is denoted by $D_{n,n}(\mathbb{F})$.

2.7 The $E_{6,1}$ Geometry

Let M_3 denote the space of 3×3 matrices over $\mathbb{F}$ and set $M = M_3 \oplus M_3 \oplus M_3$. Define $T \colon M \to \mathbb{F}$ by

$$T(m_1, m_2, m_3) = det(m_1) + det(m_2) + det(m_3) - trace(m_1 m_2 m_3).$$

T is a cubic form in 27 variables. It contains 45 monomials, with each variable in 5 such monomials. By taking (formal) partials of T with respect to the 27 variables we obtain 27 quadratic forms. The points of the $E_{6,1}(\mathbb{F})$ geometry are the points of PG(M) which are singular with respect to all of these quadratic forms and the lines are the lines which are singular with respect to all the forms.

2.8 The $E_{7,7}(\mathbb{F})$ Geometry

The $E_{7,7}(\mathbb{F})$ geometry is the Lie incidence geometry with diagram

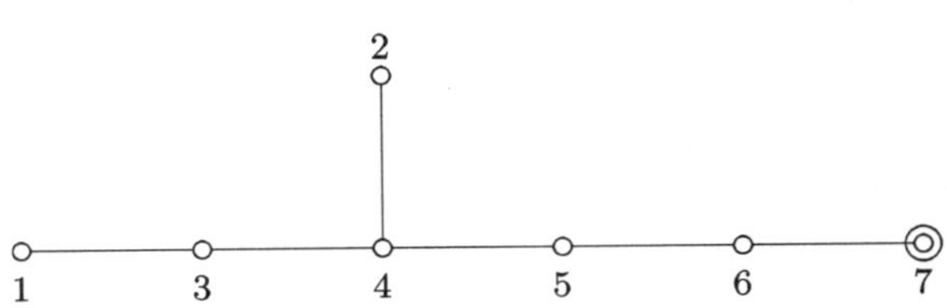

A concrete construction of this geometry, as a collection of certain points in $PG(55, \mathbb{F})$, is possible but too lengthy to describe here.

2.9 Long Root Geometries

If G is a Chevalley group or a group of Lie type over a field $\mathbb{F}$ it has a conjugacy class of "long root" subgroups $\mathcal{R}$ which are isomorphic, as groups, to the additive group of $\mathbb{F}$ (or a subfield of $\mathbb{F}$ in the case of unitary groups). Distinct long root subgroups meeting trivially (only in the identity). When $X \neq Y$ are in $\mathcal{R}$ then the group generated by X and Y has a limited set of possibilities as given in the following:

Theorem 2.1. *If $X \neq Y$ are long root subgroups of the group of Lie type G then one of the following occurs:*

(1) X and Y commute and every non-identity element in $\langle X, Y\rangle$ is contained in a long root subgroup.
(2) X and Y commute but the only long root subgroups which intersect $\langle X, Y\rangle$ non-trivially are X and Y.
(3) The center of the group $\langle X, Y\rangle$ is equal to its commutator $\langle X, Y\rangle'$ and is isomorphic to a maximal unitary subgroup of $SL_3(\mathbb{F})$.
(4) $\langle X, Y\rangle$ is isomorphic to $SL_2(\mathbb{F})$ or $PSL_2(\mathbb{F})$.

The **long root subgroup geometry** of G has $\mathcal{P} = \mathcal{R}$. Two long root subgroups X and Y are collinear when they have relation (1) and then the line on them consists of all the long root subgroups which meet $\langle X, Y\rangle$ non-trivially. We note that sometimes relation (2) does not occur, for example when G is $A_2(\mathbb{F}) = SL_3(\mathbb{F})$, $G_2(\mathbb{F})$ or ${}^3D_4(\mathbb{F})$. In this case the geometry is a generalized hexagon. Also, there are instances in which (3) does not occur, for example when G is symplectic or unitary. In these cases the geometry is the corresponding polar space.

When (V, Q) is an orthogonal space with Witt index n, $dim(V) \in \{2n, 2n+1, 2n+2\}$, and G is the group of isometries of Q, the long root subgroup geometry of G is isomorphic to the orthogonal Grassmannian with $l = 2$; that is, it is one of the geometries $D_{n,2}(\mathbb{F})$, $B_{n,2}(\mathbb{F})$, ${}^2D_{n+1,2}(\mathbb{F})$.

When $G = SL_n(V)$ where V is an n-dimensional vector space over $\mathbb{F}$ we describe the long-root geometry: The point set, $\mathcal{P}$ can be identified with (p, H) where $p \in L_1(V)$, $H \in L_{n-1}(V)$ and $p \subset H$. There are two types of lines: Let $U \in L_2(V)$, $H \in L_{n-1}(V)$ with $U \subset H$. The corresponding line is $\{(p, H) | p \in L_1(U)\}$. For the second line let $W \in L_{n-2}(V)$ and $p \in L_1(W)$. The corresponding line is $\{(p, H) | H \in S_{n-1}(V, W)\}$.

3 Generating Lie Geometries

We begin with a comprehensive statement for polar spaces.

3.1 Generating Polar Spaces

The following theorem summarizes what is known about the the generating rank of polar spaces with either maximal or nearly maximal Witt index. Parts (1) and (3) are straightforward but can be found explicitly stated in a paper of Blok and Brouwer [3]. Parts (2) and (4) are in a paper of Cooperstein and Shult [14]. (5) should be well-known but I have not found a reference and so is included in a recent paper of Blok and Cooperstein [4]. (6) can be deduced from an argument found in the above cited paper of Cooperstein and Shult.

Theorem 3.1. *(1) Assume the characteristic of* $\mathbb{F}$ *is not two. Then the polar space* $C_{n,1}(\mathbb{F})$ *is generated by any subgraph of type* $K_{2,2,\dots,2}$ *(n 2s) of its collinearity graph.*
(2) A polar space of type $B_{n,1}(\mathbb{F})$ *is generated by any subgraph of type* $K_{2,2,\dots,2,3}$ *((n − 1) 2s) of the collinearity graph of its collinearity graph*
(3) A polar space of type $D_{n,1}(\mathbb{F})$ *is generated by any subgraph of type* $K_{2,2,\dots,2}$ *(n 2s) of its collinearity graph.*
(4) A polar space of type $^2D)n(\mathbb{F})$ *can be generated by a subgraph of type* $K_{2,2,\dots,2,4}$ *(n − 1 2s) of its collinearity graph.*
(5) A polar space of type $^2A_{2n-1,1}(\mathbb{F})$ *is generated by any subgraph of type* $K_{2,2,\dots,2}$ *(n 2s) of its collinearity graph.*
(6) subgraph of type $K_{2,2,\dots,3}$ *(n − 1 2s) of the collinearity graph which generate A polar space of type* $^2A_{2n,1}(\mathbb{F})$ *can be generated by a subgraph of type* $K_{2,2,\dots,3}$ *(n − 1 2s) of its collinearity graph.*

We remark in parts (4) and (6) we are asserting that there exists subgraphs of the given type which are generating sets for the geometry not that every subgraph of this isomorphism type is a generating set.

3.2 Geometries Spanned by the Shadow of an Apartment

Apart from projective spaces and polar spaces probably the first geometries for which the generating rank was determined are those geometries which are generated by the shadow of an apartment. We will refer to these as miniscule geometries since these are precisely those geometries which have an embedding afforded by a module for the corresponding group whose high weight is miniscule. The most explicit statement and proof can be found in the aforementioned paper of Blok and Brouwer [3].

Theorem 3.2. *Let M be one of the types* A_n, B_n, C_n, D_n *or* E_n *and* $\mathbb{F}$ *a field. Let* Δ *be a building with diagram M over* $\mathbb{F}$ *and let* $i \in I$ *be the label of a node of M. Then the* $i-$*shadow geometry of* Δ *is spanned by the shadow of an apartment of the building* Δ *if and only if* M_i *is one of the following:*

(1) $A_{n,l}$ *with* $1 \le l \le n$;
(2) $D_{n,1}$, $D_{n,n-1}$, $D_{n,n}$;

(3) $E_{6,1}$, $E_{6,6}$, $E_{7,7}$;
(4) $B_{n,n}$ *provided* $char(\mathbb{F}) \neq 2$;
(5) $C_{n,1}$ *provided* $char(\mathbb{F}) \neq 2$.

The above result can also be found implicitly in Ronan and Smith [20] as well as in a paper of Cooperstein and Shult [12]. As a corollary we have the following theorem on generating ranks:

Theorem 3.3. *(1) The generating rank of* $A_{n,k}(\mathbb{F})$ *is* $\binom{n+1}{k}$.
(2) If $char(\mathbb{F}) \neq 2$ *the generating rank of* $B_{n,n}(\mathbb{F})$ *is* 2^n.
(3) The generating rank of $D_{n,n}(\mathbb{F}) \cong D_{n,n-1}(\mathbb{F})$ *is* 2^{n-1}.
(4) The generating rank of $E_{6,1}(\mathbb{F}) \cong E_{6,6}(\mathbb{F})$ *is 27.*
(5) The generating rank of $E_{7,7}(\mathbb{F})$ *is 56.*

3.3 Generating Long Root Subgroup Geometries

The situation for long root subgroups is not nearly as complete as polar or miniscule geometries. So far two main results have been obtained.

Theorem 3.4. [9] *Let* $\mathbb{F}_p$ *be a field of prime order. Then the following holds:*

(1) *For* $n \geq 3$ *the generating rank of the long root subgroup geometry for* $A_n(\mathbb{F}_p) \cong SL_{n+1}(\mathbb{F}_p)$ *is* $(n+1)^2 - 1$.
(2) *Let* $G = \Omega_n(\mathbb{F}_p)$ *be the group of isometries of a non-singular orthogonal space of dimension* $d \in \{2n, 2n+1, 2n+2\}$ *and Witt index n. If* $d = 2n+1$ *assume* $p \neq 2$. *Then the generating rank of the long root subgroup geometry of G is* $\binom{d}{2}$.

In [5] Blok and Pasini have obtained a powerful extension of this result:

Theorem 3.5. *Given a prime power q and an integer* $n \geq 3$, *let* Γ *be the long root subgroup geometry of the group* $G = SL_{n+1}(q)$, $\Omega_{2n+1}(q)$ *or* $\Omega^+_{2n}(q)$. *Then* Γ *has generating rank r with* $r_0 \leq r \leq r_0 + 1$ *where* r_0 *is as follows:*

(1) *For* $SL_{n+1}(q)$, $r_0 = n^2 - 1$.
(2) *For* $\Omega_{2n+1}(q)$, $r_0 = \binom{2n+1}{2} = 2n^2 + n$.
(3) *For* $\Omega^+_{2n}(q)$, *q odd,* $r_0 = \binom{2n}{2} = 2n^2 - n$.

The situation for characteristic two is more complicated, though Blok and Pasini obtain results in this case as well.

Open Questions for Root Subgroup Geometries
The main open question is what are the generating ranks of the exceptional long root subgroup geometries: $F_{4,4}(\mathbb{F})$, $E_{6,2}(\mathbb{F})$, $E_{7,1}(\mathbb{F})$, and $E_{8,1}(\mathbb{F})$.

3.4 Generating Ranks of Dual Polar Spaces

The situation for embeddable dual polar spaces is nearly complete (at least over finite fields). The results follow:

Theorem 3.6. *Assume* $|\mathbb{F}| > 2$ *is a finite field. Then the generating rank of the symplectic dual polar space,* $DS(2n, \mathbb{F})$ *is the* $(n+1)^{st}$ *Catalan number,* $Cat(n+1) =$

$$\frac{1}{n+2}\binom{2n+2}{n+1} = \binom{2n}{n} - \binom{2n}{n-2}.$$

This result was obtained by Cooperstein in [10] by constructing a generating set recursively whose cardinality is $Cat(n+1)$. Since it is known that $DS(2n, \mathbb{F})$ has a projective embedding with this dimension it follows that the generating set is minimal and a basis of the geometry.

Our next result is for unitary dual polar spaces:

Theorem 3.7. *Assume* $\mathbb{F} = \mathbb{F}_{q^2}$ *with* $q > 2$. *Then the generating rank of* $DU(2n, \mathbb{F})$ *is* $\binom{2n}{n}$.

This result is also due to Cooperstein [7] and proved in a manner entirely analogous with the result for symplectic dual polar spaces.

These results were extended to arbitrary fields with more than 2 elements by De Bruyn and Pasini [16].

The following results were obtained by Cooperstein and Shult [14] for finite fields. As with the previous mentioned cases they recursively construct a generating set with cardinality equal to that of an embedding. A proof was obtained holding for general fields by De Bruyn [15].

Theorem 3.8. *For* $n \geq 2$, $DO^-(2n+2, \mathbb{F})$ *has generating rank* 2^n.

For completeness we mention the case of $DO(2n+1, \mathbb{F})$ which is isomorphic to the geometry $B_{n,n}(\mathbb{F})$ whose generating rank, as previously indicated, has been determined many times:

Theorem 3.9. *Assume the characteristic of* $\mathbb{F} \neq 2$. *Then the generating rank of the dual polar space* $DO(2n, \mathbb{F})$ *is* 2^n.

Note that when the characteristic of the field $\mathbb{F}$ is two and perfect the dual polar spaces $DO(2n+1, \mathbb{F})$ and $DS(2n, \mathbb{F})$ are isomorphic. Consequently, when the characteristic of $\mathbb{F}$ is two, $|\mathbb{F}| > 2$, and $\mathbb{F}$ is perfect we know the generating rank by the aforementioned result of Cooperstein, De Bruyn-Pasini.

Open Cases for Dual Polar Spaces In most cases the generating rank of $DS(2n, \mathbb{F}_2)$ is unknown. What we do know is as follows:

Independently, Paul Li [19] and Blokhuis-Brouwer [6] proved the first Brouwer conjecture:

Theorem 3.10. *The universal embedding dimension of the dual polar space $DS(2n, \mathbb{F}_2)$ is $\frac{(2^n+1)(2^{n-1}+1)}{3}$.*

Paul Li also proved the second Brouwer conjecture [18]:

Theorem 3.11. *The universal embedding dimension of the dual polar space $DU(2n, \mathbb{F}_4)$ is $\frac{4^n-1}{3}$.*

The results of Blokhuis-Brouwer and Li give lower bounds on the generating rank of $DS(2n, \mathbb{F}_2)$ and $DU(2n, \mathbb{F}_4)$ most of which are unknown: For $n \geq 6$, the generating rank of $DS(2n, \mathbb{F}_2)$ are unknown and the actual value remains open. For $n \leq 5$ the generating ranks have been determine [8, 13]. Also, in [8] a very strong upper bound on the generating rank of $DS(12, \mathbb{F}_2)$ is obtained. The results are as follows:

Theorem 3.12. *(1) The generating rank of $DS(6, \mathbb{F}_2)$ is 15.*
(2) The generating rank of $DS(8, \mathbb{F}_2)$ is 51.
(3) The generating rank of $DS(10, \mathbb{F}_2)$ is 187.
(4) The generating rank of $DS(12, \mathbb{F}_2)$ is at most 716.

We have even less knowledge of the generating rank of $DU(2n, \mathbb{F}_4)$. The only known cases are for $n = 2$ where the dual polar space is isomorphic to the orthogonal generalized quadrangle $O^-(6, \mathbb{F}_2)$ and has generating rank 6 as previously indicated; and $n = 3$ which has been shown to have generating rank 22 by Cooperstein [11].

3.5 Generating Symplectic Grassmannians

As a consequence of work of Rieuwert Blok we know the generating rank of all the geometries $C_{n,k}(\mathbb{F})$ provided *char* $\mathbb{F} \neq 2$. This work began with the previously mentioned joint paper with [5] in which they established that over a prime field $\mathbb{F}_p$, $p > 2$ the generating rank of $C_{n,2}(\mathbb{F})$ is

$$2n^2 - n - 1 = \binom{2n}{2} - 1 = \binom{2n}{2} - \binom{2n}{0}.$$

The result was extended in [1] by Blok to all fields $\mathbb{F}$ of characteristic greater then 2. Then his program culminated in the following omnibus result [2]

Theorem 3.13. *Assume char $\mathbb{F} \neq 2$. Then the generating rank of $C_{n,k}(\mathbb{F})$ is*

$$\binom{2n}{k} - \binom{2n}{k-2}.$$

When the characteristic of $\mathbb{F}$ is two there are no results. Since induction on k plays a role in the proof and the generating rank of the symplectic polar space over a field $\mathbb{F}$ of characteristic two is $2n+1$ when $\mathbb{F}$ is a finite field and not $2n$ we can expect a different result. Moreover, we think it is likely that the cases of $\mathbb{F} = \mathbb{F}_2$ and $char(\mathbb{F}) = 2$, $|\mathbb{F}| > 2$ will be very different. Things become even wilder when $\mathbb{F}$ is a non-perfect field of characteristic two: the generating rank can be infinite as pointed out to the author by one of the referees (See [17]).

3.6 Generating Hermitian Grassmannians

Fairly recently the generating rank of the entire family of Hermitian Grassmannians has been obtained by Blok and Cooperstein [4]:

Theorem 3.14. *The generating rank of* ${}^2A_{n-1,k}(\mathbb{F})$ *is* $\binom{2n}{k}$.

3.7 Generating Generalized Hexagons

A group of type $G_2(\mathbb{F})$ acts on two dual generalized hexagons. One is a long root geometry and has an embedding in $\mathbb{PG}(13, \mathbb{F})$ and is denoted by $H^D(\mathbb{F})$ ($H^D(q)$ if $|\mathbb{F}| = q$). This is referred to as the dual Cayley generalized hexagon. The other has an embedding in $\mathbb{PG}(6, \mathbb{F})$ and is known as the Cayley generalized hexagon. It is denoted by $H(\mathbb{F})$ ($H(q)$ when $\mathbb{F} = \mathbb{F}_q$). In a few cases we know the generating rank [11, 21]:

Theorem 3.15. *(1) The generating rank of* $H(2)$ *is 14.*
(2) The generating rank of $H^D(2)$ *is 14.*

Some Further Open Problems
Virtually nothing is known about the generating rank of the analog of Grassmannians for the exceptional geometries $E_{6,1}$ and $E_{7,7}$. However, based on some numerology we put forward the following:
Conjecture If p is a prime then the generating rank of $E_{6,3}(\mathbb{F}_p)$ is $\binom{27}{2}$.
Many problems remain open, in particular, the generating rank of orthogonal Grassmannians as well as any of the geometries that arise as the shadow of a spherical building of type F_4 or E_8 over a field $\mathbb{F}$.

References

1. Blok, R.J.: The generating rank of the symplectic line-Grassmannian. Beiträge Algebra Geom. **44**(2), 575–580 (2003)
2. Blok, R.J.: The generating rank of the symplectic Grassmannians: hyperbolic and isotropic geometry. Eur. J. Comb. **28**(5), 1368–1394 (2007)
3. Blok, R.J, Brouwer, A.E.: Spanning point-line geometries in buildings of spherical type. J. Geom. **62**(1–2), 26–35 (1998)

4. Blok, R.J., Cooperstein, B.N.: The generating rank of the unitary and symplectic Grassmannians. J. Comb. Theory Ser. A **119**(1), 1–13 (2012)
5. Blok, R.J, Pasini, A.: Point-line geometries with a generating set that depends on the underlying field. In: Finite Geometries, vol. 3, pp. 1–25, Kluwer Academic Publishers Dev. Math, Dordrecht (2001)
6. Blokhuis, A., Brouwer, A.E.: The universal embedding dimension of the binary symplectic dual polar space. The 2000 Com^{2}MaC conference on association schemes, codes and designs (Pohang). Discrete Math. **264**(1–3), 3–11 (2003)
7. Cooperstein, B.N.: On the generation of dual polar spaces of unitary type over finite fields. Eur. J. Comb. **18**, 849–856 (1997)
8. Cooperstein, B.N.: On the generation of some dual polar spaces of symplectic type over GF(2). Eur. J. Comb. **18**(7), 741–749 (1997)
9. Cooperstein, B.N: Generating long root subgroup geometries of classical groups over prime fields. Bull. Belg. Math. Soc. **5**, 531–548 (1998)
10. Cooperstein, B.N.: On the generation of dual polar spaces of symplectic type over finite fields. J. Comb. Thy, Ser A. **83**, 221–232 (1998)
11. Cooperstein, B.N.: On the generation of some embeddable GF(2) geometries. J. Algebraic Comb. **13**(1), 15–28 (2001)
12. Cooperstein, B.N., Shult, E.E.: Frames and bases of lie incidence geometries. J. Geom. **60**, 17–46 (1997)
13. Cooperstein, B.N., Shult, E.E.: Combinatorial construction of some near polygons. J. Comb. Theory Ser. A **78**(1), 120–140 (1997)
14. Cooperstein, B.N., Shult, E.E.: A note on embedding and generating dual polar spaces. Adv. Geom. **1**, 37–48 (2001)
15. De Bruyn, B.: A note on the spin-embedding of the dual polar space $DQ^-(2n+1, \mathbb{K})$. Ars Comb. **99**, 365–375 (2011)
16. De Bruyn, B., Pasini, A.: Generating symplectic and Hermitian dual polar spaces over arbitrary fields nonisomorphic to $\mathbb{F}_2$. Electron. J. Comb. **14**(1), 17 (Research Paper 54) (2007)
17. De Bruyn, B., Pasini, A.: On symplectic polar spaces over non-perfect fields of characteristic 2. Linear Multilinear Algebra **57**, 567–575 (2009)
18. Li, P.: On the universal embedding of the $U_{2n}(2)$ dual polar space. J. Comb. Theory Ser. A **98**(2), 235–252 (2002)
19. Li, P.: On the universal embedding of the $Sp_{2n}(2)$ dual dolar space. J. Comb. Theory Ser. A **94**(1), 100–117 (2001)
20. Ronan, M., Smith, S.: Sheaves on buildings and modular representations of Chevalley groups. J. Algebra **96**, 319–346 (1985)
21. Thas, J.A., Van Maldeghem, H.: Embeddings of small generalized polygons. Finite Fields Appl. **12**(4), 565–594 (2006)

Chapter 6
Witt-Type Theorems for Subspaces of Lie Geometries: A Survey

Bruce N. Cooperstein

Abstract We define the notion of an independent clique in the collinearity graph of a para-projective geometry. This notion of independence is extended to arbitrary subgraphs. We then define several graphs which we will refer to as 'partial frames' which are isomorphic to subgraphs of an i-shadow of an apartment which we refer to as apartment partial frames. With this background we subsequently survey results which classify independent subgraphs of the collinear graph of a Lie geometry which are isomorphic to an apartment partial frame.

Keywords Para-projective space · s-independent subgraph of a para-projective space · Partial frame of a Lie incidence geometry · Polar Grassmannian

1 Preliminaries

We begin with several definitions:

A *point-line geometry* is a pair $\Gamma = (\mathcal{P}, \mathcal{L})$ where $\mathcal{P}$ is a set whose elements are called 'points' and $\mathcal{L}$ is a collection of subsets of $\mathcal{P}$ called 'lines' with the property that any two points belong to at most one line.

The *collinearity graph* of a point-line geometry $\Gamma = (\mathcal{P}, \mathcal{L})$ is the graph whose set of vertices is $\mathcal{P}$ and whose edges consists of distinct pairs $\{p, q\}$ such that p and q are on a line. We will denote the set of points which are collinear with a point p by $\Delta(p)$.

A *subspace* of $\Gamma = (\mathcal{P}, \mathcal{L})$ is a subset X of the point-set $\mathcal{P}$ such that any line containing at least two points of X entirely belongs to X. A subspace is *singular* if it is a clique in the collinearity graph. Clearly the intersection of subspaces is a subspace.

B. N. Cooperstein (✉)
Department of Mathematics, University of California, Santa Cruz, USA
e-mail: coop@ucsc.edu

N. S. N. Sastry (ed.), *Groups of Exceptional Type, Coxeter Groups and Related Geometries*, Springer Proceedings in Mathematics & Statistics 82,
DOI: 10.1007/978-81-322-1814-2_6, © Springer India 2014

Assume X is a subset of the point-line geometry $\Gamma = (\mathcal{P}, \mathcal{L})$. Then the *subspace generated by* X, denoted by $\langle X\rangle_\Gamma$, is the intersection of the collection of subspaces which contain X. X is said to *generate* Γ if $\langle X\rangle_\Gamma = \mathcal{P}$.
A point-line geometry $\Gamma = (\mathcal{P}, \mathcal{L})$ is said to be a *gamma space* if for every point p, $p^\perp := \{p\} \cup \Delta(p)$ is a subspace of $(\mathcal{P}, \mathcal{L})$. This means if $L \in \mathcal{L}$, $p \in \mathcal{P} \setminus L$ and p is collinear with at least two points of L then p is collinear with every point on L. We remark if a subset X of $\mathcal{P}$ is a clique of the collinearity graph that the subspace generated by X, $\langle X\rangle_\Gamma$, is singular.

Assume $\Gamma = (\mathcal{P}, \mathcal{L})$ is a gamma space. It is a *para-projective space* if the singular subspaces which properly contain lines are isomorphic to projective spaces. If Γ is para-projective and every singular subspace spanned by a clique consisting of a line and a point off the line is isomorphic to $\mathbb{PG}(2, \boldsymbol{F})$ for a fixed field $\boldsymbol{F}$ then we say that Γ is a para-projective space over $\boldsymbol{F}$.

Assume that $\Gamma = (\mathcal{P}, \mathcal{L})$ is a para-projective space over a field $\boldsymbol{F}$. A finite clique C in the collinearity graph $(\mathcal{P}, \Delta)$ is said to be *independent* if

$$\langle C\rangle_\Gamma \cong \mathbb{PG}(|C| - 1, \boldsymbol{F}).$$

A subset S of $\mathcal{P}$ is *singularly independent* or, simply, *s-independent* if every finite clique contained in S is independent.
Throughout this paper we will presume the reader is familiar with the following Lie geometries. We refer to the companion paper [9] in this volume.
The incidence geometry of l-Grassmannians in an n-dimensional vector space V over a field $\mathbb{F}$. We denote this by $\mathcal{G}_{n,l}(\mathbb{F})$ and also by $A_{n-1,l}(\mathbb{F})$.

The incidence geometry of unitary Grassmannians, denoted by ${}^2A_{2n,k}(\boldsymbol{F})$ and ${}^2A_{2n+1,k}(\boldsymbol{F})$ where $2 \le k \le n$.

The incidence geometry of symplectic Grassmannians, denoted by $C_{n,k}(\boldsymbol{F})$ where $2 \le k < n$.

The Orthogonal Grassmannians: $D_{n,k}(\boldsymbol{F})$, $B_{n,k}(\boldsymbol{F})(2 \le k \le n-2),{}^2 D_{n+1,k}(\boldsymbol{F})$ where $2 \le k \le n-2$ and $\boldsymbol{F}$ is a field.

The half-spin geometry $D_{n,n}(\boldsymbol{F})$ on a class of maximal singular subspaces of a hyperbolic orthogonal space of dimension $2n$ over the field $\boldsymbol{F}$.

The dual polar space whose points are the maximal singlar subspaces of a non-singular orthogonal space of dimension $2n + 1$ with Witt index n over the field $\boldsymbol{F}$. We denote this by $B_{n,n}(\boldsymbol{F})$.

The miniscule geometry for the group $E_6(\boldsymbol{F})$, denoted by $E_{6,1}(\boldsymbol{F})$.

The minuscule geometry for the group $E_7(\boldsymbol{F})$ denote by $E_{7,7}(\boldsymbol{F})$.

These geometries are all para-projective.

Let $\Gamma = (\mathcal{P}, \mathcal{L})$ be a Lie geometry. We will say a subspace $\mathcal{S}$ is *parabolic* if its stabilizer in $Aut(\Gamma)$ is a parabolic subgroup.

2 Some Graphs Related to an Apartment of a Spherical Building

Let Γ be a spherical building with a connected diagram M and type set I. A typical Lie incidence geometry, $\Gamma = (\mathcal{P}, \mathcal{L})$ is derived from the building by choosing some type i in I and setting $\mathcal{P}$ equal to all the objects of this type. Let J consist of the types connected to i in the diagram M and let F be a flag of type J. The lines of the geometry are in one-to-one correspondence with such flags and consists of all those objects x of type i such that $F \cup \{x\}$ is a flag. Let A be an apartment of Γ. The i-shadow of A consists of all the objects of type i which are contained in A with the graph induced by the collinearity graph of $(\mathcal{P}, \mathcal{L})$. We will denote the shadow of A by $\Gamma(A)$ and often refer to $\Gamma(A)$ as an apartment, when we mean the shadow of an apartment. We will refer to a subset of $\Gamma(A)$ as a *partial apartment*.

2.1 Frames and Partial Frames for $A_{n-1,l}(F)$

We illustrate with a building of type A_{n-1} over a field $\boldsymbol{F}$ which is just the projective space $\mathbb{PG}(V)$, where V is an n-dimensional vector space over $\boldsymbol{F}$. The objects of type $i, 1 \le i \le n-1$ are the i-dimensional subspaces of V. Suppose $1 < l < n-1$. Then in the diagram A_{n-1} the node l is connected to nodes $l-1$ and $l+1$. Therefore the relevant flags that give rise to lines are pairs (A, C) of subspaces of V with $A \subset C$, $dim(A) = l-1$, and $dim(C) = l+1$. The corresponding line then consists of the l-dimensional subspaces B of V such that $A \subset B \subset C$. We thus obtain the l-Grassmannians, $\mathcal{G}_{n,l}(\boldsymbol{F}) = A_{n-1,l}(\boldsymbol{F})$. Note that two l-subspaces B, B' are collinear when $dim(B \cap B') = l-1$.

An apartment can be described as follows. Chose a basis $(\boldsymbol{v}_1, \dots, \boldsymbol{v}_n)$ for V. Let J be a subset of $I = \{1, 2, \dots, n\}$ and set $V_J = Span(\{\boldsymbol{v}_j | j \in J\})$. Then $A = \{V_J | J \subset I\}$ is an apartment. The l-shadow of the apartment consists of those V_J where $|J| = l$. Now suppose J, J' have cardinality l. When are V_j and $V_{J'}$ adjacent in the collinearity graph? As we remarked above, when $dim(V_J \cap V_{J'}) = l-1$. However, this occurs if and only if $|J \cap J'| = l-1$. Therefore the shadow of an apartment in this geometry is isomorphic, as a graph, to the the graph whose vertices consist of the l-element subsets of $I_n = [1, n] = \{1, 2, \dots, n\}$ with two subsets α, α' adjacent if $|\alpha \cap \alpha'| = l-1$. This graph is known in the literature as a Johnson graph and denoted by $J(n, l)$. We use this notation and also the notation $A_{n-1,l}(1)$.

A *partial frame* of a Grassmannian geometry $\mathcal{G}_{n.l}(\boldsymbol{F})$ is an s-independent subgraph S which is isomorphic to $A_{m,k}(1)$ for some m and k. As an example, let $K \subset M \subset [1, n]$ such that $|K| < l-1$ and $|M| > l+1$. Denote by $s_l(M, K)$ the set of all J with $|J| = l$ and $K \subset J \subset M$. Then $\{Span(V_J) | J \in s_l(M, K)\}$ is a partial frame. Note that this is a partial apartment.

2.2 The Graph $D_{n,n}(1)$ and Partial Frames of $D_{n,n}(F)$

The graph denoted by $D_{n,n}(1)$ has as vertex set the subsets of even cardinality of $[1, n]$ (including the empty set). If α, β are vertices define $\alpha \sim \beta$ if and only if $|(\alpha \cup \beta) \setminus (\alpha \cap \beta)| = 2$. The graph $D_{n,n}(1)$ is isomorphic to the *half-cube graph* H_n which consists of the vectors of even weight in $\mathbb{F}_2^n$ with two vectors adjacent if their Hamming distance is two. A *partial frame* of $D_{n,n}(\mathbb{F})$ is a subset S of the point set which is $s-$independent and, as a graph, is isomorphic to $D_{m,m}(1)$ for some m satisfying $4 \leq m \leq n$.

As an example, let $(\boldsymbol{v}_1, \ldots, \boldsymbol{v}_n, \boldsymbol{w}_1, \ldots, \boldsymbol{w}_n)$ be a hyperbolic basis of a $2n$-dimensional non-singular orthogonal space (V, Q) (with Witt index n). Let J be a subset of $[1, n] = \{1, 2, \ldots, n\}$ of even cardinality and set $U_J = \langle \boldsymbol{v}_i | i \notin J\rangle_F \oplus \langle \boldsymbol{w}_j | j \in J\rangle_F$. Note that for $J \neq J'$

$$\begin{aligned} dim(U_J) - dim(U_J \cap U_{J'}) &= dim(U_{J'}) - dim(U_J \cap U_{J'}) \\ &= |J| - |J \cap J'| \end{aligned}$$

Therefore the graph induced on $\{U_J : J \subset [1, n], |J| \text{ even }\}$ is $D_{n,n}(1)$. By choosing a $K \subset [1, n]$ of cardinality k and only allowing those subsets J such that $J \cap K = \emptyset$ we obtain a subgraph isomorphic to $D_{n-k,n-k}(1)$. This is a partial apartment.

2.3 The Graph $B_{n,n}(1)$ and Partial Frames in $B_{n,n}(\mathbb{F})$

The graph denoted by $B_{n,n}(1)$ has vertex set the subsets of $[1, n]$ with α, β adjacent if and only if $|\alpha \setminus \alpha \cap \beta| = 1$. Alternatively, take all vectors in $\mathbb{F}_2^n$ with two adjacent if their Hamming distance is 1. This graph is also referred to as the hypercube graph and denoted by H_m. The graph is isomorphic to the n-shadow of an apartment in a spherical building of type B_n. A *partial frame* of $B_{n,n}(\mathbb{F})$ is a subset S of the point set which is s-independent and, as a graph, isomorphic to $B_{m,m}(1)$.

2.4 The $E_{6,1}(1)$ Graph

Let V consist of elements x_i, y_i for $i = 1, 2, \ldots 6$, and z_{ij} where $1 \leq i < j \leq 6$. We assume $x_i \sim x_j, y_i \sim y_j$ for all $i \neq j$ and $x_i \sim y_i$. We further assume that $x_i, y_i \sim z_{kl}$ provided $i \notin \{k, l\}$. Finally, $z_{ij} \sim z_{kl}$ if and only if $\{i, j\} \cap \{k, l\} \neq \emptyset$. This graph is isomorphic to the 1-shadow of an apartment of a spherical building of type E_6 when the diagram is labeled as below. An s-independent subset of the $E_{6,1}(\boldsymbol{F})$ geometry which is isomorphic to the $E_{6,1}(1)$ graph is a frame of the geometry.

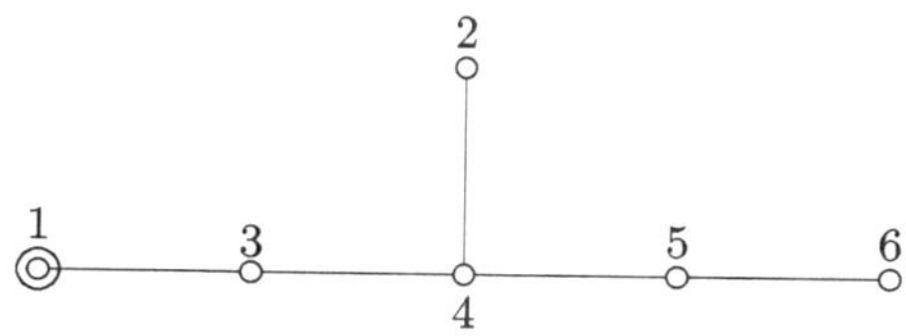

2.5 *The* $E_{7,7}(1)$ *Graph*

The $E_{7,7}(\mathbb{F})$ geometry is the Lie incidence geometry with diagram

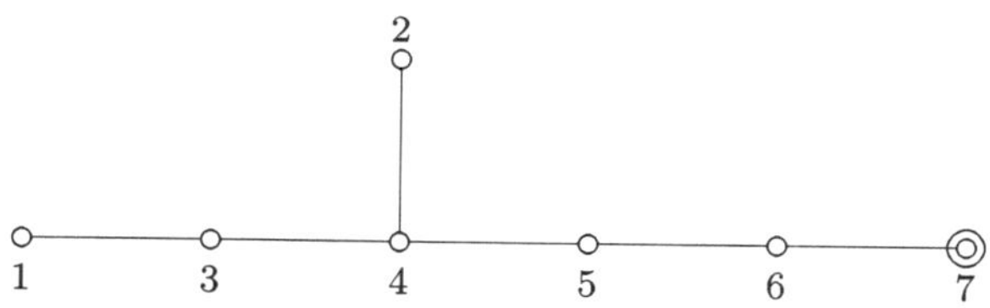

This group can be constructed as the isometry group of a four homogeneous form in 56 variables (see cf [5]). The 7-shadow of an apartment, which we refer to as the $E_{7,7}(1)$ graph, can be described as follows:

Let x_{ij}, y_{ij} be distinct objects for $1 \le i < j \le 8$. Define $x_{ij} \sim x_{kl}$ and $y_{ij} \sim y_{kl}$ if and only if $|\{i, j\} \cap \{k, l\}| = 1$ and $x_{ij} \sim y_{kl}$ if and only if $\{i, j\} \cap \{k, l\} = \emptyset$. Note that for any vertex v the graph induced on the set of vertices adjacent to v is isomorphic to $E_{6,1}(1)$. An s-independent subset of the point set $\mathcal{P}$ which is isomorphic to the graph $E_{7,7}(1)$ is called a frame.

3 Witt-Type Theorems

Let V be a finite dimensional vector space over a field $\boldsymbol{F}$ and let $B : V \times V \to \boldsymbol{F}$ be a non-degenerate reflexive bilinear or sesquilinear form (so that B is either alternate, symmetric or equivalent to a Hermitian form). Recall that an isometry of (V, B) is a linear transformation $S : V \to V$ such that $B(S(\boldsymbol{v}_1), S(\boldsymbol{v}_2)) = B(\boldsymbol{v}_1, \boldsymbol{v}_2)$. Witt's extension theorem states the following:

Theorem (Witt). Let U, U' be subspaces of V and $\sigma : U \to U'$ a vector space isomorphism such that $B(\sigma(\boldsymbol{u}_1), \sigma(\boldsymbol{u}_2)) = B(\boldsymbol{u}_1, \boldsymbol{u}_2)$ for all $\boldsymbol{u}_1, \boldsymbol{u}_2 \in U$. Then there exists an isometry S such that S restricted to U is σ.

Assume the Witt index of (V, B) is $m \geq 2$ and let $\mathcal{P}$ denote the set of singular (projective) points $\langle \boldsymbol{v} \rangle_F$ where $B(\boldsymbol{v}, \boldsymbol{v}) = 0$ and for such a point p denote by $\Delta(p)$ the set of $q \in \mathcal{P}, q \neq p$, such that $p \perp q$, that is, $B(p, q) = 0$. The following is a corollary of Witt's theorem.

Theorem Let $X, Y \subset \mathcal{P}$ and assume the full induced subgraphs of $(\mathcal{P}, \Delta)$ of X and Y are $K_{2,2,\ldots,2}$ where there are l 2's with $l \leq m$. Then there exists an isometry S of (V, B) such that $S(X) = Y$.

We loosely refer to theorems for Lie geometries such as the above as Witt-type theorems. In what follows we survey some results that have been previously obtained as well as some recent ones.

3.1 Classification of (Partial) Frames in Minuscule Geometries

The first paper with Witt-type results for Lie geometries is by Cooperstein and Shult [11]. In this paper they consider each of the minuscule geometries and classify subgraphs of the collinearity graph which are isomorphic to certain subgraphs of the shadow of an apartment and map to a set of independent points for some projective embedding of the geometry. The independence condition is, of course, inherited by subsets and so could be used in induction arguments. This result was considerably improved by Cooperstein, Kasikov, and Shult in [10] where they introduce the concept of s-independence, an intrinsic notion of independence as contrasted with the extrinsic embedding independence conditions. Their results follow:

Theorem 3.1. *Let $(\mathcal{P}, \mathcal{L})$ be one of the Lie geometries $\mathcal{G}_{n,l}(\boldsymbol{F}) = A_{n-1,l}(\boldsymbol{F})$, $B_{n,n}(\boldsymbol{F})$ or $D_{n,n}(\boldsymbol{F})$ and S a partial frame of the geometry. Then S is a partial apartment.*

As an immediate corollary they obtain the following result classifying subspaces of the geometry by their isomorphism type:

Theorem 3.2. *(1) A subspace of $\mathcal{G}_{n,l}(\boldsymbol{F})$ isomorphic to $\mathcal{G}_{m,k}(\boldsymbol{F})$ is parabolic.*
(2) A subspace of $B_{n,n}(\boldsymbol{F})$ which is isomorphic to $B_{m,m}(\boldsymbol{F})$ is parabolic.
(3) A subspace of $D_{n,n}(\boldsymbol{F})$ which is isomorphic to $D_{m,m}(\boldsymbol{F})$ is parabolic.

In another vein, Mark Pankov in [14], determines precisely when every isometric embedding of a Johnson graph $J(n, l)$ in $\mathcal{G}_{n,l}(\boldsymbol{F})$ is an apartment and, more generally, obtains a classification of all isometric embeddings of $J(n, l)$ in $\mathcal{G}_{n,l}(\boldsymbol{F})$.

3.2 Grassmannian Subspaces of Symplectic Grassmannians

Cooperstein [6] has obtained a complete classification of subspaces of $C_{n,l}(\boldsymbol{F})$ which are ordinarily Grassmannians:

Theorem 3.3. *Assume $\mathcal{S}$ is a subspace of $C_{n,l}(\boldsymbol{F})$ and $\mathcal{S}$ is isomorphic to $\mathcal{G}_{m,j}(\boldsymbol{F})$. Then either (i) $\mathcal{S}$ is parabolic or (ii) $char(\boldsymbol{F}) = 2$, $(m, j) = (4, 2)$, there is a totally singular subspace U of dimension $l-1$, and a subspace Y of dimension $l+6$ such that Y/U is non-degenerate and $\mathcal{S}$ a geometric hyperplane of $S(Y, U)$ which consists of all the l-dimensional totally isotropic subspaces containing U and contained in Y.*

Blok and Cooperstein [2] extended this result to other polar grassmannians, specifically, they classify all the subspaces of a polar Grassmannian (orthogonal, sympectic, or unitary) which are isomorphic to an ordinary Grassmannian.

Theorem 3.4. *Let $(\mathcal{P}, \mathcal{L})$ be one of the geometries in the following table.*

$M_{n,l}$	Char$(\boldsymbol{F})$	β	group	n	l
$B_{n,l}(\boldsymbol{F})$	any	parabolic orthogonal	$O_{2n+1}(\boldsymbol{F})$	≥ 3	$1 \leq l \leq n$
$C_{n,l}(\boldsymbol{F})$	$\neq 2$	symplectic	$\mathrm{Sp}_{2n}(\boldsymbol{F})$	≥ 3	$1 \leq l \leq n$
$D_{n,l}(\boldsymbol{F})$	any	hyperbolic orthogonal	$O_{2n}^{+}(\boldsymbol{F})$	≥ 3	$1 \leq l \leq n-2$
${}^2A_{2n,l}(\boldsymbol{F})$	any	σ-hermitian	$U_{2n}^{+}(\boldsymbol{F})$	≥ 3	$1 \leq l \leq n$
${}^2A_{2n+1,l}(\boldsymbol{F})$	any	σ-hermitian	$U_{2n+1}^{+}(\boldsymbol{F})$	≥ 3	$1 \leq l \leq n$
${}^2D_{n+1,l}(\boldsymbol{F})$	any	elliptic orthogonal	$O_{2n+2}^{-}(\boldsymbol{F})$	≥ 3	$1 \leq\leq n$

Assume $S \cong A_{m,j}(\boldsymbol{F})$ be a subspace of $(\mathcal{P}, \mathcal{L})$. Then

(i) *If M_n is of type C_n, ${}^2A_{2n}$, ${}^2A_{2n+1}$, or $D_{n,n-2}$, then S is parabolic.*
(ii) *If $M_{n,}$ is of type $B_{n,l}$, ${}^2D_{n+1,l}$, or $D_{n,l}$ with $1 \leq l \leq n-3$, then S is parabolic or $l \leq n-3$ and $S \cong A_{3,2}(\boldsymbol{F})$ is embedded naturally as $D_{3,1}(\boldsymbol{F})$ into the polar subspace $S_l(C^{\perp}, C)$, for some totally singular (isotropic) subspace C of dimension $l-1$.*

3.3 Symplectic Grassmannian Subspaces of Symplectic Grassmannians

In [7] Cooperstein considered the more general problem of classifying subspaces of a sympectic Grassmannian which are isomorphic to a symplectic Grassmannian. He achieved a complete answer in the case that $char(\boldsymbol{F}) \neq 2$:

Theorem 3.5. *Assume $char(\boldsymbol{F}) \neq 2$ and let $\mathcal{S}$ be a subspace of $C_{n,l}(\boldsymbol{F})$ with $\mathcal{S} \cong C_{m,k}(\boldsymbol{F})$. Then there exists a totally isotropic subspace A of dimension $l-k$ and a subspace B of dimension $2m + l = l$ such that $A \subset B \subset A^{\perp}$ and B/A non-degenerate and $\mathcal{S} = S_l(B, A)$.*

3.4 Witt-Type Theorems for the Geometry $D_{n,n}(F)$

In recent work Cooperstein [8] has obtained Witt-type results for half-spin geometries, in particular, for subspaces which are isomorphic to ordinary Grassmannians. His main theorem is

Theorem 3.6. *Let $\Gamma \cong D_{n,n}(F)$ and S and s-independent subgraph of the collinearity graph isomorphic to $A_{m,l}(1)$. Then $l = 2$ and $m \leq n-1$. Moreover, if $3 \leq m \leq n-1$ then there is a unique class under the action of $Aut(\Gamma)$. In particular, S is a partial apartment.*

As an immediate corollary we have

Corollary 3.7. *Let $\Gamma \cong D_{n,n}(F)$ and $\mathcal{S}$ a subspace isomorphic to $A_{m,l}(F)$. Then $l = 2, m \leq n-1$. Moreover, for $3 \leq m \leq n-1$ there is a unique class and it is parabolic.*

3.5 Characterization of Apartments in Polar Grassmannians and Dual Polar Spaces

Let $\Gamma = (\mathcal{P}, \mathcal{L})$ be a polar space of rank n. A frame, F, of Γ is a full subgraph which is isomorphic to $K_{2,2,\dots,2}$ (n 2's). Let $C(F)$ denote the collection of maximal cliques of F. For $C, C' \in C(F)$ say C and C' are adjacent if $|C \cap C'| = n-1$. The graph on $C(F)$ defined by this adjacency is isomorphic to the graph $B_{n,n}(1)$ which is isomorphic to the hypercube graph H_n. This graph is the shadow of an apartment of the polar space Γ in the dual polar space, $D(\Gamma)$ associated to Γ. More generally, assume $1 < m < n$ and let D be a clique of a frame F with $|D| = n-m$. Set $C_D(F)$ consist of those $C \in C(F)$ such that $D \subset C$. Then the full induced subgraph on $C_D(F)$ is isomorphic to the hypercube H_m. For our purposes we refer to $C_D(F)$ as a partial apartment of $D(\Gamma)$.

In the paper [13], Mark Pankov proves the following characterization of isometric embeddings of the hypercube graph into the collinearity graph of a dual polar space:

Theorem 3.8. *Let $\Gamma = (\mathcal{P}, \mathcal{L})$ be a polar space of rank n. The image of every isometric embedding of H_m into $D(\Gamma)$ is a partial apartment.*

In a similar vein, in the paper [15], Pankov obtains a characterization of the shadow of an apartment in non-maximal polar Grassmannians.

3.6 Witt-Type Theorems for the Geometry $E_{6,1}(F)$

In the aforementioned paper [8] Cooperstein has obtained several results on s-independent subgraphs and subspaces of the geometry $E_{6,1}(F)$. These are as follows:

Theorem 3.9. *$A_{5,3}(\boldsymbol{F})$ is not a subspace of $E_{6,1}(\boldsymbol{F})$.*

Theorem 3.10. *Let $\Gamma = (\mathcal{P}, \mathcal{L}) \cong E_{6,1}(\boldsymbol{F})$ and set $G = Aut(\Gamma)$. Assume S is an s-independent subgraph of Γ, $S \cong A_{m,l}(1)$. Then $l = 2$ and $m \leq 5$ and S is a partial apartment. In particular, G is transitive on all such S.*

Corollary 3.11. *Let $\Gamma = (\mathcal{P}, \mathcal{L}) \cong E_{6,1}(\boldsymbol{F})$ and set $G = Aut(\Gamma)$. Assume $\mathcal{S} \cong A_{m,l}(\boldsymbol{F})$. Then $l = 2$ and $m \leq 5$. Moreover, G is transitive on all such $\mathcal{S}$ for a fixed admissible pair (m, l).*

Theorem 3.12. *Let $\Gamma = (\mathcal{P}, \mathcal{L}) \cong E_{6,1}(\boldsymbol{F})$ and set $G = Aut(\Gamma)$. Assume S is an s-independent subgraph of Γ, $S \cong D_{m,m}(1)$. Then $m \leq 5$ and S is a partial apartment. In particular, G is transitive on all such S.*

Corollary 3.13. *Let $\Gamma = (\mathcal{P}, \mathcal{L}) \cong E_{6,1}(\boldsymbol{F})$ and set $G = Aut(\Gamma)$. Assume $\mathcal{S} \cong D_{m,m}(\boldsymbol{F})$. Then $m \leq 5$. Moreover, G is transitive on all such $\mathcal{S}$.*

Theorem 3.14. *Let $\Gamma = (\mathcal{P}, \mathcal{L}) \cong E_{6,1}(\boldsymbol{F})$. Assume S is an s-independent subgraph of Γ, $S \cong E_{6,1}(1)$. Then S is an apartment.*

3.7 Witt-Type Theorems for the Geometry $E_{7,7}(F)$

Finally, in [8] are nearly complete results for a geometry isomorphic to $E_{7,7}(\boldsymbol{F})$. A summary follows:

Lemma 3.15. *$A_{6,3}(\boldsymbol{F})$ is not a subspace of $E_{7,7}(\boldsymbol{F})$.*

Theorem 3.16. *Let $\Gamma = E_{7,7}(\boldsymbol{F})$ and set $G = Aut(\Gamma)$. Assume S is an s-independent subset of Γ, $S \cong A_{m,l}(1)$. Then either $(m, l) = (5, 3)$ or $l = 2$ and $m \leq 6$. Moreover, for each admissible (m, l) there is a unique class under the action of G. In particular, S is a partial apartment.*

Corollary 3.17. *Let $\Gamma = E_{7,7}(\boldsymbol{F})$ and set $G = Aut(\Gamma)$. Assume $\mathcal{S} \cong A_{m,l}(\boldsymbol{F})$. Then either $(m, l) = (5, 3)$ or $l = 2$ and $m \leq 6$. Moreover, for each admissible (m, l) there is a unique class under the action of G.*

Theorem 3.18. *Let $\Gamma = E_{7,7}(\mathbb{F})$ and set $G = Aut(\Gamma)$. Assume S is an s-independent subset of Γ, $S \cong D_{m,m}(1)$. Then $m \leq 7$. Moreover, for $3 \leq m \leq 6$ there is a unique class under the action of G. In particular, S is a partial apartment.*

Corollary 3.19. *Let $\Gamma = E_{7,7}(\boldsymbol{F})$ and set $G = Aut(\Gamma)$. Assume $\mathcal{S}$ is a subspace of Γ, $\mathcal{S} \cong D_{m,m}(\boldsymbol{F})$. Then $m \leq 7$. Moreover, for $3 \leq m \leq 7$ there is a unique class under the action of G.*

Theorem 3.20. *Let $\Gamma = E_{7,7}(\mathbb{F})$ and set $G = Aut(\Gamma)$. Assume S is an s-independent subset of Γ, $S \cong E_{6,1}(1)$. Then there is one class of such S under the action of G. Consequently, S is a partial apartment.*

Corollary 3.21. *Let $\Gamma = E_{7,7}(\boldsymbol{F})$ and set $G = Aut(\Gamma)$. Assume $\mathcal{S}$ is a subspace of Γ, $\mathcal{S} \cong E_{6,1}(\boldsymbol{F})$. Then there is a unique class under the action of G.*

Theorem 3.22. *Let $\Gamma = E_{7,7}(\mathbb{F})$ and set $G = Aut(\Gamma)$. Assume S is an s-independent subset of Γ, $S \cong E_{7,7}(1)$. Then there is one class of such S under the action of G. Consequently, S is an apartment.*

4 Methods of the Proof

We briefly discuss our methods of proof.

The main tool in demonstrating the existence of subspaces is the generation of subgroups by full root subgroups, drawing on results of Kantor [12] and Cooperstein [3, 4]. These subgroups are restricted to an appropriate vector space (which affords the universal embedding for the geometry) and a decomposition of the space is obtained.

For example, $E_7(\boldsymbol{F})$ contains a subgroup isomorphic to $A_7(\boldsymbol{F})$ generated by full root subgroups. When the 56 dimensional module of $E_7(\mathbb{F})$ is restricted to this group it has the form $\wedge^2(V) \oplus \wedge^2(V^*)$ where V is an eight dimensional space on which $A_7(\boldsymbol{F})$ acts and V^* is its dual space.

The group $E_7(\boldsymbol{F})$ also contains a subgroup $A_1(\mathbb{F}) \cdot D_6(\mathbb{F})$. When the 56 space is restricted to $A_1(\mathbb{F}) \cdot D_6(\boldsymbol{F})$ it decomposes as $[V_2 \otimes O_{12}^+] \oplus HS_{32}$ where V_2 is the standard module for $A_1(\boldsymbol{F})$, O_{12}^+ is the standard module for $D_6(\boldsymbol{F})$ and HS_{32} is a half-spin module of dimension 32 for $D_6(\boldsymbol{F})$.

As another example, the group $E_7(\boldsymbol{F})$ also contains a subgroup $A_2(\boldsymbol{F}) \cdot A_5(\boldsymbol{F})$. When the 56 space is restricted to $A_2(\boldsymbol{F}) \cdot A_5(\boldsymbol{F})$ it decomposes as $\wedge^3(V_6) \oplus V_6 \otimes V_3 \oplus V_6^* \otimes V_3^*$ where V_6 is the standard module for $A_5(\boldsymbol{F})$, V_6^* is its dual, V_3 is the standard module for $A_2(\boldsymbol{F})$ and V_3^* is its dual.

The existence of s-independent subsets of the required isomorphism type follows from results of Blok-Brouwer [1], Cooperstein-Shult [11], and Ronan-Smith [16].

Non-existence aspects of the proof rely heavily on analyzing the residue at a point as well as particular properties of the geometries. For example, in $D_{5,5}(\boldsymbol{F})$ if l is a line and M is a singular subspace isomorphic to $PG(2, \boldsymbol{F})$, then there is a line $m \subset M$ such that $l \cup m$ generates a grid. As a consequence of this we prove that $A_{5,3}(\boldsymbol{F})$ is not a subspace of $E_{6,1}(\boldsymbol{F})$.

When subgraphs exist they are classified inductively by building them up by rank. For example, in $E_{6,1}(\boldsymbol{F})$ there is a unique s-independent subgraph isomorphic to $A_{3,2}(1)$ since this is contained in a symp. We then show that any s-independent subgraph S isomorphic to $A_{4,2}(1)$ must have $S^\perp$ a line (and then a classification in $D_{5,5}(\boldsymbol{F})$ can be used). This is then used to classify the s-independent subgraphs $S' \cong A_{5,2}(1)$.

References

1. Blok, R.J., Brouwer, A.E.: Spanning point-line geometries in buildings of spherical type. J. Geom. **62**(1–2), 26–35 (1998)
2. Blok, R.J., Cooperstein, B.N.: Projective sub-Grassmannians of polar Grassmannians. Bull. Belg. Math. Soc. Simon Stevin **17**(4), 675–691 (2010)
3. Cooperstein, B.N.: The geometry of root subgroups in exceptional groups. I. Geom. Dedicata **8**(3), 317–381 (1979)
4. Cooperstein, B.N.: The geometry of root subgroups in exceptional groups. II. Geom. Dedicata **15**(1), 1–45 (1983)
5. Cooperstein, B.N.: The fifty-six-dimensional module for E7. I. A four form for E7. J. Algebra **173**(2), 361–389 (1995)
6. Cooperstein, B.N.: Classical subspaces of symplectic Grassmannians. Bull. Belg. Math. Soc. Simon Stevin **12**(5), 719–725 (2005)
7. Cooperstein, B.N.: Symplectic subspaces of symplectic Grassmannians. European J. Combin. **28**(5), 1442–1454 (2007)
8. Cooperstein, B.N.: Witt-type theorems for exceptional lie geometries, in preparation
9. Cooperstein, B.N.: Generation of lie incidence geometries: a survey. In: Proceedings ISI Workshop and Conference, Bangalore, 2013
10. Cooperstein, B.N., Kasikova, A., Shult, E.E.: Witt-type theorems for Grassmannians and lie incidence geometries. Adv. Geom. **5**(1), 15–36 (2005)
11. Cooperstein, B.N., Shult, E.E.: Frames and bases of lie incidence geometries. J. Geom. **60**, 17–46 (1997)
12. Kantor, W.M.: Subgroups of classical groups generated by long root elements. Trans. Amer. Math. Soc. **248**(2), 347–379 (1979)
13. Pankov, Mark: Metric characterization of apartments in dual polar spaces. J. Combin. Theorey Ser. A **118**(4), 1313–1321 (2011)
14. Pankov, Mark: Isometric embeddings of Johnson graphs in Grassman graphs. J. Algebraic Combin. **33**(4), 555–570 (2011)
15. Pankov, Mark: Characterization of apartments in polar Grassmannians. Bull. Belg. Math. Soc. Simon Stevin **19**(2), 345–366 (2012)
16. Ronan, M., Smith, S.: Sheaves on buildings and modular representations of Chevalley groups. J. Alg. **96**, 319–346 (1985)

Chapter 7
Embeddings of Cotriangular Spaces

Hans Cuypers and Jos in 't panhuis

Abstract A cotriangular space is a partial linear space with three points per line, such that a point is collinear to 0 or 2 points on the line. Such a space is called irreducible if it is connected, and no two points are collinear with the same set of points. Results by Hall and Shult imply that irreducible cotriangular spaces can be subdivided into three classes: those of triangular type, those of symplectic type, and those of orthogonal type. In these spaces any two intersecting lines are in a dual affine plane. We consider embeddings of cotriangular spaces in which any two intersecting lines are in a dual affine plane into projective spaces over arbitrary fields. We show that if such a cotriangular space admits an embedding over a field $\mathbb{F}$, it admits a universal embedding over the field $\mathbb{F}$. Moreover we classify all universal embeddings of irreducible cotriangular spaces. For the spaces of symplectic or orthogonal type we describe, if the characteristic $\mathbb{F}$ is two, the embedding using the associated quadratic forms. For other characteristics the universal embeddings only exist for cotriangular spaces associated to the root systems of type E_6, E_7, and E_8 and to root systems of type A_n, where $n > 4$ or $n = \infty$.

Keywords Embedding · Cotriangular spaces · Root lattices

The first author likes to thank Jonathan Hall for some fruitful discussions on the topics of this paper.

H. Cuypers (✉) · J. in 't panhuis
Department of Mathematics and Computer Science, Technische Universiteit Eindhoven, P.O. Box 513, 5600 MB Eindhoven, The Netherlands
e-mail: hansc@win.tue.nl

J. in 't panhuis
e-mail: josintpanhuis@gmail.com

N. S. N. Sastry (ed.), *Groups of Exceptional Type, Coxeter Groups and Related Geometries*, Springer Proceedings in Mathematics & Statistics 82,
DOI: 10.1007/978-81-322-1814-2_7,

1 Introduction

In this paper, we consider partial linear spaces in which each line contains three points and in which a point not on a line is collinear with no or two points of that line, that is, we consider *cotriangular spaces*.

A cotriangular spaces is called *irreducible* if it is connected and has the property that no two points are collinear to the same set of points.

Results by Hall [3] and Shult [7] make that the irreducible cotriangular spaces can be subdivided into three classes: those of triangular type, those of symplectic type, and those of orthogonal type. In particular, in an irreducible cotriangular space any two intersecting lines are contained in a subspace isomorphic to a dual affine plane with 6 points and 4 lines.

Moreover, Hall also considered embeddings of cotriangular spaces into a projective space over the field $\mathbb{F}_2$, see [2, 3]. He showed that for a cotriangular space in which any two intersecting lines are in a dual affine plane, there is a universal embedding over $\mathbb{F}_2$ such that each embedding over $\mathbb{F}_2$ is a quotient of this embedding. In this paper we generalize this result to arbitrary fields:

Theorem 1.1. *A cotriangular space in which any two intersecting lines are contained in a dual affine plane admitting an embedding over an arbitrary field* $\mathbb{F}$ *has a universal embedding over* $\mathbb{F}$.

We classify all universal embeddings of the irreducible cotriangular spaces.

Theorem 1.2. *Let* Π *be an irreducible cotriangular space and* ϕ *a universal embedding of* Π *into a vector space* V *over a field* $\mathbb{F}$. *Then, up to isomorphism and equivalence of the embedding, we have one of the following:*

1. Π *is* $\mathcal{NO}(V, Q)$ *or* $\mathcal{HS}p(V, f)$ *for some nondegenerate orthogonal space* (V, Q) *or symplectic space* (V, f) *over* $\mathbb{F}_2$ *and* ϕ *is its natural orthogonal embedding over* $\mathbb{F}$, *where* $\mathbb{F}$ *has even characteristic.*
2. Π *is* $\mathcal{T}_\Omega$ *for some set* Ω *of size at least* 5, *and* ϕ *its embedding of type* A;
3. Π *is* $\mathcal{NO}_6^-(2)$ *and* ϕ *its embedding of type* E_6;
4. Π *is* $\mathcal{NO}_7(2)$ *and* ϕ *its embedding of type* E_7;
5. Π *is* $\mathcal{NO}_8^+(2)$ *and* ϕ *its embedding of type* E_8.

The paper is organized as follows. In the next section we introduce the basic notions and describe the cotriangular spaces appearing in the above result. Moreover, we introduce embeddings and provide examples of embeddings of cotriangular spaces. The proof of Theorem 1.1 is given in Sect. 3, while Sect. 4 is devoted to the proof of Theorem 1.2.

This paper is partly based on the Ph.D.-thesis of the second author, see [6].

2 Preliminaries

Let Π be a pair $(\mathcal{P}, \mathcal{L})$ consisting of a set $\mathcal{P}$ of *points* and a set $\mathcal{L}$ of *lines*. Each line in $\mathcal{L}$ is a subset of $\mathcal{P}$ of size at least two. The pair $(\mathcal{P}, \mathcal{L})$ is called a *point-line space*. If any two distinct points are on at most one line, then $(\mathcal{P}, \mathcal{L})$ is called a *partial linear space*.

Let $\mathcal{X}$ be a subset of $\mathcal{P}$. Then it is a *subspace* of $(\mathcal{P}, \mathcal{L})$ if any line intersecting $\mathcal{X}$ in at least two points is completely contained in $\mathcal{X}$. If we define $\mathcal{K}$ to be the set of lines in $\mathcal{L}$ completely contained in $\mathcal{X}$, then, assuming $\mathcal{X}$ is a subspace, $(\mathcal{X}, \mathcal{K})$ is a point-line space which we identify with $\mathcal{X}$.

Next, consider the intersection of all subspaces of $(\mathcal{P}, \mathcal{L})$ containing $\mathcal{X}$. This is again a subspace and we denote it by $\langle\mathcal{X}\rangle$. The elements of $\mathcal{X}$ are called the *generators* of $\langle\mathcal{X}\rangle$ and $\langle\mathcal{X}\rangle$ is said to be *generated* by $\mathcal{X}$. Suppose n is the minimal cardinality of a generating set of $(\mathcal{P}, \mathcal{L})$, then n is said to be the *generating rank* of $(\mathcal{P}, \mathcal{L})$.

A subspace of Π generated by two intersecting lines is called a *plane*.

The *collinearity graph* of a point-line space $(\mathcal{P}, \mathcal{L})$ is the graph where two (distinct) points in $\mathcal{P}$ are adjacent if and only if there is a line in $\mathcal{L}$ containing both of them. If the collinearity graph is connected, then $(\mathcal{P}, \mathcal{L})$ is called *connected*. Two points in $(\mathcal{P}, \mathcal{L})$ are called *collinear* if they are adjacent in the collinearity graph.

A *cotriangular space* is a partial linear space in which any line contains exactly three points and any point not on a line is collinear with either no or all but one of the points of that line. If two points x and y are equal or non-collinear we write $x \perp y$. Otherwise, we write $x \sim y$. Moreover, given a point x we write Δ_x for the set of points non-collinear with x, that is, $\Delta_x = x^{\perp} \setminus \{x\}$. A connected cotriangular space is called *irreducible* if no two distinct points x and y exist with $\Delta_x = \Delta_y$.

Examples 2.1–2.3 describe some important cotriangular spaces

Example 2.1. If Ω is a set, then the partial linear space $(\mathcal{P}, \mathcal{L})$ with

$$\mathcal{P} = \{\{i, j\} \mid i \neq j \in \Omega\} \ \wedge\ \mathcal{L} = \{\{x, y, z\} \mid x, y, z \in \mathcal{P} \wedge x \neq y \neq z \neq x \wedge |x \cup y \cup z| = 3\}$$

is denoted by $\mathcal{T}(\Omega)$. We write $\mathcal{T}_n$ instead of $\mathcal{T}(\Omega)$ if $\Omega = [n]$ for a certain $n \in \mathbb{N}$. (By $[n]$ we denote the set $\{1, \ldots, n\}$ and by $[m, n]$ for $m < n$ the set $\{m, m+1, \ldots, n\}$.)

Example 2.2. Suppose (V, f) is a symplectic space over the field $\mathbb{F}_2$. Then the partial linear space $(\mathcal{P}, \mathcal{L})$ with

$$\mathcal{P} = V \setminus \{0\} \ \wedge\ \mathcal{L} = \{\{x, y, x + y\} \mid x, y \in \mathcal{P} \wedge f(x, y) = 1\}$$

is denoted by $\mathcal{HS}p(V, f)$.

If $V = \mathbb{F}_2^{2n}$ ($n \in \mathbb{N}$), and f nondegerenate, then we can take f to be the symplectic form

$$((x_1, \ldots, x_{2n}), (y_1, \ldots, y_{2n})) \mapsto \sum_{i=1}^{n} (x_{2i-1}y_{2i} + y_{2i-1}x_{2i}),$$

and we write $\mathcal{HS}p_{2n}(2)$ instead of $\mathcal{HS}p(V, f)$.

Example 2.3. Suppose (V, Q) is an orthogonal space over the field $\mathbb{F}_2$. Moreover, let f be the symplectic form associated to Q. Then the partial linear space $(\mathcal{P}, \mathcal{L})$ with

$\mathcal{P} = \{x \mid x \in V \setminus \mathrm{Rad}(f) \wedge Q(x) = 1\} \ \wedge \ \mathcal{L} = \{\{x, y, x+y\} \mid x, y, x+y \in \mathcal{P}\}$

is denoted by $\mathcal{NO}(V, Q)$.

If $V = \mathbb{F}_2^{2n+1}$ $(n \in \mathbb{N})$ and Q nondegenerate, then we can assume Q to be the quadratic form

$$(x_1, \ldots, x_{2n+1}) \mapsto \sum_{i=1}^{n} x_{2i-1}x_{2i} + x_{2n+1}^2,$$

and we write $\mathcal{NO}_{2n+1}(2)$ instead of $\mathcal{NO}(V, Q)$.

If $V = \mathbb{F}_2^{2n}$ $(n \in \mathbb{N})$, then there are two possibilities for Q. Either we can take Q to be the quadratic form

$$(x_1, \ldots, x_{2n}) \mapsto \sum_{i=1}^{n} x_{2i-1}x_{2i},$$

and we write $\mathcal{NO}_{2n}^{+}(2)$ instead of $\mathcal{NO}(V, Q)$, or we can take Q to be the form

$$(x_1, \ldots, x_{2n}) \mapsto \sum_{i=1}^{n} x_{2i-1}x_{2i} + x_{2n-1}^2 + x_{2n}^2,$$

and write $\mathcal{NO}_{2n}^{-}(2)$ instead of $\mathcal{NO}(V, Q)$.

In fact, the cotriangular spaces from Examples 2.1–2.3 are all what is needed to give a complete classification of the irreducible cotriangular spaces.

Theorem 2.4. [3, 7] *Each irreducible cotriangular space is isomorphic to*

- $\mathcal{T}(\Omega)$ *for a set* Ω *of size at least 5,*
- $\mathcal{HS}p(V, f)$ *for a non-degenerate symplectic space* (V, f) *of dimension at least 6 over* $\mathbb{F}_2$,
- $\mathcal{NO}(V, Q)$ *for a non-degenerate orthogonal space* (V, Q) *of dimension at least 6 over* $\mathbb{F}_2$.

Moreover, each plane in an irreducible cotriangular space is isomorphic to a dual affine plane of order two.

The following lemma describes some well known isomorphisms amongst the different cotriangular spaces occurring in Theorem 2.4.

Lemma 2.5.

$$\mathcal{T}_6 \cong \mathcal{HS}p_4(2) \cong \mathcal{NO}_5(2)$$

$$\mathcal{NO}_6^{+}(2) \cong \mathcal{T}_8 \ \textit{and} \ \forall_{n \in \mathbb{N}} : \ \mathcal{HS}p_{2n}(2) \cong \mathcal{NO}_{2n+1}(2).$$

Another way to obtain cotriangular spaces is related to root systems of types A and E. We refer to [5] for the relevant definitions regarding root systems.

Example 2.6. Let X_m be one of the root systems E_m with $m \in [6, 8]$ or A_m with $m \geq 3$ an integer. Let Φ be the set of simple roots in X_m. Then the partial linear space $(\mathcal{P}, \mathcal{L})$ with

$$\mathcal{P} = \{\mathbb{R}x \mid x \in \Phi\} \quad \wedge \quad \mathcal{L} = \{\{\mathbb{R}x, \mathbb{R}y, \mathbb{R}z\} \mid x, y, z \in \Phi \wedge x + y + z = 0\}$$

is denoted by $\mathcal{R}(X_m)$. Cotriangular spaces isomorphic to $\mathcal{R}(X_m)$ are said to be of *type* X_m.

Of course, the definition of cotriangular spaces of type A_m can easily be generalized to obtain infinite examples. Indeed, let V be a real vector space and $\mathcal{B}$ a basis consisting of vectors e_ω with ω in some index set Ω. Then for Φ we take the set of vectors $e_\omega - e_{\omega'}$, where $\omega \neq \omega' \in \Omega$ (which we will also call the simple roots). Points are now the 1-space spanned by simple roots and lines are the triples of points $\{\mathbb{R}(e_{\omega_1} - e_{\omega_2}), \mathbb{R}(e_{\omega_2} - e_{\omega_3}), \mathbb{R}(e_{\omega_3} - e_{\omega_1})\}$, where ω_1, ω_2 and ω_3 are three distinct elements from Ω. These spaces are denoted by $\mathcal{R}(A_\Omega)$ and are also said to be of type A.

The following lemma provides some well known isomorphisms.

Lemma 2.7. *We have the following isomorphisms:*

(a) $\mathcal{T}_\Omega \cong \mathcal{R}(A_\Omega)$ *for any set* Ω*;*
(b) $\mathcal{NO}_6^-(2) \cong \mathcal{R}(E_6)$*;*
(c) $\mathcal{NO}_7(2) \cong \mathcal{R}(E_7)$*;*
(d) $\mathcal{NO}_8^+(2) \cong \mathcal{R}(E_8)$.

Suppose $\Pi = (\mathcal{P}, \mathcal{L})$ is a cotriangular space. Let ϕ be a map from the point set $\mathcal{P}$ of Π into the point set $\mathcal{P}(\mathbb{P})$ of a projective space $\mathbb{P} = \mathbb{P}(V)$ with V a vector space over a field $\mathbb{F}$. Then ϕ is called a *embedding* of Π over $\mathbb{F}$ into $\mathbb{P}$ if

(a) ϕ is injective;
(b) ϕ maps the three points of any line $\ell \in \mathcal{L}$ into a line of $\mathbb{P}(V)$, which intersects $\phi(\mathcal{P})$ in precisely $\phi(\ell)$;
(c) $\phi(\mathcal{P})$ generates $\mathbb{P}$.

Example 2.8. Let (V, Q) be a (nondegenerate) quadratic space over $\mathbb{F}_2$, the field with two elements. The cotriangular space $\Pi = \mathcal{NO}(V, Q)$ naturally embeds into V as well as into the space which we obtain by tensoring V with a field $\mathbb{F}$ of even characteristic. These embeddings will be called the natural (orthogonal) embeddings of Π over $\mathbb{F}$.

Example 2.9. Let (V, f) be a (nondegenerate) symplectic space over the field with 2 elements and suppose $\Pi = \mathcal{HS}p(V, f)$ is the cotriangular space of hyperbolic lines in (V, f). Then, as Hall shows in [3], Sect. 5, there is an orthogonal space $(\tilde{V}, \tilde{Q})$

with 1-dimensional non-singular radical, such that $\mathcal{NO}(\tilde{V}, \tilde{Q})$ is isomorphic to Π. The isomorphism induces an embedding of Π into $(\tilde{V}, \tilde{Q})$ which, as Hall shows, is the universal embedding of Π over $\mathbb{F}_2$.

Of course we can tensor this embedding (just as the embedding into (V, f)) with any field $\mathbb{F}$ of characteristic 2 to get an embedding over that field. This embedding will be referred to as the orthogonal (symplectic) embedding of Π over $\mathbb{F}$.

Example 2.10. Let Π be a cotriangular space of type A or E, as described in 2.6 corresponding to a root lattice (or its infinite version of type A). Then the set Φ of simple roots and the lattice they generate are defined over $\mathbb{Z}$. Thus, Π not only embeds in a real vector space, but in any vector space over a field of characteristic 0. Moreover, as we can consider the lattice modulo p for any prime p, we can obtain an embedding of the points and lines of Π over any field.

So, let $\mathbb{F}$ be an arbitrary field. As $\mathcal{T}_\Omega \cong \mathcal{R}(A_\Omega)$, the above provides us with an embedding of $\mathcal{T}_\Omega$ into the hyperplane of the $\mathbb{F}$-vector with a basis e_ω $(\omega \in \Omega)$ spanned by the vectors $e_\omega - e_{\omega'}$, where $\omega, \omega' \in \Omega$. This embedding will be called the *embedding of $\mathcal{T}_\Omega$ of type A over $\mathbb{F}$.*

The above together with the isomorphisms $\mathcal{NO}_6^-(2) \simeq \mathcal{R}(E_6)$, $\mathcal{NO}_7(2) \simeq \mathcal{R}(E_7)$, and $\mathcal{NO}_8^+(2) \simeq \mathcal{R}(E_8)$, provide us with embeddings of $\mathcal{NO}_6^-(2)$, $\mathcal{NO}_7(2)$ and $\mathcal{NO}_8^+(2)$, respectively, into 6-, 7- and 8-dimensional vector spaces, respectively, over $\mathbb{F}$. We refer to these embedding as the *embedding of type E_n* over $\mathbb{F}$ of $\mathcal{NO}_6^-(2)$, $\mathcal{NO}_7(2)$ and $\mathcal{NO}_8^+(2)$, where $n = 6, 7$ or 8, respectively.

For the field $\mathbb{F}_2$ the natural embeddings as introduced in Examples 2.8–2.10 coincide with the embeddings as introduced by Hall [2].

Since we want to classify embeddings of a cotriangular space, we need a notion of equivalence. Therefore, let Π be a cotriangular space admitting two embeddings ϕ and ψ over $\mathbb{F}$ into the projective spaces $\mathbb{P}(U)$ and $\mathbb{P}(V)$, respectively. Then ϕ and ψ are called *equivalent* if there is an invertible semi-linear transformation $g : U \to V$ satisfying $\psi = g \circ \phi$, that is, the diagram

$$\begin{array}{ccc} \Pi & \xrightarrow{\ \phi\ } & \mathbb{P}(U) \\ & \searrow^{\psi} & \downarrow{g} \\ & & \mathbb{P}(V) \end{array}$$

commutes.

The embedding ϕ is called the *universal embedding of Π* over $\mathbb{F}$ if, for every embedding φ of Π over $\mathbb{F}$ into a projective space $\mathbb{P}(W)$, there is a semi-linear transformation $g : U \to W$ which satisfies $\varphi = g \circ \phi$, that is, the diagram

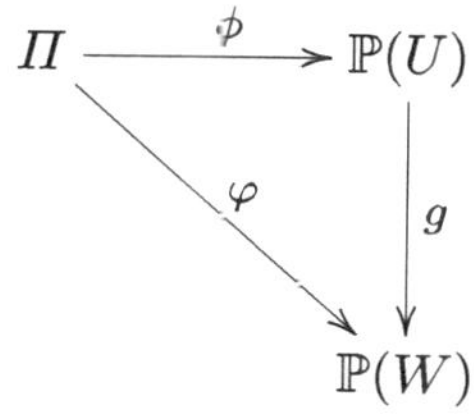

commutes. Note that, in this case, g is not necessarily bijective.

We close this section with a result relating the dimension of an embedding ϕ of a partial linear space Π with the generating rank of Π.

Proposition 2.11. *Let ϕ be an embedding of a partial linear space $\Pi = (\mathcal{P}, \mathcal{L})$ into a projective space $\mathbb{P}$. If the generating rank of Π is n, then* $\dim(\phi) \leq n$.

Proof. Identify Π with its image under ϕ. Let X be a generating set for Π of size n. Then X generates a subspace of $\mathbb{P}$ meeting $\mathcal{P}$ in a subspace of Π. As X generates Π, this subspace of $\mathbb{P}$ equals $\mathbb{P}$. So, the dimension of ϕ is at most n.

3 The Universal Embeddings of Cotriangular Spaces

This section is devoted to the proof of Theorem 1.1:

Theorem 3.1. *Let $\mathbb{F}$ be a field and let Π be a connected cotriangular space in which any two intersecting lines are in a dual affine plane.*

If Π admits an embedding over $\mathbb{F}$, it admits a universal embedding over $\mathbb{F}$. This universal embedding can be realized over the prime field of $\mathbb{F}$.

Proof. Suppose $\Pi = (\mathcal{P}, \mathcal{L})$ admits an embedding ϕ. Fix a point $x \in \mathcal{P}$ and a vector $0 \neq v_x \in \phi(x)$. For each line $\{x, y, z\}$ through x we can fix vectors $0 \neq v_y \in \phi(y)$ and $0 \neq v_z \in \phi(z)$ such that $v_x + v_y + v_z = 0$. (Notice that these vectors are uniquely defined by v_x.)

Now suppose $u \neq x$ is a point not collinear to x. Then u is inside a dual affine plane π with x. Suppose this plane consists of the lines $\{x, y_1, z_1\}$, $\{x, y_2, z_2\}$, $\{u, y_1, y_2\}$ and $\{u, z_1, z_2\}$. As $-v_x = v_{y_1} + v_{z_1} = v_{y_2} + v_{z_2}$ we find $v_{y_1} - v_{y_2} = v_{z_2} - v_{z_1}$. So, $\phi(u)$ equals $\mathbb{F}(v_{y_1} - v_{y_2}) = \mathbb{F}(v_{z_2} - v_{z_1})$. In particular, there is a vector $0 \neq v_u \in \phi(u)$ with $v_u = v_{y_1} - v_{y_2} = v_{z_2} - v_{z_1}$. (Here we make a choice, we could also have chosen $v'_u = -v_u = v_{y_2} - v_{y_1} = v_{z_1} - v_{z_2}$.)

Now suppose π' is a second dual affine plane on u and x consisting of the lines $\{x, y'_1, z'_1\}$, $\{x, y'_2, z'_2\}$, $\{u, y'_1, y'_2\}$ and $\{u, z'_1, z'_2\}$. Then as above, we find that $\phi(u)$ equals $\mathbb{F}(v_{y'_1} - v_{y'_2}) = \mathbb{F}(v_{z'_2} - v_{z'_1})$. So, there is a vector $0 \neq w_u \in \phi(u)$ with $w_u = v_{y'_1} - v_{y'_2} = v_{z'_2} - v_{z'_1} = 0$.

We claim that $w_u = \pm v_u$. Of course, there is a $\lambda \in \mathbb{F}$ with $w_u = \lambda v_u$.

Suppose that y_1 and y'_1 are collinear. As

$$w_u = v_{y'_1} - v_{y'_2} = \lambda v_u = \lambda(v_{y_1} - v_{y_2}),$$

we have $\lambda v_{y_1} - v_{y'_1} = \lambda v_{y_2} - v_{y'_2}$, which spans $\phi(w)$, where w is the third point on the line through y_1 and y'_1.

Inside the dual affine plane spanned by x, y_1 and y'_1, however, we find $\phi(w)$ to be spanned by $v_{y_1} - v_{y'_1}$. This implies that $\lambda = 1$, and indeed, we find $v_u = w_u$. If y_1 and y'_1 are not collinear, we find, by similar arguments, that $w_u = -v_u$.

Up till now we have seen that we can attach to each point p of Π a vector v_p spanning $\phi(p)$ such that for each line $\ell = \{p, q, r\}$ on x we have $-v_p = v_q + v_r$ and for each line $\ell = \{p, q, r\}$ not on x nor in $x^\perp$ we have $v_p = v_q + v_r$ (up to a permutation of p, q, r). (Notice that this permutation is determined by the choice of v_u for points $u \perp x$.)

Now suppose that $\ell = \{p, q, r\}$ is a line in $x^\perp$. Then ℓ is contained in a dual affine plane of Π meeting $x^\perp$ in exactly ℓ. Denote the three points of this plane not on ℓ by p', q', r', where $a' \perp a$ for $a = p, q$ or r. By the above we then have $\pm v_p = v_{q'} - v_{r'}$, $\pm v_q = v_{r'} - v_{p'}$ and $\pm v_r = v_{p'} - v_{q'}$. So, up to the choice of the sign, the sum of the three vectors is 0. Notice that this relation is forced by the relations on the three lines of the plane different from ℓ.

Let $\mathbb{F}\mathcal{P}$ be the vector space with as basis the points in $\mathcal{P}$. Denote by R the subspace of $\mathbb{F}\mathcal{P}$ spanned by the elements of the form $p - q - r$ or $p + q + r$ where p, q, r are points of a line for which $v_p - v_q - v_r = 0$, or $v_p + v_q + v_r = 0$, respectively. Clearly,

$$\begin{aligned} \psi : \mathcal{P} &\to \mathbb{F}\mathcal{P}/R \\ p &\mapsto p + R \end{aligned}$$

defines an embedding of Π and ϕ is a quotient of ψ. As ϕ was arbitrary, we find that any embedding is a quotient of ψ and we have found a universal embedding of Π.

Moreover, as all relations only involve coefficients ± 1, this universal embedding can be realized over the prime field of $\mathbb{F}$.

4 Universal Embeddings of Irreducible Cotriangular Spaces

Let Π be an irreducible cotriangular space. Then, as follows from Theorem 2.4, any two intersecting lines of Π are contained in a dual affine plane. Thus, by Theorem 3.1, every embedding of Π is a quotient of a universal embedding. In this section we will determine the universal embeddings of Π.

First we consider the case of embeddings over a field $\mathbb{F}$ of characteristic 2. By Theorem 3.1, we can realize any universal embedding over $\mathbb{F}$ already over the field $\mathbb{F}_2$ with two elements. This implies that we can apply the results of [3] to obtain the following.

Theorem 4.1. *Suppose $\mathbb{F}$ is a field of even characteristic. Let Π be an irreducible cotriangular space and ϕ its universal embedding over $\mathbb{F}$. Then, up to equivalence we have the following:*

(a) $\Pi \cong \mathcal{T}_\Omega$ *for some set* Ω *of size at least 5 and* ϕ *the embedding of* Π *of type A over* $\mathbb{F}$*;*

(b) $\Pi \cong \mathcal{NO}(V, Q)$ *or* $\mathcal{HS}p(V, f)$ *for some nondegenerate orthogonal space* (V, Q) *or symplectic space* (V, f) *over* $\mathbb{F}_2$ *and* ϕ *is its natural orthogonal embedding over* $\mathbb{F}$*.*

Now we concentrate on embeddings over fields of characteristic not two. The following result implies that we only have to consider triangular spaces and symplectic and orthogonal spaces of small dimension.

Proposition 4.2. *If the cotriangular space* $\mathcal{NO}_8^-(2)$ *does admit an embedding over the field* $\mathbb{F}$*, then the characteristic of* $\mathbb{F}$ *is even.*

Proof. Suppose the cotriangular space $\Pi = (\mathcal{P}, \mathcal{L}) = \mathcal{NO}_8^-(2)$ does admit an embedding ϕ into $\mathbb{P}(V)$ for some vector space V over the field $\mathbb{F}$ of characteristic different from two. Then, as follows from the proof of Theorem 3.1, there is a set Δ of vectors of V such that

- for all $v \in \Delta$ there is a point $x \in \mathcal{P}$ with $\langle v \rangle = \phi(x)$;
- for all $x \in \mathcal{P}$ the 1-space $\phi(x)$ meets Δ in two opposite vectors $\pm v_x$;
- for every line $\ell = \{x, y, z\}$ there are vectors v_x, v_y and v_z in Δ with $\langle v_x \rangle = \phi(x)$, $\langle v_x \rangle = \phi(x)$, $\langle v_x \rangle = \phi(x)$, and $v_x + v_y + v_z = 0$.

It is well known that the collinearity graph of Π is strongly regular with

$$(v, k, \lambda, \mu) = (136, 72, 26, 40).$$

We can now define the following relations between two vectors $v, w \in \Delta$ (we identify each point of $\mathcal{P}$ with its image $\phi(x)$):

R_0: $v = w$;
R_1: $\langle v \rangle$ and $\langle w \rangle$ are collinear and $v + w \in \Delta$;
R_2: $\langle v \rangle$ and $\langle w \rangle$ are not collinear;
R_3: $\langle v \rangle$ and $\langle w \rangle$ are collinear and $v - w \in \Delta$;
R_4: $v = -w$.

Let $\ell = \{x, y, z\}$ be a line of Π and v_x, v_y and v_z vectors in Δ spanning the points $\phi(x)$, $\phi(y)$ and $\phi(z)$, respectively, with $v_x + v_y + v_z = 0$. Let u be a point of Π not on ℓ but collinear with x and $v_u \in \Delta$ be a vector spanning $\phi(u)$. Suppose u is not collinear to z. If $(v_u, v_x) \in R_1$, then $\pm(v_u + v_x) \in \Delta$. So, $(v_y - v_u) + (v_u + v_x) = v_x + v_y = v_z$ and $(v_y - v_u) = v_z - (v_u + v_x) \in \Delta$. So, $(v_y, v_u) \in R_3$.

Using this observation, it is straightforward to check that the relations $R_0, \ldots, R_4$ define a 4-class association scheme on Δ. Indeed, denote by p_{ij}^k the number of elements z such that $(x, z) \in R_i$ and $(z, y) \in R_j$ for some pair $(x, y) \in R_k$. Then we have:

$$(p^k_{0,j}) = \begin{pmatrix} 1&0&0&0&0\\ 0&1&0&0&0\\ 0&0&1&0&0\\ 0&0&0&1&0\\ 0&0&0&0&1 \end{pmatrix}, \quad (p^k_{1,j}) = \begin{pmatrix} 0&1&0&0&0\\ 72&1&20&35&0\\ 0&35&32&35&0\\ 0&35&20&1&72\\ 0&0&0&1&0 \end{pmatrix},$$

$$(p^k_{2,j}) = \begin{pmatrix} 0&0&1&0&0\\ 0&35&32&35&0\\ 126&56&60&56&126\\ 0&35&20&35&0\\ 0&0&1&0&0 \end{pmatrix}, \quad (p^k_{3,j}) = \begin{pmatrix} 0&0&0&1&0\\ 0&35&20&1&72\\ 0&35&32&35&0\\ 72&1&20&35&0\\ 0&1&0&0&0 \end{pmatrix},$$

$$(p^k_{4,j}) = \begin{pmatrix} 0&0&0&0&1\\ 0&0&0&1&0\\ 0&0&1&0&0\\ 0&1&0&0&0\\ 1&0&0&0&0 \end{pmatrix}.$$

A common eigenvector for these matrices is

$$\begin{pmatrix} 1\\ 2\\ 0\\ -2\\ -1 \end{pmatrix}.$$

The dimension of the corresponding common eigenspace of the adjacency matrices of the association scheme is then

$$\frac{2 \cdot 136}{\frac{1^2}{1} + \frac{2^2}{72} + \frac{0^2}{126} + \frac{(-2)^2}{72} + \frac{1^2}{1}} = \frac{136}{76}.$$

(See for example [[1], Proposition 2.2.2]). As this is not an integer, we obtain a contradiction. So, Π does not admit an embedding over $\mathbb{F}$.

We are now in a position to prove the following:

Theorem 4.3. *Let Π be an irreducible cotriangular space and ϕ a universal embedding of Π into a vector space V over a field of characteristic different from 2. Then, up to isomorphism and equivalence of the embedding, we have one of the following:*

(a) Π is $\mathcal{T}_\Omega$ for some set Ω of size at least 5, and ϕ its embedding of type A;
(b) Π is $\mathcal{NO}^-_6(2)$ and ϕ its embedding of type E_6;
(c) Π is $\mathcal{NO}_7(2)$ and ϕ its embedding of type E_7;
(d) Π is $\mathcal{NO}^+_8(2)$ and ϕ its embedding of type E_8.

Proof. Suppose Π is an irreducible cotriangular space which admits an embedding over a field $\mathbb{F}$ of characteristic not two. Then by Proposition 4.2, Π can not contain a subspace isomorphic to $\mathcal{NO}_8^-(2)$. So, by the classification of irreducible cotriangular spaces, see Theorem 2.4, we find Π to be isomorphic to $\mathcal{T}_\Omega$ for some set Ω of size at least three or to one of the three spaces $\mathcal{NO}_6^-(2)$, $\mathcal{NO}_7(2)$ or $\mathcal{NO}_8^+(2)$.

First consider the case that Π is isomorphic to $\mathcal{NO}_6^-(2)$, $\mathcal{NO}_7(2)$ or $\mathcal{NO}_8^+(2)$. Then Π can be generated by $n = 6, 7$ or 8 elements, respectively [3]. But that implies, by 2.11, that any embedding of Π is at most n-dimensional. Thus, the embedding of type E_n is universal.

Now consider the case that Π is isomorphic to $\mathcal{T}_n$ for some $n \geq 2$. Then Π can be generated by $n-1$ elements (see [3]) and, by Proposition 2.11, all its embeddings are of dimension at most $n-1$. As the embedding of type A of Π is of dimension $n-1$, it is universal.

Now consider the case that $\Pi \cong \mathcal{T}_\Omega$ for some infinite set Ω. If the embedding ψ of Π of type A is not universal, there is a finite subspace X of Π for which $\psi|_X$ is not universal. However, as each connected component of X is isomorphic to $\mathcal{T}_n$ for some n, this contradicts the above. So, also in case Ω is infinite, the embedding of type A is universal.

Theorem 1.2 is now obtained by combining the results from Theorems 4.1 and 4.3.

References

1. Brouwer, A.E., Cohen, A.M., Neumaier, A.: Distance-Regular Graphs. Springer, Berlin (1989)
2. Hall, Jonathan I.: Linear representations of cotriangular spaces. Linear Algebra Appl. **49**, 257–273 (1983)
3. Hall, J.I.: Graphs, geometry, 3-transpositions, and symplectic F_2-transvection groups. Proc. London Math. Soc. **58**(3), 89–111 (1989)
4. Higman, D.G.: Admissible graphs, in finite geometries (Pullman, Wash., 1981). In: Lecture Notes in Pure and Applied Mathematics, vol. 82, pp. 211–222. Dekker, New York (1983)
5. Humphreys, J.E.: Introduction to lie algebras and representation theory. In: Graduate Texts in Mathematics, vol. 9. Springer, New York (1978) (Second printing, revised)
6. In 't panhuis, J.: Lie algebras, extremal elements, and geometry. Ph.D. Dissertation, Technische Universiteit Eindhoven, Eindhoven (2009)
7. Shult, E.E.: Groups, polar spaces and related structures. In: Combinatorics, Part 3: combinatorial group theory (Proceedings of Advanced Study Institute, Breukelen, 1974), Math. Centrum, pp. 130–161. Math. Centre Tracts, No. 57, Amsterdam (1974)
8. Taylor, D.E.: The geometry of the classical groups. In: Sigma Series in Pure Mathematics, vol. 9. Heldermann, Berlin (1992)

Chapter 8
Unipotent Overgroups in Simple Algebraic Groups

Iulian I. Simion and Donna M. Testerman

Abstract In this article we survey recent results on the structure of centers of centralizers of unipotent elements u in simple linear algebraic groups G. We bring forth the case of bad characteristic treated by the first author as well as a new case-free proof in characteristic 0 of the second author giving a lower bound for $\dim Z(C_G(u))$ in case u is an even element. We also point out properties of the group $Z(C_G(u)^\circ)$.

Keywords Semisimple groups · Unipotent overgroups · Double centralizers · Unipotent · Nilpotent

1 Introduction

Throughout G will denote a simple linear algebraic group defined over an algebraically closed field k of characteristic $p \geq 0$, u will denote a unipotent element of G and e a nilpotent element in the Lie algebra $\mathrm{Lie}(G)$. We will say p is *bad for G* if $p = 2$ and G is not of type A_n or if $p = 3$ and G is of exceptional type or if $p = 5$ and G is of type E_8; otherwise we say p is *good for G*. Characteristic 0 is considered to be good for all G.

It was shown in [19, 27] that if $p > 0$ and $u \in G$ has order p, with the exception of precisely one conjugacy class of elements, u lies in an A_1-type subgroup of G. In particular, u lies in a closed connected 1-dimensional subgroup of G. The exception is the $\tilde{A}_1^{(3)}$ class in G of type G_2 when $\mathrm{char}(k) = 3$. Even in the one exceptional

The first author would like to acknowledge the support of the Swiss National Science Foundation through grants numbers 200021-122267 and 200020-135144.

I. I. Simion · D. M. Testerman (✉)
École Polytechnique Fédérale de Lausanne, Lausanne, Switzerland
e-mail: donna.testerman@epfl.ch

I. I. Simion
e-mail: simion@mathematik.uni-kl.de

N. S. N. Sastry (ed.), *Groups of Exceptional Type, Coxeter Groups and Related Geometries*, Springer Proceedings in Mathematics & Statistics 82,
DOI: 10.1007/978-81-322-1814-2_8, © Springer India 2014

case, u lies in a 1-dimensional closed connected subgroup of G. Moreover, if p is a good prime for G, it follows from [20, Theorem 1.2] that there exists a 1-dimensional subgroup U containing u which has particularly nice properties, for example:

$$C_G(u) = C_G(U) = C_G(\text{Lie}(U)) \tag{1}$$

It is natural to ask: does there exist a canonically defined overgroup of u satisfying the above equalities, either in bad characteristic, or when u no longer has order p? The condition $C_G(U) = C_G(u)$ means that the subgroup U lies in $Z(C_G(u)) = C_G(C_G(u))$. The structure of abelian algebraic groups, and in particular abelian connected unipotent groups, shows that one should aim for a t-dimensional group if $o(u) = p^t$.

The first work in this direction was done by Richard Proud in [18], who showed:

Theorem 1.1. *[18, Main Theorem] Let G be a simple algebraic group defined over an algebraically closed field k and assume* $\text{char}(k) = p > 0$ *is good for G. Let* $u \in G$ *be of order* p^t, $t > 1$. *Then there exists a closed connected abelian t-dimensional unipotent subgroup* $W \leq G$ *with* $u \in W$.

This existence result does not point to any particularly canonical properties of the overgroup, whereas the group $C_G(C_G(u))$ is a canonically defined abelian overgroup. But is it unipotent? And what about the connected component, which is also a canonically defined group, associated to u - does it even contain u? These questions were addressed in [17, 21]. They showed (independently) the following result, where for an abelian group H we denote by H_u the subgroup of unipotent elements.

Proposition 1.2. *Let G and u be as above. Then* $Z(C_G(u)) = Z(G) \times Z(C_G(u))_u$. *Moreover, if p is good for G, then* $Z(C_G(u))_u = \big(Z(C_G(u))\big)^\circ$. *In particular, if p is good, the group* $\big(Z(C_G(u))\big)^\circ$ *is a (canonically defined) connected abelian unipotent overgroup of u.*

Continuing with the case of good characteristic, we establish here the following corollary of [13], showing that we have Equality (1) above when we replace U by $Z(C_G(u))^\circ$.

Proposition 1.3. *Let G be a simple algebraic group defined over an algebraically closed field of characteristic p. Let* $u \in G$ *be unipotent and set* $Z = Z(C_G(u))$. *If p is good for G then*

$$C_G(u) = C_G(Z^\circ) = C_G(\text{Lie}(Z^\circ)).$$

In what follows we discuss some recent work on describing $Z(C_G(u))$. The analysis is different depending on whether the characteristic is a bad prime for G or not. We consider as well the case where the field is of characteristic 0, as we obtain new results even in this setting.

2 Good Characteristic

In this section we consider the case where $\mathrm{char}(k)$ is good for G. For some statements we will need the notion of *very good prime*: p is *very good* for G if it is good and in addition does not divide $n+1$ when G is of type A_n.

A certain number of very powerful tools are available under our current assumption on $\mathrm{char}(k)$:

- Springer maps: Given G and u, we fix a Springer map, a G-equivariant homeomorphism $\varphi : \mathcal{U} \to \mathcal{N}$, between the variety of unipotent elements in G and the variety of nilpotent elements in $\mathrm{Lie}(G)$ (for a discussion of this see [12, Proposition 29]). Such a bijection exists as long as $\mathrm{char}(k)$ is good for G (and in fact is an isomorphism of varieties as long as the characteristic is very good for G). So we have $C_G(u) = C_G(\varphi(u))$, which allows us to reduce questions about centralizers of unipotent elements to the study of centralizers of nilpotent elements in $\mathrm{Lie}(G)$.
- Smoothness of centralizers. We can use the result of Slodowy [23, p. 38]: if $\mathrm{char}(k)$ is a very good prime for G, then $\mathrm{Lie}(C_G(x)) = C_{\mathrm{Lie}G}(x)$ where x is either a unipotent element of G or a nilpotent element in $\mathrm{Lie}(G)$.
- The Bala-Carter-Pommerening classification of unipotent classes in G and nilpotent orbits in $\mathrm{Lie}(G)$ (see [1, 2, 14, 15]).
- For each nilpotent element $e \in \mathrm{Lie}(G)$, there exists an associated cocharacter. The definition is given below and the existence is given by [14, Satz (3.1)] or [16, Theorem A].

We can now prove Proposition 1.3.

Proof. (of Proposition 1.3.) The first equality follows from Proposition 1.2. The containment

$$C_G(Z^\circ) \subset C_G(\mathrm{Lie}(Z^\circ))$$

is clear. For the reverse inclusion, we fix a Springer map φ as above and set $e = \varphi(u)$; so $C_G(e) = C_G(u)$. If G is of type A_n then it follows from the explicit description of Z° given in [9, Sect. 4.1] that $e \in \mathrm{Lie}(Z)$. If G is not of type A_n, then using [13, Theorem A], we again have $e \in \mathrm{Lie}(Z)$. In all cases $e \in \mathrm{Lie}(Z^\circ)$, so $C_G(\mathrm{Lie}(Z^\circ)) \subset C_G(e)$.

In order to state some results from [9], we will require the following two definitions.

Definition 2.1. *Let H be a connected reductive algebraic group. We say that a nilpotent element $e \in \mathrm{Lie}(H)$ is a* distinguished *nilpotent element in* $\mathrm{Lie}(H)$ *if $C_H(e)^\circ$ contains no noncentral semisimple elements or, equivalently, each torus of $C_H(e)$ lies in $Z(H)$.*

Note that taking S to be a maximal torus of $C_G(e)$, we have that e is distinguished in the Lie algebra of the reductive subgroup $C_G(S)$ (a Levi subgroup of G).

Definition 2.2. *Let $e \in \mathrm{Lie}(G)$ be nilpotent. A morphism $\tau : k^* \to G$ is said to be an* associated cocharacter for e *if*

i. $\tau(c)e = c^2 e$ *for all* $c \in k^*$, *and*
ii. $\mathrm{im}(\tau) \subseteq [L, L]$, *for some Levi subgroup of* G, *such that* e *is distinguished in* $\mathrm{Lie}(L)$.

One can show that any two cocharacters associated to e are conjugate by an element of $C_G(e)^\circ$ (see [5, Lemma 5.3]). Moreover, an associated cocharacter determines a unique *weighted Dynkin diagram* where each node is labelled with an integer $i_\alpha \in \{0, 1, 2\}$. (This is analogous to the usual Kostant-Dynkin theory in characteristic 0.) We will call this the *weighted Dynkin diagram of* e.

Assume for the moment that the characteristic is very good for G. Since $\mathrm{im}(\tau)$ normalizes $C_G(e)$ it acts on the subspace $\mathrm{Lie}(C_G(e))$ with a certain set of (integral) weights. The assumption on $\mathrm{char}(k)$ implies that $\mathrm{Lie}(C_G(e)) = C_{\mathrm{Lie}G}(e)$. We denote by $C_{\mathrm{Lie}G}(e)_+$ the subalgebra spanned by the set of $\mathrm{im}(\tau)$-weight spaces associated to strictly positive weights. We also have $C_G(e) = CR$, a semi-direct product of $R = R_u(C_G(e))$ and a reductive (not necessarily connected) group C. In fact, $C = C_G(e) \cap C_G(\mathrm{im}(\tau))$ (see [5, Proposition 5.10]) and $\mathrm{Lie}(R) = C_{\mathrm{Lie}G}(e)_+$. The following description of $\mathrm{Lie}(Z(C_G(e)))$ allows one to determine this object computationally.

Theorem 2.3. *[9, Theorem 3.9] Let* e, τ, *and* C *be as above and assume* $\mathrm{char}(k)$ *is very good for* G. *Then* $\mathrm{Lie}(Z(C_G(e))) = \left(Z(C_{\mathrm{Lie}G}(e)_+)\right)^C$, *that is, the fixed points of* C *acting on* $Z(C_{\mathrm{Lie}G}(e)_+)$.

In order to apply Theorem 2.3 one has to: find a basis for $Z(C_{\mathrm{Lie}G}(e)_+)$ (if G is classical, this can be deduced from the basis for $Z(C_{\mathrm{Lie}G}(e))$ given by Yakimova in [28, Sect. 2]), determine the fixed point space of the connected reductive group C° acting there, and find representatives for the component group C/C° and let them act as well. In the exceptional groups, lengthy case-by-case considerations are required for most results.

In order to state one of the main results we will need the following additional definition.

Definition 2.4. *We write* Δ_e *for the weighted Dynkin diagram of* e *and* $n_2(\Delta_e)$ *for the number of weights equal to* 2 *in* Δ_e. *We say that* e *is* even *if all weights of* τ *are even, so the weighted Dynkin diagram has all labels either 0 or 2. (For example, distinguished nilpotent elements are even.)*

Theorem 2.5. *[9, Theorem 2] Let* G *be as above and assume* $\mathrm{char}(k)$ *is good for* G. *Let* $e \in \mathrm{Lie}(G)$ *be an even nilpotent element. Then*

$$\dim Z(C_G(e)) = n_2(\Delta_e) = \dim Z(C_G(\mathrm{im}(\tau))).$$

In [9], the dimension of $Z(C_G(e))$ was determined for e in any nilpotent orbit. Indeed, [9, Theorem 4] gives a formula for $\dim Z(C_G(e))$ as a function of the weighted Dynkin diagram Δ_e. The formula for noneven elements is more technical and we refer the reader to the original article for the precise statement.

In the same article, the authors establish the following connection between the degrees of the invariant polynomials of the Weyl group for G and the weights of an associated cocharacter τ for e on $C_{\mathrm{Lie}G}(e)_+$. Again, the proof for the exceptional groups follows from lengthy case-by-case analysis.

Theorem 2.6. *[9, Theorem 1] Let $e \in \mathrm{Lie}(G)$ be a distinguished nilpotent element, with associated cocharacter τ. Let $d_1, \cdots, d_\ell$ be the degrees of the invariant polynomials of the Weyl group of G, ordered such that d_ℓ is ℓ if G is of type D_ℓ, and otherwise d_ℓ is $\max\{d_i\}$, and $d_i < d_j$ if $i < j < \ell$. Then the weights of $\mathrm{im}(\tau)$ on $\mathrm{Lie}(Z(C_G(e)))$ are the $n_2(\Delta_e)$ integers $2d_i - 2$ for $i \in S_\Delta$, where*

$$S_\Delta = \begin{cases} \{1, \ldots, n_2(\Delta_e) - 1, \ell\} & \text{if } G \text{ is of type } D_\ell \text{ and } \Delta = \cdots{}^2_2; \\ \{1, \ldots, n_2(\Delta_e) - 1, n_2(\Delta_e)\} & \text{otherwise.} \end{cases}$$

Here $\cdots{}^2_2$ stands for any weighted Dynkin diagram of type D_ℓ for which the last two nodes (in Bourbaki notation) have label 2.

Open Problems 1

1. *Find a proof in characteristic 0 (free of case-by-case analysis) that for an even nilpotent element e we have $\dim Z(C_G(e)) = n_2(\Delta_e)$. In what follows we give a proof for the inequality $\dim Z(C_G(e)) \geq n_2(\Delta_e)$.*
2. *Give a case-free proof of the more general formula for $\dim Z(C_G(e))$ in [9, Theorem 4].*
3. *Theorem 2.6 was deduced from the case-by-case considerations in [9]. If $k = \mathbb{C}$ and e is regular then this result was established in [6] (see Sect. 4 for the definition of regular). Give a case-free proof of Theorem 2.6, at least over fields of characteristic 0.*

2.1 A Proof of an Inequality in Characteristic 0

Here we assume that e is a non-zero *even* element and that k is the field of complex numbers $\mathbb{C}$. With this assumption, there exist h and f in $\mathrm{Lie}(G)$ such that $[h, e] = 2e$, $[h, f] = -2f$ and $[e, f] = h$ and in particular e, h, f span an $\mathfrak{sl}_2$-subalgebra of $\mathrm{Lie}(G)$ which we denote by $\mathfrak{a}$ (see [4]). Moreover, $C_{\mathrm{Lie}G}(e) = C_{\mathrm{Lie}G}(\mathfrak{a}) \oplus \mathfrak{r}$ where $\mathfrak{r}$ is nilpotent and $C_{\mathrm{Lie}G}(\mathfrak{a})$ is a reductive subalgebra and h spans the Lie algebra of $\mathrm{im}(\emptyset)$ where τ is an associated cocharacter for e. So $C_G(\mathrm{im}(\tau)) = C_G(h)$.

Lemma 2.7. *Let $e \in \mathrm{Lie}(G)$ be a non-zero even nilpotent element. Then $\dim C_G(e) = \dim C_{\mathrm{Lie}G}(e)$ is equal to the dimension of the 0 weight space for $\mathrm{ad}(h)$, and hence, for all $t \in \mathbb{C}^*$ we have*

$$\begin{aligned}\dim C_{\mathrm{Lie}G}(e) &= \dim C_{\mathrm{Lie}G}(h) = \dim C_G(h) = \dim C_G(th) = \dim C_{\mathrm{Lie}G}(th) \\ &= \dim C_{\mathrm{Lie}G}(th + e).\end{aligned}$$

Proof. Since we are assuming that e is an even element, in a decomposition of Lie(G) into a direct sum of irreducible $\mathfrak{a}$-submodules, each irreducible summand has an even highest weight and the zero weight occurs with multiplicity 1 in each irreducible summand. Moreover, $C_{\mathrm{Lie}G}(e)$ is precisely the set of fixed points for ad(e) acting on Lie(G), and there is a 1-dimensional subspace of such vectors in each irreducible summand. This establishes the first equality. The second, third and fourth equalities are clear. The last equality follows from the fact that th and $th+e$ are conjugate under the action of the closed connected (P)SL$_2$-subgroup $A \subseteq G$ with Lie(A) $= \mathfrak{a}$.

In characteristic 0 it is not difficult to prove the Lie algebra version of Proposition 1.2. In fact, we have $Z(C_{\mathrm{Lie}G}(e)) \subseteq \mathfrak{r}$. Using this and some basic facts about centralizers and centers, we can show:

Lemma 2.8. $\mathrm{Lie}(Z(C_G(e))) = \big(Z(C_{\mathrm{Lie}G}(e))\big)^C$ *where* $C = C_G(e) \cap C_G(\mathrm{im}(\tau))$.

For the proof of the following inequality we use an argument based upon ideas in [6, Theorem 5.7].

Proposition 2.9. *If* $e \in \mathrm{Lie}G$ *is a non-zero even element then* $\dim Z(C_G(e)) \geq n_2(\Delta_e) = \dim Z(C_G(h))$.

Proof. Set $f = \dim Z(C_{\mathrm{Lie}G}(th+e))$ and $d = \dim C_{\mathrm{Lie}G}(h) = \dim C_{\mathrm{Lie}G}(e) = \dim C_{\mathrm{Lie}G}(th+e)$, for all $t \in \mathbb{C}$ by Lemma 2.7. Choose a sequence $\{t_n\} \subset \mathbb{C}$ with $t_n h + e$ converging to e. We consider the Grassmannian of d-dimensional subspaces of Lie(G), which is a compact space. Hence, eventually after refining $\{t_n\}$, we may assume that the sequence of d-dimensional subalgebras $C_{\mathrm{Lie}G}(t_n h + e)$ converges to a d-dimensional subspace $\mathfrak{u}$ in the Grassmannian. Let $w_1, \ldots, w_d$ be a basis of $\mathfrak{u}$, and for each $1 \leq i \leq d$, choose $\{w_i^n \in C_{\mathrm{Lie}G}(t_n h+e)\}_{n\in\mathbb{N}}$ such that $\{w_i^n\}$ converges to w_i. Since $[w_i^n, t_n h + e] = 0$ for all n, we have $[w_i, e] = 0$ for all i. Hence $\mathfrak{u} \subseteq C_{\mathrm{Lie}G}(e)$. But, by our assumption that e is an even element and Lemma 2.7, we have that $d = \dim C_{\mathrm{Lie}G}(e)$ and so $\mathfrak{u} = C_{\mathrm{Lie}G}(e)$.

After again extracting a subsequence, we may assume that the sequence of f-dimensional subspaces $Z(C_{\mathrm{Lie}G}(t_n h+e))$ converges to an f-dimensional subspace $\mathfrak{z}$ of Lie(G). As before we have that $\mathfrak{z} \subseteq C_{\mathrm{Lie}G}(e)$ and we claim that $\mathfrak{z} \subseteq Z(C_{\mathrm{Lie}G}(e))^C$. Take a basis $\{z_1, \ldots, z_f\}$ of $\mathfrak{z}$ and for each j a sequence $\{z_j^n \in Z(C_{\mathrm{Lie}G}(t_n h+e))\}_{n\in\mathbb{N}}$ such that $\{z_j^n\}$ converges to z_j. In particular, $[z_j^n, w_i^n] = 0$ for all n, since $w_i^n \in C_{\mathrm{Lie}G}(t_{n_k} h + e)$. So $[z_j, w_i] = 0$ for all i, j which shows that $z_j \in Z(C_{\mathrm{Lie}G}(e))$. Moreover

$$Z(C_{\mathrm{Lie}G}(t_n h + e)) = \mathrm{Lie}(Z(C_G(t_n h + e))),$$

since $C_G(t_n h+e)$ is connected, and this latter is equal to $\mathrm{Lie}(C_G(t_n h+e))^{C_G(t_n h+e)}$. But, since

$$C = C_G(e) \cap C_G(h) \subseteq C_G(t_n h + e) \text{ for all } t_n,$$

we also have that C fixes all elements in $Z(C_{\mathrm{Lie}G}(t_n h+e))$ and so C fixes the z_j^n and hence fixes z_j for all j. This shows that the f-dimensional subspace $\mathfrak{z}$ indeed lies in

$Z(C_{\mathrm{Lie}G}(e))^C$. By Lemma 2.8, it follows that $\mathfrak{z} \subseteq \mathrm{Lie}(Z(C_G(e))$ and we conclude that $\dim Z(C_G(e)) \geq n_2(\Delta_e)$.

3 Bad Characteristic

In this section we focus on the case where G is of exceptional type and the characteristic of k is bad for G. The most recent reference on unipotent and nilpotent classes, with an extensive treatment of bad characteristic, is [10]. In this setting almost all main tools used in the analysis for good characteristic fail. The difficulties are:

- No Springer isomorphism. Liebeck and Seitz show that if the characteristic is not 2 then there is a bijective correspondence between nilpotent orbits and unipotent classes which comes close to a Springer map (see [10, Theorem 1]). They also show that in the case of characteristic 2 there is an injective map from unipotent to nilpotent orbits satisfying some useful properties.
- Centralizers are not smooth, that is, for $u \in G$ unipotent, we do not always have $\mathrm{Lie}(C_G(u)) = C_{\mathrm{Lie}G}(u)$. So now studying $C_G(u)$ and $Z(C_G(u))$ cannot be 'linearized'.
- For u unipotent, we do not necessarily have u in $C_G(u)^\circ$ and so $Z(C_G(u))^\circ$ will not work as a canonically defined connected abelian overgroup of u. Springer showed in [25] that for $u \in G$ regular and $\mathrm{char}(k)$ a bad prime for G, then $u \notin C_G(u)^\circ$. All classes u for which $u \notin C_G(u)^\circ$ were determined in [10, Corollary 4].
- Associated cocharacters still exist but several of the useful properties of their good-characteristic counterparts are lost. For more details see [10, Theorem 9.1].

In order to complete the analysis of $Z(C_G(u))$ by extending it to bad characteristic, the following description of $Z(C_G(u))^\circ$ can be used to algorithmically determine this group.

Theorem 3.1. *[22, Theorem A] Let B be a Borel subgroup of a simple algebraic group G defined over an algebraically closed field and set $U = R_u(B)$. Let $u \in G$ be a unipotent element and suppose that B contains a Borel subgroup of $C_G(u)$. Then*

$$\left(Z(C_G(u))\right)^\circ = C_{Z(C_U(u)^\circ)^\circ}(T_u, \tilde{A})^\circ$$

where T_u is a maximal torus of $C_B(u)$ and $\tilde{A}$ is a set of coset representatives for $C_G(u)^\circ$ in $C_G(u)$.

A consequence of this is

Corollary 3.2. *[22, Corollary 2.8] With notation as in Theorem 3.1 and under the assumption that the characteristic of the field is 0 we have* $\mathrm{Lie}Z(C_G(u)) = Z(C_{\mathrm{Lie}G}(u))^{\tilde{A}}$.

Using a Springer map one has a similar statement for nilpotent elements which gives a proof of Lemma 2.8. This corollary shows that the component group of the centralizer plays an important role in determining the double centralizer and can be viewed as a justification for the difficulty of showing dim $Z(C_G(e)) \leq n_2(\Delta_e)$ in case e is even, other than through a case-by-case analysis.

In order to apply Theorem 3.1, one first needs to find a Borel subgroup which contains a Borel subgroup of the centralizer. Once this is determined one can computationally obtain $Z(C_U(u)^\circ)^\circ$. This was carried out case by case in [22] for each class of unipotent elements in the exceptional algebraic groups defined over fields of bad characteristic:

Theorem 3.3. *[22, Theorem D] Suppose that G is of exceptional type and that the characteristic of k is bad for G. Let $u \in G$ be a unipotent element. Then* dim $Z(C_G(u))$ *is explicitly determined; the tables are given in [22, Sect. 9].*

Moreover, the analysis determines if u lies in $Z(C_G(u))^\circ$, and when u lies in $Z(C_G(u)^\circ)^\circ$. In particular, we find that u does not necessarily lie in $Z(C_G(u))^\circ$, even if u does lie in $C_G(u)^\circ$. Note that in all cases, we have

$$Z(C_G(u))^\circ \subset Z(C_G(u)^\circ)^\circ \subset Z(C_G(u)^\circ) \subset C_G(u)^\circ \subset C_G(u).$$

Clearly, when $u \notin C_G(u)^\circ$, we have $u \notin Z(C_G(u))^\circ$. But in fact, there exist u with $u \in C_G(u)^\circ$, but $u \notin Z(C_G(u))^\circ$. There are examples which show that each of the above inclusions may be proper:

Corollary 3.4. *Let G be an exceptional algebraic group and suppose that the characteristic is 2. Except for the classes A_2 and A_2A_1 in E_6, an element u lies in $Z(C_G(u))^\circ$ if and only if it has order* 2.

Open Problems 2 *The problem of determining* dim $Z(C_G(u))$, *for u a unipotent element in an orthogonal or symplectic group over a field of characteristic* 2, *is still open.*

4 Related Questions

The results mentioned so far can be used to address related questions. The questions considered here revolve around several characterizations of regular unipotent elements.

Definition 4.1. *We call an element x of G* regular *if* dim $C_G(x)$ *is minimal.*

If u is unipotent, then there are various characterizations of the condition that dim $C_G(u)$ is minimal. More precisely, if the characteristic is good for G then we have

$$
\begin{aligned}
u \text{ regular unipotent} &\Leftrightarrow \dim C_G(u) = \operatorname{rank}(G) && (2)\\
&\Leftrightarrow G.u \text{ dense in the variety of unipotent elements of } G && (3)\\
&\Leftrightarrow C_G(u) \text{ abelian} && (4)\\
&\Leftrightarrow C_G(u)^\circ \text{ abelian} && (5)\\
&\Leftrightarrow \dim Z(C_G(u)) = \operatorname{rank}(G) && (6)\\
&\Leftrightarrow \dim Z(C_G(u)^\circ) = \operatorname{rank}(G) && (7)
\end{aligned}
$$

whereas in bad characteristic only the first three equivalences remain true.

For the first two equivalences we refer the reader to [26, Chapter III]. In [24], Springer proved that for a regular unipotent element u, $C_G(u)^\circ$ is abelian. Then Lou showed in [11] that for regular unipotent elements the full centralizer $C_G(u)$ is abelian. In [7] Kurtzke showed that in good characteristic a unipotent element $u \in G$ is regular if and only if $C_G(u)^\circ$ is abelian. Lawther extended these results to cover bad characteristics, so the equivalence in (4) is true:

Theorem 4.2. *[8, Theorem 2] Let u be a unipotent element in G. Then u is regular if and only if $C_G(u)$ is abelian.*

The result of Kurtzke does not generalize however. In (5) we only have '$\Rightarrow$' in general. Our analysis gives a second proof for exceptional groups in bad characteristic of the following result.

Theorem 4.3. *[8, Theorem 1] For $u \in G$ unipotent, with $C_G(u)^\circ$ abelian, then either u is regular, or $u \in G = G_2$, $p = 3$ and u lies in the class of subregular elements.*

The implications '$\Rightarrow$' in (6) and (7) follow from (4), (5) and (2). That the reversed implications do not hold in general can be deduced from Theorem 3.3.

Corollary 4.4. *Let G be of exceptional type and suppose that* $\operatorname{char}(k)$ *is bad for G. If $u \in G$ is unipotent then*

i. $\dim Z(C_G(u)^\circ) \leq \operatorname{rank}(G)$ *unless $p = 2$, G is of type F_4 and u is in the $F_4(a_3)$ class; in this case* $\dim Z(C_G(u)^\circ) = 6$.
ii. $\dim Z(C_G(u)) = \operatorname{rank}(G)$ *if and only if u is in the regular class or (G, p, class of u) is one of the triples $(G_2, 3, \tilde{A}_1^{(3)})$, $(F_4, 2, F_4(a_1))$ and $(F_4, 2, C_3(a_1)^{(2)})$; in these cases* $\dim Z(C_G(u)) = \dim Z(C_G(u)^\circ)$.

In fact, it is not known whether the inequality $\dim Z(C_G(u)) \leq \operatorname{rank}(G)$ holds in general. From [9, Theorem 4] and Theorem 3.3 we deduce the following

Corollary 4.5. *Assume $p \neq 2$ if G is of type B_n, C_n or D_n. Then* $\dim Z(C_G(u)) \leq \operatorname{rank}(G)$.

Open Problems 3 *1. Give a case-free proof of the inequality* $\dim Z(C_G(u)) \leq \operatorname{rank}(G)$, *in characteristic 0, or even in good positive characteristic.*

2. *Determine whether* $\dim Z(C_G(u)) \leq \text{rank}(G)$ *in the cases which are excluded in Corollary 4.5.*

We conclude by showing that if the characteristic is good for G then the implications '$\Leftarrow$' in (6) and (7) hold. The implication '$\Leftarrow$' in (6) follows from [9, Theorem 4]. We therefore turn our attention to $Z(C_G(u)^\circ)$.

For exceptional groups, if the characteristic is good, the dimension of $Z(C_G(u)^\circ)$ can be deduced from the tables in [9], where a basis for $\mathcal{Z}^\natural = Z(C_{\text{Lie}G}(e)_+)^{C^\circ}$ is given. We need the following lemma.

Lemma 4.6. *Assume* $\text{char}(k)$ *is a very good prime for* G*. Then*

$$\text{Lie}(Z(C_G(e)^\circ)) = Z(C_{\text{Lie}G}(e)_+)^{C^\circ}.$$

Proof. By [22, Lemma 2.6], $Z(C_G(u)^\circ) = Z(C_G(u)^\circ)_u \times (Z(G) \cap C_G(u)^\circ)$, for all unipotent elements $u \in G$; so via a Springer map, the analogous statement is true for all nilpotent elements $e \in \text{Lie}G$. In particular, setting $C_G(e) = CR$, where $R = R_u(C_G(e))$ as in Sect. 2, we have $Z(C_G(e)^\circ)^\circ \subseteq R$. So

$$\begin{aligned} \text{Lie}Z(C_G(e)^\circ) &= \text{Lie}(C_G(e)^\circ)^{C^\circ R} \cap \text{Lie}R \\ &= (\text{Lie}C \oplus \text{Lie}R)^{C^\circ R} \cap \text{Lie}R \\ &= (\text{Lie}R)^{RC^\circ} \\ &= (\text{Lie}Z(R))^{C^\circ}. \end{aligned}$$

Now we argue as in the proofs of Propositions [9, 3.7, 3.8, 3.9], to see that $Z(\text{Lie}R) = \text{Lie}(Z(R))$. So finally we have $\text{Lie}(Z(C_G(e)^\circ)) = Z(\text{Lie}R)^{C^\circ} = Z(C_{\text{Lie}G}(e)_+)^{C^\circ} = \mathcal{Z}^\natural$.

In particular, we now deduce from [9, Sect. 11] that for the exceptional groups defined over fields of good characteristic p, we have $\dim Z(C_G(u)^\circ) = \text{rank}(G)$ if and only if u is regular.

We finish by establishing this result for the classical groups.

Lemma 4.7. *Let* G *be a classical group defined over a field of characteristic* 0 *or of good characteristic* p*. Let* $u \in G$ *be unipotent. Then* $\dim Z(C_G(u)^\circ) = \text{rank}(G)$ *if and only if* u *is regular.*

Proof. Using a Springer map, we will consider $e \in \text{Lie}G$ nilpotent. If e is regular nilpotent, then by Theorem 4.2, $C_G(e)$ is abelian and so we have the result.

Now take $e \in \text{Lie}G$ a non-regular nilpotent element.

If G is of type A_n, then $C_G(e)$ is connected and so $Z(C_G(e)^\circ) = Z(C_G(e))$ and Theorem [9, Theorem 4] shows that the latter is of dimension strictly less than $\text{rank}(G)$.

For G of type B_n, C_n and D_n, set $Y = \langle e, e^3, \dots \rangle$, the subspace of Lie$G$ spanned by the odd powers of e. By [28, Theorem 2.3], $Z(C_{\mathrm{Lie}G}(e)) = Y$, unless G is of type B_n or D_n and e has at least 3 Jordan blocks on the natural module for G, 2 of which are blocks of odd sizes strictly bigger than 1. In the exceptional cases, $Z(C_{\mathrm{Lie}G}(e)) = Y \oplus \langle x \rangle$ for some $0 \neq x \in \mathrm{Lie}G$. Using the known Jordan block structure of nilpotent elements in Lie$G$ acting on the natural module for G (for example as given in [3, Chapter 13]), we deduce that dim $Y < \mathrm{rank}(G)$ for all non-regular nilpotent elements e, and dim $Y + 1 < \mathrm{rank}(G)$, unless G has type B_n and e has a Jordan block of size $2n - 1$ on the natural kG-module. In the latter case, e has exactly three blocks of sizes $2n-1$, 1 and 1, and so [28, Theorem 2.3] implies that $Z(C_{\mathrm{Lie}G}(e)) = Y$. Since $\mathrm{Lie}(Z(C_G(e)^\circ)) \subset Z(\mathrm{Lie}(C_G(e)^\circ)) = Z(\mathrm{Lie}C_G(e)) = Z(C_{\mathrm{Lie}G}(e))$, we have that dim $Z(C_G(e)^\circ) < \mathrm{rank}(G)$.

References

1. Bala, P., Carter, R.W.: Classes of unipotent elements in simple algebraic groups I. Math. Proc. Cambridge Philos. Soc. **79**(3), 401–425 (1976)
2. Bala, P., Carter, R.W.: Classes of unipotent elements in simple algebraic groups II. Math. Proc. Cambridge Philos. Soc. **80**(1), 1–18 (1976)
3. Carter, R.W.: Finite groups of lie type: conjugacy classes and complex characters. In: Pure and Applied Mathematics (New York). Wiley, New York (1985)
4. Jacobson, N.: A note on three dimensional simple Lie algebras. J. Math. Mech. **7**, 823–831 (1958)
5. Jantzen, J.C.: Nilpotent orbits in representation theory. In: Lie Theory. Progr. Math. vol. 228, pp. 1–211. Birkhäuser Boston, Boston, MA (2004)
6. Kostant, B.: The principal three-dimensional subgroup and the Betti numbers of a complex simple Lie group. Amer. J. Math. **81**, 973–1032 (1959)
7. Kurtzke Jr, J.F.: Centralizers of irregular elements in reductive algebraic groups. Pacific J. Math. **104**(1), 133–154 (1983)
8. Lawther, R.: Elements in reductive algebraic groups with abelian connected centralizers. J. Algebra **359**, 1–34 (2012)
9. Lawther, R., Testerman, D.M.: Centres of centralizers of unipotent elements in simple algebraic groups. Mem. Amer. Math. Soc. **210**(988), vi+188 (2011)
10. Liebeck, M.W., Seitz, G.M.: Unipotent and nilpotent classes in simple algebraic groups and lie algebras. In: Mathematical Surveys and Monographs. Amer. Math. Soc., vol. 180 (2012)
11. Lou, B.: The centralizer of a regular unipotent element in a semisimple algebraic group. Bull. Amer. Math. Soc **74**, 1144–1146 (1968)
12. McNinch, G.J.: Sub-principal homomorphisms in positive characteristic. Math. Z. **244**(2), 433–455 (2003)
13. McNinch, G.J., Testerman, D.M.: Nilpotent centralizers and Springer isomorphisms. J. Pure Appl. Algebra **213**(7), 1346–1363 (2009)
14. Pommerening, K.: Über die unipotenten Klassen reduktiver Gruppen. J. Algebra **49**(2), 525–536 (1977)
15. Pommerening, K.: Über die unipotenten Klassen reduktiver Gruppen. II. J. Algebra **65**(2), 373–398 (1980)
16. Premet, A.: Nilpotent orbits in good characteristic and the Kempf-Rousseau theory (Special issue celebrating the 80th birthday of Robert Steinberg). J. Algebra **260**(1), 338–366 (2003)
17. Proud, R.: On centralizers of unipotent elements in algebraic groups. Unpublished manuscript

18. Proud, R.: Witt groups and unipotent elements in algebraic groups. Proc. London Math. Soc. 82(3), 647–675 (2001)
19. Proud, R., Saxl, J., Testerman, D.M.: Subgroups of type A_1 containing a fixed unipotent element in an algebraic group. J. Algebra **231**(1), 53–66 (2000)
20. Seitz, G.M.: Unipotent elements, tilting modules, and saturation. Invent. Math. **141**(3), 467–502 (2000)
21. Seitz, G.M.: Unipotent centralizers in algebraic groups. J. Algebra **279**(1), 226–259 (2004)
22. Simion, I.I.: Centers of Centralizers of Unipotent Elements in Exceptional Algebraic Groups. PhD thesis, EPF Lausanne (2013)
23. Slodowy, P.: Simple singularities and simple algebraic groups. Lecture Notes in Mathematics, vol. 815. Springer, Berlin (1980)
24. Springer, T.A.: A note on centralizers in semi-simple groups. Nederl. Akad. Wetensch. Proc. Ser. A 69=Indag. Math. 28, 75–77 (1966)
25. Springer, T.A.: Some arithmetical results on semi-simple Lie algebras. Inst. Hautes Études Sci. Publ. Math. **30**, 115–141 (1966)
26. Springer, T.A., Steinberg, R.: Conjugacy classes. In: Seminar on Algebraic Groups and Related Finite Groups (The Institute for Advanced Study, Princeton, N.J., 1968/69), Lecture Notes in Mathematics, vol. 131, pp. 167–266. Springer, Berlin (1970)
27. Testerman, D.M.: A_1-type overgroups of elements of order p in semisimple algebraic groups and the associated finite groups. J. Algebra **177**(1), 34–76 (1995)
28. Yakimova, O.: Surprising properties of centralisers in classical Lie algebras. Ann. Inst. Fourier (Grenoble) **59**(3), 903–935 (2009)

Chapter 9
The Axes of a Majorana Representation of A_{12}

A. Castillo-Ramirez and A. A. Ivanov

Abstract This paper contributes to an axiomatic approach (which goes under the name of *Majorana theory*) to the largest sporadic simple group $\mathbf{M}$, known as the Monster, and to its 196, 884-dimensional algebra $V_{\mathbf{M}}$. We study the axiomatic version V of the subalgebra V_A of $V_{\mathbf{M}}$ generated by the Majorana axes corresponding to the $2A$-involutions of an alternating subgroup A_{12} of $\mathbf{M}$ whose centraliser is isomorphic to A_5. Although the dimension of $V_A^{(2A)}$, the linear span of the generating Majorana axes of V_A, has been known for a while to be 3498, the dimension of V_A itself remains unknown. For each $3 \leq N \leq 5$, let $V^{(NA)}$ be the linear span of the NA-axes of V. In this paper, we examine the space

$$V^{\circ} = \left\langle V^{(NA)} : 2 \leq N \leq 5 \right\rangle.$$

We prove that $V^{(2A)}$ contains the $3A$-axes corresponding to 3-cycles in A_{12}, but it does not contain any of those corresponding to products of two 3-cycles. By considering a 21-dimensional space determined by a $2B$-involution, we also show that no $4A$-axis of V belongs to $V^{(2A)}$. An argument due to Á. Seress enables us to show further that any $5A$-axis of V is linearly expressible in terms of Majorana axes and $3A$-axes of V. When $V = V_A$, our results, enhanced by information about the characters of $\mathbf{M}$, allow us to deduce that

---

A Castillo-Ramirez—Funded by the *Universidad de Guadalajara* and an Imperial College International Scholarship.

A. Castillo-Ramirez (✉) · A. A. Ivanov
Department of Mathematics, Imperial College London, 180 South Kensington Campus, SW7 2AZ London, UK
e-mail: alonso.castillo-ramirez09@imperial.ac.uk

A. A. Ivanov
e-mail: a.ivanov@imperial.ac.uk

N. S. N. Sastry (ed.), *Groups of Exceptional Type, Coxeter Groups and Related Geometries*, Springer Proceedings in Mathematics & Statistics 82,
DOI: 10.1007/978-81-322-1814-2_9, © Springer India 2014

$$\left\langle V_A^{(2A)}, V_A^{(3A)} \right\rangle = \left\langle V_A^{(2A)}, V_A^{(4A)} \right\rangle = V_A^{\circ},$$

and that V_A° is the the direct sum of $V_A^{(2A)}$ and a 462-dimensional irreducible $\mathbb{R}A$-module; hence, we conclude that $\dim(V_A^{\circ}) = 3960$. The whole algebra V_A may still contain V_A° properly, but its codimension is bounded by 1191.

Keywords Majorana representation · Monster group · Conway-Griess-Norton algebra · Alternating group of degree 12

1 Introduction

The definition of a *Majorana representation* appeared for the first time in [4] and was refined in [5]. Explicitly, it is a tuple

$$(G, T, V, \cdot, (,), \varphi, \psi) \tag{1}$$

where G is a finite group, T is a G-stable set of generating involutions of G, $(V, \cdot, (,))$ is a real commutative (not necessarily associative) algebra with inner product, $\varphi : G \to GL(V)$ is a linear representation on V that preserves both products, and $\psi : T \to V \setminus \{0\}$ is an injective mapping such that $\psi(t^g) = \psi(t)^{\varphi(g)}$. Furthermore, the tuple satisfies the axioms **M1-M7** stated in [5]. The elements of $\psi(T)$, which are idempotents of length 1, are called *Majorana axes*, while the automorphisms in $\varphi(T)$ are called *Majorana involutions*. It is standard notation to write a_t for the Majorana axis corresponding to $t \in T$.

The fundamental example of a Majorana representation is when G is isomorphic to $\mathbf{M}$, the sporadic simple Monster group, T is the conjugacy class of $2A$-involutions of $\mathbf{M}$, and V is the 196, 884-dimensional Conway-Griess-Norton algebra $V_{\mathbf{M}}$. In this situation, the Majorana involutions coincide with the restrictions of the *Miyamoto involutions* of the Moonshine module defined in [9].

In general, we say that the Majorana representation (1) is *based on an embedding in the Monster* if G is a subgroup of $\mathbf{M}$ generated by $2A$-involutions and V is isomorphic to the subalgebra of $V_{\mathbf{M}}$ generated by the Majorana axes $\psi(T)$.

The definition of a Majorana representation was inspired by Sakuma's Theorem [13], originally established in the context of Vertex Operator Algebras, which states that the product of any two Majorana involutions has order at most six, and that there are at most eight possibilities for the isomorphism type of an algebra generated by two distinct Majorana axes. It was shown in [5] that axioms **M1-M7** are sufficiently strong to prove Sakuma's theorem.

When the representation is based on an embedding in the Monster, the algebras generated by pairs of Majorana axes had been described by Simon Norton in [11]. In this context, it turns out that the isomorphism type of the algebra generated by a_t and a_g, $t, g \in T$, $t \neq g$, may be labeled by

$$2A,\ 2B,\ 3A,\ 3C,\ 4A,\ 4B,\ 5A \text{ and } 6A,$$

since it is uniquely determined by the conjugacy class of tg in $\mathbf{M}$.

In view of Sakuma's theorem and Norton's description, we axiomatise that the algebra generated by any two Majorana axes a_t and a_g in an abstract Majorana representation is isomorphic to one of the two-generated subalgebras in $V_{\mathbf{M}}$ of type NX, where $N = |\varphi(tg)|$ and $X \in \{A, B, C\}$. In this context, these algebras are called the *Norton-Sakuma algebras*. Note that, when $N \leq 4$, there is a choice for the type of the Norton-Sakuma algebra generated by a_t and a_g. For this reason, we define the *shape of a Majorana representation* as the rule which specifies the type of the Norton-Sakuma algebra generated by every pair of Majorana axes; this rule must be stable under conjugation and must respect the inclusions of the algebras:

$$2A \hookrightarrow 4B,\ 2B \hookrightarrow 4A,\ 2A \hookrightarrow 6A,\ 3A \hookrightarrow 6A.$$

Suppose that a_t and a_g generate a Norton-Sakuma algebra of type NA and let $\rho = \varphi(tg)$. If N is 3, 4 or 5, the basis of the algebra contains a vector u_ρ, v_ρ or $\pm w_\rho$ defined as a linear combination of products of Majorana axes, which is called a $3A$-, $4A$-, or $5A$-axis, respectively. It follows from their definition that these axes satisfy the relations

$$u_\rho = u_{\rho^{-1}},\ v_\rho = v_{\rho^{-1}} \text{ and } w_\rho = -w_{\rho^2} = -w_{\rho^3} = w_{\rho^4}.$$

In particular, the NA-axes of $V_{\mathbf{M}}$ are in invariant correspondence with the cyclic groups $\langle\rho\rangle \leq \mathbf{M}$ of order N, where $3 \leq N \leq 5$. We axiomatise further this property for an abstract Majorana representation (cf. **M8** in Sect. 2).

Majorana representations of many groups have already been studied. It is implied by the axioms that the Majorana representations of the dihedral group D_{2N}, with $2 \leq N \leq 6$, coincide with the Norton-Sakuma algebras of types NX, $X \in \{A, B, C\}$. In [5], it was shown that the symmetric group S_4 has exactly four Majorana representations, all of them based on embeddings on the Monster. After this, Majorana representations of the projective linear groups $L_3(2)$ and $L_2(11)$, and of the alternating groups A_5, A_6 and A_7, have been discussed in [2, 6, 8, 9], respectively. Several groups with 2-closed Majorana representations were studied [16] using computer calculations. Although most of the representations obtained before are based on embeddings in the Monster, there are some counterexamples, such as the representation of A_6 analyzed in [7].

In the present paper, we study a Majorana representation of $A \cong A_{12}$, the largest of the alternating groups involved in $\mathbf{M}$. We define the shape of our representation according to the embedding of A_{12} into the Monster as the centraliser of an A_5-subgroup. Hence, in particular, T is the set of involutions of A with cycle types 2^2 and 2^6, and so V is the algebra generated by the $11,880$ Majorana axes corresponding to these involutions. This is the Majorana representation involving the highest number of Majorana axes studied so far, besides, of course, the case of the Monster itself.

It is clear that, for each $3 \leq N \leq 5$, the algebra V must contain the linear span $V^{(NA)}$ of the NA-axes contained in all the Norton-Sakuma algebras generated by pairs of Majorana axes. Therefore, it is important to study the space

$$V^{\circ} := \left\langle V^{(NA)} : 2 \leq N \leq 5 \right\rangle,$$

where $V^{(2A)}$ is the linear span of the Majorana axes.

In the following sections, we study each one of the quotient spaces

$$Q^{(NA)} := \left\langle V^{(2A)}, V^{(NA)} \right\rangle / V^{(2A)},$$

for $3 \leq N \leq 5$. We start in Sect. 2 by introducing some notation and results that will be required in the rest of the paper. In Sect. 3, we use the inner product structure of V together with *Pasechnik's relation* [6, p. 11] to show that $Q^{(3A)}$ is non-trivial and that $\dim(Q^{(3A)}) \leq 9{,}240$. Section 4 is devoted to prove that, for each involution z of A not in T, there exists an alternating sum of $4A$-axes, preserved by $C_A(z)$ up to sign, that is not linearly expressible in terms of Majorana axes. We also show that the dimension of the quotient space $Q^{(4A)}$ has bounds:

$$1 \leq \dim\left(Q^{(4A)}\right) \leq [A : C_A(z)] = 51{,}975.$$

After this, in Sect. 4, we use the argument written by Á. Seress in [14] to deduce that every $5A$-axis of V is linearly expressible in terms of Majorana axes and $3A$-axes. Finally, in Sect. 5, we refine all our results by considering that $V = V_A$ is the subalgebra of $V_{\mathbf{M}}$ generated by the Majorana axes $\psi(T)$; in this situation, we show that

$$Q_A^{(3A)} = Q_A^{(4A)} \text{ and } \dim(Q_A^{(3A)}) = \dim(Q_A^{(4A)}) = 462.$$

Therefore, Dima Pasechnik's calculation of $\dim(V_A^{(2A)})$ enable us to conclude that the dimension of V_A° is 3,960.

2 A Majorana Representation of A_{12}

First of all, we introduce some notation. For any group G, let $G^{(n)}$ be the set of elements of G of order $n \in \mathbb{N}$. For any subset Ω of an algebra V, denote by $\langle\langle \Omega \rangle\rangle$ the subalgebra of V generated by Ω, and by $\langle \Omega \rangle$ the linear subspace of V spanned by Ω. Whenever G is finite and V is an $\mathbb{R}G$-module, define, for any $v \in V$, the sum

$$S_v^G := \frac{1}{|G_v|} \sum_{g \in G} v^g,$$

where G_v is the stabiliser in G of v.

For the rest of the paper, we assume there exists a Majorana representation

$$\mathbf{R} := (A, T, V, \cdot, (,), \varphi, \psi),$$

where $A \cong A_{12}$ and T is the union of the conjugacy classes of involutions with cycle types 2^2 and 2^6. Because of the action of A on T, φ is a faithful linear representation, so we may consider A as a group of automorphisms of V.

The following are the axioms introduced in [5] that **R** must satisfy:

M1 The inner product associates in $(V, \cdot)$ in the sense that

$$(u, v \cdot w) = (u \cdot v, w) \text{ for all } u, v, w \in V.$$

M2 The *Norton inequality*,

$$(u \cdot u, v \cdot v) \geq (u \cdot v, u \cdot v),$$

holds for every $u, v \in V$.

M3 The elements of $\psi(T)$ are idempotents of length 1.

M4 For any $a \in \psi(T)$, the adjoint transformation of a is diagonalisable with spectrum contained in Sp $:= \{0, 1, \frac{1}{2^2}, \frac{1}{2^5}\}$.

M5 For any $a \in \psi(T)$, 1 is a simple eigenvalue of a.

M6 For any $a \in \psi(T)$, the endomorphism $\tau(a)$ of V defined by

$$u^{\tau(a)} := (-1)^{2^5\mu}v, \text{ with } \mu \in \text{Sp and } v \in V_{\mu}^{(a)},$$

preserves the algebra product of V. Furthermore, $\tau(\psi(t)) = \varphi(t)$ for every $t \in T$.

M7 For any $a \in \psi(T)$, the endomorphism $\sigma(a)$ of $C_V(\tau(a))$ defined by

$$u^{\sigma(a)} := (-1)^{2^2\mu}v, \text{ with } \mu \in \text{Sp} \setminus \{\frac{1}{2^5}\} \text{ and } v \in V_{\mu}^{(a)},$$

preserves the algebra product of $C_V(\tau(a))$.

In this paper, we only use axioms **M1** and **M3-M5**.

In the following sections, we will consider various subalgebras generated by Majorana axes of **R**. As we mentioned in the introduction, every subalgebra $\langle\langle a_t, a_g\rangle\rangle$, $t, g \in T$, $t \neq g$, is isomorphic to a Norton-Sakuma algebra of type NX, where $N = |\rho|$, $\rho := \varphi(tg) = tg$. For $i \in \mathbb{Z}$, let $g_i := t\rho^i$. In Table 1, we describe the basis and inner product of each Norton-Sakuma algebra, using the same scaling as Table 3 in [5].

The missing inner products between the basis vectors of Table 1 may be reconstructed by applying the symmetries of the algebras and their mutual inclusions. We

Table 1 Bases and inner products of the Norton-Sakuma algebras

Type	Basis	Inner products
$2A$	a_t, a_g, a_ρ	$(a_t, a_g) = (a_t, a_\rho) = (a_g, a_\rho) = \frac{1}{2^3}$
$2B$	a_t, a_g	$(a_t, a_g) = 0$
$3A$	$a_t, a_g, a_{g^{-1}}, u_\rho$	$(a_t, a_g) = \frac{13}{2^8}, (a_t, u_\rho) = \frac{1}{2^2}, (u_\rho, u_\rho) = \frac{2^3}{5}$
$3C$	$a_t, a_g, a_{g^{-1}}$	$(a_t, a_g) = \frac{1}{2^6}$
$4A$	$a_t, a_g, a_{g^{-1}}, a_{g^2}, v_\rho$	$(a_t, a_g) = \frac{1}{2^5}, (a_t, a_{g^2}) = 0, (a_t, v_\rho) = \frac{3}{2^3}, (v_\rho, v_\rho) = 2$
$4B$	$a_t, a_g, a_{g^{-1}}, a_{g^2}, a_{\rho^2}$	$(a_t, a_g) = \frac{1}{2^6}, (a_t, a_{g^2}) = (a_t, a_{\rho^2}) = \frac{1}{2^3}$
$5A$	$a_t, a_g, a_{g^{-1}}, a_{g^2}, a_{g^{-2}}, w_\rho$	$(a_t, a_g) = \frac{3}{2^7}, (a_t, w_\rho) = 0, (w_\rho, w_\rho) = \frac{5^3 7}{2^{19}}$
$6A$	$a_t, a_g, a_{g^{-1}}, a_{g^2}, a_{g^{-2}}, a_{g^3}, a_{\rho^3}, u_{\rho^2}$	$(a_t, a_g) = \frac{5}{2^8}, (a_t, a_{g^2}) = \frac{13}{2^8}, (a_t, a_{g^3}) = \frac{1}{2^3}, (a_{\rho^3}, u_{\rho^2}) = 0$

do not describe the algebra products of the Norton-Sakuma algebras as they are not explicitly required for the arguments used in this paper.

We assume that the Norton-Sakuma subalgebras of V satisfy the following axioms:

M8 Let $t_i \in T$ and $a_i := \psi(t_i)$ for $1 \leq i \leq 4$.

(a) If $t_1t_2t_3 = 1$ then the Norton-Sakuma algebra $\langle\langle a_1, a_2 \rangle\rangle$ has type $2A$ and $a_3 = a_{\varphi(t_1t_2)}$.
(b) If $\langle\langle a_1, a_2 \rangle\rangle$ has type $2A$, $4B$ or $6A$, then t_1t_2, $(t_1t_2)^2$ or $(t_1t_2)^3$ belongs to T, and $\psi(t_1t_2)$, $\psi((t_1t_2)^2)$ or $\psi((t_1t_2)^3)$ coincides with $a_{\varphi(t_1t_2)}$, $a_{\varphi(t_1t_2)^2}$ or $a_{\varphi(t_1t_2)^3}$, respectively.
(c) Suppose that $t_1t_2 = t_3t_4$, and that $\langle\langle a_1, a_2 \rangle\rangle$ and $\langle\langle a_2, a_3 \rangle\rangle$ have both type $3A$, $4A$, or $5A$. Then $u_{\varphi(t_1t_2)} = u_{\varphi(t_3t_4)}$, $v_{\varphi(t_1t_2)} = v_{\varphi(t_3t_4)}$, or $w_{\varphi(t_1t_2)} = w_{\varphi(t_3t_4)}$, respectively.

In order to define the shape of **R**, and to motivate the choice of T, we use the following result, which is a consequence of Lemma 6 in [10].

Proposition 2.1. *Denote the conjugacy classes of* **M** *as in* [1]. *Suppose that $A \cong A_{12}$ is the centraliser of an A_5-subgroup of* **M** *of type* $(2A, 3A, 5A)$. *Then,*

(i) *The set $A \cap 2A$ contains precisely the involutions of A with cycle types 2^2 and 2^6.*
(ii) *The set $A \cap 3A$ contains the elements of A with cycle types 3, 3^2 and 3^4.*
(iii) *The sets $(A \cap 4A)^2$ and $(A \cap 4B)^2$ are equal to $A^{(2)} \setminus 2A$ and $A \cap 2A$, respectively.*

For $2 \leq N \leq 4$, define subsets $A \cap NA$ of A as in Lemma 2.1. It is clear, with this notation, that $T = A \cap 2A$ and that $A^{(2)} \setminus T$ coincides with set of involutions of A with cycle type 2^4.

The shape of **R** is defined as follows: let $\langle\langle a_t, a_g \rangle\rangle$, $t, g \in T$, be of type NA if $tg \in A \cap NA$, where $2 \leq N \leq 4$. Define the spaces

$$V^{(2A)} := \langle a_t : t \in T \rangle, \qquad V^{(3A)} := \langle u_\rho : \rho \in A \cap 3A \rangle,$$
$$V^{(4A)} := \langle v_\rho : \rho \in A \cap 4A \rangle, \quad V^{(5A)} := \langle w_\rho : \rho \in A^{(5)} \rangle.$$

We say that an NA-axis has *type* $N_1^{k_1} N_2^{k_2} \ldots N_r^{k_r}$ if this is the cycle type of its corresponding elements in A.

Several of the arguments that we use in this paper are based on the following observation.

Lemma 2.2. *Let C be a finite group and W an $\mathbb{R}C$-module. Let Ω be a finite spanning C-invariant subset of W. Suppose that $H \leq C$ is an index-two subgroup such that there exits a non-zero element $\omega \in W$ with*

$$\omega^h = \omega \textit{ for } h \in H \textit{ and } \omega^c = -\omega \textit{ for } c \in C \setminus H.$$

Then, the following assertions hold:

(i) *There exists a C-orbit on Ω that splits into two H-orbits.*
(ii) *For any $c \in C \setminus H$, we have that $\omega \in \langle S_a^H - S_{a^c}^H : a \in \Omega \rangle$.*

Proof. Let $\mathbf{O} := \{O_i : 1 \leq i \leq r\}$ be the set of H-orbits on Ω. Since $\omega \in W = \langle \Omega \rangle$, we may write

$$\omega = \sum_{i=1}^{r} \sum_{a \in O_i} \lambda_a a,$$

for some $\lambda_a \in \mathbb{R}$. For each $1 \leq i \leq r$, let a_i be a representative of O_i. Since $\omega^h = \omega$ for all $h \in H$, we must have that

$$\omega = \frac{1}{|H|} \sum_{h \in H} \omega^h = \sum_{i=1}^{r} \lambda_i S_{a_i}^H, \tag{2}$$

where $\lambda_i = \frac{1}{|O_i|} \sum_{a \in O_i} \lambda_a$.

Let $c \in C \setminus H$. Observe that c acts on $\mathbf{O}$ by fixing the H-orbits that are equal to a C-orbit, and by permuting the pairs of H-orbits whose union is a C-orbit. By relabeling if necessary, assume that $\{O_i : 1 \leq i \leq m\}$ is a complete set of representatives of the $\langle c \rangle$-orbits on $\mathbf{O}$ of size 2. This set is non-empty because ω is a non-zero vector inverted by c; hence, (i) is established. Furthermore,

$$\omega = \frac{1}{2}(\omega - \omega^c) = \frac{1}{2} \sum_{i=1}^{m} (\lambda_i - \lambda_i^c)(S_{a_i}^H - S_{a_i^c}^H),$$

where $\lambda_i^c \in \mathbb{R}$ is the $S_{a_i^c}^H$-coefficient of ω in (4.6). Part (ii) follows.

For the following sections, consider Ω to be the set of Majorana axes of $\mathbf{R}$. In order to simplify notation, if $H \leq A$ and $t \in T$, denote by S_t^H the sum $S_{a_t}^H$.

Throughout our exposition, we will use results obtained in [5] about the Majorana representations of $S \cong S_4$ of shapes $(2B, 3A)$ and $(2A, 3A)$; we give a brief description of the algebras involved in these representations:

- The S_4-algebra of shape $(2B, 3A)$ has dimension 13 and basis

$$\left\{a_t : t \in ((1,2))^S\right\} \cup \left\{u_r : r \in S^{(3)}\right\} \cup \left\{v_f : f \in S^{(4)}\right\}.$$

  The inner and algebra product rules are described in Sect. 4.1 of [5].
- The S_4-algebra of shape $(2A, 3A)$ has dimension 13 and basis

$$\left\{a_t : t \in S^{(2)}\right\} \cup \left\{u_r : r \in S^{(3)}\right\}.$$

  The inner and algebra product rules are described in Sects. 4.2 and 5 of [5].

Recall that, in the bases given above, the $3A$- and $4A$-axes u_r and v_f are determined by the cyclic groups generated by r and f, respectively.

An appendix that gives some inner product relations between distinct axes of the Norton-Sakuma and S_4-algebras introduced above concludes the paper.

3 The 3*A*-Axes

The $3A$-axes of the Majorana representation **R** defined in Sect. 2 may have types 3^1, 3^2 and 3^4. The following lemmas deal with each one of these cases.

Lemma 3.1. *Let $u_r \in V^{(3A)}$ be a $3A$-axis of type 3^1 and consider $H := N_A(\langle r \rangle)$. Then, we have that*

$$u_r = \frac{2}{135}\left(5S^H_{x_1} - 16S^H_{x_2} + S^H_{x_3} - 2S^H_{x_5}\right), \tag{3}$$

where every $x_i \in T$ has cycle shape 2^2 and $|rx_i| = i + 1$.

Proof. By the positive-definiteness of the inner product, it is enough to show that

$$(u_r, u_r) - 2(u_r, S) + (S, S) = 0,$$

where S is the expression on the right-hand side of (3). In order to make calculations explicit, let $r = (1, 2, 3)$ and consider

$$x_1 = (1,2)(4,5),\ x_2 = (1,2)(3,4),\ x_3 = (1,4)(5,6),\ \text{and } x_5 = (4,5)(6,7).$$

Observe that $\langle\langle u_r, a_{x_1}\rangle\rangle$ is a Norton-Sakuma algebra of type $3A$, while $\langle\langle u_r, a_{x_3}\rangle\rangle$ is an S_4-algebra of shape $(2A, 3A)$. By Table 1, and Table 9 in [5], we know that $(u_r, u_r) = \frac{8}{5}$, $(u_r, a_{x_1}) = \frac{1}{4}$ and $(u_r, a_{x_3}) = \frac{1}{36}$.

The Norton-Sakuma algebras

$$\langle\langle u_r, a_{(1,2)(5,6)}\rangle\rangle \text{ and } \langle\langle a_{x_2}, a_{(1,2)(5,6)}\rangle\rangle$$

have types $3A$ and $2A$, respectively, so by Table 3 in the Appendix, we deduce that $(u_r, a_{x_2}) = \frac{1}{9}$. Similarly,

$$\langle\langle u_r, a_{(2,3)(8,9)}\rangle\rangle \text{ and } \langle\langle a_{x_5}, a_{(2,3)(8,9)}\rangle\rangle$$

have types $3A$ and $2B$, respectively, so $(u_r, a_{x_5}) = 0$. Therefore, by the H-invariance of the inner product, we have that $(u_r, S) = \frac{8}{5}$.

In order to compute (S, S), we introduce some notation. Consider

$$\mathbf{Ty} := \{2A, 2B, 3A, 3C, 4A, 4B, 5A, 6A\}.$$

For every $t, g \in T$ and $NX \in \mathbf{Ty}$, define

$$\lambda_{NX} := (a_t, a_g) \text{ where } \langle\langle a_t, a_g\rangle\rangle \text{ has type } NX,$$
$$H_t^{NX}(g) := \left\{h \in g^H : \langle\langle a_t, a_h\rangle\rangle \text{ has type } NX\right\}.$$

Now, observe that for any $i \neq j$, we have that

$$(S_{x_i}^H, S_{x_j}^H) = \left|x_i^H\right| \sum_{NX \in \mathbf{Ty}} \lambda_{NX} \left|H_{x_i}^{NX}(x_j)\right|,$$
$$(S_{x_i}^H, S_{x_i}^H) = \left|x_i^H\right| + \left|x_i^H\right| \sum_{NX \in \mathbf{Ty}} \lambda_{NX} \left|H_{x_i}^{NX}(x_i)\right|,$$

where x_i^H is the H-orbit of x_i on T. Using the above identities, we calculate that $(S, S) = \frac{8}{5}$. The result follows as

$$(u_r, u_r) - 2(u_r, S) + (S, S) = \frac{8}{5} - 2 \cdot \frac{8}{5} + \frac{8}{5} = 0.$$

Corollary 3.2. *Every $3A$-axis of type 3^1 is linearly expressible in terms of Majorana axes.*

Our aim now is to show that no $3A$-axis of type 3^2 belongs to the space $V^{(2A)}$. In order to achieve this, we are going to apply Lemma 2.2 with a suitable choice of groups.

Let $a, b \in A$ be disjoint 3-cycles. Define

$$\mathbf{u} := u_{ab} - u_{ab^{-1}} \in V^{(3A)},$$

and consider the groups

$$C := N_A\left(\left\langle ab, ab^{-1}\right\rangle\right) \text{ and } H := N_A(\langle ab\rangle).$$

With this choice, H is an index-two subgroup of C such that $\mathbf{u}^h = \mathbf{u}$ for every $h \in H$, and $\mathbf{u}^c = -\mathbf{u}$ for every $c \in C \setminus H$.

Lemma 3.3. *With the notation defined above, there are exactly three C-orbits on Ω that split into two H-orbits.*

Proof. By Cauchy-Frobenius lemma, and the fact that $[C : H] = 2$, we know that the number of C-orbits on Ω that split into two H-orbits is equal to the inner product (ξ, χ), where ξ is the permutation character of C on Ω and χ is the lift of the non-trivial character of C/H to C (so χ is the linear character of C such that $\chi(c) = 1$ for $c \in H$ and $\chi(c) = -1$ for $c \in C \setminus H$). Computations in [3] show that $(\xi, \chi) = 3$.

Alternatively, the previous lemma may be established by combinatorial computer-free calculations.

In order to make computations explicit, assume that $a = (1, 2, 3)$ and $b = (4, 5, 6)$. In general, for $t \in T$, a C-orbit a_t^C splits into two H-orbits if and only if the centraliser of t in C is contained in H. Using this fact, we check that the Majorana axes corresponding to the involutions

$$\begin{aligned} t_1 &:= (2, 4)(3, 5), \\ t_2 &:= (1, 4)(2, 5)(3, 6)(7, 8)(9, 10)(11, 12), \\ t_3 &:= (1, 4)(2, 5)(3, 7)(6, 8)(9, 10)(11, 12), \end{aligned}$$

are representatives of the three distinct splitting C-orbits on Ω.

Before applying Lemma 2.2, we need to calculate some inner products.

Lemma 3.4. *With the notation defined above, the following assertions hold:*

(i) $(\mathbf{u}, a_{t_1}) = \frac{1}{12}$.
(ii) $(\mathbf{u}, a_{t_2}) = -\frac{1}{4}$.
(iii) $(\mathbf{u}, a_{t_3}) = \frac{1}{36}$.

Proof. First, by the C-invariance of the inner product, we have that

$$(\mathbf{u}, a_{t_i}) = (u_r, a_{t_i} - a_{c^{-1}t_ic}),$$

where $r := ab$, $1 \leq i \leq 3$ and $c \in C \setminus H$.

The Norton-Sakuma algebras

$$\langle\langle a_{t_1}, a_{(2,3)(4,5)}\rangle\rangle \text{ and } \langle\langle u_r, a_{(2,3)(4,5)}\rangle\rangle$$

have types $2A$ and $3A$, respectively, so by Table 3 in the Appendix, we calculate that $(u_r, a_{t_1}) = \frac{1}{9}$. Since $\langle\langle u_r, a_{c^{-1}t_1c}\rangle\rangle$ is an S_4-algebra of shape $(2A, 3A)$, we have that $(u_r, a_{c^{-1}t_1c}) = \frac{1}{36}$. This establishes (i).

The Norton-Sakuma algebras $\langle\langle a_{t_2}, a_g\rangle\rangle$ and $\langle\langle u_r, a_g\rangle\rangle$ have types $2B$ and $3A$, respectively, where a_g corresponds to

$$g := (1,4)(2,6)(3,5)(7,8)(9,11)(10,12) \in T.$$

Again by Table 3 in the Appendix, we deduce that $(u_r, a_{t_2}) = 0$. Since $\langle\langle u_r, a_{c^{-1}t_2c}\rangle\rangle$ is a Norton-Sakuma algebra of type $3A$, we have that $(u_r, a_{c^{-1}t_2c}) = \frac{1}{4}$. Part (ii) follows.

In order to show part (iii), note that $\langle\langle u_r, a_{t_3}\rangle\rangle$ is an S_4-algebra of shape $(2B, 3A)$, so $(u_r, a_{t_3}) = \frac{13}{180}$ by Table 11 in [5]. The Norton-Sakuma algebras

$$\langle\langle a_{c^{-1}t_3c}, a_{(1,2)(4,6)}\rangle\rangle \text{ and } \langle\langle u_r, a_{(1,2)(4,6)}\rangle\rangle$$

have types $2A$ and $3A$, respectively. Therefore, $(u_r, a_{c^{-1}t_3c}) = \frac{2}{45}$.

Now we are ready to show that $\mathbf{u} \notin V^{(2A)}$.

Lemma 3.5. *Let $a, b \in A$ be disjoint 3-cycles. The difference of $3A$-axes $\mathbf{u} = u_{ab} - u_{ab^{-1}}$ is not linearly expressible in terms of Majorana axes.*

Proof. We are going to use the notation introduced in the previous paragraphs. By Lemma 3.3, there are exactly three splitting C-orbits on Ω with representatives a_{t_i}, $1 \le i \le 3$. Assuming that $\mathbf{u} \in V^{(2A)}$, Lemma 2.2 (ii) shows that

$$\mathbf{u} = \lambda_1(S^H_{t_1} - S^H_{t_1^c}) + \lambda_2(S^H_{t_2} - S^H_{t_2^c}) + \lambda_3(S^H_{t_3} - S^H_{t_3^c}), \tag{4}$$

for $c \in C \setminus H$ and some scalars $\lambda_i \in \mathbb{R}$.

For each $1 \le i \le 3$, by taking the inner product of a_{t_i} with both sides of (4), we deduce the following linear equation on the scalars:

$$(\mathbf{u}, a_{t_i}) = \lambda_1(S^H_{t_1} - S^H_{t_1^c}, a_{t_i}) + \lambda_2(S^H_{t_2} - S^H_{t_2^c}, a_{t_i}) + \lambda_3(S^H_{t_3} - S^H_{t_3^c}, a_{t_i}).$$

We calculate the coefficients of these equations. The inner products $(S^H_{t_j}, a_{t_i})$, $1 \le i \le j \le 3$, may be calculated using the same technique as in the proof of Lemma 3.1, while the products $(\mathbf{u}, a_{t_i})$ are given in Lemma 3.4. Thus, we have the following system of linear equations:

$$
\begin{aligned}
\frac{1}{12} &= \frac{3^2 5}{2^6}\left(\frac{3}{2}\lambda_1 - \frac{5}{2}\lambda_2 - 15\lambda_3\right),\\
-\frac{1}{4} &= \frac{3^2 5}{2^6}\left(-\frac{1}{2}\lambda_1 + \frac{7}{2}\lambda_2 - 3\lambda_3\right),\\
\frac{1}{36} &= \frac{15}{2^6}\left(-\frac{1}{2}\lambda_1 - \frac{1}{2}\lambda_2 + 9\lambda_3\right).
\end{aligned}
$$

It follows that $\mathbf{u} \notin V^{(2A)}$ as the above system has no solutions.

The next result uses the Majorana representation of the group $P \cong 3^2 : 2$ examined in Lemma 3.3 of [6, p. 11].

Lemma 3.6. *Every $3A$-axis is linearly expressible in terms of Majorana axes and $3A$-axes of type 3^2.*

Proof. Because of Corollary 3.2, it is enough to show that every $3A$-axis of type 3^4 is linearly expressible in terms of Majorana axes and $3A$-axes of type 3^2. Denote by $V_{3^2}^{(3A)}$ the linear span of the $3A$-axes of type 3^2 in V. Let $a, b, c, d \in A$ be disjoint 3-cycles. Note that there always exists an involution $t \in T$ that inverts both ab and cd. Hence $\langle ab, cd, t\rangle \cong 3^2 : 2$, so we may use Pasechnik's relation [6, p. 11] to deduce that

$$u_{abcd} + u_{abc^{-1}d^{-1}} \in \left\langle V^{(2A)}, V_{3^2}^{(3A)}\right\rangle. \tag{5}$$

Since $u_\rho = u_{\rho^{-1}}$ for any $\rho \in A \cap 3A$, we obtain that

$$
\begin{aligned}
u_{abcd} &= \frac{1}{2}\left(u_{abcd} + u_{abc^{-1}d^{-1}}\right) - \frac{1}{2}\left(u_{abc^{-1}d^{-1}} + u_{ab^{-1}cd^{-1}}\right)\\
&\quad + \frac{1}{2}\left(u_{ab^{-1}cd^{-1}} + u_{a^{-1}b^{-1}c^{-1}d^{-1}}\right).
\end{aligned}
$$

By (5), each sum in brackets on the right-hand side of the above identity belongs to $\left\langle V^{(2A)}, V_{3^2}^{(3A)}\right\rangle$, so the result follows.

We conclude this section by establishing a bound for the codimension of $V^{(2A)}$ in $\langle V^{(2A)}, V^{(3A)}\rangle$.

Proposition 3.7. *Consider the quotient space*

$$Q^{(3A)} := \left\langle V^{(2A)}, V^{(3A)}\right\rangle / V^{(2A)}.$$

The following assertions hold:

(i) $Q^{(3A)} = \langle u_{ab} - u_{ab^{-1}} + V^{(2A)} : a, b \in (1,2,3)^A, ab = ba\rangle$.
(ii) $1 \leq \dim(Q^{(3A)}) \leq 9{,}240$.

Proof. If $a, b \in A$ are disjoint 3-cycles, Pasechnik's relation and Corollary 3.2 imply that

$$u_{ab} + u_{ab^{-1}} \in V^{(2A)}.$$

Hence, (i) follows by Lemma 3.6.

Clearly, $Q^{(3A)}$ is non-trivial by Lemma 3.5. In order to deduce an upper bound for the dimension, fix two disjoint 3-cycles $a, b \in A$. Consider the $\mathbb{R}C$-module $M := \langle u_{ab} - u_{ab^{-1}} \rangle$, with $C = N_A(\langle ab, ab^{-1} \rangle)$. Observe that the character of this module is the same one defined in the proof of Lemma 3.3. Now, by part (i), $Q^{(3A)}$ may be embedded into the induced $\mathbb{R}A$-module $M \uparrow^A$, so

$$\dim(Q^{(3A)}) \leq \dim\left(M \uparrow^A\right) = \frac{1}{4}\left|((1,2,3)(4,5,6))^A\right| = 9{,}240.$$

4 The $4A$-Axes

Our aim in this section is to show that the set of $4A$-axes is not contained in the linear span of the Majorana axes. The most direct way to reach our goal is to show that the Gram matrix of a linearly independent subset of Ω and a $4A$-axis of V has non-zero determinant. However, this plan is computationally too involved; instead, we are going to use a different strategy. We start by taking an involution $z \in A^{(2)} \setminus T$ and defining a vector $\omega_z \in V^{(4A)}$ as an alternating sum of $4A$-axes v_ρ with $\rho^2 = z$. Then, we show that there are two $C_A(z)$-orbits on Ω that split into two H-orbits, where H is an index-two subgroup of $C_A(z)$ that centralises ω_z. The rest of the section consists in proving that the assumption $\omega_z \in V^{(2A)}$ contradicts Lemma 2.2 (ii).

All of the constructions of this section depend on the choice of an arbitrary involution of cycle shape 2^4. In order to make explicit computations, we take

$$z := (1,2)\,(3,4)\,(5,6)\,(7,8) \in A^{(2)} \setminus T.$$

However, it should be noted that all our calculations are invariant under conjugation, so different choices of z lead to equivalent results.

Consider the group

$$C := C_A(z).$$

In order to find $H \leq C$ and $\omega_z \in V^{(4A)}$ with the required properties, observe that the 2-core of C is the internal direct product of the groups $E \cong 2_+^{1+4}$ and $R \cong 2^2$. Hence the quotient space

$$X := O_2(C)/\langle z \rangle \cong 2^6$$

carries the natural structure of an orthogonal $GF(2)$-space with symplectic form induced by the commutator map and orthogonal form induced by the squaring map.

The space $\langle z \rangle R / \langle z \rangle$ is the radical of X while $\overline{E} = E/\langle z \rangle$ is its non-singular complement in X. Observe that the non-singular vectors of X precisely correspond to cosets with representatives in $A \cap 4A$. Since $\overline{E}$ is a 4-dimensional orthogonal $GF(2)$-space of plus type, we know it contains six non-singular vectors; let $E^{(4A)}$ be a set of coset representatives for such vectors. Since $\rho z = \rho^{-1}$ for any $\rho \in E^{(4A)}$, it is clear that there is a bijection between $E^{(4A)}$ and the set of $4A$-axes v_σ of type 4^2 such that $\sigma^2 = z$.

A possible choice for the elements of $E^{(4A)}$ is:

$$\begin{aligned} \rho_1 &:= (1,3,2,4)(5,7,6,8), \; \rho_4 := (1,4,2,3)(5,7,6,8), \\ \rho_2 &:= (1,8,2,7)(3,6,4,5), \; \rho_5 := (1,6,2,5)(3,8,4,7), \\ \rho_3 &:= (1,6,2,5)(3,7,4,8), \; \rho_6 := (1,8,2,7)(3,5,4,6). \end{aligned}$$

The extraspecial group $E \cong 2_+^{1+4}$ may be written as the central product of two quaternion groups

$$E = Q_8^{(0)} * Q_8^{(1)},$$

where each $Q_8^{(i)}$ is generated by precisely three elements of $E^{(4A)}$. Define

$$E_i^{(4A)} := E^{(4A)} \cap Q_8^{(i)},$$

for $i \in \{0, 1\}$. In particular, with the above choice of elements of $E^{(4A)}$, we have that

$$E_i^{(4A)} := \{\rho_{k+3i} : 1 \leq k \leq 3\}.$$

Let H be the stabilizer in C of the central product decomposition of E and define $\omega_z \in V^{(4A)}$ to be the following alternating sum of $4A$-axes:

$$\omega_z := \sum_{\rho \in E_0^{4A}} v_\rho - \sum_{\sigma \in E_1^{4A}} v_\sigma.$$

Since C is transitive on $E^{(4A)}$, it is clear that H is an index-two subgroup such that $\omega_z^h = \omega_z$ for every $h \in H$ and $\omega_z^c = -\omega_z$ for every $c \in C \setminus H$. Explicitly, we have that

$$H = N_A\left(Q_8^{(0)}\right) = N_A\left(Q_8^{(1)}\right) = \left\langle C', (9,10), (1,3)(2,4)(7,8) \right\rangle.$$

Lemma 4.1. *With the notation defined above, there are exactly two C-orbits on Ω that split into two H-orbits.*

Proof. The result follows because $(\xi, \chi) = 2$, where ξ is the permutation character of C on Ω, and χ is the lift of the non-trivial character of C/H to C.

In particular, the Majorana axes corresponding to

$$x := (1, 8)(2, 9)(3, 7)(4, 6)(5, 11)(10, 12),$$
$$y := (1, 8)(2, 9)(3, 6)(4, 10)(5, 11)(7, 12),$$

are representatives of the two H-splitting C-orbits on Ω.

If $\omega_z \in V^{(2A)}$, Lemma 2.2 implies that

$$\omega_z = \lambda_1 \left(S_x^H - S_{x^c}^H\right) + \lambda_2 \left(S_y^H - S_{y^c}^H\right), \tag{6}$$

for $c \in C \setminus H$ and some scalars $\lambda_1, \lambda_2 \in \mathbb{R}$.

We are going to show that relation (6) does not hold for any $\lambda_1, \lambda_2 \in \mathbb{R}$ by calculating several inner products. The next lemma is the first step in this plan.

Lemma 4.2. *With the notation defined above, the following assertions hold for any* $c \in C \setminus H$:

(i) $\left(S_x^H, S_x^H - S_{x^c}^H\right) = 3240.$
(ii) $\left(S_x^H, S_y^H - S_{y^c}^H\right) = 0.$
(iii) $\left(S_y^H, S_y^H - S_{y^c}^H\right) = 1620.$

Proof. The result follows by the formulas given in the proof of Lemma 3.1 and computations in [3].

Our goal now is to calculate the inner products (ω_z, ω_z), (ω_z, a_x) and (ω_z, a_y). In order to find the first product, we look at various subalgebras containing the $4A$-axes v_ρ, with $\rho \in E^{(4A)}$, and deduce some relations using the orthogonality between eigenvectors of Majorana axes. These relations will eventually allow us to compute (ω_z, ω_z). In order to find (ω, a_x) and (ω, a_y), we simplify calculations by considering the space V_z spanned by the sums of Majorana axes corresponding to the non-trivial singular vectors of X and the $4A$-axes corresponding to the non-singular vectors of X.

4.1 The Inner Product (ω_z, ω_z)

Let $t\langle z\rangle \in X$ be a non-trivial singular vector. Say that $t\langle z\rangle$ has type $2A$ if $t \in T$, or type $2B$ otherwise. We may check directly that this definition does not depend on the coset representative. Alternatively, when the Majorana representation is based on an embedding in the Monster, it is known that for any involution $g \in O_2(C_{\mathbf{M}}(z)) \cong 2_+^{1+24}$, $g \neq z$, the elements g and gz are always conjugated in $O_2(C_{\mathbf{M}}(z))$ (see Sect. 2.3 in [4]).

Let $E^{(2A)}$ and $E^{(2B)}$ be sets of representatives for the vectors in $\overline{E}$ of type $2A$ and $2B$, respectively. For $NX \in \{2A, 2B, 4A\}$, denote by $\overline{E}^{(NX)}$ the image of $E^{(NX)}$ in $\overline{E}$. It is clear that

$$\overline{E} = \{\langle z \rangle\} \cup \overline{E}^{(2A)} \cup \overline{E}^{(2B)} \cup \overline{E}^{(4A)},$$

where the union is disjoint.

In this subsection we show the following result.

Proposition 4.3. *Let $\rho, \sigma \in E^{(4A)}$. Then, the inner product (v_ρ, v_σ) is completely determined by the type of the vector $\rho\sigma \langle z \rangle \in \overline{E}$. In particular,*

(i) $\rho\sigma \langle z \rangle = \langle z \rangle$ *if and only if* $(v_\rho, v_\sigma) = 2$.
(ii) $\rho\sigma \langle z \rangle \in \overline{E}^{(4A)}$ *if and only if* $(v_\rho, v_\sigma) = \frac{1}{2}$.
(iii) $\rho\sigma \langle z \rangle = \overline{E}^{(2A)}$ *if and only if* $(v_\rho, v_\sigma) = \frac{8}{9}$.
(iv) $\rho\sigma \langle z \rangle \in \overline{E}^{(2B)}$ *if and only if* $(v_\rho, v_\sigma) = \frac{2}{9}$.

Corollary 4.4. *With $\omega_z \in V^{(4)}$ as defined in this section, $(\omega_z, \omega_z) = 10$.*

Observe that the non-diagonal orbits of $A \cong A_{12}$ on $\overline{E}^{(4A)} \times \overline{E}^{(4A)}$ coincide with the sets

$$\{(\rho \langle z \rangle, \sigma \langle z \rangle) : \rho\sigma \langle z \rangle \in \overline{E}^{(NX)}\},$$

for $NX \in \{2A, 2B, 4A\}$. Since

$$\rho_1\rho_4 \langle z \rangle \in \overline{E}^{(2A)}, \ \rho_1\rho_5 \langle z \rangle \in \overline{E}^{(2B)}, \text{ and } \rho_1\rho_2 \langle z \rangle \in \overline{E}^{(4A)},$$

the next three lemmas complete the proof of Proposition 4.3.

Lemma 4.5. *With the notation defined in this section, $(v_{\rho_1}, v_{\rho_5}) = \frac{2}{9}$.*

Proof. Let

$$t = (1,7)(2,8)(3,5)(4,6)(9,10)(11,12) \in T.$$

Then $\langle\langle a_t, a_h \rangle\rangle$ and $\langle\langle a_t, a_{h'} \rangle\rangle$ are both Norton-Sakuma algebras of type $4A$, with $h = t\rho_1$ and $h' = t\rho_5$. Since $\rho_1^2 = z = \rho_5^2$, we have that $a_{t\rho_1^2} = a_{t\rho_5^2}$. By Lemma A.2 in the Appendix,

$$(v_{\rho_1}, v_{\rho_5}) = \frac{1}{3} + \frac{2^2}{3}(v_{\rho_1}, a_{h'}) - 2^2 (v_{\rho_5}, a_h).$$

Since $\langle\langle a_h, a_{h'} \rangle\rangle$ has type $2B$, we use Table 3 to calculate the inner products on the right-hand side of the above relation:

$$(v_{\rho_1}, a_{h'}) = (v_{\rho_5}, a_h) = \frac{1}{2^3 3}.$$

Lemma 4.6. *With the notation defined in this section, $(v_{\rho_1}, v_{\rho_2}) = \frac{1}{2}$.*

Proof. Let $U_1 := \langle\langle a_t, a_h \rangle\rangle$ be a Norton-Sakuma algebra of type $4A$, where

$$t := (1,7)(2,8)(3,5)(4,6)(9,10)(11,12)$$

and $h := t\rho_1$. Let $U_2 := \langle\langle a_t, a_g, a_{g_2} \rangle\rangle$ be an S_4-algebra of shape $(2B, 3A)$, where

$$g := (1,2)(3,4)(5,11)(6,12)(7,9)(8,10),$$

and $g_2 := t\rho_1^2$. It is clear that U_1 and U_2 contain the $4A$-axes v_{ρ_1} and v_{ρ_2}, respectively.
Using the orthogonality between the eigenvectors of a_t in U_1 and U_2 stated in Tables 2 and 4 in the Appendix, we obtain the following equations:

$$\begin{aligned} 3 &= 48(v_{\rho_2}, v_{\rho_1}) + 15[2^2(u_r, a_h) - 3(u_r, v_{\rho_1})] + 2^6[(a_g, v_{\rho_1}) - (v_{\rho_2}, a_h)], \\ 281 &= 2^6[9(v_f, v_{\rho_1}) - (a_g, v_{\rho_1})] + 315[3(u_r, v_{\rho_1}) - 4(u_r, a_h)], \\ 31 &= 12(v_{\rho_2}, v_{\rho_1}) + 45[(u_r, v_{\rho_1}) + 4(u_r, a_h)] + 48[(v_{\rho_2}, a_h) - (v_f, v_{\rho_1})]. \end{aligned}$$

where u_r and v_f are the $3A$- and $4A$-axes of U_2 corresponding to

$$r := (1,7,9)(2,8,10)(3,5,11)(4,6,12) \text{ and } f := (1,9,2,10)(3,11,4,12).$$

Using Table 3 and Lemma A.1 in the Appendix A, we are going to calculate some inner products in the above equations. Let

$$\begin{aligned} b_1 &:= (1,3)(2,4)(5,8)(6,7)(9,10)(11,12), \\ b_2 &:= (1,5)(2,6)(3,7)(4,8)(9,11)(10,12), \\ b_3 &:= (1,8)(2,7)(3,6)(4,5)(9,10)(11,12), \end{aligned}$$

be involutions of T. Since the Norton-Sakuma algebras $\langle\langle a_{b_1}, v_{\rho_2} \rangle\rangle$ and $\langle\langle a_{b_1}, a_h \rangle\rangle$ have types $4A$ and $2B$, respectively, it follows that $(a_h, v_{\rho_2}) = \frac{1}{2\cdot 3}$. Similarly, since the algebras $\langle\langle a_{b_2}, u_r \rangle\rangle$ and $\langle\langle a_{b_2}, a_h \rangle\rangle$ have types $3A$ and $2B$, respectively, it follows that $(a_h, u_{r_1}) = \frac{1}{2^2 3^2}$.
Finally, the algebras $\langle\langle a_{b_3}, u_r \rangle\rangle$ and $\langle\langle a_{b_3}, v_{\rho_1} \rangle\rangle$ have types $3A$ and $4A$, respectively, so we deduce that

$$(v_{\rho_1}, u_{r_1}) = \frac{11}{2\cdot 3\cdot 5} - 2\left(a_{b_3\rho_1}, u_{r_1}\right).$$

By the A-invariance of the inner product, we know that $(a_{b_3\rho_1}, u_{r_1}) = (a_h, u_{r_1}) = \frac{1}{2^2 3^2}$. Hence $(v_{\rho_1}, u_{r_1}) = \frac{2\cdot 7}{3^2 5}$.
With the previous computations, we reduce the system of equations obtained above:

$$0 = 1 - 2^2 3 \left(v_{f_2}, v_{\rho_1}\right) - \left(v_{\rho_2}, v_{\rho_1}\right),$$
$$0 = 1 + 2^2 3 \left(v_{f_2}, v_{\rho_1}\right) - 3 \left(v_{\rho_2}, v_{\rho_1}\right).$$

The result follows by solving this system.

Lemma 4.7. *With the notation defined in this section,* $(v_{\rho_1}, v_{\rho_4}) = \frac{8}{9}$.

Proof. Let $Y_1 := \langle\langle a_t, a_g, a_{t_2} \rangle\rangle$ and $Y_2 := \langle\langle a_t, a_g, a_{t_3} \rangle\rangle$ be S_4-algebras of shape $(2B, 3A)$, where

$$t := (2,3)(6,7),\ g := (1,4)(5,8),$$
$$t_2 = (1,2)(5,6),\ t_3 := (3,4)(5,6).$$

Note that Y_1 and Y_2 contain the $4A$-axes v_{ρ_1} and v_{ρ_4}, respectively.

Define

$$a = a_{t_2} + a_{(3,4)(7,8)} \in Y_1 \text{ and } u = u_{r_2} + u_{tr_2t} \in Y_2,$$

where $r_2 = (1,2,4)(5,8,7)$. By the orthogonality between the eigenvectors of a_t on Y_1 and Y_2 stated in Table 4 in the Appendix, we obtain the equations:

$$\begin{aligned} 2^6 241 &= 96[15(a,u) - 2^4(a, v_{\rho_4} + v_{t\rho_4 t})] + 2^6 3^3 [2^4 (v_{\rho_1}, v_{\rho_4} + v_{t\rho_4 t}) - 15(v_{\rho_1}, u)] \\ &\quad + 105[2^4 3^3 (u_r, v_{\rho_4} + v_{t\rho_4 t}) - 3^4 5 (u_r, u)], \\ 7 &= 2^5[2^4(a, v_{\rho_4} + v_{t\rho_4 t}) - 15(a,u)] + 45[15(u_r, u) - 2^4(u_r, v_{\rho_4} + v_{t\rho_4 t})], \end{aligned}$$

where $u_r \in Y_1$ is the $3A$-axis corresponding to $r := (1,2,4)(5,6,8)$.

Now, by looking at the various S_4-algebras generated by the pair of axes involved in the previous equations, it is possible to reduce the above system to obtain that

$$(v_{\rho_1}, v_{\rho_4}) = \frac{7^2}{2^3 3^2} + \frac{3 \cdot 5}{2^3}(v_{\rho_1}, u_{r_2}). \tag{7}$$

In order to calculate the inner product (v_{ρ_1}, u_{r_2}), we use the orthogonality between the $\frac{1}{4}$-eigenvector of a_t in $\langle\langle a_g, a_h \rangle\rangle$, with $h := (1,2)(7,8)$, and one of the 0-eigenvectors of a_t in Y_1 stated in Table 4. We deduce that:

$$\begin{aligned} 2^3 113 &= 45[2^6 9 (v_{\rho_1}, u_{r_2}) + 945(u_{r_3}, u_{r_2}) - 2^5(a, u_{r_2})] \\ &\quad - 2^5 3^2 [105(u_{r_3}, a_h + a_{tt_3t}) + 2^6(v_{\rho_1} + v_{t\rho_1 t}, a_h)], \end{aligned} \tag{8}$$

where $u_{r_3} \in Y_1$ is the $3A$-axis corresponding to $r_3 := (2,3,4)(6,7,8)$.

We are going to compute some of the inner products in (8). The S_4-algebra $\langle\langle a_{t_3}, a_h, a_{(2,4)(6,7)} \rangle\rangle$ of shape $(2B, 3A)$ contains v_{ρ_1}, so $(a_h, v_{\rho_1}) = \frac{1}{2^3 3}$ and $(a_h, v_{t\rho_1 t}) = \frac{31}{2^6 3}$. On the other hand, the algebras $\langle\langle a_h, u_{r_3} \rangle\rangle$ and $\langle\langle a_{tt_3t}, u_{r_3} \rangle\rangle$ are both S_4-algebras of shape $(2A, 3A)$, so that

$$(u_{r_3}, a_h) = (u_{r_3}, a_{tt_3t}) = \frac{1}{2^2 3^2}.$$

Finally, in order to calculate (u_{r_2}, u_{r_3}), observe that $\langle\langle a_{(2,4)(7,8)}, u_{r_2}\rangle\rangle$ and $\langle\langle a_{(2,4)(7,8)}, u_{r_3}\rangle\rangle$ are both Norton-Sakuma algebras of type $3A$. The orthogonality between eigenvectors of $a_{(2,4)(7,8)}$ in these algebras allow us to obtain that $(u_{r_2}, u_{r_3}) = \frac{2^3 7}{3^3 5^2}$.

With the previous results, we calculate in (8) that $(v_{\rho_1}, u_{r_2}) = \frac{1}{3^2}$, so by (7) we have that $(v_{\rho_1}, v_{\rho_4}) = \frac{2^3}{3^2}$.

4.2 The Inner Products (ω, a_x) and (ω, a_y)

In this section, we construct a subspace of V that will considerably simplify the calculations of (ω, a_x) and (ω, a_y), where x and y are representatives in T of the two H-splitting C-orbits on Ω. Let $X^{(2A)}$ be the set of the twenty-four non-trivial singular vectors of type $2A$ in $X \cong 2^6$. Define the sets

$$B_1 := \left\{a_t + a_{tz} : t\langle z\rangle \in X^{(2A)}\right\},$$

$$B_2 := \left\{v_\rho : \rho \in E^{(4A)}\right\},$$

and consider the space

$$V_z := \langle B_1, B_2\rangle .$$

It is clear that the group $C = C_A(z)$ acts on V_z because both B_1 and B_2 are C-invariant sets.

It was shown by S. Norton in private communication how the space V_z embeds in $S^2(\Lambda)$, the symmetric square algebra of the Leech lattice, which is an important 300-dimensional subalgebra of $V_{\mathbf{M}}$. Using this embedding, it is straightforward to compute that $\dim V_z = 21$ and $\dim V_z/\langle B_1\rangle = 1$. We are going to prove that this result follows as well from the Majorana axioms.

In order to maintain explicit the computations, consider the following representatives of the cosets of $X^{(2A)}$:

$$\begin{aligned}
t_1 &= (1,2)(3,4), & t_{7+i} &= (1,3)(2,4)(5,7)(6,8)t_{4+i},\\
t_2 &= (1,2)(5,6), & t_{10+i} &= (1,4)(2,3)(5,7)(6,8)t_{4+i},\\
t_3 &= (3,4)(5,6), & t_{13+i} &= (1,5)(2,6)(3,7)(4,8)t_{4+i},\\
t_4 &= (9,10)(11,12), & t_{16+i} &= (1,6)(2,5)(3,7)(4,8)t_{4+i},\\
t_5 &= (9,11)(10,12), & t_{19+i} &= (1,7)(2,8)(3,5)(4,6)t_{4+i},\\
t_6 &= (9,12)(10,11), & t_{22+i} &= (1,8)(2,7)(3,5)(4,6)t_{4+i},
\end{aligned}$$

for $0 \leq i \leq 2$. Define $b_i := a_{t_i} + a_{t_i z}$.

Lemma 4.8. *With the above notation,* $\dim(\langle B_1\rangle) = 20$.

Proof. Since the type of a vector in X is independent of the choice of representative, we have that $g \in T \cap O_2(C)$ if and only if $gz \in T \cap O_2(C)$. Hence,

$$\left(a_{t_i}, a_{t_j}\right) = \left(a_{t_i z}, a_{t_j z}\right) = \left(a_{t_i z}, a_{t_j}\right) = \left(a_{t_i}, a_{t_j z}\right),$$

which implies that $(b_i, b_j) = 4(a_{t_i}, a_{t_j})$ and $(b_i, b_i) = 2$ for $1 \leq i < j \leq 24$. With this observation, we calculate in [3] that the Gram matrix of B_1 has rank 20, so the result follows.

In order to obtain the dimension of V_z, first we need to compute some inner products.

Lemma 4.9. *With the notation defined above, the following assertions hold:*

(i) $\left(b_i, v_{\rho_1}\right) = \frac{3}{2^2}$ *for* $16 \leq i \leq 21$.
(ii) $\left(b_i, v_{\rho_1}\right) = \frac{1}{3}$ *for* $i = 1$ *or* $13 \leq i \leq 15$ *or* $22 \leq i \leq 24$.
(iii) $\left(b_i, v_{\rho_1}\right) = 0$ *for* $4 \leq i \leq 6$.
(iv) $\left(b_i, v_{\rho_1}\right) = \frac{1}{2^2 3}$ *for* $i = 2, 3$ *or* $7 \leq i \leq 12$.

Proof. Let $\rho = \rho_1$. Direct calculations show that the orbits of the group $N_A(\langle\rho\rangle)$ on $T \cap O_2(C)$ are

$$\{t_1, t_1 z\}, \{t_2, t_3, t_2 z, t_3 z\}, \{t_4, t_5, t_6\}, \{t_4 z, t_5 z, t_6 z\},$$

$$\{t_i, t_i z : 7 \leq i \leq 12\}, \{t_i, t_i z : 16 \leq i \leq 21\},$$

$$\text{and } \{t_i, t_i z : 13 \leq i \leq 15 \text{ or } 22 \leq i \leq 24\}.$$

For this reason, it is enough to calculate the inner products (v_ρ, a_{t_i}), for $i \in \{1, 2, 4, 10, 16, 22\}$, and $(v_\rho, a_{t_4 z})$.

Since $\langle\langle a_{t_{16}}, a_{t_{16}\rho}\rangle\rangle$ is a Norton-Sakuma algebra of type $4A$ that contains v_ρ, we have that $(a_{t_{16}}, v_\rho) = \frac{3}{2^3}$.

Observe that $\langle\langle a_{t_{16}}, a_{t_i}\rangle\rangle$, for $i = 4, 10, 22$, and $\langle\langle a_{t_{16}}, a_{t_4 z}\rangle\rangle$ are Norton-Sakuma algebras of type $2B$. Hence, Table 3 in the Appendix allows us to deduce that

$$(a_{t_4 z}, v_\rho) = (a_{t_4}, v_\rho) = 0, \;\; (a_{t_{10}}, v_\rho) = \frac{1}{2^3 3} \text{ and } (a_{t_{22}}, v_\rho) = \frac{1}{2 \cdot 3}.$$

Similarly, $\langle\langle a_{t_{16}}, a_{t_2}\rangle\rangle$ is a Norton-Sakuma algebra of type $2A$, so by Table 3, $(a_{t_2}, v_\rho) = \frac{1}{2^3 3}$.

Now we calculate $\left(a_{t_1}, v_\rho\right)$. In this case, $\langle\langle a_{t_1}, a_{t_2}\rangle\rangle$ is an algebra of type $2A$ and $\langle\langle a_{t_2}, a_{g_2}, a_{g_3}\rangle\rangle$ is an S_4-algebra of shape $(2B, 3A)$, where $g_2 = (3,4)(7,8)$ and $g_3 = (2,3)(6,7)$. Since ρ is inverted by t_2, Lemma A.3 implies that

$$\left(a_{t_1}, v_\rho\right) = -\frac{2}{3}\left(a_{t_1} - a_{t_1 t_2}, a_{g_3} + a_{g_5}\right) + \frac{15}{2^4}\left(a_{t_1} - a_{t_1 t_2}, u_{r_1}\right) + \left(a_{t_1 t_2}, v_\rho\right), \quad (9)$$

where $g_5 = (2,4)(6,8)$ and $r_1 = (2,3,4)(6,7,8)$. We are going to calculate the inner products in the right-hand side of (9).

By the action of $N_A(\langle\rho\rangle)$, we have that $(a_{t_1t_2}, v_\rho) = (a_{t_2}, v_\rho) = \frac{1}{2^3 3}$, as can be read in Table 11 of [5]. Observe that

$$\langle\langle a_{t_1}, a_{g_3}\rangle\rangle \text{ and } \langle\langle a_{t_1t_2}, a_{g_3}\rangle\rangle$$

are both Norton-Sakuma algebras of types $2A$, while $\langle\langle a_{g_3}, a_{g_2}\rangle\rangle$ is a Norton-Sakuma algebra of type $3A$ that contains u_{r_1}. Hence, by Table 3,

$$(a_{t_1t_2}, u_{r_1}) = \frac{1}{2^2 3^2} \text{ and } (a_{t_1}, u_{r_1}) = \frac{1}{3^2}.$$

Substituting the above computations in (9), we obtain that $(a_{t_1}, v_\rho) = \frac{1}{2\cdot 3}$.

Corollary 4.10. *With the notation defined above,* $\dim(V_z) = 21$.

Proof. Using Lemma 4.3 and the fact that C is transitive on X^{4A}, it is straightforward to obtain the inner products (b_i, v_{ρ_j}) for $1 \leq i \leq 24$ and $1 \leq j \leq 6$. The inner products between the $4A$-axes in B_2 were obtained in Lemma 4.3. Therefore, we may calculate in [3] that the rank of the Gram matrix of $B_1 \cup B_2$ is 21.

Lemma 4.11. *The following relations hold for the* $4A$*-axes in* V_z*:*

$$\begin{aligned}
v_{\rho_2} =& \frac{2}{3}(b_3 + b_8 + b_{14} + b_{17} + b_{23}) + \frac{2^2}{3}(b_5 + b_{11}) \\
&- \frac{2}{3}(b_1 + b_4 + b_6 + b_{16} + b_{18} + b_{19} + b_{21} + b_{22} + b_{24}) + v_{\rho_1}, \\
v_{\rho_3} =& \frac{2}{3}(b_2 + b_7 + b_9 + b_{11} + b_{23}) - \frac{2}{3}(b_1 + b_{14} + b_{16} + b_{18} + b_{20}) + v_{\rho_1}, \\
v_{\rho_4} =& \frac{2}{3}(b_1 + b_8 + b_{16} + b_{18} + b_{22} + b_{24}) + \frac{2^2}{3}(b_{14} + b_{20}) \\
&- \frac{2}{3}(b_2 + b_3 + b_4 + b_6 + b_7 + b_9 + b_{11}) - v_{\rho_1}, \\
v_{\rho_5} =& \frac{2}{3}(b_{11} + b_{17} + b_{19} + b_{20} + b_{21} + b_{23}) - \frac{2}{3}(b_3 + b_4 + b_6) - v_{\rho_1}, \\
v_{\rho_6} =& \frac{2}{3}(b_8 + b_{14} + b_{16} + b_{17} + b_{18} + b_{20}) - \frac{2}{3}(b_2 + b_4 + b_6) - v_{\rho_1}.
\end{aligned}$$

Proof. These relations may be checked directly using the positive-definiteness of the inner product.

Now are ready to calculate (ω, a_x) and (ω, a_y).

Lemma 4.12. *Consider* $\omega_z \in V^{(4)}$ *as defined in this section, and let* $x, y \in T$ *be representatives of the* H*-splitting* C*-orbits on* Ω*. Then, the following assertions hold:*

(i) $(\omega_z, a_x) = -\frac{11}{2^6 3}$.
(ii) $\left(\omega_z, a_y\right) = 0$.

Proof. We start by computing (v_{ρ_1}, a_x) and (v_{ρ_1}, a_y). Consider

$$t_1 := (1,6)(2,5)(3,7)(4,8)(9,11)(10,12),$$
$$t_2 := (1,8)(2,7)(3,6)(4,5)(9,12)(10,11).$$

The Norton-Sakuma algebras $\langle\langle a_{t_1}, a_x\rangle\rangle$ and $\langle\langle a_{t_2}, a_y\rangle\rangle$ have both type $2B$, while $\langle\langle a_{t_1}, v_{\rho_1}\rangle\rangle$ and $\langle\langle a_{t_2}, v_{\rho_1}\rangle\rangle$ have both type $4A$. Hence, by Table 3 in the Appendix,

$$\left(v_{\rho_1}, a_x\right) = \frac{19}{2^7 3} \text{ and } \left(v_{\rho_1}, a_y\right) = \frac{11}{2^8}.$$

The result follows using the relations of Lemma 4.11.

4.3 The Quotient Space $Q^{(4A)}$

With the previous inner product calculations, we are able to show that the alternating sum of $4A$-axes ω_z does not belong to the space $V^{(2A)}$.

Lemma 4.13. *Consider $\omega_z \in V^{(4A)}$ as defined in this section. Then ω_z is not linearly expressible in terms of Majorana axes.*

Proof. As established before, if $\omega_z \in V^{(2A)}$, then the relation

$$\omega_z = \lambda_1\left(S_x^H - S_{x^c}^H\right) + \lambda_2\left(S_y^H - S_{y^c}^H\right) \tag{10}$$

holds by Lemma 2.2 (ii), where $c \in C \setminus H$ and $x, y \in T$ are the representatives of the H-splitting C-orbits on Ω defined before. Using Lemmas 4.2 and 4.12, we calculate the inner product of a_y with both sided of (10) in order to deduce that $\lambda_2 = 0$. Hence, by the positive-definiteness and invariance of the inner product, (10) holds if and only if

$$(\omega_z, \omega_z) - 4\lambda_1\left(\omega_z, S_x^H\right) + 2\lambda_1^2\left(S_x^H, S_x^H - S_{x^c}^H\right) = 0.$$

By Corollary 4.4, and Lemmas 4.2 and 4.12, the above relation is equivalent to the quadratic equation

$$5 + 132\lambda_1 + 3240\lambda_1^2 = 0,$$

which has no real solutions. This proves that $\omega_z \notin V^{(2A)}$.

Lemmas 4.11 and 4.13 imply that no $4A$-axis of type 4^2 is linearly expressible in terms of Majorana axes. The next result deals with the case of the $4A$-axes of type $4^2 2^2$.

Lemma 4.14. *Consider* $\rho_1 = (1,3,2,4)(5,7,6,8)$ *and*

$$q := (1,3,2,4)(5,8,6,7)(9,10)(11,12) \in O_2(C).$$

The following relation holds:

$$\begin{aligned} v_q = {} & \frac{2}{3}(b_2 + b_3 + b_5 + b_6 + b_8 + b_9 + b_{10} + b_{14} + b_{15}) \\ & - \frac{2}{3}(b_1 + b_4 + b_7 + b_{17} + b_{18}) - \frac{4}{3}(b_{13} + b_{19}) + v_{\rho_1}. \end{aligned}$$

Proof. Observe that $\langle\langle a_{(1,2)(5,6)}, v_{\rho_1} \rangle\rangle$ is an S_4-algebra of shape $(2B, 3A)$ and $\langle\langle a_{(1,2)(5,6)}, v_q \rangle\rangle$ is a Norton-Sakuma algebra of type $4A$. Thus, we calculate, using orthogonality between various eigenvectors, that $(v_{\rho_1}, v_q) = 0$. The inner products (b_i, v_q), $1 \leq i \leq 19$, may be computed with a similar technique as in the proof of Lemma 4.9, and so the above relation is verified using the positive-definiteness of the inner product.

Corollary 4.15. *Every* $4A$*-axis of type* $4^2 2^2$ *is linearly expressible in terms of Majorana axes and* $4A$*-axes of type* 4^2.

The next result is the equivalent of Proposition 3.7 for $4A$-axes.

Proposition 4.16. *Define the quotient space*

$$Q^{(4A)} := \left\langle V^{(2A)}, V^{(4A)} \right\rangle / V^{(2A)}.$$

Then, the following assertions hold:

(i) $Q^{(4A)} = \langle \omega_z + V^{(2A)} : z \in A^{(2)} \setminus T \rangle$.
(ii) $1 \leq \dim(Q^{(4A)}) \leq 51{,}975$.

Proof. Let $z \in A^{(2)} \setminus T$. By Lemmas 4.11 and 4.14, the 21-dimensional space V_z contains exactly twenty-four $4A$-axes of $V^{(4A)}$, which correspond to all the elements $\rho \in A^{(4)}$ such that $\rho^2 = z$. By Lemma 4.3 and Corollary 4.5 we have that

$$\left\langle V^{(2A)}, V_z \right\rangle / V^{(2A)} = \left\langle \omega_z + V^{(2A)} \right\rangle.$$

Part (i) follows since every $4A$-axis of $V^{(4A)}$ corresponds to an element whose square is an involution in $A^{(2A)} \setminus T$.

It is clear that $Q^{(4A)}$ is non-trivial because of Lemma 4.13. Fix $z \in A^{(2)} \setminus T$, and consider the one-dimensional $\mathbb{R}C$-module $M := \langle \omega_z \rangle$; note that the character of this module is the same one considered in Lemma 4.1. Now, by part (i), $Q^{(4A)}$ may be embedded into the induced $\mathbb{R}A$-module $M \uparrow^A$. This implies that

$$1 \leq \dim(Q^{(4A)}) \leq \dim\left(M \uparrow^A\right) = \left|A^{(2)} \setminus T\right| = 51,975.$$

5 The 5A-Axes

The argument in the following lemma was found by Á. Seress in [15].

Lemma 5.1. *Every 5A-axis of type* 5^1 *is linearly expressible in terms of Majorana axes and 3A-axes.*

Proof. Let $H \cong A_5$. Recall that H has two conjugacy classes K and R of elements of order 5, where $f \in K$ if and only if $f^2 \in R$. Since

$$w_f = -w_{f^2} = -w_{f^3} = w_{f^4},$$

for any $f \in A^5$, we deal with only one of the conjugacy classes. By *Norton's relation* (see Lemma 4.5 in [8]), for every $f \in K$, we have that

$$6w_f - w \in \left\langle V^{(2A)}, V^{(3A)} \right\rangle, \tag{11}$$

where

$$w := \frac{1}{2} \sum_{f \in K} w_f. \tag{12}$$

Let $I = \{1, 2, 3, 4, 5, 6, 7\}$ and let $G_k \cong A_6$ be the stabiliser in A_7 of $k \in I$. The group G_k has as well two conjugacy classes of elements of order 5, but these classes are fused together in A_7. Denote by $H_i^k \cong A_5$ the stabiliser in G_k of $i \in I \setminus \{k\}$; these groups are all conjugated in G_k.

For H_2^1, choose a conjugacy class K of elements of order 5 and define a vector w_2^1 as in (12). For H_i^1, $3 \leq i \leq 7$, define w_i^1 by choosing the conjugacy classes that fuse together with K in G_1. Consider elements $\phi_{1,k} \in A_7$ that transpose 1 and k, $2 \leq k \leq 7$. Now, for $i \neq k$, define w_i^k using the image of the class used to define $w^1_{\phi_{1,k}^{-1}(i)}$ under $\phi_{1,k}$.

There exists another conjugacy class of 6 twisted A_5-subgroups of G_k; denote them by $\{L_s : 1 \leq s \leq 6\}$. Define vectors w_s as in (12) for each $1 \leq s \leq 6$. Since the intersection $H_i^k \cap L_s$ always contains an element of order 5, we deduce from (11) that

$$w_i^k - w_s \in \left\langle V^{(2A)}, V^{(3A)} \right\rangle.$$

Therefore,

$$w_i^k - w_j^k = w_i^k - w_s + w_s - w_j^k \in \left\langle V^{(2A)}, V^{(3A)} \right\rangle. \tag{13}$$

for every $i, j \in I \setminus \{k\}$.

Define

$$\mathbf{w_k} = \sum_{i \in I \setminus \{k\}} w_i^k. \tag{14}$$

Using (13), we may see that

$$13 w_i^k + \mathbf{w_k} \in \left\langle V^{(2A)}, V^{(3A)} \right\rangle$$

for every $i \in I \setminus \{k\}$. It follows by (11) that

$$78 w_f + \mathbf{w_k} \in \left\langle V^{(2A)}, V^{(3A)} \right\rangle, \tag{15}$$

for every element f of order 5 in G_k.

By construction, we know that

$$(\mathbf{w_k})^{\phi_{k,i}} = \mathbf{w_i},$$

where $\phi_{k,i} = (\phi_{1,n})^{\phi_{1,i}} \in A_7$, $n = \phi_{1,i}^{-1}(\kappa)$. As $\phi_{k,i}$ induces an odd permutation on $I \setminus \{k, i\}$, we have that $(w_f)^{\phi_{k,i}} = -w_f$ for every $f \in H_i^k$. Therefore, by (15) we obtain that

$$78 (w_f)^{\phi_{k,i}} + (\mathbf{w_k})^{\phi_{k,i}} = -78 w_f + \mathbf{w_i} \in \left\langle V^{(2)}, V^{(3)} \right\rangle.$$

This implies that

$$\mathbf{w_k} + \mathbf{w_i} \in \left\langle V^{(2A)}, V^{(3A)} \right\rangle$$

for every $i, k \in I$, $i \neq k$, and so, by symmetry,

$$\mathbf{w_k} - \mathbf{w_i} = \mathbf{w_k} + \mathbf{w_j} - \mathbf{w_j} - \mathbf{w_i} \in \left\langle V^{(2A)}, V^{(3A)} \right\rangle.$$

Therefore,

$$\mathbf{w_i} \in \left\langle V^{(2A)}, V^{(3A)} \right\rangle,$$

and again by (15), we conclude that

$$w_f \in \left\langle V^{(2A)}, V^{(3A)} \right\rangle,$$

for every element of order 5 in A_7.

The result follows since any element of A with cycle shape 5^1 is contained in an A_7-subgroup.

Lemma 5.2. *Every $5A$-axis of type 5^2 is linearly expressible in terms of Majorana axes and $3A$-axes.*

Proof. Consider the conjugacy class K in $A \cong A_{12}$ of elements with cycle shape 5^2. Let $V_{5^2}^{(5A)}$ be the linear span of the $5A$-axes corresponding to the elements of K, and define the $\mathbb{R}A$-module

$$Q := \left\langle V^{(2A)}, V^{(3A)}, V_{5^2}^{(5A)} \right\rangle / \left\langle V^{(2A)}, V^{(3A)} \right\rangle.$$

Let $\Delta(K)$ be the graph on K where two vertices are adjacent if they are contained in a common A_5-subgroup of A. Using the conjugacy classes of the chain of subgroups $L_2(11) < M_{11} < M_{12} < A$, we may show that $\Delta(K)$ is a connected graph. It follows by Norton's relation that the dimension of Q is at most 1.

If $\dim(Q) = 1$, then the action of A on Q must be trivial because A is simple. However, for any $f \in K$ we may find $g \in A$ such that $f^g = f^2$, so g negates the $5A$-axis corresponding to f. This implies that Q must be trivial and the result follows.

The argument in the proof of Lemma 5.2 also works for Lemma 5.1; however, Seress' approach has the advantage of finding explicit expressions for the $5A$-axes of type 5^1 in terms of Majorana axes and $3A$-axes.

6 The Majorana Representation Based on an Embedding in the Monster

As we mentioned in the introduction, the results of this paper may be considerably refined when our representation is based on an embedding in the Monster. Let $A \cong A_{12}$ be a subgroup of $\mathbf{M}$ such that $G := C_{\mathbf{M}}(A) \cong A_5$, and consider the subalgebra V_A of $V_{\mathbf{M}}$ generated by the Majorana axes corresponding to the $2A$-involutions of A.

Proposition 6.1. *With the notation defined above, the following statements hold:*

(i) *V_A is contained in the $4,689$-dimensional space $C_{V_{\mathbf{M}}}(G)$.*
(ii) *The space $C_{V_{\mathbf{M}}}(G)$ has the following irreducible constituents as $\mathbb{R}A$-module:*

$$4689 = 4 \cdot 1 + 11 + 54 + 2 \cdot 132 + 154 + 275 + 462_1 + 2 \cdot 462_2 + 616 + 1925.$$

Proof. Part (i) may be calculated using the character table of the Monster in [1], as

$$\dim\left(C_{V_{\mathbf{M}}}(G)\right) = (\psi\downarrow_G, 1_G) + 1,$$

where ψ is the irreducible character of **M** of degree 196, 883 and 1_G is the trivial character of G.

Part (ii) was derived by finding the fusion of the character table of $A_{12} \times A_5$ into the character table of the Monster. This calculation was performed in [3] by S. Shpectorov.

Since the inner products (a_t, a_s), $t, s \in T$, are determined by the class of the product ts, the dimension of the linear span $V_A^{(2A)}$ of the Majorana axes may be obtained by calculating the rank of the symmetric matrix

$$M = [(a_t, a_s)]_{s \times t}$$

which has 11, 880 rows and columns. Dima Pasechnik calculated in [12] that this rank is 3498.[1] Moreover, he found the irreducible constituents of $V_A^{(2A)}$ as $\mathbb{R}A$-module :

$$3498 = 1 + 11 + 54 + 154 + 275 + 462_2 + 616 + 1925.$$

It is clear that the above calculations also hold in the general setting of the abstract Majorana representation of $A \cong A_{12}$.

We use the previous information to prove the following result.

Theorem 6.2 *With the above notation, define*

$$V_A^{\circ} := \left\langle V_A^{(NA)} : 2 \leq N \leq 5 \right\rangle.$$

Then,

$$\dim(V_A^{\circ}) = 3960.$$

Proof. Let $Q_A^{(3A)}$ and $Q_A^{(4A)}$ be the quotient spaces defined in Propositions 3.7 and 4.16, respectively. By the results of Sects. 3, 4 and 5, we know that

$$V_A^{\circ}/V_A^{(2A)} = \left\langle Q_A^{(3A)}, Q_A^{(4A)} \right\rangle.$$

Recall that $Q_A^{(3A)}$ and $Q_A^{(4A)}$ may be embedded in $\mathbb{R}A$-modules, induced from linear characters, of dimensions 9, 240 and 51, 975, respectively. We calculated in [3] that the irreducible constituents of this 9, 240-dimensional module are

$$9240 = 330 + 462_1 + 616 + 1050_1 + 1050_2 + 1728 + 1925 + 2079, \qquad (16)$$

while the irreducible constituents of this 51, 975-dimensional module are

---

[1] This number was independently approximated, with an error of 8×10^{-10}, by Alexander Balikhin in his final year project – 'On a Majorana Representation of A_{12}', Imperial College London, (2009). The error in Balikhin's computation means that 3498 is the rank of a matrix that differs from M on each entry by at most 8×10^{-10}.

Table 2 Eigenvectors of Norton-Sakuma algebras

Type	0	$\frac{1}{2^2}$
$2A$	$a_g + a_{tg} - \frac{1}{2^2} a_t$	$a_g - a_t g$
$3A$	$u_{\rho_1} - \frac{2.5}{3^3} a_t + \frac{2^5}{3^3}(a_s + a_{s_{-1}})$	$u_{\rho_1} - \frac{2^3}{3^2 5} a_t - \frac{2^5}{3^2 5}(a_s + a_{s_{-1}})$
$4A$	$v_{\rho_2} - \frac{1}{2} a_t + 2(a_h + a_{h_{-1}}) + a_{h_2}$	$v_{\rho_2} - \frac{1}{3} a_t - \frac{2}{3}(a_h + a_{h_{-1}}) - \frac{1}{3} a_{h_2}$

$$\begin{aligned} 51975 = {} & 330 + 462_1 + 1050_1 + 1050_2 + 1728 + 2 \cdot 1925 + 2079 \\ & + 2 \cdot 3520 + 2 \cdot 3564 + 3696 + 3850_1 + 3850_2 + 4455 \\ & + 5632 + 5775. \end{aligned} \tag{17}$$

By Proposition 6.1 (ii) and Pasechnik's calculation of the irreducible constituents of $V_A^{(2)}$, we deduce that $\left\langle Q_A^{(3A)}, Q_A^{(4A)} \right\rangle$ is contained in an $\mathbb{R}A$-module with irreducible constituents

$$1191 = 3 \cdot 1 + 2 \cdot 132 + 462_1 + 462_2. \tag{18}$$

Only the constituent 462_1, which corresponds to χ_{12} in the character table of A_{12} in [1], appears in all decompositions (16), (17) and (18). It follows that $Q_A^{(3A)} = Q_A^{(4A)}$ and $\dim(Q_A^{(3A)}) = \dim(Q_A^{(4A)}) = 462$. Thus, we conclude that

$$\dim(V_A^\circ) = 3498 + 462 = 3960.$$

Appendix: Orthogonality Relations

Let $(G, T, V, (\,,\,), \cdot, \varphi, \psi)$ be a Majorana representation. It follows directly from axiom **M1** that eigenvectors of the adjoint transformation of a Majorana axis corresponding to distinct eigenvalues are orthogonal. This fact may be used to obtain several inner product relations involving distinct $3A$- and $4A$-axes. This appendix is devoted to state some of these relations.

Let $t, g, q, s, h \in T$ and suppose that the Norton-Sakuma algebras $\langle\langle a_t, a_g \rangle\rangle$, $\langle\langle a_t, a_q \rangle\rangle$, $\langle\langle a_t, a_s \rangle\rangle$ and $\langle\langle a_t, a_h \rangle\rangle$ have types $2A$, $2B$, $3A$ and $4A$, respectively. Let $\rho_1 := ts$ and $\rho_2 := th$, and define $s_i := t\rho_1^i$ and $h_i := t\rho_2^i$, for $i \in \mathbb{Z}$. Table 2 describes some of the eigenvectors of the adjoint transformation of a_t contained in the previous algebras. These eigenvectors were obtained in [5].

We summarise in Table 3 some of the relations obtained using the orthogonality between the eigenvectors of Table 2. In Table 3, the expression of each entry equals to the inner product between the axes labeling the row and column.

The following lemmas state further useful relations.

Lemma A.1. *Consider the algebras $\langle\langle a_t, a_s \rangle\rangle$ and $\langle\langle a_t, a_h \rangle\rangle$ of types $3A$ and $4A$, respectively, as defined above. Then,*

Table 3 Inner product relations using Norton-Sakuma algebras

	$3A$-axis u_{ρ_1}	$4A$-axis v_{ρ_2}
a_q	$\frac{2^6}{3^25}(a_q, a_s)$	$\frac{2^2}{3}(a_q, a_h) + \frac{1}{3}(a_q, a_{h_2})$
a_g	$\frac{1}{3^25} - \frac{2^6}{3^35}(a_g, a_s) + \frac{2^8}{3^35}(a_{tg}, a_s)$	$\frac{1}{2^33} - \frac{2^2}{3}(a_g, a_h) + \frac{2^3}{3}(a_{tg}, a_h) + \frac{1}{3}(a_g, a_{h_2})$

Table 4 Eigenvectors of $a_{(ij)}$ in the S_4-algebra of shape $(2B, 3A)$

Ev.	Eigenvector
0	$-\frac{1}{2^33}a_{(ij)} + \frac{1}{3}(a_{(ik)} + a_{(il)} - a_{(jk)} + a_{(jl)}) - \frac{3.5}{2^5}(u_i + u_j) + v_j$
0	$-\frac{79}{2^43^2}a_{(ij)} + \frac{1}{2.3^2}(a_{(ik)} + a_{(il)} + a_{(jk)} + a_{(jl)}) - \frac{3.5.7}{2^6}(u_i + u_j) + v_k + v_l$
$\frac{1}{4}$	$\frac{1}{2.3}(a_{(kl)} - a_{(ij)}) - \frac{3.5}{2^4}(u_i + u_j) - \frac{1}{2}v_j + v_k + v_l$

$$
\begin{aligned}
\left(a_{h_2}, u_{\rho_1}\right) &= \frac{2^6}{3^25}\left(a_{h_2}, a_s\right),\\
\left(v_{\rho_2}, u_{\rho_1}\right) &= \frac{11}{2^23^3} - 2\left(a_h, u_{\rho_1}\right) + \frac{2^6}{3^3}\left[\left(a_h, a_s + a_{s_{-1}}\right) + \frac{1}{5}\left(a_{h_2}, a_s\right)\right],\\
\left(v_{\rho_2}, a_s\right) &= \frac{7}{2^83} + \frac{3^25}{2^5}\left(a_h, u_{\rho_1}\right) + \frac{1}{3}\left[\left(a_{h_2}, a_s\right) - \left(a_h, a_s + a_{s_{-1}}\right)\right].
\end{aligned}
$$

Lemma A.2. *Let $\langle\langle a_t, a_h\rangle\rangle$ be the algebra of type $4A$ defined above, and suppose that $\langle\langle a_t, a_{h'}\rangle\rangle$ is also an algebra of type $4A$ with $\rho_2' := th'$ and $h_i' := t\left(\rho_2'\right)^i$, $i \in \mathbb{Z}$. Suppose that $h_2 = h_2'$. Then*

$$
\left(v_{\rho_2}, v_{\rho_2'}\right) = \frac{1}{3} + \frac{2^2}{3}\left(v_{\rho_2}, a_{h'}\right) - 2^2\left(a_h, v_{\rho_2'}\right) + \frac{2^3}{3}\left(a_h + a_{h_{-1}}, a_{h'}\right).
$$

For future reference, Table 4 contains some eigenvectors of $a_{(i,j)}$ in the S_4-algebra of shape $(2B, 3A)$, using the notation of the basis introduced in Sect. 5 of [5].

Lemma A.3. *Let t, (ij), (kl), $(ik) \in T$. Suppose that*

$$
\langle\langle a_t, a_{(ij)}\rangle\rangle \text{ and } \langle\langle a_{(ij)}, a_{(kl)}, a_{(ik)}\rangle\rangle
$$

are a Norton-Sakuma algebra of type $2A$ and an S_4-algebra of shape $(2B, 3A)$, respectively. Then, we have that

$$
\left(a_t - a_{t\cdot(ij)}, v_j\right) = -\frac{2}{3}\left(a_t - a_{t\cdot(ij)}, a_{(ik)} + a_{(il)}\right) + \frac{3\cdot 5}{2^4}\left(a_t - a_{t\cdot(ij)}, u_i\right),
$$

where v_j is the $4A$-axis corresponding to (i, k, j, l), and u_i is the $3A$-axis corresponding to (j, k, l).

Proof. The result follows by the orthogonality between the first 0-eigenvector of $a_{(i,j)}$ given in Table 4 and the $\frac{1}{4}$-eigenvector of $a_{(ij)}$ in $\langle\langle a_{(ij)}, a_{(kl)}, a_{(ik)} \rangle\rangle$.

References

1. Conway, J.H., Curtis, R.T., Norton, S.P., Parker, R.A., Wilson, R.A.: Atlas of Finite Groups. Clarendon Press, Oxford (1985)
2. Decelle, S.: The $L_2(11)$-subalgebra of the monster algebra. J. Ars Math. Contemp. **7**(1), 83–103 (2014)
3. The GAP Group, Gap—Groups Algorithms and Programming, Version 4.4.12, http://www.gap-system.org (2008)
4. Ivanov, A.A.: The Monster Group and Majorana Involutions. Cambridge Tracts in Mathematics 176. Cambridge University Press, Cambridge (2009)
5. Ivanov, A.A., Pasechnik, D.V., Seress, Á., Shpectorov, S.: Majorana representations of the symmetric group of degree 4. J. Algebra **324**, 2432–2463 (2010)
6. Ivanov, A.A.: On Majorana representations of A_6 and A_7. Comm. Math. Phys. **307**, 1–16 (2011)
7. Ivanov, A.A.: Majorana representations of A_6 involving $3C$-algebras. Bull. Math. Sci. **1**, 356–378 (2011)
8. Ivanov, A.A., Seress, Á.: Majorana representations of A_5. Math. Z. **272**, 269–295 (2012)
9. Ivanov, A.A., Shpectorov, S.: Majorana representations of $L_3(2)$. Adv. Geom. **14**, 717–738 (2012)
10. Miyamoto, M.: Griess algebras and conformal vectors in vertex operator algebras. J. Algebra **179**, 523–548 (1996)
11. Norton, S.P.: 'The uniqueness of the monster', in finite simple groups, coming of age. In: McKay J. (ed.) Contemporary Mathematical, vol. 45, pp. 271–285, American Mathematical Society, Providence, Rhode Island (1982)
12. Norton, S.P.: 'The monster algebra: some new formulae', in moonshine, the monster and related topics, Contemporary Mathematics, vol. 193, pp. 297–306. American Mathematical Society, Providence, Rhode Island (1996)
13. Pasechnik, D.: private communication (2008)
14. Sakuma, S.: 6-Transposition property of τ-involutions of vertex operator algebras. Int. Math. Res. Not. **rnm030**, 19 (2007)
15. Seress, Á.: private communication, 3 Feb 2011
16. Seress, Á.: 'Construction of 2-closed M-representations' in Proceedings ISAAC12, pp. 311–318. American Mathematical Society, New York (2012)

Chapter 10
GIT Related Problems of the Flag Variety for the Action of a Maximal Torus

S. Senthamarai Kannan

Abstract Let G be a simple adjoint group over the field $\mathbb{C}$ of complex numbers. Let T be a maximal torus of G. Let P be a parabolic subgroup of G. In this article, we give a survey on the Geometric Invariant Theory related problems for the left action of T on G/P.

Keywords Schubert varieties · Line bundles · Semi-stable points

1 Introduction

Let V denote a two dimensional vector space over $\mathbb{C}$. Consider the natural action of $SL(2, \mathbb{C})$ on the nth symmetric power $S^n(V^*)$. Classical problem in Invariant Theory is the study of the ring of $SL(2, \mathbb{C})$-invariants of $\mathbb{C}[S^n(V^*)]$; in particular, the problem is to describe a set of generators of this algebra of invariants and the relations among the generators. In the modern context, this is a part of Geometric Invariant Theoretic (GIT) study in the sense of Mumford (cf. [25]) for the action of $SL(2, \mathbb{C})$ on $S^n(V^*)$ (or the corresponding projective space).

Recall that the GIT study of the projective space $\mathbb{P}(S^n(V^*))$ for the action of $SL(2, \mathbb{C})$ is equivalent to the GIT study of the Grassmannian $G_{2,n}$ of all two dimensional subspaces of $\mathbb{C}^n$ for the action of the normaliser N of the diagonal matrices T in $SL(n, \mathbb{C})$ (cf. proofs of Theorems 1 and 4 in [30]). Note that T is a normal subgroup of N such that N/T is the symmetric group on n symbols. Hence, the GIT study of $G_{2,n}$ for the action of T is also an interesting problem. Note that $G_{2,n}$ is

---

Dedicated to Professor N.S.N. Sastry on the occasion of his 60th birth day.

S. S. Kannan (✉)
Chennai Mathematical Institute, Plot H1, SIPCOT IT Park, Siruseri,
Kelambakkam 603103, Tamilnadu, India
e-mail: kannan@cmi.ac.in

N. S. N. Sastry (ed.), *Groups of Exceptional Type, Coxeter Groups and Related Geometries*, Springer Proceedings in Mathematics & Statistics 82,
DOI: 10.1007/978-81-322-1814-2_10,

a homogeneous space $SL(n, \mathbb{C})/P$, where P is a "maximal parabolic subgroup" of $SL(n, \mathbb{C})$. A more general question in this setting is the study of GIT related problems on G/P for the action of N or T, where G is a semisimple algebraic group, T a maximal torus of G, N the normaliser of T in G and P a parabolic subgroup of G.

Further, we have an interesting class of sub varieties which are known as Schubert varieties. These are the Zariski closures of the B-orbits of T-fixed points in G/P, where B is a "Borel subgroup" of G containing T. In the case of $G = SL(n, \mathbb{C})$, B is the subgroup of all upper triangular matrices in $SL(n, \mathbb{C})$. These varieties are stable for the left action of T on G/P. Note that these varieties are not necessarily N-stable.

Of course, a most general problem to this is the study of the GIT related problems on G/P for the H-linearised line bundles, where G is a semisimple algebraic group, P a parabolic subgroup of G and H a reductive subgroup of G.

There is a very interesting connection between GIT quotients of G/P for the action of T with the Kostant multiplicity formula.

By the Borel-Weil theorem, for any line bundle $\mathcal{L}$ on G/B associated to a "dominant character" of B, the space of global sections $H^0(G/B, \mathcal{L})$ is an irreducible G-module (cf. [4, p. 436]). We recall that the homogeneous coordinate ring of G/B for the line bundle $\mathcal{L}$ on G/B is $\bigoplus_{d\in\mathbb{Z}_{\geq 0}} H^0(G/B, \mathcal{L}^{\otimes d})$. Hence, we see that the vector subspace of the T-invariants of $H^0(G/B, \mathcal{L}^{\otimes d})$ is equal to the zero weight space of $H^0(G/B, \mathcal{L}^{\otimes d})$. Hence, its dimension is given by Kostant multiplicity formula (cf. [10, Theorem in p. 138]).

We now set up some notation as follows:

Let G be a semisimple algebraic group over $\mathbb{C}$. Let T be a maximal torus of G and B be a Borel subgroup of G containing T. Let G/B denote the variety of all Borel subgroups of G. Now, we consider the left action of T on G/B.

We have the G-linearised line bundle $\mathcal{L}_\chi$ on G/B corresponding to an arbitrary character χ of B. So, in particular the line bundle $\mathcal{L}_\chi$ is T-linearised. For the notion of G-linearisation and the GIT quotients, see [25, p. 30, 40].

Let w be an element of the Weyl group W of G with respect to T. Let $X(w)$ denote the closure of the B-orbit of the point wB in G/B. We recall that $X(w)$ is known as the Schubert variety corresponding to w.

We have the restriction of $\mathcal{L}_\chi$ to Schubert varieties $X(w)$, $w \in W$. Since each $X(w)$ is T-stable for the left action of T on G/B, the restriction of $\mathcal{L}_\chi$ to $X(w)$ is T-linearised as well.

Since T is reductive, by using the Geometric Invariant Theory of Mumford (cf. [25, p. 36, Definition 1.7] and [26]), we can talk about the semi-stable points $X(w)^{ss}_T(\mathcal{L}_\chi)$ of $X(w)$ with respect to the T-linearised line bundle $\mathcal{L}_\chi$ on $X(w)$.

The aim of this article is to give a survey on GIT related problems for the action of a maximal torus T on G/P, where P is any parabolic subgroup of G.

The problems that we discuss in this article are the following:

1. Existence of some smooth varieties as GIT quotients of G/P for T-linearised line bundles.

2. Projective normality of the GIT quotients of G/P for T-linearised ample line bundles.
3. Automorphism group of the GIT quotients of G/P for T-linearised ample line bundles.
4. GIT equivalence of T-linearised ample line bundles on G/B.
5. A connection between the ring of T-invariants of the homogeneous coordinate ring $\oplus_{d\in\mathbb{Z}_{\geq 0}} H^0(G/B, \mathcal{L}_\chi^{\otimes d})^T$ being a polynomial ring and the semi-stable points of $X(c)$ for a suitable T-linearised line bundle and a Coxeter element c.
6. A connection between the automorphism group of a smooth Schubert variety $X(w)$ and semi-stable points in $X(w)$ for a suitable T-linearised line bundle.
7. A cohomological characterisation of a Coxeter element c in W for which $X(c)$ contains semi-stable points for a suitable T-linearised line bundle.

Dabrowski, Howard, Shrawan Kumar, Skorobogatov, Strickland, Zhgoon have also worked on these problems.

Hilbert-Mumford criterion (cf. [25, Theorem 2.1]) and a combinatorial formula of Seshadri involving one parameter subgroups of T lying in the fundamental dominant chamber (cf. [31, Lemma (2.1)]) are the key tools to obtain some of the above mentioned results.

The organisation of this article is as follows:

Section 2 consists of preliminaries and notation.

In Sect. 3, we discuss the study of smooth varieties as GIT quotients of G/P for T-linearised line bundles.

In Sect. 4, we discuss the study of projective normality problems related to GIT quotients of G/P for T-linearised ample line bundles.

In Sect. 5, we discuss the study of the automorphism group of the GIT quotients of G/P for T-linearised ample line bundles.

In Sect. 6, we discuss the study of the GIT equivalence of T-linearised ample line bundles on the full flag variety G/B.

In Sect. 7, we discuss the study of the connection between the ring of T-invariants of the homogeneous coordinate ring $\oplus_{d\in\mathbb{Z}_{\geq 0}} H^0(G/B, \mathcal{L}_\chi^{\otimes d})$ being a polynomial ring and the semi-stable points of $X(c)$ for a suitable T-linearised line bundle and a Coxeter element c.

In Sect. 8, we discuss the study of a connection between the automorphism group of a smooth Schubert variety $X(w)$ and semi-stable points in $X(w)$ for a suitable T-linearised line bundle.

In Sect. 9, we discuss the study of a cohomological characterisation of a Coxeter element c in W for which $X(c)$ contains semi-stable points for a suitable T-linearised line bundle.

2 Preliminaries and Notation

In this section, we set up further notation.

The books [10, 11] of Humphreys and the book [13] of Jantzen are good references for this.

We now assume G to be a semisimple group of rank n over $\mathbb{C}$. Let T be a maximal torus of G, B be a Borel subgroup of G containing T. Let $N_G(T)$ denote the normaliser of T in G. Let $W = N_G(T)/T$ denote the Weyl group of G with respect to T.

We denote by $\mathfrak{g}$ the Lie algebra of G. We denote by $\mathfrak{h} \subseteq \mathfrak{g}$ the Lie algebra of T. Let R denote the roots of G with respect to T. Let $R^+ \subset R$ be the set of positive roots with respect to B. Let $S = \{\alpha_1, \alpha_2, \ldots, \alpha_n\} \subset R^+$ denote the set of simple roots with respect to B. Let $\langle ., . \rangle$ denote the restriction of the Killing form to $\mathfrak{h}$. Let $\check{\alpha}_i$ denote the coroot corresponding to α_i. Let $\varpi_1, \varpi_2, \ldots, \varpi_n$ denote the fundamental weights corresponding to S.

Let s_i denote the simple reflection corresponding to the simple root α_i. For any subset J of $\{1, 2, \ldots, n\}$, we denote by W_J the subgroup of W generated by s_j, $j \in J$. We denote the complement of J in $\{1, 2, \ldots, n\}$ by J^c. For each $w \in W$, we choose an element n_w in $N_G(T)$ such that $n_w T = w$. We denote the parabolic subgroup of G containing B and $\{n_w\colon w \in W_{J^c}\}$ by P_J. In particular, we denote the maximal parabolic subgroup of G generated by B and $\{n_{s_j};\ j \neq i\}$ by P_i.

Let w_0 denote the longest element of W corresponding to B. Let $B^- = w_0 B w_0^{-1}$ denote the Borel subgroup of G opposite to B with respect to T.

Let $\chi \in X(B)$, we have an action of B on $\mathbb{C}$, namely $b.k = \chi(b^{-1})k, b \in B, k \in \mathbb{C}$. Consider the equivalence relation $\sim$ on $G \times \mathbb{C}$ defined by $(gb, b.k) \sim (g, k), g \in G, b \in B, k \in \mathbb{C}$. The set of all equivalence classes is the total space of a line bundle over G/B. We denote this G-linearised line bundle associated to χ by $\mathcal{L}_\chi$.

We recall the notion of semi-stable and stable points introduced by Mumford.

A point x in $X(w)$ is said to be semi-stable with respect to the T-linearised line bundle $\mathcal{L}_\chi$ if there is a positive integer d and a T-invariant section s of $H^0(X(w), \mathcal{L}_\chi^{\otimes d})$ such that $s(x) \neq 0$. For more details, see [25]. We denote the set of all semi-stable points in $X(w)$ with respect to the T-linearised line bundle $\mathcal{L}_\chi$ by $X(w)^{ss}_T(\mathcal{L}_\chi)$.

A point x in $X(w)$ is said to be stable with respect to the T-linearised line bundle $\mathcal{L}_\chi$ if x is in $X(w)^{ss}_T(\mathcal{L}_\chi)$, the T-orbit of x in $X(w)^{ss}_T(\mathcal{L}_\chi)$ is closed and the isotropy subgroup of x in T is finite. We denote the set of all stable points in $X(w)$ with respect to the T-linearised line bundle $\mathcal{L}_\chi$ by $X(w)^{s}_T(\mathcal{L}_\chi)$.

3 Smooth GIT Quotients of G/P for T-Linearised Line Bundles

In this section, we give some examples of smooth projective varieties as GIT quotients of G/P for some T-linearised line bundles.

We first recall some notation from Sect. 2. Let G be a semisimple algebraic group over $\mathbb{C}$. Let T be a maximal torus of G. Let B be a Borel subgroup of G containing

T. Let P be a parabolic subgroup of G containing B. We recall that P is of the form $P = \bigcap_{r \in J} P_r$ for some subset J of S, where P_r is the maximal parabolic subgroup of G corresponding to the simple root α_r. In the case of $G = SL(n, \mathbb{C})$, S is indexed by $\{1, 2, \ldots n-1\}$.

We denote the greatest common divisor of two positive integers r and n by (r, n). Then, we have:

Theorem 3.1 [15, Theorems 3.1 and 3.5]

1. *Let $G = SL(n, \mathbb{C})$. Let $P = \bigcap_{r \in J} P_r$ for some subset J of $\{1, 2, \ldots, n-1\}$. Then, there is a line bundle $\mathcal{L}$ on G/P such that $G/P_T^{ss}(\mathcal{L}) = G/P_T^{s}(\mathcal{L})$ if and only if the least common multiple of $\{n/(r, n) \colon r \in J\}$ is n.*
2. *Let G be a simple algebraic group of type different from A. Let P be a parabolic subgroup of G containing B. Then, there is a line bundle $\mathcal{L}$ on G/P such that $G/P_T^{ss}(\mathcal{L}) = G/P_T^{s}(\mathcal{L})$ if and only if $P = B$.*

The special case of a maximal parabolic subgroup of $SL(n, \mathbb{C})$ is studied in [14].

We now give an example of a smooth projective variety as a GIT quotient of G/P for some choice of a T-linearised line bundle.

To describe the example with details, we need the following lemma on the root system of type A_{n-1}.

Let R denote the root system of type A_{n-1}. Let R^+ denote the set of positive roots. Let $S := \{\alpha_1, \alpha_2, \ldots \alpha_{n-1}\} \subset R^+$ denote the set of simple roots. Here, we choose the ordering of simple roots as in Humphreys [10]. Let $\{\beta_1, \beta_1, \ldots \beta_m\}$ be a subset of R^+ such that the $\mathbb{Q}$ spans $\sum_{j=1}^{m} \mathbb{Q}\beta_j$ and $\sum_{i=1}^{n-1} \mathbb{Q}\alpha_i$ are equal. Then, we have

Lemma 3.2. $\sum_{j=1}^{m} \mathbb{Z}\beta_j = \sum_{i=1}^{n-1} \mathbb{Z}\alpha_i$.

Proof. Proof is by induction on the rank $= n-1$. If $n - 1 = 1$, $m = 1$ and $\beta_1 = \alpha_1$. Hence, we are done.

In the general case, since $\alpha_1 \in \sum_{j=1}^{m} \mathbb{Q}\beta_j$, we see that there is a $j \in \{1, 2, \ldots m\}$ such that $\alpha_1 \leq \beta_j$. Let J be the set of integers j in $\{1, 2, \ldots m\}$ such that $\alpha_1 \leq \beta_j$. By the description of positive roots in type A_{n-1}, for every $j \in J$, there is a $a_j \in \{1, 2, \ldots n-1\}$ such that $\beta_j = \sum_{i=1}^{a_j} \alpha_i$. Let $k \in J$ be such that a_k is the minimum in the set $\{a_j : j \in J\}$.

Now, for each $j \in J$ different from k, let $\beta'_j = \beta_j - \beta_k$ and let $\beta'_k = \beta_k$. Also, let $\beta'_j = \beta_j$ for every $j \in \{1, 2, \ldots m\}$ but not in J. Clearly, every β'_j is a positive root. Further, since $\sum_{j=1}^{m} \mathbb{Q}\beta_j$ and $\sum_{i=1}^{n-1} \mathbb{Q}\alpha_i$ are equal, we have $\sum_{j=1}^{m} \mathbb{Q}\beta'_j$ and $\sum_{i=1}^{n-1} \mathbb{Q}\alpha_i$ are equal. We note that β'_j's need not be distinct. However, we can take all the distinct elements among them and work with them.

By replacing each β_j by β'_j, we may assume that there is a unique k in $\{1, 2, \ldots m\}$ such that $\alpha_1 \leq \beta_k$. Hence, the $\mathbb{Q}$ spans $\sum_{j \neq k} \mathbb{Q}\beta_j$ and $\sum_{i=2}^{n-1} \mathbb{Q}\alpha_i$ are equal. By indunction on rank (using the fact that $\{\alpha_2, \alpha_3, \ldots \alpha_{n-1}\}$ is the set of simple roots for type A_{n-2}), we conclude that $\{\alpha_i : i \geq 2\}$ is in the lattice $\sum_{j \neq k} \mathbb{Z}\beta_j$. Since

$\beta_k = \sum_{i=1}^{a_k} \alpha_i$, α_1 is in the lattice spanned by $\{\alpha_i : i \geq 2\}$ and β_k. Thus, α_1 is in the lattice generated by β_j's.

This completes the proof.

Example 3.3. Let $G = SL(n, \mathbb{C})$. Let J be a subset of $\{1, 2, \ldots, n-1\}$ such that the least common multiple of $\{n/(r, n) : r \in J\}$ is n. Let $P = \bigcap_{r \in J} P_r$. Let $\mathcal{L}$ be a line bundle G/P such that $G/P_T^{ss}(\mathcal{L}) = G/P_T^{s}(\mathcal{L})$ (exists by [15, Theorem 3.1]). Then, the GIT quotient $T\backslash\backslash G/P_T^{ss}(\mathcal{L})$ is a smooth projective variety.

Proof. Since $G/P_T^{ss}(\mathcal{L}) = G/P_T^{s}(\mathcal{L})$, for any point $x \in G/P_T^{ss}(\mathcal{L})$, the T-orbit of x is closed in $G/P_T^{ss}(\mathcal{L})$ and the isotropy subgroup T_x of x in T is finite.

Now, let x be a semi-stable point of G/P lying in the cell BwP/P. Let $R^+(w^{-1})$ denote the set of all positive roots made negative by w^{-1}. Choose a subset $\{\beta_1, \beta_2, \beta_3, \ldots \beta_m\}$ of $R^+(w^{-1})$ such that $x = u_{\beta_1}(t_1)u_{\beta_2}(t_2)\ldots u_{\beta_m}(t_m)wP/P$, where each $u_{\beta_j}(t_j)$ is an element of the root group U_{β_j} and $t_j \neq 0$ for every $j = 1, 2, \ldots m$. We note that the isotropy subgroup T_x is equal to $\bigcap_{j=1}^{m} ker(\beta_j)$.

Since the isotropy subgroup T_x is finite, the kernel of the homomorphism ϕ: $T \longrightarrow (\mathbb{C}^\times)^m$ defined by $\phi(t) = (\beta_1(t), \beta_2(t), \ldots, \beta_m(t))$ is finite. Let $T' = \phi(T)$. Then, the group $X(T')$ of the characters of T' is a subgroup of the group $X(T)$ of the characters of T of finite index. Hence, the subset $\{\beta_1, \beta_2, \beta_3, \ldots \beta_m\}$ of $R^+(w^{-1})$ generates the $\mathbb{Q}$-vector space spanned by S. By previous Lemma, S is in the lattice $\sum_{j=1}^{m} \mathbb{Z}\beta_j$. Hence, we have

$$T_x = \bigcap_{j=1}^{m} ker(\beta_j) = \bigcap_{i=1}^{n-1} ker(\alpha_i) = Z(G).$$

Hence, by working with the adjoint group of G, we may assume that the isotropy subgroup is trivial. Thus, $T\backslash\backslash G/P_T^{ss}(\mathcal{L})$ is smooth.

In this example, the GIT quotient of any smooth Schubert variety in G/P is also smooth. A good reference for the study of smooth Schubert varieties is the book of Billey-Lakshmibai [3].

Strickland gave a short proof of the above Theorem. The key idea of her proof is [36, Theorem 1.1, p. 2].

We recall some notation to describe her result.

Let $Y(T)$ denote the group of all one parameter subgroups of G. That is, the group of all homomorphisms of algebraic groups from $\mathbb{C}^\times$ into T.

Let $\langle , \rangle : X(T) \otimes Y(T) \longrightarrow \mathbb{Z}$ be given by $(\chi \cdot \lambda)(a) = a^{\langle \chi, \lambda \rangle}$ for every $a \in \mathbb{C}^\times$.

For every simple root α, there is a fundamental one parameter subgroup λ_α such that $\langle \alpha, \lambda_\alpha \rangle = 1$ and $\langle \beta, \lambda_\alpha \rangle = 0$ for every simple root β different from α. With this notation, we have

Theorem 3.4 [36, Theorem 1.1] *Let G be a semisimple algebraic group. Let P be a parabolic subgroup of G containing B. Let χ be a character of P such that corresponding line bundle $\mathcal{L}_\chi$ is ample. Then, $G/P_T^{ss}(\mathcal{L}_\chi) = G/P_T^{s}(\mathcal{L}_\chi)$ if and only* *if* $\langle \chi, w\lambda \rangle \neq 0$ *for every fundamental one parameter subgroup* λ *of* T *and for every* w *in* W.

We now present another example of a smooth GIT quotient as follows:

Let $G = SL(n+1, \mathbb{C})$. Let B_{n+1} denote the upper triangular matrices in G. Let T be the diagonal matrices in G. Consider the natural inclusion of $SL(n, \mathbb{C})$ in G. Let $B_n = SL(n, \mathbb{C}) \bigcap B_{n+1}$. Then, B_n is a Borel subgroup of $SL(n, \mathbb{C})$.

By [17, Theorem 5.2], there is an ample line bundle $\mathcal{L}$ on $SL(n+1, \mathbb{C})/B_{n+1}$ such that $T\backslash\backslash(SL(n+1, \mathbb{C})/B_{n+1})^{ss}_T(\mathcal{L})$ is isomorphic to $SL(n, \mathbb{C})/B_n$. Thus, we have

Example 3.5. Let $G = SL(n+1, \mathbb{C})$. $T\backslash\backslash(SL(n+1, \mathbb{C})/B_{n+1})^{ss}_T(\mathcal{L})$ is smooth.

4 Projective Normality of the GIT Quotients of G/P for T-Linearised Ample Line Bundles

In this section, we discuss the problem of projective normality related to torus action.

We first recall that in [7], Dabrowski has proved that for any parabolic subgroup P of G, the Zariski closure of a generic T-orbit in G/P is normal. For a precise statement, see [7, Theorem 3.2, p. 327].

In the context of a problem on projective normality for torus actions, we recall that Howard proved that for any parabolic subgroup P of $SL(n, \mathbb{C})$, the Zariski closure $\overline{T \cdot x}$ of the T-orbit of any point x in $SL(n, \mathbb{C})/P$ is projectively normal for the choice of any ample line bundle $\mathcal{L}$ on $SL(n, \mathbb{C})/P$. For a precise statement, see [9, Theorem 5.4, p. 540].

We now let $G = SL(6, \mathbb{C})$. Let T, B as usual. Let P be the maximal parabolic subgroup of G corresponding to the simple root α_2. Here, G/P is the Grassmannian $G_{2,6}$. Clearly, three times the fundamental weight ω_2 is in the root lattice. Hence, by a Theorem of Shrawan Kumar, $\mathcal{L}_{3\omega_2}$ descent to the GIT quotient $T\backslash\backslash G/P^{ss}_T(\mathcal{L}_{3\omega_2})$ (cf. [22, Theorem 3.10, p. 758]).

We let $X = T\backslash\backslash G/P^{ss}_T(\mathcal{L}_{3\omega_2})$ and we denote the descent of $\mathcal{L}_{3\omega_2}$ to X by $\mathcal{M}$.

Counter example The polarised variety $(X, \mathcal{M})$ *is not projectively normal.*

Proof. Let i and j be two integers from $\{1, 2, \ldots 6\}$ such that $i < j$.

Let p_{ij} denote the Plucker coordinate corresponding to the pair (i, j).

Then, using the Standard Monomial Theory for Grassmannian (cf. [32, Theorem 3.7]),we see that $p_{12}p_{14}p_{24}p_{35}p_{36}p_{56}$ is in $H^0(X, \mathcal{M}^{\otimes 2})$ but is not in the image of the multiplication map $H^0(X, \mathcal{M}) \bigotimes H^0(X, \mathcal{M}) \longrightarrow H^0(X, \mathcal{M}^{\otimes 2})$.

Besides, there are also examples of projective normal GIT quotients. For instance, the example 3.5 in Sect. 3 is $SL(n, \mathbb{C})/B_n$. This is obtained as a GIT quotient of $SL(n+1, \mathbb{C})/B_{n+1}$. Hence, by a Theorem of Ramanan-Ramanathan, this quotient is projectively normal (cf. [28, Theorem 2]).

In this context, we pose the following problem:

Problem 1: Given a parabolic subgroup P of G containing B, describe all ample line bundles $\mathcal{L}$ which descend to the GIT quotient of G/P for the action of T such that the polarised variety $(G/P^{ss}_T(\mathcal{L}), \mathcal{M})$ is projectively normal.

Here $\mathcal{M}$ is the descent of $\mathcal{L}$ to the GIT quotient $T\backslash\backslash G/P^{ss}_T(\mathcal{L})$.

5 Automorphism Group of the GIT Quotients of G/P for T-Linearised Ample Line Bundles

In this section, we discuss the problem of studying the automorphism group of the GIT quotient of G/P, where P is a parabolic subgroup of G, for the action of T.

Skorobogatov has described the automorphism group of the GIT quotients for maximal parabolic subgroups with some conditions on the isotropy subgroups of stable points. He considers a semisimple algebraic over a field k and describes his result over its algebraic closure $\overline{k}$.

Since the base field in this article is $\mathbb{C}$, we present his result with our usual notation.

Let G, T, B as usual. Recall from Sect. 2, P_i denote the maximal parabolic subgroup of G containing B corresponding to the simple root α_i.

Let Aut(S) denote the automorphism of the Dynkin diagram of G with respect to T and B. Let Aut(S, α_i) denote the subgroup of all elements of Aut(S) fixing α_i. Let $\mathcal{L}$ denote the ample generator of the Picard group of G/P_i.

Since the normaliser N of T acts on G/P, the Weyl group W acts on the quotient $T\backslash\backslash(G/P_i)^s_T(\mathcal{L})$.

Then, we have

Theorem 5.1 [34, Theorem 2.2] *Assume that the complement of the set of stable points in G/P_i with trivial isotropy subgroups in T has codimension at least two and the pair (G, α_i) is not of the form (B_n, α_n) and (G_2, α_1). Then, the automorphism group of $T\backslash\backslash(G/P_i)^s_T(\mathcal{L})$ is isomorphic to the semi direct product $W \rtimes Aut(S, \alpha_i)$.*

Up to our knowledge, the automorphism group of these GIT quotients are not studied when P is not a maximal parabolic subgroup of G. It is interesting to study the automorphism groups of the GIT quotients of Schubert varieties in G/P.

In this context, we pose the following problem:

Problem 2: Let G be a semisimple algebraic group over $\mathbb{C}$. Let T, B as usual. Let P be a parabolic subgroup of G containing B. Let $\mathcal{L}$ be an ample line bundle. Let $w \in W/W_P$. Describe the automorphism group of $T\backslash\backslash X(w)^{ss}_T(\mathcal{L})$.

6 GIT Equivalence of T-Linearised Ample Line Bundles on G/B

In this section, we discuss the work of Zhgoon on the GIT equivalence of T-linearised ample line bundles on G/B.

Let G, T, B as usual. Let χ_1 and χ_2 be two regular dominant characters of T. χ_1 and χ_2 are said to be equivalent if $G/B_T^{ss}(\mathcal{L}_{\chi_1}) = G/B_T^{ss}(\mathcal{L}_{\chi_2})$.

Let A denote the cone generated by all simple roots. Given a χ in the interior C^0 of the fundamental Weyl chamber C, the cone σ_χ is defined by $\sigma_\chi = \bigcap_{w \in W \chi \in (-wA)} (-wA) \bigcap C$.

We then have:

Theorem 6.1 [40, Theorem 1.9] *The set* $\{\sigma_\chi : \chi \in C^0\}$ *of cones is finite. These cones form a fan with support C. The interior points of these cones corresponds to GIT equivalence classes.*

When $G = SL(n+1, \mathbb{C})$, let C_1 denote the set of all regular dominant characters χ of the form $\chi = \sum_{j=1}^n a_j\alpha_j$, where $a_n > a_{n-1} \ldots a_2 > a_1 \geq 1$. Let $\chi \in C_1$. Let $J = \{1, 2, \ldots n-1\}$. Recall from Sect. 2 that W_J is the subgroup of W generated by s_j such that $j \in J$. By [17, Lemma 5.1], we see that $\chi \in -w(A)$ if and only if w is in the left coset $W_J s_n s_{n-1} \ldots s_1$. Hence, we have $C_1 = \sigma_\chi$ for some $\chi \in C_1$.

Further, by [17, Theorem 5.2], we see that the GIT quotient $T\backslash\backslash(SL(n+1, \mathbb{C})/B_{n+1})_T^{ss}(\mathcal{L}_\chi)$ is isomorphic to $SL(n, \mathbb{C})/B_n$ for any $\chi \in C_1$. Therefore, C_1 is a good example of the interior of a cone in the sense of Zhgoon.

Shrawan Kumar and Zhgoon have studied the Picard group of $T\backslash\backslash(SL(n+1, \mathbb{C})/B_{n+1})_T^{ss}(\mathcal{L}_\chi)$ for regular dominant characters χ (cf. [41, Sect. 3] and [22, Theorem 3.10]).

Hence, it is interesting to study the GIT quotient $T\backslash\backslash G/B_T^{ss}(\mathcal{L}_\chi)$ corresponding to every regular dominant character χ in the interior of every cone explicitly.

7 Connection Between the Algebra $\oplus_{d\in\mathbb{Z}_{\geq 0}} H^0(G/B, \mathcal{L}_\chi^{\otimes d})^T$ and the Semi-stable Points $X(c)_T^{ss}(\mathcal{L}_\chi)$

In this section, we discuss the connection between the set $X(c)_T^{ss}(\mathcal{L}_\chi)$ of semi-stable points for a Coxeter element c and the algebra $\oplus_{d\in\mathbb{Z}_{\geq 0}} H^0(G/B, \mathcal{L}_\chi^{\otimes d})^T$.

The ring of invariants of a finitely generated K algebra being a polynomial ring over K is an interesting problem in Invariant Theory.

In Chevalley [6], Serre [29], Shephard and Todd [33] have proved that for any finite dimensional faithful representation V of a finite group H over $\mathbb{C}$, the ring $\mathbb{C}[V]^H$ of H- invariants of $\mathbb{C}[V]$ is a polynomial algebra if and only if H is generated by pseudo reflections in $GL(V)$.

Chevalley also proved that for any semisimple algebraic group G over $\mathbb{C}$, the ring $\mathbb{C}[\mathfrak{g}]^G$ of G invariants of the coordinate ring $\mathbb{C}[\mathfrak{g}]$ of the adjoint representation $\mathfrak{g}$ of G is a polynomial algebra (cf. [10, p. 127]).

In [35], Steinberg proved that for any semisimple simply connected algebraic group G (over any algebraically closed field K) acting on itself by conjugation, the ring of G-invariants $K[G]^G$ is a polynomial ring.

For instance, see [12, Theorem in p. 41].

In [38], Wehlau obtained a complete characterisation of those rational representations V of a torus S such the algebra of S- invariants in the coordinate ring $K[V]$ is a polynomial algebra (cf. [38, Theorem 5.8]).

So, the above mentioned results tempt the authors in [19] to ask the following question:

For which dominant character χ of T the algebra $\oplus_{d\in\mathbb{Z}_{\geq 0}} H^0(G/B, \mathcal{L}_\chi^{\otimes d})^T$ of the T-invariants of the homogeneous coordinate ring $\oplus_{d\in\mathbb{Z}_{\geq 0}} H^0(G/B, \mathcal{L}_\chi^{\otimes d})$ is a polynomial ring ?

We use the following concepts in stating Theorem 7.1.

First, we recall from [12, p. 56] that an element w of W is said to be a Coxeter element if it has a reduced expression of the form $w = s_{i_1}s_{i_2}\dots s_{i_l}$, where $i_j \neq i_k$ when ever $j \neq k$ and l is the rank of G.

Since the group $X(T)$ of characters of T is a free abelian group, we use additive notation for the group law in $X(T)$ in defining indecomposable dominant characters as follows:

We say that a dominant character χ is said to be indecomposable if it can not be expressed as a sum of two non zero dominant characters.

We note that since G is adjoint, the group $X(T)$ of characters of T is the same as the root lattice. In particular, a fundamental weight is not necessarily a character of T. So, the notion of an indecomposable dominant character is not the same as the notion of a fundamental weight.

For instance, the indecomposable dominant characters of T in the case when G is $\mathbb{P}GL(3, \mathbb{C})$ are $2\alpha_1 + \alpha_2$, $\alpha_1 + \alpha_2$ and $\alpha_1 + 2\alpha_2$.

The following Theorem is an illustration of connecting the set $X(w)_T^{ss}(\mathcal{L}_\chi)$ of semi-stable points with freeness of the algebra $\oplus_{d\in\mathbb{Z}_{\geq 0}} H^0(G/B, \mathcal{L}_\chi^{\otimes d})^T$.

Theorem 7.1 [19, Theorem 4.8] *Let χ be an indecomposable dominant character of T such that there is a Coxeter element w in the Weyl group W for which $X(w)_T^{ss}(\mathcal{L}_\chi) \neq \emptyset$.*

Then, the graded $\mathbb{C}$- algebra $\oplus_{d\in\mathbb{Z}_{\geq 0}} H^0(G/B, \mathcal{L}_\chi^{\otimes d})^T$ is a polynomial ring if and only if $dim(H^0(G/B, \mathcal{L}_\chi)^T) \leq rank(G)$.

The special case of this result applied to $\chi = \alpha_0$, the highest root is of some interest.

We discuss this briefly.

Since $H^0(G/B, \mathcal{L}_{\alpha_0})$ is an irreducible self dual G module with highest weight α_0, the G modules $H^0(G/B, \mathcal{L}_{\alpha_0})$ and $\mathrm{Hom}(\mathfrak{g}, \mathbb{C})$ are isomorphic.

On the other hand, the natural T-invariant projection from $\mathfrak{g}$ to $\mathfrak{h}$ induces an isomorphism $Hom(\mathfrak{h}, \mathbb{C}) \to Hom(\mathfrak{g}, \mathbb{C})^T$.

Thus we have a homomorphism $f\colon \mathbb{C}[\mathfrak{h}] \longrightarrow \oplus_{d\in\mathbb{Z}_{\geq 0}} H^0(G/B, \mathcal{L}_{\alpha_0}^{\otimes d})^T$ of $\mathbb{C}$ algebras.

Further, we also have an isomorphism $Hom(\mathfrak{h}, \mathbb{C}) \to H^0(G/B, \mathcal{L}_{\alpha_0})^T$. Thus, we have $\dim(H^0(G/B, \mathcal{L}_{\alpha_0})^T) = \mathrm{rank}(G)$.

By [18, Theorem 4.2], there is a Coxeter element w in W such that $X(w)_T^{ss}(\mathcal{L}_{\alpha_0}) \neq \emptyset$ if and only if G is of type A_n, B_2 or C_n.

Hence, we can apply Theorem 3.1 only if G is of type A_n, B_2 or C_n to conclude that the graded algebra $\oplus_{d\in\mathbb{Z}_{\geq 0}} H^0(G/B, \mathcal{L}_{\alpha_0}^{\otimes d})^T$ is a polynomial ring.

It is a natural question to ask whether we can describe this polynomial ring explicitly, for instance in terms of the homomorphism $f\colon \mathbb{C}[\mathfrak{h}] \longrightarrow \oplus_{d\in\mathbb{Z}_{\geq 0}} H^0(G/B, \mathcal{L}_{\alpha_0}^{\otimes d})^T$.

In this direction, the following Theorem is proved in [19].

Theorem 7.2 [19, Theorem 3.3] *The homomorphism $f\colon \mathbb{C}[\mathfrak{h}] \longrightarrow \oplus_{d\in\mathbb{Z}_{\geq 0}} H^0(G/B, \mathcal{L}_{\alpha_0}^{\otimes d})^T$ is an isomorphism if and only if $X(w)_T^{ss}(\mathcal{L}_{\alpha_0})$ is non-empty for some Coxeter element w in W.*

Is the first assumption of Theorem 7.1 necessary?

In this context, we pose the following problem:

Problem 3: Is it true that if the algebra $\oplus_{d\in\mathbb{Z}_{\geq 0}} H^0(G/B, \mathcal{L}_{\chi}^{\otimes d})^T$ is a polynomial ring, then, there is a Coxeter element w in W such that $X(w)_T^{ss}(\mathcal{L}_{\chi})$ is non-empty ?

The above Theorem describes the ring $\oplus_{d\in\mathbb{Z}_{\geq 0}} H^0(G/B, \mathcal{L}_{\alpha_0}^{\otimes d})^T$ only when G is of type A_n, B_2 or C_n.

So, it is interesting to study the following problem:

Let G be a simple algebraic group of type different from A_n, B_2 and C_n.

Problem 4: What are the generators and relations of the algebra $\oplus_{d\in\mathbb{Z}_{\geq 0}} H^0(G/B, \mathcal{L}_{\alpha_0}^{\otimes d})^T$?

8 Connection with Automorphism Group of a Smooth Schubert Variety

Studying the automorphism group of a variety is an interesting problem. For instance, Onishchik in [27] and Tits in [37] have studied the automorphism groups of some homogeneous spaces.

In this section, we state a result from [20] which connects the automorphism group of a smooth Schubert variety with the semi-stable points.

Again, we assume that G is semisimple. We follow the convention in [8, 13]. So, let B be the Borel subgroup of G containing T corresponding to negative roots.

In [5], Bott proved that for any semisimple algebraic group G, for any Borel subgroup B of G, all the higher cohomologies $H^i(G/B, \mathcal{T}_{G/B})$ of the tangent bundle $\mathcal{T}_{G/B}$ on the flag variety G/B vanishes. He further showed that the G- module of global sections $H^0(G/B, \mathcal{T}_{G/B})$ is the adjoint representation $\mathfrak{g}$ of G.

Consider the restriction of $\mathcal{T}_{G/B}$ to a Schubert variety $X(w)$. It is interesting to know whether all higher cohomologies $H^i(X(w), \mathcal{T}_{G/B})$ vanish. It is also interesting to study the B- module $H^0(X(w), \mathcal{T}_{G/B})$ of global sections.

The tangent space of idB in G/B is the B-module $\mathfrak{g}/\mathfrak{b}$. Since B corresponds to negative roots, the T-module $\mathfrak{g}/\mathfrak{b}$ is a direct sum of weight spaces each of whose weight is not dominant except the highest root α_0 if G is simply laced.

Since we are dealing with line bundles $\mathcal{L}_\chi$ on Schubert varieties corresponding to non-dominant characters χ of T, we would like to mention the developments in the study of cohomology of line bundles corresponding to non-dominant characters on Schubert varieties.

Demazure character formula gives the Euler characteristic of the line bundle $\mathcal{L}_\chi$ corresponding to a character χ of T (see [13, II, 14.18]) or (see [1, Theorem in p. 617]). When χ is dominant, using Frobenius splitting methods, Mehta-Ramanathan proved that $H^i(X(w), \mathcal{L}_\chi) = (0)$ for all $w \in W$ and for all $i \geq 1$ (cf. [24]). Hence, the character of the T-module $H^0(X(w), \mathcal{L}_\chi)$ is given by the Demazure character formula. Further, Ramanan-Ramanathan use Frobenius splitting methods to prove some vanishing results and deduce the projective normality of Schubert varieties (cf. [28, Theorem 2]).

Using Standard Monomial Theory methods, Lakshmibai-Musili-Seshadri have proved cohomology vanishing results for Schubert varieties. They deduce geometric properties like Arithmetic Cohen Macaulay using the vanishing results (cf. [32, Theorem 1.4.19]). Further, Littelmann has constructed a basis of $H^0(X(w), \mathcal{L}_\chi)$ corresponding to the set of all LS paths of shape χ which is compatible with the geometry of Schubert varieties. He used such a basis to deduce the projective normality property of Schubert varieties (cf. [23]).

Demazure character formula for non dominant characters is an important formula for the study of cohomology of line bundles on a Schubert variety corresponding to non dominant characters. For recent study in the non dominant cases, see [2, 16].

Therefore, study of the cohomology modules of line bundles on Schubert varieties is interesting.

We state the following theorem from [20] which relates the cohomological problem on the Schubert variety $X(w)$ with the set $X(w)^{ss}_T(\mathcal{L}_{\alpha_0})$ of semi-stable points. Here, we assume that G is simple and simply laced.

Theorem 8.1. *1.* $H^i(X(w), \mathcal{L}(\mathfrak{g}/\mathfrak{b})) = (0)$ *for every* $i \geq 1$.
2. $H^0(X(w), \mathcal{L}(\mathfrak{g}/\mathfrak{b})) = \mathfrak{g}$ *if and only if the set* $X(w^{-1})^{ss}_T(\mathcal{L}_{\alpha_0})$ *of semi-stable points is non-empty.*

Since $\mathcal{L}(\mathfrak{g}/\mathfrak{b})$ is the tangent bundle of G/B , this Theorem can be formulated as follows:

Theorem 8.1 [20, Theorem 3.8]

1. $H^i(X(w), \mathcal{T}_{G/B}) = (0)$ *for every* $i \geq 1$.
2. $H^0(X(w), \mathcal{T}_{G/B})$ *is the adjoint representation* $\mathfrak{g}$ *of* G *if and only if the set* $X(w^{-1})^{ss}_T(\mathcal{L}_{\alpha_0})$ *of semi-stable points is non-empty.*

Proof of the above Theorem essentially uses the idea of Demazure's short proof of Borel-Weil-Bott's theorem in [8]. Using the above Theorem, the following corollary is obtained in [20].

Let G, B, T, χ, w be as above. Further, we assume that $X(w)$ is smooth. Let $\mathrm{Aut}(X(w))^0$ denote the connected component of the identity element of the automorphism group $\mathrm{Aut}(X(w))$.

The following Corollary is an illustration of the connection between semi-stable points and the automorphism group.

Corollary 8.3 [20, Corollary 3.9] *$Aut(X(w))^0$ is a parabolic subgroup of G containing B if and only if the set $X(w^{-1})^{ss}_T(\mathcal{L}_{\alpha_0})$ of semi-stable points is non-empty.*

9 Connection with a Cohomological Characterisation of a Coxeter Element

We first recall the dot action of W on $X(T)$. Let $w \in W$ and let $\chi \in X(T)$. Let ρ denote the half sum of positive roots. Then, $w \cdot \chi$ is defined by $w \cdot \chi = w(\chi + \rho) - \rho$.

In Sect. 3, we have discussed the connection between $X(w)^{ss}_T(\mathcal{L}_{\alpha_0})$ and $H^0(G/B, \mathcal{L}_{\alpha_0}^{\otimes m})^T$.

In Sect. 4, we have discussed the connection between $X(w)^{ss}_T(\mathcal{L}_{\alpha_0})$ and $H^0(X(w), \mathcal{T}_{G/B})$.

In this section, we discuss the connection between $X(w)^{ss}_T(\mathcal{L}_{\alpha_0})$ and the cohomology modules $H^i(X(w), \mathcal{L}_{w^{-1}\cdot 0})$.

Let $w \in W$. Consider the weight $\chi = w^{-1} \cdot 0$.

By Borel-Weil-Bott's theorem (cf. [5]), all cohomologies $H^i(G/B, \mathcal{L}_\chi)$ vanish except $i = l(w)$ and $H^{l(w)}(G/B, \mathcal{L}_\chi)$ is the trivial one dimensional representation of G.

This special case of Borel-Weil-Bott's theorem tempts the authors in [21] to ask the following question:

Given $w \in W$, is it true that $H^i(X(w), \mathcal{L}_{w^{-1}\cdot 0})$ vanish except $i = l(w)$ and $H^{l(w)}(X(w), \mathcal{L}_{w^{-1}\cdot 0})$ is the trivial one dimensional representation of B.

In this context, we state the following partial result from [21]:

Theorem 9.1.
1. *Let w be an element of W. The cohomology module $H^{l(w)}(X(w), \mathcal{L}_{w^{-1}\cdot 0})$ is the one dimensional trivial representation of B.*
2. *Let c be a Coxeter element of W. Then, $H^i(X(c), \mathcal{L}_{c^{-1}\cdot 0})$ is zero for every $i \neq l(c)$ if and only if both $X(c)^{ss}_T(\mathcal{L}_{\alpha_0})$ and $X(c^{-1})^{ss}_T(\mathcal{L}_{\alpha_0})$ are non-empty.*

Let c be a Coxeter element of W. Let C be the cyclic subgroup of W generated by c.

The following Proposition of Yang-Zelevinsky is useful to calculate the weights in the C-orbit of a fundamental character in the proof of the above Theorem:

Proposition 9.2. [39, Proposition 1.3] *Let c be a Coxeter element. Let α be a simple root. Then, there is a $j \in \mathbb{N}$ such that $c^j(\omega_\alpha) = w_0(\omega_\alpha)$.*

We note that there are Coxeter elements c such that either $X(c)^{ss}_T(\mathcal{L}_{\alpha_0})$ is empty or $X(c^{-1})^{ss}_T(\mathcal{L}_{\alpha_0})$ is empty. Hence, by Theorem 6.1, there is at least one $i \neq l(c)$ such that $H^i(X(c), \mathcal{L}_{c^{-1}.0})$ is non zero. So, for such c, the character of the T-module $H^i(X(c), \mathcal{L}_{c^{-1}.0})$ is not completely understood.

In this connection, we pose the following question:

Problem 5: Let w be an element of W.

1. What are the integers i (in terms of a combinatorial description of w) for which $H^i(X(w), \mathcal{L}_{w^{-1}.0})$ is non-zero ?
2. What are the characters of these T-modules ?

Acknowledgments We thank the referees for the useful comments. We thank Professor C.S. Seshadri for useful discussions

References

1. Andersen, H.H.: Schubert varieties and Demazure's character formula. Invent. Math. **79**, 611–618 (1985)
2. Balaji, V., Kannan, S.S., Subrahmanyam, K.V.: Cohomology of line bundles on Schubert varieties-I. Transform. Groups **9**(2), 105–131 (2004)
3. Billey, S.C., Lakshmibai, V.: Singular Loci of Schubert Varieties, Progress in Mathematics, vol. 182. Birkhäuser, Boston (2000)
4. Borel, A.: Linear representations of semi-simple algebraic groups. Proc. Symp. Pure Math. **29**, 421–440 (1975)
5. Bott, R.: Homogeneous vector bundles. Ann. Math. Ser. **2**(66), 203–248 (1957)
6. Chevalley, C.: Invariants of finite groups generated by reflections. Amer. J. Math. **77**, 778–782 (1955)
7. Dabrowski, R.: On normality of the closure of a generic torus orbit in G/P. Pacific J. Math. **172**(2), 321–330 (1996)
8. Demazure, M.: A very simple proof of Bott's theorem. Invent. Math. **33**, 271–272 (1976)
9. Howard, B.J.: Matroids and geometric invariant theory of torus actions on flag spaces. J. Algebra **312**, 527–541 (2007)
10. Humphreys, J.E.: Introduction to Lie Algebras and Representation Theory. Springer, Berlin (1972)
11. Humphreys, J.E.: Linear Algebraic Groups. Springer, New York (1975)
12. Humphreys, J.E.: Conjugacy classes in semisimple algebraic groups, Math. Surveys Monographs. Amer. Math. Soc. **43**, 2961–2988 (1995)
13. Jantzen, J.C.: Representations of Algebraic Groups, Pure and Applied Mathematics. Academic Press, Orlando (1987)
14. Kannan, S.S.: Torus quotients of homogeneous spaces. Proc. Indian Acad. Sci. (Math. Sci.) **108**(1), 1–12 (1998)
15. Kannan, S.S.: Torus quotients of homogeneous spaces-II, Proc. Indian Acad. Sci. (Math. Sci.) **109**(1), 23–39 (1999)
16. Kannan, S.S.: Cohomology of line bundles on Schubert varieties in the Kac-Moody setting. J. Algebra **310**, 88–107 (2007)
17. Kannan, S.S., Sardar, P.: Torus quotients of homogeneous spaces of the general linear group and the standard representation of certain symmetric groups. Proc. Indian Acad. Sci. (Math. Sci.) **119**(1), 81–100 (2009)

18. Kannan, S.S., Pattanayak, S.K.: Torus quotients of homogeneous spaces-minimal dimensional Schubert varieties admitting semi-stable points. Proc. Indian Acad. Sci. (Math. Sci.) **119**(4), 469–485 (2009)
19. Kannan, S.S., Chary, B.N., Pattanayak, S.K.: Torus invariants of the homogeneous coordinate ring of G/B—connection with Coxeter elements. Commun. Algebra **42**(5), 1880–1895 (2014)
20. Kannan, S.S.: On the automorphism group of a smooth Schubert variety (2014, preprint), http://www.arxiv.org/abs/1312.7066
21. Kannan, S.S., Chary, B.N.: Top cohomology of a Schubert variety with trivial B-module. preprint (2014)
22. Kumar, S.K.: Descent of line bundles to GIT quotients of flag varieties by maximal torus. Transform. Groups **13**(3–4), 757–771 (2008)
23. Littelmann, P.: Contracting modules and standard monomial theory for symmetrizable Kac-Moody algebras. J. Amer. Math.Soc. **11**(3), 551–567 (1998)
24. Mehta, V.B., Ramanathan, A.: Frobenius splitting and cohomology vanishing for Schubert varieties. Ann. Math. (2) **122**(1), 27–40 (1985)
25. Mumford, D., Fogarty, J., Kirwan, F.: Geometric Invariant Theory, 3rd edn. Springer, New York (1994)
26. Newstead, P.E.: Introduction to Moduli Problems and Orbit Spaces, TIFR Lecture Notes, Tata Institute of Fundamental Research, Bombay (1978)
27. Onishchik, A.L.: On compact Lie groups transitive on certain manifolds. Sov. Math. Dokl. **1**, 1288–1291 (1961)
28. Ramanan, S., Ramanathan, A.: Projective normality of flag varieties and Schubert varieties. Invent. Math. **79**(2), 217–224 (1985)
29. Serre, J.P.: Groupes finis d'automorphismes d'anneaux locaux reguliers, pp. 1–11. Colloq. d'Alg. Ecole Norm. de Jeunes Filles, Paris (1967)
30. Seshadri, C.S.: Mumford's conjecture for $GL(2)$ and applications, International Colloquim on Algebraic Geometry, Bombay, (1968)
31. Seshadri, C.S.: Quotient Spaces modulo reductive algebraic groups. Ann. Math. **95**, 511–556 (1972)
32. Seshadri, C.S.: Introduction to the Theory of Standard Monomials, Texts and Readings in Mathematics 46. Hindustan Book Agency, New Delhi (2007)
33. Shephard, G.C., Todd, J.A.: Finite unitary reflection groups. Can. J. Math. **6**, 274–304 (1954)
34. Skorobogatov, A.N.: Automorphisms and forms of toric quotients of homogeneous spaces (Russian). Mat. Sb. **200**(10), 107–122 (2009); translation inSb. Math. **200**(9–10), 1521–1536 (2009)
35. Steinberg, R.: Regular elements of semisimple algebraic groups. Inst. des Hautes Etudes Sci. Publ. Math. **25**, 49–80 (1965)
36. Strickland, E.: Quotients of flag varieties by a maximal torus. Math. Z. **234**(1), 1–7 (2000)
37. Tits, J.: Espaces homogenes Complexes Compact. Comm. Helv. Math. **37**, 111–120 (1962)
38. Wehlau, D.: When is a ring of torus invariants a polynomial ring. Manuscripta Math. **82**, 161–170 (1994)
39. Yang, S.W., Zelevinsky, A.: Cluster algebras of finite type via Coxeter elements and principal minors. Transform. Groups **13**(3–4), 855–895 (2008)
40. Zhgoon, V.S.: Variation of Mumford quotients by torus actions on full flag varieties. I (Russian). Izv. Ross. Akad. Nauk Ser. Mat. **71**(6), 29–45 (2007); translation inIzv. Math. **71**(6), 1105–1122 (2007)
41. Zhgoon, V.S.: Variation of Mumford quotient by torus actions on full flag varieties. II (Russian). Mat. Sb. **199**(3), 25–44 (2008); translation inSb. Math. **199**(3–4), 341–359 (2008)

Chapter 11
Characterizations of Trialities of Type I_{id} in Buildings of Type D_4

Hendrik Van Maldeghem

Abstract A triality of type I_{id} in a building Δ of type D_4 is a type rotating automorphism of order 3 whose structure of fixed flags is the building of type G_2 related to Dickson's simple groups (in geometric term, this building is the split Cayley generalized hexagon over the field in question). Such a triality exists over any field and is unique up to conjugacy. In this paper, we present two characterizations of such trialities among all type rotating automorphisms (hence not necessarily of order 3). We prove that, if for a type rotating automorphism θ of Δ, no non-fixed line and its image are contained in adjacent chambers, and θ fixes at least one line, then θ is a triality of type I_{id} (here, lines are vertices of type 2, with Bourbaki labeling). Also, if a type rotating automorphism θ of Δ never maps a line to an opposite line, then it is also a triality of type I_{id}. We moreover show that this condition is equivalent with θ not mapping any chamber to an opposite one. The latter completes the programme for type rotating automorphisms of buildings of type D_4 of determining all domestic automorphisms of spherical buildings.

Keywords Triality quadric · Spherical building · Opposition · Generalized hexagon

1 Introduction

A *domestic* automorphism of a spherical building is an automorphism which does not map any chamber to an opposite one. This notion arose from work of Abramenko and Brown [2], who proved that in (thick) non-spherical buildings, only the identity has bounded displacement on the set of chambers. In the spherical case, they proved that only the identity maps no flag to an opposite flag. This led Temmermans, Thas and the present author [8–10] to the more refined definition of *J-domestic automorphism*:

H. Van Maldeghem (✉)
Department of Mathematics, Ghent University, Krijgslaan 281-S22, B-9000 Ghent, Belgium
e-mail: hvm@cage.ugent.be

N. S. N. Sastry (ed.), *Groups of Exceptional Type, Coxeter Groups and Related Geometries,* Springer Proceedings in Mathematics & Statistics 82,
DOI: 10.1007/978-81-322-1814-2_11, © Springer India 2014

this is an automorphism of a spherical building over the type set $I \supseteq J$, with J stable under the opposition map on types, mapping no flag of type J to an opposite flag. For $J = I$, we obtain domesticity as defined above. It is clear that J-domesticity for any (self-opposite) $J \subseteq I$ implies domesticity. Hence in order to understand all J-domestic automorphisms, one has to classify the domestic automorphisms. This was done for projective spaces in [8], for generalized quadrangles in [9], for type interchanging automorphisms of buildings of type E_6 in [14], and for generalized hexagons in work to be published by Parkinson and the author. Partial results for polar spaces are contained in [10]. In a lot of cases, it seems that domesticity is intimately related to the notion of geometric hyperplane in an appropriate point–line geometry arising from the building. For instance, in a projective space, the domestic dualities fix a hyperplane of the line-Grassmannian, namely a linear complex, also known as the set of lines fixed under a symplectic polarity. In a polar pace, an automorphism is line domestic if and only if it pointwise fixes a geometric hyperplane. Also, the domestic dualities of buildings of type E_6 fix pointwise a building of type F_4, the standard point set of which is a geometric hyperplane of the standard point set of the building of type E_6. In the situation of the present paper, this phenomenon will also show up again. But I do not know of an abstract reason for it.

We investigate domestic type rotating automorphisms of (thick) spherical buidings Δ of type D_4 and we show that there is only one conjugacy class of such automorphisms, namely, the trialities of type I_{id} (as defined by Tits in [11]), which fix a split Cayley hexagon. Now, it is well known (see Paragraph 5.2.2 of [11]) that the standard point set of this hexagon is the standard point set of a subbuilding of type B_3, and thus again constitutes a geometric hyperplane of the standard point set of Δ.

We will proceed by showing that a domestic automorphism maps no line to an opposite line, and then use this to prove that it is a triality of type I_{id}. But domesticity also implies that, if we view Δ as a hyperbolic quadric, no line and its image are contained in a common singular plane. We also show that a partial converse holds, in particular, that the fact that no non-fixed line and its image under a type rotating automorphism θ are contained in a plane, and some line is fixed, implies that θ is a triality of type I_{id}. We can rephrase these conditions with no reference to the hyperbolic quadric using the following observation. Two distinct lines of the quadric are contained in a common plane if and only if they are contained in respective chambers which are adjacent (these chambers contain the intersection point and the two maximal singular subspaces through the plane).

Geometrically, the split Cayley hexagons and the symplectic quadrangles are very similar objects, sharing a number of properties and characterizations, see Chap. 6 of [13]. The present paper adds one more such characterization to this list: they both arise from the unique type rotating domestic automorphism of their ambient building.

In fact, the same is true for all symplectic polar spaces. Note that the symplectic polar spaces and the split Cayley hexagons are *split* buildings whose diagram contains a multiple bond. There are two more classes of split buildings with a double bond in the diagram, namely the polar spaces (of type B_n) related to quadrics of maximal Witt index in even dimensional projective spaces, and the standard buildings of type F_4. The latter also arise from domestic polarities, in these cases in buildings of type E_6, and these polarities are (again) unique as domestic type rotating automorphisms. For the former (the polar spaces of type B_n), a characterization has yet to be proved, but there are "polarities" of buildings of type D_{n+1} producing them, and these polarities are domestic. For the moment it is not clear whether these are the only domestic dualities (for low ranks, $n \in \{2, 3\}$, it can be deduced from the results in [10] that they are unique). Hence it seems that the split spherical buildings with a double bond in the diagram can be generally characterized by domesticity of type rotating automorphisms of spherical buildings with only single bonds in the diagram. This would be a rather beautiful and remarkable synthetic characterization of such buildings.

For a relation between domesticity and the Freudenthal–Tits Magic Square, see the introduction of [14].

Finally, our result shows that the *Phan-geometry* of a type rotating automorhism of a building of type D_4 is empty if and only if it is a triality of type I_{id}. Phan-geometries (which are the geometries of chambers mapped on opposite ones by a certain automorphism) play a prominent role in Phan–Curtis–Tits theory, see [6]. Note that in this specific case, the *almost-Phan geometry* (roughly defined as the geometry induced on the chambers mapped "as far away as possible", but not opposite) is studied in [4], where its diagram is established and where it is proved that it is simply connected.

Remark 1.1. Although the general setting of the problem treated in this paper concerns spherical buildings, we are here only concerned about buildings of type D_4, which is a "classical" type. Hence we will always take the natural point–line geometry point of view to approach these buildings, i.e., we consider these buildings as non-thick polar spaces. Then the elements (vertices) of the building (viewed as a simplicial complex) are the singular subspaces except for the next-to-maximal ones (and the maximal ones represent two types of vertices, see below). In fact, in this way, we can identify the building with a hyperbolic quadric of Witt index 4, and we will do so explicitly below. Note that this is closer to the original approach [12] than to a more recent approach [15]. When we mention flags, vertices, chambers, then we mean the terminology of [12]; in the terminology of [15] these are all residues. But we will not use the terminology of [15] in the present paper.

2 Preliminaries and Main Results

Let Δ be a spherical building, i.e., a building with a finite Weyl group. For the definition of buildings, we refer to [1]. We always consider *thick* buildings, if not explicitly stated otherwise. Let $I = \{1, 2, 3, 4\}$ be the set of types of Δ (and we use Bourbaki labeling [3] for the nodes of the corresponding Coxeter diagram). As mentioned in the introduction, for a subset $J \subseteq I$, a *J-domestic* automorphism is an automorphism which does not map any flag of type J onto an opposite one. If

$J = I$, then we simply talk about a *domestic* automorphism. If J is a singleton, then we sometimes replace the "J" in J-domestic by the name of the type of elements of type J. If J is not stable under the opposition relation, then every automorphism is automatically J-domestic, hence we assume from now on that J is fixed under the opposition relation. It is clear that in this case J-domesticity follows from J'-domesticity if $J' \subseteq J$. Hence the most general situation is that of domesticity. In the present paper, we concentrate on domestic type rotating automorphisms of buildings of type D_4. Our main result will classify all such automorphisms and show that, here, domesticity is equivalent to $\{2\}$-domesticity.

So let Δ be a building of type D_4. It can be identified with a hyperbolic quadric H in a projective space $\mathsf{PG}(7, \mathbb{K})$ of dimension 7 over some (commutative) field $\mathbb{K}$. Hence the elements of Δ are the *points*, the *lines*, and the two systems of maximal isotropic or singular subspaces of dimension 3 of H (the oriflamme complex of H). Each such system plays the same role as the point set, and hence there is the *principle of triality*, which just means that there exist type rotating automorphisms of Δ, inducing period 3 on the types. A *triality*, in the sense of Tits [11], is then a type rotating automorphism of order 3, the smallest possible order it can have. Note that sometimes a type rotating automorphism itself is already called a triality, but we will follow Tits's terminology here, and call a general type rotating automorphism a *tairality* (after a Welsh expression "tair" for three). Since lines are the elements of type 2 in Δ, we also speak about *line-domesticity* instead of $\{2\}$-domesticity, as explained in the previous paragraph.

Tits classified in [11] all trialities which admit at least one fixed flag of type $\{1, 3, 4\}$. The most interesting trialities are the ones of type I_σ, where σ is an automorphism of the underlying field $\mathbb{K}$. The fixed flags of such trialities define a building of type G_2, a so-called *generalized hexagon*. If $\sigma = \mathsf{id}$, the identity, then the standard point set of such hexagon is exactly the standard point set of a subbuilding of type B_3—in geometric terms, a parabolic quadric obtained from H by intersecting it with a non-tangent hyperplane of $\mathsf{PG}(7, \mathbb{K})$. The hexagons related to trialities of type I_{id} are called *split Cayley hexagons*, because they can be constructed directly using the split Cayley algebra over $\mathbb{K}$, see Schellekens [7].

We can now state our first main result.

Main Result 2.1 *A type rotating automorphism θ of a building Δ of type D_4 is a triality of type I_{id} if and only if θ is domestic if and only if θ is line-domestic.*

Note that two lines L, M in H can be in six distinct possible mutual positions:

1. L could be equal to M;
2. L could be concurrent with M and $L \cup M$ contained in a singular subspace of H;
3. L could be concurrent with L but some point of L is not collinear with some point of M;
4. $L \cup M$ could be contained in a singular subspace of H but not in a plane;
5. there could be a unique point on L collinear to all points of M (and then there is a unique point on M collinear with all points of L);
6. L and M are opposite, so each point of L is collinear to a unique point of M.

Viewing L and M as elements of Δ, the cases (3) and (4) are the same, since both say that L and M are incident with a unique common element of Δ. In case (5), we say that (L, M) is a *special pair*. Case (2) is equivalent to saying that L and M are distinct and contained in adjacent chambers.

Now we can state our second main result.

Main Result 2.2 *A type rotating automorphism θ of a building Δ of type D_4 is a triality of type I_{id} if and only if no non-fixed line and its image under θ are contained in adjacent chambers, and θ fixes some line.*

Concerning Main Result 2.2, it also seems to be a general phenomenon for the split buildings with a multiple bond in the diagram that a certain mutual position other than opposition of a vertex and its image of the corresponding type rotating automorphism of the ambient building with only single bonds in the diagram cannot occur, and that this automorphism is characterized by that property. For instance, [14] shows that symplectic polarities θ of buildings of type E_6 are characterized by the absence of pairs (p, p^θ), where p is a point and p^θ is a "symp" not opposite and not incident with p, and the fact that there is at least one pair (p, p^θ) with the point p incident with p^θ. A similar thing holds for symplectic polarities in projective space; we will show in a short appendix the following characterization.

Proposition 2.3. *Let θ be a duality of a projective space $\mathsf{PG}(n, \mathbb{K})$, $n \geq 3$, for some skew field $\mathbb{K}$. Then θ is a symplectic polarity if and only if for no line L the image L^θ intersects L in precisely one point, and there is some line L included in its image L^θ if and only if for no line L the image L^θ intersects L in precisely one point, and there is some absolute point.*

The restriction $n \geq 3$ stems from the fact that for $n = 2$, the hypotheses are self-contradictory (and there is indeed no symplectic polarity in a plane). Note also that the last criterion is the most general because if a line is included in its image, then every point of that line is absolute (a point being *absolute* if it is contained in its image).

3 Proof of Main Result 2.1

Let θ be a type rotating automorphism of a building Δ of spherical type D_4. We repeat that we call θ a *tairality* (see the introduction). If θ has order 3, then we call it a *triality*. Let H be the corresponding hyperbolic quadric in $\mathsf{PG}(7, \mathbb{K})$, for some well determined field $\mathbb{K}$. Then H can be viewed as a non-thick polar space of rank 4 containing two systems Σ_1 and Σ_2 of maximal singular subspaces (all of dimension 3; they are also called *generators*) with the property that two maximal singular subspaces intersect in a singular subspace of dimension 0 or 2 if and only if they belong to a different system. Each system of generators can be identified with one type of vertices of Δ. Without loss of generality, we assume that θ maps points to elements of Σ_1, maps Σ_1 to Σ_2 and maps Σ_2 to the point set of H. Notice that the planes of H correspond with flags of Δ containing an element of Σ_1 and one of Σ_2.

We will identify Δ with H. In particular, we will view each element of Δ as a set of points (the points incident with that element).

In order to prove Main Result 2.1, it suffices to prove that, if θ is a triality of type $\mathrm{I_{id}}$, then it is line-domestic, and that, if θ is a domestic tairality, it is a triality of type $\mathrm{I_{id}}$. We begin with the former.

Let θ be a triality of type $\mathrm{I_{id}}$, and let L be a line of H. Then L contains at least one absolute point p (as the set of absolute points consists of a parabolic subquadric, see Paragraph 5.2.2 of [11]), and $L^\theta \subseteq p^\theta$. It follows that L and L^θ are not opposite, hence θ is line-domestic. Note that this has also been observed in [4] (Sect. 3) using a coordinate calculation.

So from now on we assume that θ is a domestic tairality, and we aim to show that it is a triality of type $\mathrm{I_{id}}$. We first show that θ is line-domestic.

Lemma 3.1. *If θ is domestic, then it is line-domestic.*

Proof. Suppose, for a contradiction, that some line L is mapped onto an opposite line L^θ. Let p be an arbitrary point on L. Then there is a unique member of Σ_1 incident with L not opposite p^θ, hence we can choose $U_1 \in \Sigma_1$ opposite p^θ. Likewise, there is a unique member of Σ_2 incident with L and not opposite U_1^θ, and there is a unique point q on L^θ not opposite p. Since there are at least three elements of Σ_2 incident with L, we can choose $U_2 \in \Sigma_2$ incident with L such that U_2 is opposite U_1^θ and such that $U_2^\theta \neq q$. Since we now have that p is opposite U_2^θ, that U_1 is opposite p^θ, and that U_2 is opposite U_1^θ, we conclude that $\{p, L, U_1, U_2\}$ is a chamber that is mapped onto an opposite chamber, a contradiction to the domesticity.

We call a point p of H *absolute* if $p \in p^\theta$. Likewise, a member U of $\Sigma_1 \cup \Sigma_2$ is *absolute* if U^θ is incident with U, i.e., if $\{U, U^\theta\}$ is a flag of Δ, or, equivalently, if $U \cap U^\theta$ is a (singular) plane of H if $U \in \Sigma_1$, and $U^\theta \in U$ if $U \in \Sigma_2$. Note that $U \in \Sigma_1 \cup \Sigma_2$ being absolute is equivalent with U^θ being absolute.

Here is a fundamental observation.

Lemma 3.2. *Let L be a line such that (L, L^θ) is a special pair. Then the unique point p on L collinear with all points of L^θ is absolute.*

Proof. Suppose, for a contradiction, that p is not absolute. Then p^θ does not share any point with L. Let π be the plane in p^θ all of whose points are collinear with p. Note that $L^\theta \subseteq \pi$. Let M^θ be a line in p^θ intersecting L^θ in a point x, and not contained in π. We choose x different from the unique point on L^θ collinear with all points of L. Then M is a line through p contained in a plane α together with L. By varying M^θ, we may assume that α does not meet L^θ. It follows that M and M^θ are not concurrent. Since p is not collinear to every point of M^θ, we see that M and M^θ are not contained in a common singular subspace. By domesticity of θ, the pair (M, M^θ) is special. Since p is not collinear to all points of M^θ, but it is collinear to x, it follows that x is collinear to all points of M. Since x is not collinear with all points of L, it is not collinear with all points of a line K in α through p, with $L \neq K \neq M$. But K^θ contains x and is not contained in π. Hence K and K^θ are opposite, a contradiction to the assumption that θ is a domestic tairality. Hence p is absolute.

We now investigate properties of the set A of absolute points. Our first main aim in this respect is to show that θ has order 3 when restricted to A.

If p is an absolute point of θ, then we denote by π_p the intersection of p^θ with p^{θ^2}.

Lemma 3.3. *Let p be an absolute point of θ. Then all points of π_p are absolute.*

Proof. Let x be an arbitrary point in π_p. Let L be any line in p^θ through x but not through p. Then L^θ is a line in p^{θ^2}, but not in p^θ. If $x \in L^\theta$, then, since $L^\theta \subseteq x^\theta$, clearly x is absolute. If $x \notin L^\theta$, then Lemma 3.2 implies that x is absolute. Hence every point x of $\pi := p^\theta \cap p^{\theta^2}$ is absolute.

Lemma 3.4. *Let p be an absolute point of θ. Then $p \in p^{\theta^2}$.*

Proof. Suppose by way of contradiction that $p \notin p^{\theta^2}$. Let x be an arbitrary point of π_p. By Lemma 3.3, we know that x is absolute. Then x^θ intersects π_p in a line L_x through x (since it intersects p^{θ^2} in a plane). It follows that $(px)^\theta = L_x$ is a line of π_p through x. But the map sending a line in p^θ through p to a line in π_p induces an isomorphism of projective planes; hence the mapping from π_p to π_p mapping the point x to the line L_x is the restriction to the points of a duality of π_p. But since $x \in L_x$, no flag can be mapped to an opposite, a contradiction to Theorem 3.1 of [8] and the fact that there are no symplectic polarities in a projective plane.

This contradiction proves the lemma.

Lemma 3.5. *Let p be an absolute point of θ. Then every line in π_p through p is fixed under θ. Also, $p = p^{\theta^3}$.*

Proof. Suppose, for a contradiction, that some line L through p in π_p is not fixed under θ. Since $p \in L$ and $L \subseteq p^\theta$, the image L^θ is contained in π_p. Put $z = L \cap L^\theta$. Let $y \in L \setminus \{p, z\}$. Since, by Lemma 3.3, y is absolute, $y \in y^\theta$. Since $y \in L$, we also have $L^\theta \subseteq y^\theta$. Hence $\pi_p \in y^\theta$, implying $p^\theta = y^\theta$, a contradiction. Hence L is fixed.

Now suppose $p \neq p^{\theta^3}$. Lemma 3.4 implies $p \in p^{\theta^2}$; hence p^θ contains p^{θ^3}. It follows that $p^{\theta^3} \in \pi_p$. The image of a line through p in π_p is a line through p^{θ^3} in π_p. Now, since $p \neq p^{\theta^3}$, there exists a line L through p in π_p not through p^{θ^3}, and so L cannot be fixed, a contradiction. We conclude that $p = p^{\theta^3}$.

Remark 3.6. Lemma 3.5 implies that no non-fixed line L is mapped onto a line L^θ such that L and L^θ are contained in a plane. Indeed, suppose that L and L^θ are contained in some plane, and $L \neq L^\theta$. Let $x = L \cap L^\theta$, then $x \in L^\theta \subseteq x^\theta$ and so x is absolute. Then $L \subseteq x^{\theta^{-1}} = x^{\theta^2}$. Since no point of $x^\theta \setminus \pi_x$ is collinear with any point of $x^{\theta^2} \setminus \pi_x$, one of L, L^θ is contained in π_x and hence fixed, a contradiction.

Lemma 3.7. *Let p be an absolute point of θ. If q is an absolute point of θ collinear with p and not contained in π_p, then there is a unique point $r \in \pi_p$ such that $\{q, p\} \subseteq \pi_r$. In particular, all points of the line pq are absolute.*

Proof. First we claim that q is not contained in $p^\theta \cup p^{\theta^2}$. Indeed, suppose $q \in p^\theta$. The line $(pq)^\theta$ is contained in p^{θ^2} but not in π_p. Then q^θ contains $(pq)^\theta$, and so, if q were absolute, q would be collinear with a point of $p^{\theta^2} \setminus \pi_p$. Hence q would be collinear with all points of p^{θ^2}, a contradiction. The same argument with θ substituted by θ^2 (and noting that $(p^{\theta^2})^{\theta^2} = p^\theta$) shows that $q \notin p^{\theta^2}$. Our claim is proved.

Hence there is a unique line L in π_p through p all of whose points are collinear with q, and there is a unique plane π containing q and L. Let U be the unique member of Σ_2 containing π. Since $L^\theta = L$, we see that the point $r := U^\theta$ is contained in L, and is absolute. Now, since q is contained in $U = r^{\theta^2}$, it must be contained in π_r by our claim above. Hence $\pi = \pi_r$.

Clearly r is unique as $\pi_s = \pi_r$ for every other point $s \in \pi_p$ such that $\{p, s\} \subseteq \pi_s$.

Lemma 3.8. *The set of absolute points is a geometric hyperplane of H.*

Proof. We start by showing that every line L contains at least one absolute point. If $L^\theta \cap L \neq \emptyset$, then for any $x \in L \cap L^\theta$, the singular subspace x^θ contains L^θ, and hence x; so x is absolute and contained in L. If $L \cap L^\theta = \emptyset$, then, by Lemma 3.1, either (L, L^θ) is a special pair, in which case the assertion follows from Lemma 3.2, or L and L^θ are skew lines in a singular three-space U. If $U \in \Sigma_1$, then, since $L^\theta \subseteq U$, we see that $U^{\theta^{-1}} \in L$ and so $U^{\theta^{-1}}$ is contained in U and is an absolute point on L. If $U \in \Sigma_2$, then U^θ is a point on L^θ, and hence $p := U^\theta$ is an absolute point and $L \subseteq p^{\theta^{-1}} = p^{\theta^2}$. Also, $L = (L^\theta)^{\theta^{-1}} \subseteq U^{\theta^{-1}}$ and the latter is by Lemma 3.5 equal to p^θ. Hence $L \subseteq \pi_p$ and so every point on L is absolute by Lemma 3.3.

Hence every line of H contains at least one absolute point. Lemmas 3.5 and 3.7 complete the proof.

Lemma 3.9. *The tairality θ is a triality.*

Proof. If p is absolute, then $p^{\theta^3} = p$ by Lemma 3.5. If p is not absolute, then we consider the three-space p^θ. Since, by Lemma 3.8, the set of absolute points is a geometric hyperplane of H, it follows that there is a plane π in p^θ consisting of absolute points. Moreover, p^θ is the unique member of Σ_1 containing π. But, by Lemma 3.5, π^{θ^3} (pointwise image) coincides with π, hence $(p^\theta)^{\theta^3} = p^\theta$, implying that $p^{\theta^3} = p$. Consequently θ has order 3 and is a triality.

We can now finish the proof of Main Result 2.1. Since θ is a triality, and since every absolute point is incident with at least three fixed lines, it follows from the second part of Paragraph 5.2.9 of [11] that θ is a triality of type I_σ, where σ is a field automorphism of order 3. However, since for every absolute point p, every line through p inside π_p is fixed, the first part of Paragraph 5.2.9 of [11] implies that $\sigma = \mathrm{id}$.

The proof of Main Result 2.1 is complete.

4 Proof of Main Result 2.2

In this section, we let θ be a tairality of a building Δ of spherical type D_4 (which we again view as the hyperbolic quadric H just like in Sect. 3) with the property that, for every line L which is not fixed by θ, the lines L and L^θ are not contained in adjacent chambers of Δ, and such that there exists at least one fixed line. We must show that θ is a triality of type I_{id}. Hence, by Main Result 2.1, it suffices to show that θ maps no line to an opposite line. Note that, by Remark 3.6 and the fact that a triality of type I_{id} is domestic, we know that trialities of type I_{id} do not map any non-fixed line L to a line which is contained in a chamber adjacent to some chamber containing L.

The outline of the proof is as follows. We set A to be the the union of all fixed lines (viewed as sets of points of H). Our first goal is the show that A is a subspace, i.e., if $x, y \in A$ are collinear in H, then all points of the line xy belong to A. Then we show that every line meets A nontrivially. Finally, we show that every line containing a point of A is not mapped onto an opposite line.

We again assume that θ maps points to elements of Σ_1, and its inverse maps points to elements of Σ_2. Note that every element of A is an absolute point. In fact, θ/A has order 3:

Lemma 4.1. *If $x \in A$, then $x^{\theta^3} = x$.*

Proof. Let L be a fixed line containing x. Then $L = L^\theta \subseteq x^\theta$, and similarly $L \subseteq x^{\theta^2}$. Replacing x by x^{θ^3} and θ by θ^{-1}, we see that $x^{\theta^3} \in x^\theta \cap x^{\theta^2} =: \pi_x$. If $x \neq x^{\theta^3}$, we consider a line $M \subseteq \pi_x$ through x but not through x^{θ^3}. Then $M^\theta \subseteq \pi_x$ and M^θ contains x^{θ^3}. Hence M and M^θ are distinct and contained in a common plane, a contradiction.

We again introduce the notation π_x for $x^\theta \cap x^{\theta^2}$.

Lemma 4.2. *If $x \in A$, then all points of π_x are in A and every line in π_x through x is fixed under θ.*

Proof. If L is a line through x in π_x, then L^θ is again a line through x in π_x. Our main assumption now implies that $L = L^\theta$. It follows that every point in π_x belongs to A.

Lemma 4.3. *The union A of all fixed lines is a non-empty subspace.*

Proof. By our main assumption, A is non-empty. Now let $x, y \in A$ with L a line of H containing x and y. If L is fixed, then we are done. Now suppose that L is not fixed. We claim that $y \notin x^\theta$. Indeed, suppose the contrary. Then $x \in L \subseteq x^\theta$ and so $L^\theta \subseteq x^\theta \cap x^{\theta^2} = \pi_x$. Also $L^\theta = x^\theta \cap y^\theta \ni y$. Hence $y \in \pi_x$ and so L is fixed due to Lemma 4.2. Our claim is proved.

Similarly $x \notin y^\theta$. Hence L^θ, which is the intersection of x^θ with y^θ, does not contain x or y. Consider the three-space S determined by the points x and y and the line L^θ. Since S intersects x^θ and y^θ in planes, it belongs to Σ_2. So S^θ is a point

p. Since x and y belong to S, we see that p is contained in $x^\theta \cap y^\theta = L^\theta$. Since S intersects x^θ and y^θ in planes (and hence is incident with them as elements of the building Δ), we see that $p \in x^{\theta^2} \cap y^{\theta^2}$. Hence $p \in \pi_x \cap \pi_y$ and Lemma 4.2 implies that the lines px and py are fixed. Consequently, these lines belong to π_p, and so $L \subseteq \pi_p$. Lemma 4.2 implies that all points of L belong to A.

Lemma 4.4. *The only points of A in $x^\theta \cup x^{\theta^2}$, for a point $x \in A$, are the points of π_x.*

Proof. Let $y \in x^\theta \setminus \pi_x$ and suppose $y \in A$. Let L be a fixed line through y. Then y^θ intersects x^{θ^2} in a plane, and contains L; hence L contains a point z of x^{θ^2}. Obviously, $z \in \pi_x$, and so L is contained in x^θ. Applying θ, we see that $L = L^\theta$ is contained in x^{θ^2}, hence in π_x, a contradiction since y is not contained in π_x. Similarly if $y \in x^{\theta^2} \setminus \pi_x$ and $y \in A$ (interchange θ and θ^{-1}).

Lemma 4.5. *Every line of H contains a point of A.*

Proof. Choose a point $p \in A$ arbitrarily. We can identify the lines of H through p with the points of a quadric H_p in $\mathsf{PG}(5, \mathbb{K})$ isomorphic to the Klein quadric over the field $\mathbb{K}$. The points of A collinear with p induce a subset A_p of the points of H_p. The plane π_p corresponds with a line L in A_p. Each point $x \in \pi_p$ produces a plane π_x through p and this plane π_x corresponds with a line L_x entirely contained in A_p and meeting L in a point. Note that by Lemma 4.4 L_x and L are not contained in a plane of H_p. We can consider three points x_1, x_2, x_3 in π_p, with x_1 and x_2 on distinct lines through p, and x_3 on the line px_1. Then the lines L, L_{x_1}, L_{x_2} generate a three-space of $\mathsf{PG}(5, \mathbb{K})$ intersecting H_p in a hyperbolic quadric Q, which is entirely contained in A_p, by Lemma 4.3. The subspace generated by Q and L_{x_3} either is a cone over Q entirely contained in A_p, or intersects H_p in a quadric Q' isomorphic to a parabolic quadric $Q(4, \mathbb{K})$ in $\mathsf{PG}(4, \mathbb{K})$. In the former case, A_p contains a plane through L, contradicting Lemma 4.4. In the latter case, the set A_p induces a subspace of $Q(4, \mathbb{K})$ containing a grid and an extra line through one of the points of the grid. Now, a subspace of a generalized quadrangle containing an ordinary quadrangle is a (full) subquadrangle, and so A_p induces in $Q(4, \mathbb{K})$ a thick (full) subquadrangle. But Proposition 5.9.4 of [13] says exactly that every thick full subquadrangle of $Q(4, \mathbb{K})$ coincides with $Q(4, \mathbb{K})$ itself. It follows that A_p contains Q'. Note that Q' is a geometric hyperplane of H_p.

Now let K be any line of H and let a be any point of A. Since A contains π_a, there is at least one point p of A collinear with all points of K. Now the plane determined by p and K defines a line in H_p. That line intersects the geometric hyperplane Q' in at least one point, and so there is a line through p contained in A and intersecting K. The lemma is proved.

One could now go on and show that A is a geometric hyperplane consisting of all points of a parabolic subquadric of H, and then show that the fixed lines define a generalized hexagon. But we can end the proof quicker by noting that θ is domestic.

Indeed, let L be any line of H, and let x be a point in $L \cap A$. Then $L^\theta \subseteq x^\theta \ni x$. Hence the point x of L is collinear to all points of L^θ and so L cannot be opposite L^θ. Consequently θ is domestic and Main Result 2.2 now follows from Main Result 2.1.

Appendix: A Characterization of Symplectic Polarities in Projective Spaces

Here we show Proposition 2.3. It suffice to show that the duality θ of the projective space $\mathsf{PG}(n, \mathbb{K})$, $n \geq 3$, maps every point x to a hyperplane x^θ containing x, as soon as θ maps no line to a subspace intersecting the line in precisely one point, and some point is absolute. Then the result follows from Lemma 3.2 of [8].

So let, by way of contradiction, x be a point with $x \notin x^\theta$. Let p be an absolute point. If $p \in x^\theta$, then the line px is mapped onto $x^\theta \cap p^\theta$, which contains p but not x, a contradiction. Hence all points of x^θ are non-absolute. Let L be a line in x^θ. Since every hyperplane meets L, every point, and hence also p, is contained in some y^θ, with $y \in L$. Now the line py is mapped onto a subspace containing p but not y, a contradiction again. The proposition is proved.

It is an open problem whether we can strengthen Main Result 2.2 to the assumptions that no line is mapped onto a line in a common plane, and there exists at least one absolute point. In the finite case, this works (and the proof, which we omit, uses the fact that every duality of a finite projective plane has at least one absolute point, see [5]).

References

1. Abramenko, P., Brown, K.: Buildings: theory and applications. In: Graduate Texts in Mathematics. Springer, New York (2008)
2. Abramenko, P., Brown, K.: Automorphisms of non-spherical buildings have unbounded displacement. Innov. Incid. Geom. **10**, 1–13 (2009)
3. Bourbaki, N.: Groupes et Algèbres de Lie, Chapters 4, 5 and 6. Actu. Sci. Ind., vol. 1337. Hermann, Paris. (1968)
4. Cohen, A., Cuypers, H., Gramlich, R.: Opposition in triality. J. Group Theor. **12**, 491–510 (2009)
5. Devillers, A., Parkinson, J., Van Maldeghem, H.: Automorphisms and opposition in twin buildings. Bull Austr. Math. Soc. **94**, 189–201 (2013)
6. Gramlich, R.: Developments in finite Phan theory. Innov. Incid. Geom. **9**, 123–175 (2009)
7. Schellekens, G.L.: On a hexagonic structure, I. and II. Indag. Math. **24**, 201–217, 218–234 (1962)
8. Temmermans, B., Thas, J.A., Van Maldeghem, H.: Domesticity in projective spaces. Innov. Incid. Geom. **12**, 141–149 (2011)
9. Temmermans, B., Thas, J.A., Van Maldeghem, H.: Domesticity in generalized quadrangles. Ann. Combin. **16**, 905–916 (2012)
10. Temmermans, B., Thas, J.A., Van Maldeghem, H.: Collineations of polar spaces with restricted displacements. Des. Codes Cryptogr. **64**, 61–80 (2012)

11. Tits, J.: Sur la trialité et les groupes qui s'en déduisent. Inst. Hautes Études Sci. Publ. Math. **2**, 13–60 (1959)
12. Tits. J.: Buildings of spherical type and finite BN-pairs. In: Springer Lecture Notes Series, vol. 386. Springer-Verlag, Berlin (1974)
13. Van Maldeghem,H.: Generalized polygons. In: Monographs in Mathematics, vol. 93. Birkhäuser Verlag, Basel, Boston, Berlin (1998)
14. Van Maldeghem, H.: Symplectic polarities in buildings of type E_6. Des. Codes Cryptogr. **65**, 115–125 (2012)
15. Weiss, R.M.: The Structure of Spherical Buildings. Princeton University Press (2004)

Chapter 12
On the Isomorphism Problem for Coxeter Groups and Related Topics

Koji Nuida

Abstract The isomorphism problem, which is a problem of deciding whether or not given two elements in a class of some kind of mathematical objects are isomorphic to each other, is a ubiquitous problem in mathematics. This paper gives a survey of recent developments for the isomorphism problem for (finitely generated or general) Coxeter groups and some related topics.

Keywords Coxeter group · Isomorphism problem

1 Introduction

In general, the *isomorphism problem* signifies the problem of determining, for a given class of mathematical objects, when two objects in the class are isomorphic. This is obviously a fundamental problem in mathematics and has been studied in various settings, and its difficulty depends highly on the choice of the underlying class of objects. For example, for the case of finitely generated groups, the complete classification of the isomorphism classes of abelian groups is quite famous; while it is also important but less elementary that the isomorphism classes of free groups are completely determined by the cardinalities of their free generating sets. On the other hand, the isomorphism problem for general finitely generated groups is unsolvable (more precisely, uncomputable by Turing machines), since the well-known halting problem for Turing machines, which is proven to be uncomputable, can be reduced to the isomorphism problem for certain kinds of finitely generated groups (see e.g., Sect. 9.3.7 of [49]).

K. Nuida (✉)
Innovative Security Research Group, Research Institute for Secure Systems, National Institute of Advanced Industrial Science and Technology, Ibaraki, Japan
e-mail: k.nuida@aist.go.jp

N. S. N. Sastry (ed.), *Groups of Exceptional Type, Coxeter Groups and Related Geometries*, Springer Proceedings in Mathematics & Statistics 82,
DOI: 10.1007/978-81-322-1814-2_12,

The aim of this survey paper is to briefly review the history and developments of the isomorphism problem for *Coxeter groups*. We recall that a group W is called a Coxeter group, if there exists a generating set S of W, referred to as a *Coxeter generating set* of W in this paper, for which the group W has the following presentation:

$$\langle S \mid (s,t)^{m(s,t)} = 1 \text{ for } s,t \in S \text{ with } m(s,t) < \infty \rangle, \tag{1}$$

where $m\colon S \times S \to \{k \in \mathbb{Z} \mid k \geq 1\} \cup \{\infty\}$ is a symmetric mapping satisfying that $m(s,t) = 1$ if and only if $s = t$. The pair (W,S) of W and S as above is called a *Coxeter system*. Here we summarize some fundamental definitions and facts about Coxeter groups; the reader may refer to the standard book [24] for those omitted in the text. The presentation (1), or equivalently the matrix $(m(s,t))_{s,t \in S}$ (called a *Coxeter matrix*), is in one-to-one correspondence to the following undirected labelled graph $\Gamma = \Gamma(W,S)$, called a *Coxeter graph*.[1] The vertex set of Γ is the set S, while two vertices $s,t \in S$ are joined in Γ by an edge if and only if $m(s,t) \geq 3$. Each edge joining vertices s and t in Γ has label $m(s,t)$; in a standard convention, the label is omitted in the case $m(s,t) = 3$, since the case $m(s,t) = 3$ appears quite often when the Coxeter group W is finite. We call the isomorphism class of the Coxeter graph $\Gamma(W,S)$ the *type* of (W,S) (or of W and of S, when the underlying Coxeter system (W,S) is unambiguously understood from the context). On the other hand, another kind of graph is sometimes more convenient to describe the structure of a Coxeter system; we define a *presentation diagram* $\Delta = \Delta(W,S)$ of a Coxeter system (W,S) in almost the same manner as the Coxeter graph Γ, but two distinct vertices s,t are now joined if and only if $m(s,t) < \infty$ instead of $m(s,t) \geq 3$ (the graph Δ has various names depending on the papers; here we adopt the terminology of [46]). We write the Coxeter system defined by a Coxeter graph Γ or a presentation diagram Δ as $(W(\Gamma), S(\Gamma))$ or $(W(\Delta), S(\Delta))$, respectively. The decompositions of Γ and Δ into connected components correspond to a direct sum decomposition and a free product decomposition of W, respectively; each direct factor of the decomposition according to Γ is called an *irreducible component* of W. The subgroup W_I of W generated by a subset I of S is called a *standard parabolic subgroup* of W, while a subgroup which is conjugate to some W_I is called a *parabolic subgroup* of W.

As explained below, most of the existing works on the isomorphism problem for Coxeter groups are restricted to the case that the *rank* (i.e., the cardinality of the Coxeter generating set) of the Coxeter group is finite. In contrast, in this paper, *we do not assume (unless otherwise specified) that the rank is finite.*

Here we show some examples of Coxeter groups.

Example 1.1. Let $W = \mathcal{D}_m$, $3 \leq m < \infty$, be the dihedral group of order $2m$, i.e., the symmetry group of a regular m-gon in the Euclidean plane. This is the finite Coxeter group of type $I_2(m)$ by following the classification of finite irreducible Coxeter

[1] It is not obvious but well-known that the Coxeter matrix is uniquely determined by the pair (W,S); namely, $m(s,t)$ is precisely the order of the element st in W.

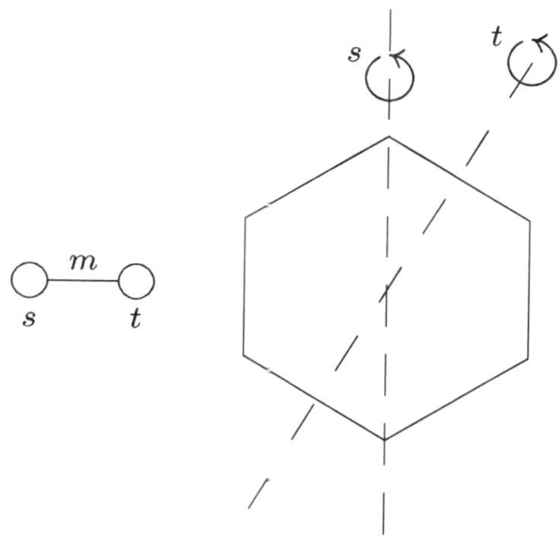

Fig. 1 *Left side* The Coxeter graph of type $I_2(m)$; *Right side* Two reflections generating $W = \mathcal{D}_m = W(I_2(m))$, when $m = 6$

groups summarized in Sect. 2.4 of [24].[2] The group is generated by two involutions s, t, with additional fundamental relation $(st)^m = 1$. Its Coxeter graph is shown in the left side of Fig. 1. For the choice of the generators in the symmetry group, we notice that, when m is odd, there are m lines of symmetry that pass through a vertex and the center of its antipodal edge; and when m is even, there are $m/2$ lines of symmetry that pass through a pair of antipodal vertices and there are $m/2$ lines of symmetry that pass through a pair of centers of antipodal edges. Then the two generators can be defined as reflections with respect to two adjacent lines of symmetry; see the right side of Fig. 1 for the case $m = 6$ (where $W(\mathcal{T})$ denotes the Coxeter group of type $\mathcal{T}$).

Example 1.2. Let $W = \mathcal{D}_\infty$ be the infinite dihedral group, which is a Coxeter group generated by two involutions s, t and has no additional fundamental relations. In Sect. 2.5 of [24], the group is classified as the affine Coxeter group of type $\widetilde{A}_1$; the left side of Fig. 2 shows its Coxeter graph. This group also has a natural realization as a reflection group on the Euclidean plane, which is generated by reflections with respect to two parallel lines; see the right side of Fig. 2.

Example 1.3. Let $W = S_n$ be the symmetric group on n letters, $1 \leq n < \infty$. It is an elementary fact that the group is generated by $n - 1$ adjacent transpositions $s_i := (i \;\; i+1)$, $1 \leq i \leq n-1$. These generators are involutions ($s_i^2 = 1$ for $1 \leq i \leq n-1$) and have relations $(s_i s_{i+1})^3 = 1$ for $1 \leq i \leq n-2$ and $(s_i s_j)^2 = 1$ for any pair of distinct and non-adjacent indices i, j. Moreover, it is less obvious but well-known that any relation in S_n is a consequence of the above relations; hence S_n

[2] Usually, when $m = 3$, the type of the group is classified as A_2 rather than $I_2(3)$. Similarly, when $m = 4$, the type of the group is classified as B_2 (also called C_2) rather than $I_2(4)$.

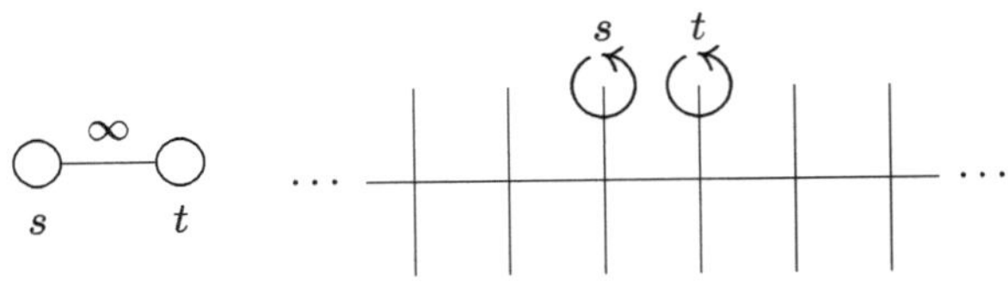

Fig. 2 *Left side* The Coxeter graph of type $\widetilde{A}_1$; *Right side* Two reflections generating $W = \mathcal{D}_\infty = W(\widetilde{A}_1)$

Fig. 3 The Coxeter graph of type A_{n-1}

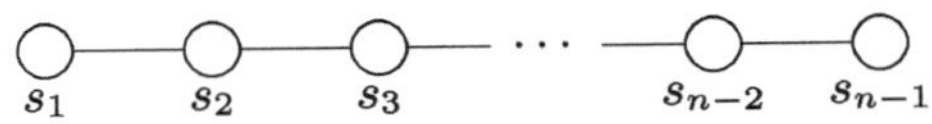

Fig. 4 The Coxeter graphs of type A_∞ (*upper part*) and type $A_{\infty,\infty}$ (*lower part*)

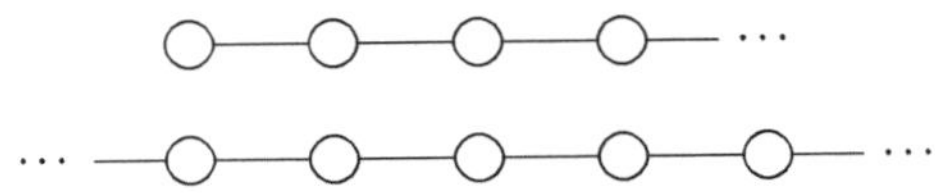

is a Coxeter group, which is classified as the finite Coxeter group of type A_{n-1}, with Coxeter graph shown in Fig. 3.

Example 1.4. Here we give two examples of Coxeter groups of (countably) infinite ranks. First, the straightforward embeddings $\{1, 2, \dots, n\} \hookrightarrow \{1, 2, \dots, n, n+1\}$ for $n \geq 1$ induce the sequence of embeddings of finite symmetric groups; $S_1 \hookrightarrow S_2 \hookrightarrow \cdots \hookrightarrow S_n \hookrightarrow S_{n+1} \hookrightarrow \cdots$. As the " limit" of the sequence, we obtain the infinite group $S_\infty := \bigcup_{n=1}^{\infty} S_n$ generated by the adjacent transpositions s_i in Example 1.3 for all $i \geq 1$. In the same way as Example 1.3, the group is also a Coxeter group, with Coxeter graph as in the upper part of Fig. 4. We call the group the Coxeter group of type A_∞ in this paper. [3] In other words, this Coxeter group of infinite rank is obtained by extending one (say, the right) end point of the Coxeter graph of type A_n to infinity, while fixing the other end point. This suggests another construction of the "limit" of finite symmetric groups, obtained by extending *both* end points of the Coxeter graph of type A_n to infinity; see the lower part of Fig. 4. We call the latter group the Coxeter group of type $A_{\infty,\infty}$ in this paper.

Example 1.5. We give one more example, for which the proof of the fact that the group under consideration forms a Coxeter group is non-trivial. We consider the projective general linear group $W = \mathrm{PGL}(2, \mathbb{Z}) = \mathrm{GL}(2, \mathbb{Z})/\{\pm 1\}$ of matrix size two over integers. We choose three elements[4] $s_1 := \begin{pmatrix} -1 & 0 \\ 0 & 1 \end{pmatrix}$, $s_2 := \begin{pmatrix} -1 & 1 \\ 0 & 1 \end{pmatrix}$, $s_3 := \begin{pmatrix} 0 & 1 \\ 1 & 0 \end{pmatrix}$ of W. Then it can be proven that W is a Coxeter group with Coxeter

[3] The author does not know whether or not there exists a standard name for the type; the same also holds for type $A_{\infty,\infty}$ below.

[4] For simplicity, we identify a matrix in $\mathrm{GL}(2, \mathbb{Z})$ with its equivalence class in $\mathrm{PGL}(2, \mathbb{Z})$.

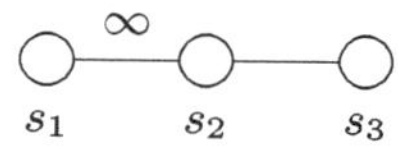

Fig. 5 The Coxeter graph for $W = \mathrm{PGL}(2, \mathbb{Z})$ in Example 1.5

generating set $\{s_1, s_2, s_3\}$ and Coxeter graph shown in Fig. 5 (see e.g., Sect. 5.1 of [24]).

In the literature, Coxeter groups have been well studied from various viewpoints such as geometry, representation theory, combinatorics, computation theory, etc. On the other hand, compared to the long history of Coxeter groups, the history of the isomorphism problem for Coxeter groups is relatively short. Indeed, before Cohen's survey [13] in 1991 on the theory of Coxeter groups, which mentioned the isomorphism problem as one of the open problems, there had been only few works related to the problem (one of the rare examples is Bannai's work [4] in the late 1960s on the automorphism groups of finite irreducible Coxeter groups). In contrast, after Cohen's survey, especially since the late of 1990s, the study of the isomorphism problem for Coxeter groups has been active and several interesting results were provided. In particular, Mühlherr's remarkable example [33] of an isomorphism between two infinite irreducible Coxeter groups of rank four (see Sect. 7) suggests that the isomorphism problem for infinite Coxeter groups would involve rich mathematical structures. Currently, especially in the finite rank case, many significant results are given in recent years and even complete solutions have been obtained for some subclasses of Coxeter groups, as explained in the following sections.

2 The Problem and Related Notions

Based on the terminology introduced in Sect. 1, the isomorphism problem for Coxeter groups can be stated as follows:

Problem 2.1. Given two Coxeter graphs Γ_1 and Γ_2, decide whether or not $W(\Gamma_1)$ and $W(\Gamma_2)$ are isomorphic as abstract groups.

Note that the same problem can be stated by using presentation diagrams Δ instead of Coxeter graphs Γ. There are two closely related problems:

Problem 2.2. Given a Coxeter group W, determine all Coxeter graphs Γ for which $W(\Gamma)$ is isomorphic to W as abstract groups.

Problem 2.3. Given a Coxeter group W, determine all subsets S of W for which (W, S) forms a Coxeter system.

In fact, the difficulty and the variety of solutions to Problems 2.2 and 2.3 highly depend on the given Coxeter group W. For the "easiest" cases of the problem, the following terminologies were introduced:

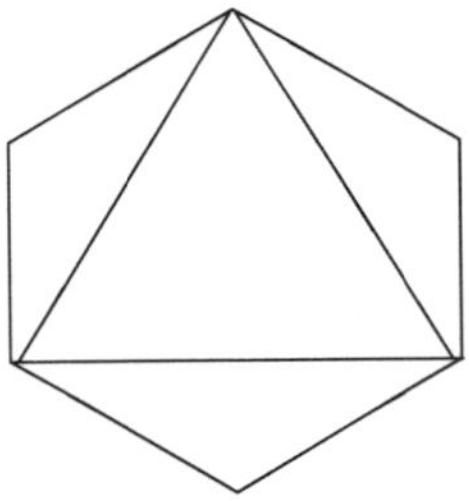

Fig. 6 *Regular triangle* inside *regular hexagon*, which induces the isomorphism $\mathcal{D}_6 \simeq S_2 \times S_3$ (here the half-turn of the *hexagon* generates the direct factor S_2)

Definition 2.4 ([5][5]). *We say that a Coxeter group W is* rigid, *if $W \simeq W(\Gamma) \simeq W(\Gamma')$ implies $\Gamma \simeq \Gamma'$.*

Definition 2.5 ([5]). *We say that a Coxeter group W is* strongly rigid, *if all Coxeter generating sets S of W are conjugate to each other.*

We notice that if W is strongly rigid then W is rigid, since conjugation in W does not change the type of a Coxeter generating set S of W. For any rigid Coxeter group W, the possible Coxeter generating sets of W correspond to the automorphisms of W (for two Coxeter generating sets S, S' of W, a graph isomorphism from $\Gamma(W, S)$ to $\Gamma(W, S')$ induces an automorphism of W), while the Coxeter generating sets appearing in Definition 2.5 correspond to the inner automorphisms of W. Therefore, it would be possible to think that the outer automorphism group $\mathrm{Out}(W) = \mathrm{Aut}(W)/\mathrm{Inn}(W)$ of a rigid Coxeter group W measures how far W is from being strongly rigid.[6] The structure of the (outer) automorphism groups of (not necessarily rigid) Coxeter groups has been studied well, e.g., [11, 16–18, 23, 25, 32, 50]. However, in the present paper we do not discuss those results on the automorphism groups in detail.

By Definition 2.4, Problem 2.2 becomes trivial if the Coxeter group W is rigid. Therefore, the isomorphism problem for Coxeter groups would have a trivial solution if all Coxeter groups were rigid. But there are in fact non-rigid Coxeter groups and the isomorphism problem is not trivial at all. A "folklore" example of non-rigid Coxeter groups is given by the pair of dihedral group $\mathcal{D}_6$ (see Example 1.1) and the direct product $S_2 \times S_3$ of two symmetric groups (see Example 1.3). We consider a pair of a regular hexagon and a regular triangle that are nested as in Fig. 6. Since any symmetry of the inner regular triangle extends uniquely to that of the outer regular hexagon, it gives an inclusion $S_3 \simeq \mathcal{D}_3 \hookrightarrow \mathcal{D}_6$. This and the fact that the half-turn of the regular hexagon is a central element of $\mathcal{D}_6$ of order two and does not belong to the S_3 implies (by comparing the orders of groups) the above-mentioned isomorphism $\mathcal{D}_6 \simeq S_2 \times S_3$.

The same argument also implies a generalization of the isomorphism:

$$W(I_2(4k+2)) \simeq W(A_1) \times W(I_2(2k+1)), \tag{2}$$

[5] To the author's best knowledge, the term "rigid" for Coxeter groups was first introduced in the title of Radcliffe's preprint [43], though the term does not appear in its text.

[6] For example, W is strongly rigid if and only if W is rigid and $\mathrm{Out}(W)$ is trivial.

where k is any positive integer and we write $I_2(3) = A_2$ (we can recover the above isomorphism $\mathcal{D}_6 \simeq S_2 \times S_3$ by putting $k = 1$ and using the properties $\mathcal{D}_6 \simeq W(I_2(6))$, $S_2 \simeq W(A_1)$ and $S_3 \simeq W(A_2)$). On the other hand, we also have another kind of isomorphism between finite Coxeter groups of different types:

$$W(B_{2k+1}) \simeq W(A_1) \times W(D_{2k+1}) \ , \tag{3}$$

where k is any positive integer as well and we write $D_3 = A_3$. These are the only cases of isomorphisms from finite irreducible Coxeter groups to Coxeter groups of different types (see e.g., [37]).

3 Direct and Free Decompositions

As mentioned above, the decomposition of a Coxeter graph Γ into connected components induces a direct product decomposition of $W(\Gamma)$ into irreducible components. Here we write W_{fin} to denote the product of all finite irreducible components of a Coxeter group W, and call it the *finite part* of W. Similarly, we write W_{inf} to denote the product of all infinite irreducible components of W, and call it the *infinite part* of W. We have $W = W_{\mathrm{fin}} \times W_{\mathrm{inf}}$ by the definition. Now we notice that the infinite part of a Coxeter group W is in general dependent on the choice of the Coxeter generating set S. On the other hand, although the definition of W_{fin} seems to be dependent similarly on the choice of S as well, it has been proven in [37] that W_{fin} is in fact *not* dependent on the choice of S (this holds even in the infinite rank case $|S| = \infty$).

The relations (2) and (3) mentioned at the last of Sect. 2 show that a finite irreducible Coxeter group is not always directly indecomposable as an abstract group. On the other hand, the following property holds for infinite irreducible Coxeter groups[7]:

Theorem 3.1. *If an infinite Coxeter group W is irreducible, then W is directly indecomposable as an abstract group.*

This property was first proven by Paris in a preprint version in 2004 of the paper [42] in the finite rank case, and then proven by the author in [37] for a general case.[8] In the finite rank case, a main tool for the proof is *essential elements* of a Coxeter group W, which are elements not belonging to any proper parabolic subgroup of W (e.g., [42]). For example, every Coxeter element of a Coxeter group is an essential element (see Theorem 3.1 of [42]). However, by definition, no essential elements exist in the infinite rank case. The proof in [37] used a different tool available in a general case as

[7] A "free product decomposition" version of Theorem 3.1 was also shown in a preprint version of the paper [30] by Mihalik, Ratcliffe and Tschantz.

[8] We note that, in the present case, the paper [42] generalizes the result on direct indecomposability in the finite rank case to a wider class consisting of subgroups of finite index of those Coxeter groups, therefore the result of [42] does not follow from the results in [37].

well, which is the centralizers of normal subgroups of a Coxeter group W generated by involutions; given any decomposition $W = H_1 \times H_2$ of an infinite irreducible Coxeter group W, we have $H_1 \subset Z_W(H_2)$ and $H_2 \subset Z_W(H_1)$, and the structural results in [37] on such centralizers imply that either H_1 or H_2 should be trivial. This example illustrates the situation that, when we want to extend an existing result on the finite rank case to a general case, the original strategy may be not effective and new techniques have to be developed in order to treat the infinite rank case.

For the relation between the isomorphism problem and direct product decompositions, the author proved a property which is an analogy to Krull–Remak–Schmidt Theorem in group theory:

Theorem 3.2 (informal; see [37] for the details). *Any group isomorphism $f : W \to W'$ between two Coxeter groups W and W' of arbitrary ranks is "approximately decomposed" into a group isomorphism $W_{\mathrm{fin}} \to W'_{\mathrm{fin}}$ between the finite parts and isomorphisms between irreducible components of W_{inf} and W'_{inf}.*

By virtue of Theorem 3.2, the isomorphism problem for Coxeter groups has been "essentially" reduced to the case of infinite irreducible Coxeter groups. For the "remaining" infinite irreducible cases, one may easily find an example of two infinite irreducible Coxeter groups which are isomorphic but have different types, by considering the free product $W(\mathcal{T}) * W'$ of a finite irreducible Coxeter group of type $\mathcal{T} = I_2(4k+2)$ or $\mathcal{T} = B_{2k+1}$, with k being a positive integer, and any non-trivial Coxeter group W'. Indeed, the non-trivial decompositions (2) and (3) for $W(\mathcal{T})$ yield another presentation for $W(\mathcal{T}) * W'$ of different type. Now we notice that, when W' has finite rank, the two presentations of $W(\mathcal{T}) * W'$ have different ranks. By avoiding such "trivial" counterexamples, the following sub-problem of the isomorphism problem was posed by Cohen [13] in 1991:

Problem 3.3. Does the isomorphism problem for *irreducible* Coxeter groups of *finite equal ranks* have a trivial solution? In other words, does $W(\Gamma) \simeq W(\Gamma')$, with Γ and Γ' being connected Coxeter graphs having the same finite number of vertices, imply $\Gamma \simeq \Gamma'$?

In fact, the answer of Cohen's question is "No"; Mühlherr [33] discovered in 2000 a counterexample for Problem 3.3, which is a pair of non-isomorphic connected Coxeter graphs with four vertices that define isomorphic infinite Coxeter groups. His counterexample is actually not a kind of "isolated singularity"; indeed, it was observed in [5] that this example is a special instance of a general construction which will be discussed in Sect. 7.

4 Reflections

For a Coxeter system (W, S), an element $w \in W$ is called a *reflection* of (W, S) (or of W, if the set S is unambiguous from the context), if it is conjugate to an element of S. From a geometric viewpoint, the condition is equivalent to that w acts as a

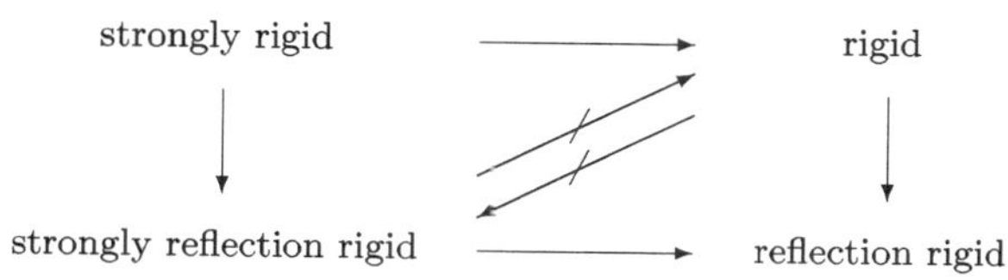

Fig. 7 Relations between variants of rigidity (here A $\rightarrow$ B means that property A implies property B, and A $\not\rightarrow$ B means that property A does not imply property B)

reflection (in the usual sense) with respect to a hyperplane in the standard geometric representation of W. We define

$$\mathrm{Ref}(W) = \mathrm{Ref}_S(W) := \{wsw^{-1} \mid s \in S, w \in W\} \tag{4}$$

to be the set of reflections of (W, S). It is naively expected that any group isomorphism between two Coxeter groups which preserves reflections would have some geometric meanings. We can consider a "reflection-preserving version" of the isomorphism problem for Coxeter groups; for example, the following relaxations of (strong) rigidity were introduced:

Definition 4.1 ([5]). *We say that a Coxeter group W with Coxeter generating set S is* reflection rigid, *if we have $\Gamma \simeq \Gamma(W, S)$ for any Coxeter graph Γ with group isomorphism $W(\Gamma) \rightarrow W$ mapping* $\mathrm{Ref}(W(\Gamma))$ *onto* $\mathrm{Ref}(W)$.

Definition 4.2 ([5]). *We say that a Coxeter group W with Coxeter generating set S is* strongly reflection rigid, *if every Coxeter generating set S' of W satisfying* $\mathrm{Ref}_{S'}(W) = \mathrm{Ref}_S(W)$ *is conjugate to S.*

By definition, strong rigidity implies both of rigidity and strong reflection rigidity; rigidity (as well as strong reflection rigidity) implies reflection rigidity. On the other hand, for rigidity and strong reflection rigidity, implication in any of the two directions does not hold in general (see Fig. 7 for a summary). Indeed, $\mathcal{D}_6 = W(I_2(6))$ is strongly reflection rigid but not rigid, while $\mathcal{D}_5 = W(I_2(5))$ is rigid but not strongly reflection rigid (see Remark 3.4 of [5]). This implies further that all of the four properties are indeed different.

The following notion which is "complementary" to (strong) reflection rigidity relative to (strong) rigidity was introduced as well:

Definition 4.3 ([1]). *We say that a Coxeter group W is called* reflection independent, *if the set* $\mathrm{Ref}_S(W)$ *of reflections is uniquely determined solely from W, independent of the choice of the Coxeter generating set S.*

It is straightforward by definition that (strongly) reflection rigid and reflection independent Coxeter groups are (strongly) rigid. In other words, a proof strategy for (strong) rigidity can be in principle divided into two steps; proof of (strong) reflection rigidity and proof of reflection independence. In fact, such a two-step strategy can be in some sense extended to more general cases of the isomorphism problem; see Sect. 9 for details.

To study the above properties, the following fact plays a significant role:

Proposition 4.4 (Brady–McCammond–Mühlherr–Neumann; see Lemma 3.7 of [5]**).** *For two Coxeter generating sets S, S' of a Coxeter group W, $S \subset \mathrm{Ref}_{S'}(W)$ implies $\mathrm{Ref}_S(W) = \mathrm{Ref}_{S'}(W)$.*

For example, by virtue of Proposition 4.4, to prove that a Coxeter group W with Coxeter generating set S is reflection independent, it suffices to show that any other Coxeter generating set of W is a subset of $\mathrm{Ref}_S(W)$.

There are some more results on relations between the above-mentioned properties. We recall some terminology: We say that a Coxeter system (W, S) is *even*, if $m(s,t)$ is not an odd integer for any pair of distinct generators $s, t \in S$ [1].

Theorem 4.5 (Bahls [1]**).** *Let (W, S) be a Coxeter system for which no pairs of generators $s, t \in S$ satisfy $m(s,t) \equiv 2 \pmod 4$. Then W is reflection independent.*

Theorem 4.6 (Bahls–Mihalik [3]**).** *Let (W, S) be an even Coxeter system of finite rank. If W is reflection independent, then W is rigid.*

We note that, when the evenness condition is removed, Mühlherr's counterexample in Sect. 3 becomes a counterexample for Theorem 4.6 as well.[9]

5 Finite Rank Cases: Geometric Arguments

In this section, we assume unless otherwise stated that any Coxeter system has finite rank. An important family of existing results on strongly rigid Coxeter groups has come from geometric arguments. An early result belonging to this family is the following one given by Charney and Davis [12] in 2000:

Theorem 5.1 (Charney–Davis). *Let W be a Coxeter group, and suppose that W is capable of acting effectively, properly and cocompactly on some contractible manifold. Then W is strongly rigid.*

In their argument, several geometric objects such as CAT(0) spaces played a central role. One of the reasons of the importance of this result is that affine Weyl groups satisfy the hypothesis of their theorem, hence affine Weyl groups (regarded as Coxeter groups) are strongly rigid. Another class to which their result is applicable is the class of compact hyperbolic Coxeter groups.

To the author's best knowledge, the state-of-the-art result in this direction is the one given by Caprace and Przytycki [10] in 2011: Recall that a Coxeter system (W, S) is *bipolar*, if for any $s \in S$, any tubular neighbourhood of the s-invariant wall separates the Cayley graph of (W, S) into exactly two connected components (see [10] for precise definitions).

[9] We note that such a counterexample cannot be found in finite Coxeter groups, since finite Coxeter groups are known to be reflection rigid; see Theorem 3.10 of [5].

Theorem 5.2 (Caprace–Przytycki). *If (W, S) is bipolar, then W is strongly rigid.*

We note that, besides the geometric definition above, a characterization of bipolar Coxeter systems in terms of Coxeter graphs is given: We say that $I \subset S$ is an *odd component* of S, if it is the vertex set of a connected component of the subgraph of $\Gamma(W, S)$ obtained by removing all edges of $\Gamma(W, S)$ having even integers or ∞ as labels (e.g., [34]).

Proposition 5.3 (Theorem 1.2 of [10]**).** *A Coxeter system (W, S) (of finite rank) is bipolar if and only if the following three conditions are all satisfied:*

- *(W, S) has no finite irreducible components.*
- *There exist no pairs of non-empty subsets $I \subset T$ of S satisfying that T is irreducible, W_I is finite, and the subset $S \setminus (I \cup T^{\perp})$ of the vertex set is not connected in the presentation diagram $\Delta(W, S)$, where we put*

$$T^{\perp} := \{s \in S \setminus T \mid s \text{ is not adjacent in} \Gamma(W, S) \text{to any } t \in T\}. \tag{5}$$

- *If $T \subset S$ is irreducible, W_T is finite and an odd component O of S is contained in $T^{\perp}$, then we have $m(t, t') < \infty$ for some $t \in O$ and $t' \in S \setminus (T \cup T^{\perp})$.*

Theorem 5.2 includes wide classes of Coxeter groups such as the ones covered by the above result of Charney and Davis (Theorem 5.1), virtually Poincaré duality Coxeter groups, and infinite irreducible 2-*spherical* Coxeter groups, where a Coxeter group W with Coxeter generating set S is called 2-spherical if and only if $m(s, t) < \infty$ for every pair of generators $s, t \in S$ (e.g., [19]). In fact, the strong rigidity for the last 2-spherical cases was already proven by a combination of the results of Franzsen, Howlett and Mühlherr [19] and of Caprace and Mühlherr [8] (the result for a subclass was given by Kaul [26]). Therefore, Theorem 5.2 can be seen as a generalization of several existing sufficient conditions for strong rigidity.

6 Finite Rank Cases: Maximal Finite Subgroups

Another important family of existing results has been obtained by considering maximal finite subgroups and maximal finite parabolic subgroups of Coxeter groups. First, we summarize some useful properties of those subgroups:

Proposition 6.1 (Franzsen–Howlett; Corollary 7 of [17]**).** *Any finite intersection of parabolic subgroups of a Coxeter group is again a parabolic subgroup.*

Proposition 6.2 (see e.g., Proposition 2 of [41]**).** *Let W be a Coxeter group of finite rank. Then any finite subgroup of W is contained in a maximal finite subgroup of W, which is a parabolic subgroup of W.*

Proposition 6.3 (Franzsen–Howlett; Lemma 8 of [17]**).** *Let W be a Coxeter group. Then any maximal finite standard parabolic subgroup of W is a maximal finite subgroup of W.*

By virtue of these properties, we can consider the following strategy to show that a group isomorphism $f\colon W \to W'$ between two Coxeter groups W, W' with finite Coxeter generating sets S, S', respectively, maps a given generator $s \in S$ of W into $\mathrm{Ref}_{S'}(W')$. First, we try to find a finite number of subsets I_j of S for which W_{I_j} are maximal finite standard parabolic subgroups of W and we have $\bigcap_j I_j = \{s\}$. If such subsets $I_j \subset S$ are found, then we have $\langle s \rangle = \bigcap_j W_{I_j}$, therefore (by Proposition 6.3) $\langle s \rangle$ is the intersection of a finite number of maximal finite subgroups of W. Since f is a group isomorphism, $\langle f(s) \rangle$ is also the intersection of a finite number of maximal finite subgroups $f(W_{I_j})$ of W'. Now Proposition 6.2 implies that each $f(W_{I_j})$ is a parabolic subgroup of W', so is their intersection $\langle f(s) \rangle$ by Proposition 6.1. This means that the involution $f(s)$ is a reflection in W', as desired.

There are some existing results based on similar properties of maximal finite subgroups. We say that a Coxeter group W with Coxeter generating set S is *right-angled*, if $m(s,t) \in \{2, \infty\}$ for any pair of distinct generators $s, t \in S$ (e.g., [44]).

Theorem 6.4 (Radcliffe [44]**).** *If W is a right-angled Coxeter group of finite rank,*[10] *then W is rigid.*

There are also some subsequent works: Radcliffe [45] extended the above result to the case of graph products of directly indecomposable groups. Hosaka [21] gave a generalization to a wider class. Moreover, Castella [11] gave a new proof for the above result, together with some structural results on $\mathrm{Aut}(W)$.

The above-mentioned idea of making use of maximal finite subgroups to prove that the image of $s \in S$ by a group isomorphism is a reflection has been refined in [19]. There the main tool is the *finite continuation* of an element w of finite order of a Coxeter group W of finite rank, which is defined in [19] by

$$\mathrm{FC}_W(w) := \bigcap \{H \mid H \subset W \text{ is a maximal finite subgroup containing } w\}. \tag{6}$$

We emphasize that the definition of $\mathrm{FC}_W(w)$ depends on the fact (Proposition 6.2) that such a w is contained in at least one maximal finite subgroup of W. Franzsen, Howlett and Mühlherr [19] determined the structure of $\mathrm{FC}_W(s)$ for every generator $s \in S$. In particular, they proved that $\mathrm{FC}_W(s) = \langle s \rangle$ for every $s \in S$ if (W,S) is an infinite, irreducible and 2-spherical Coxeter system of finite rank. Using this property, they gave the following result (as mentioned in Sect. 5):

Theorem 6.5 ([19])**.** *If (W,S) is an infinite, irreducible and 2-spherical Coxeter system of finite rank, then W is reflection independent.*

As mentioned in Sect. 5, it follows that any Coxeter group W as in Theorem 6.5 is strongly rigid, by combining Theorem 6.5 with the following result:

Theorem 6.6 (Caprace–Mühlherr [8]**).** *If (W,S) is an infinite, irreducible and 2-spherical Coxeter system of finite rank, then W is strongly reflection rigid.*

---

[10] In the same paper, Radcliffe mentioned (without proof) that the condition for the finiteness of ranks can be removed.

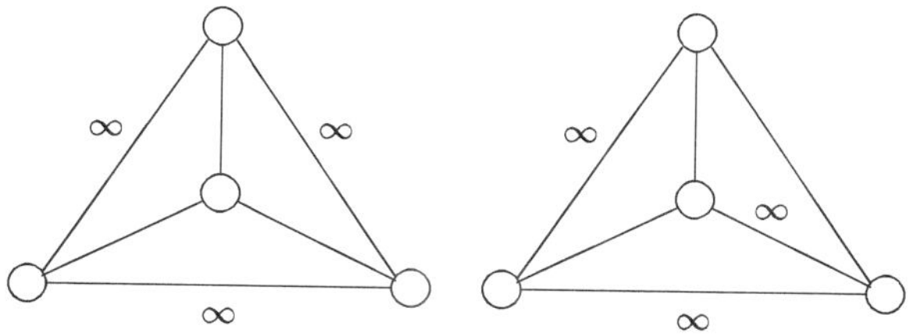

Fig. 8 Two Coxeter graphs in Mühlherr's counterexample

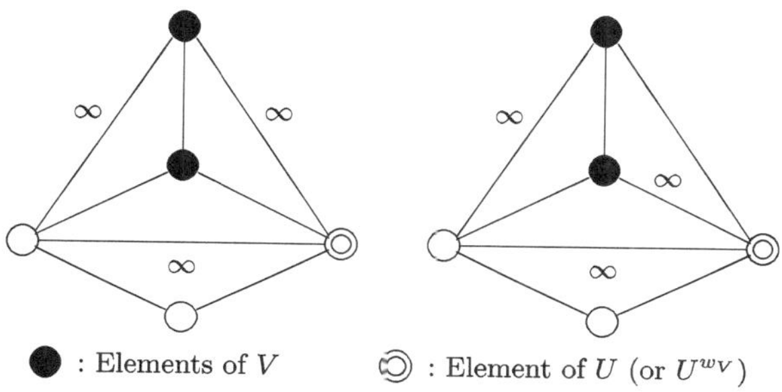

Fig. 9 Example of diagram twisting (here the graph in the *right side* is obtained from the *left side* by a diagram twisting).

7 Diagram Twisting

We revisit Mühlherr's counterexample [33] for Cohen's question mentioned in Sect. 3. The counterexample consists of two non-isomorphic connected Coxeter graphs as in Fig. 8, which define isomorphic Coxeter groups. In [5], Brady, McCammond, Mühlherr and Neumann presented a general method to obtain a whole family of such pairs of Coxeter graphs, called *diagram twisting*. We describe the definition of diagram twisting:

Definition 7.1 (Brady et al. [5]**).** *Let (W, S) be a Coxeter system. Let U and V be disjoint subsets of S satisfying the following conditions:*

- *W_V is finite, with longest element denoted here by w_V.*
- *If $s_1 \in S \setminus (U \cup V)$ is adjacent to V in $\Gamma(W, S)$, then $m(s_1, s_2) = \infty$ for every $s_2 \in U$.*

In this setting, another subset S' of W and another Coxeter graph Γ' are constructed as follows:

- *$S' := (S \setminus U) \cup U^{w_V}$, where $U^{w_V} := \{u^{w_V} \mid u \in U\}$ and $u^{w_V} := w_V u w_V^{-1}$ for each $u \in U$.*
- *Γ' is obtained from $\Gamma(W, S)$ by replacing each edge from a vertex $u \in U$ to a vertex $v \in V$ with an edge from $u^{w_V} \in U^{w_V}$ to $v^{w_V} \in V$, having the same edge label as the original.*

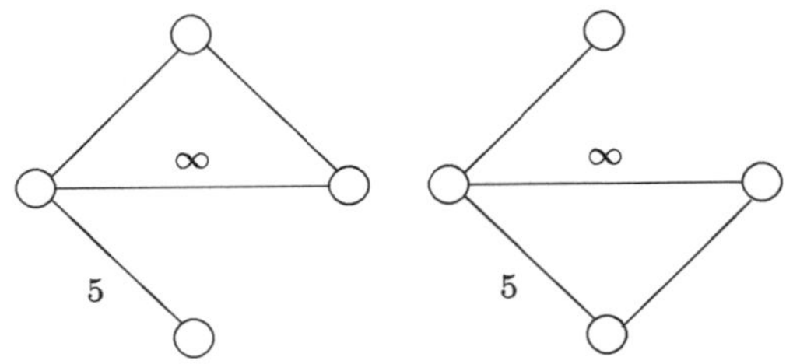

Fig. 10 Counterexample for Conjecture 7.3

For example, the Coxeter graph in the right side of Fig. 8 can be obtained from the one in the left side by a diagram twisting. Another example of diagram twisting is shown in Fig. 9. Then the result of Brady et al. is the following:

Theorem 7.2 (Brady et al. [5]**).** *In this setting, (W, S') is a Coxeter system with Coxeter graph Γ'. Moreover, we have* $\mathrm{Ref}_{S'}(W) = \mathrm{Ref}_S(W)$.

By Theorem 7.2, W is not reflection rigid if the Coxeter graph $\Gamma(W, S)$ admits a diagram twisting that yields a non-isomorphic Coxeter graph. Based on this result, Brady et al. formulated the following conjecture:

Conjecture 7.3 (Conjecture 8.1 of [5]*).* Coxeter systems are "reflection rigid up to diagram twistings", namely, if $\mathrm{Ref}_S(W) = \mathrm{Ref}_{S'}(W)$, then $\Gamma(W, S)$ can be converted to $\Gamma(W, S')$ by consecutive diagram twistings.

There are some positive results on the conjecture. Recall that a Coxeter system (W, S) is called *skew-angled*, if we have $m(s, t) \geq 3$ for any pair of distinct generators $s, t \in S$ [36].

Theorem 7.4 (Brady et al. [5]**).** *If the presentation diagram $\Delta(W, S)$ of (W, S) is a finite tree, then W is reflection rigid up to diagram twistings.*

Theorem 7.5 (Mühlherr–Weidmann [36]**).** *If (W, S) is skew-angled and of finite rank, then W is reflection rigid up to diagram twistings.*

We notice that Mühlherr and Weidmann [36] also gave a characterization of reflection independent Coxeter groups and a sufficient condition for strongly rigid ones among skew-angled Coxeter groups of finite ranks.

In spite of those positive data, it has been shown that Conjecture 7.3 is in fact false. A counterexample shown in Fig. 10 is constructed by using a result of Ratcliffe and Tschantz [46]; the two non-isomorphic Coxeter graphs define isomorphic Coxeter groups. In fact, the counterexample is a special case of another kind of transformation for Coxeter graphs, called 5-*edge angle deformation*. See [46] for the precise definition of this transformation. Based on their result, it is now natural to ask whether or not Conjecture 7.3 becomes true when we consider the new transformations introduced in [46] together with the diagram twistings. We notice that, by the results in [27], angle deformations are well understood. This yields in particular to a revised version of Conjecture 7.3 which is still open.

8 Some Solved Cases

In this section, we assume unless otherwise stated that any Coxeter system has finite rank. Here we describe some more cases of the isomorphism problem for Coxeter groups which have been completely solved.

First we summarize some results relevant to even Coxeter systems:

Theorem 8.1 (Radcliffe [44]**).** *Let* (W, S) *be a Coxeter system satisfying that* $m(s, t) \in \{2, \infty\} \cup 4\mathbb{Z}$ *for any pair of distinct generators* $s, t \in S$. *Then* W *is rigid.*

Theorem 8.2 (Brady et al. [5]**).** *Let* (W, S) *be an even Coxeter system. Then* W *is reflection rigid.*

Bahls and Mihalik [3] gave characterizations of reflection independent cases among even Coxeter systems, and of even Coxeter systems having other non-even Coxeter generating set. Moreover, Mihalik [28] gave the following result:

Theorem 8.3 (informal; see [28] **for details).** *An algorithm to determine the possible types of Coxeter generating sets of any even Coxeter group* W *is given. Hence the isomorphism problem for even Coxeter groups has been solved.*

For the second case here, we recall that a Coxeter system (W, S) is called *chordal*, if the presentation diagram $\Delta(W, S)$ is chordal, i.e., every cycle in $\Delta(W, S)$ of length at least 4 has a shortcutting edge [46]. It was shown by Ratcliffe and Tschantz (Theorem 2.2 of [46]) that the property of being chordal is independent of the choice of S. Moreover, they gave the following result:

Theorem 8.4 (informal; see [46] **for details).** *An algorithm to decide whether or not two chordal Coxeter groups are isomorphic is given. Hence the isomorphism problem for chordal Coxeter groups has been solved.*

For the last case in the section, we recall that a Coxeter system (W, S) is called *twist-rigid*, if $\Gamma(W, S)$ admits no diagram twistings that yield non-isomorphic Coxeter graphs [9]. It was shown by Caprace and Przytycki (Corollary 1.3 of [9]) that twist-rigidity is independent of the choice of S. Moreover, they gave the following result:

Theorem 8.5 (informal; see [9] **for details).** *An algorithm to output all possible Coxeter graphs that define Coxeter groups isomorphic to a given twist-rigid Coxeter group is given. Hence the isomorphism problem for twist-rigid Coxeter groups has been solved.*

9 Finite Rank Cases: Reductions

In this section, we assume unless otherwise stated that any Coxeter system has finite rank. There are results which aim at reducing the isomorphism problem for a class of Coxeter groups to some sub-case. First, we recall that a Coxeter system (W, S)

is called 2-*dimensional*, if we have $|W_I| = \infty$ for every $I \subset S$ with at least three elements[11] [2]. Then Hosaka [20] gave the following result:

Theorem 9.1 ([20]). *If (W, S) and (W, S') are both 2-dimensional, then S can be converted to another Coxeter generating set S'' of W in such a way that $\Gamma(W, S'') \simeq \Gamma(W, S)$ and $\mathrm{Ref}_{S''}(W) = \mathrm{Ref}_{S'}(W)$.*

Intuitively speaking, this result shows that the isomorphism problem for 2-dimensional Coxeter groups is reduced to "reflection-preserving" cases.

Such a reduction result for the case of general Coxeter groups (of finite ranks) has been given by an unpublished preprint of Howlett and Mühlherr [22] (see also a survey paper [34] for a description). In their argument, the following notion plays a central role: A generator $s \in S$ is called a *pseudo-transposition*, if there exists a subset J of S containing s with the following properties [22, 34];

- for each $t \in S \setminus J$, we have either $m(s, t) = \infty$, or $m(t, t') = 2$ for every $t' \in J$;
- (W_J, J) is either of type $I_2(4k + 2)$ with k being a positive integer, or of type B_{2k+1} with k being a positive integer where s is the end vertex of $\Gamma(W_J, J)$ that is adjacent to the unique edge with label 4.

A Coxeter system (W, S) is called *reduced*, if it has no pseudo-transpositions [22, 34]. They showed that, roughly speaking, a pseudo-transposition can be removed by "locally" applying the relations $W(I_2(4k + 2)) \simeq W(A_1 \times I_2(2k + 1))$ and $W(B_{2k+1}) \simeq W(A_1 \times D_{2k+1})$ mentioned in Sect. 2. Iterating the process, the following result is derived:

Theorem 9.2 (Howlett–Mühlherr). *There is a canonical way to convert the Coxeter generating set S of any Coxeter system (W, S) into another Coxeter generating set S' of W for which (W, S') is reduced.*

Moreover, for reduced Coxeter systems, they proved the following:

Theorem 9.3 (Howlett–Mühlherr). *For any reduced Coxeter system (W, S), there exists a finite subgroup Σ of the automorphism group of W, determined by using finite continuations, satisfying the following property: If (W', S') is reduced and $f: W \to W'$ is a group isomorphism, then we have $f(\sigma(S)) \subset \mathrm{Ref}_{S'}(W')$ for some $\sigma \in \Sigma$.*

By virtue of the two results above, the isomorphism problem for Coxeter groups (of finite ranks) can be reduced to "reflection-preserving" cases in the following manner: Given two Coxeter graphs Γ and Γ', to check whether or not $W(\Gamma) \simeq W(\Gamma')$,

1. first convert Γ and Γ' (as in Theorem 9.2) into Γ_* and Γ'_*, respectively, for which $(W(\Gamma_*), S(\Gamma_*))$ and $(W(\Gamma'_*), S(\Gamma'_*))$ are both reduced;
2. then for each (finitely many) $\sigma \in \Sigma$ associated to $(W(\Gamma_*), S(\Gamma_*))$ (as in Theorem 9.3), decide whether or not there exists a group isomorphism $\sigma(W(\Gamma_*)) \to W(\Gamma'_*)$ which maps $\mathrm{Ref}(\sigma(W(\Gamma_*)))$ onto $\mathrm{Ref}(W(\Gamma'_*))$.

---

[11] This condition means that the dimension of the Davis complex $\Sigma(W, S)$ for (W, S) is at most two.

The above result on the reduction to "reflection-preserving" cases has been further improved by Marquis and Mühlherr [27]. For two Coxeter generating sets S, S' of W, we say that S' is *sharp-angled* with respect to S, if for any $s, t \in S$ with $m(s,t) < \infty$, the subset $\{s, t\} \subset S$ is conjugate in W to a subset of S' [27].

Theorem 9.4 (informal; see [27] **for details).** *The isomorphism problem for Coxeter groups is reduced to the following problem: Given a Coxeter system* (W, S), *determine all subsets* $S' \subset \mathrm{Ref}_S(W)$ *with the property that* (W, S') *is a Coxeter system and* S' *is sharp-angled with respect to* S.

We note that the result of Marquis and Mühlherr is in fact used in the above-mentioned complete solution (Theorem 8.5) by Caprace and Przytycki for twist-rigid cases.

We also note that Mühlherr presented in [34] a modification of Conjecture 7.3, where the set S' in the statement is supposed to be sharp-angled with respect to S. By virtue of Theorem 9.4, this slightly weakened conjecture is still enough for solving the isomorphism problem for Coxeter groups of finite ranks. For the conjecture, Ratcliffe and Tschantz [47] recently studied a special kind of decomposition of Coxeter groups as graphs of groups, called a *JSJ decomposition*, and showed that the conjecture is reduced to the case that the Coxeter group is indecomposable with respect to a certain class of JSJ decompositions (see Theorem 5.3 of [47] for the details). See also e.g., a preprint by Mihalik [29] and a paper by Mihalik and Tschantz [31] for other results on JSJ decompositions and similar decompositions of Coxeter groups.

10 Arbitrary Rank Cases: Centralizers

In the previous sections we mainly focused on the finite rank case of the isomorphism problem. A lot of powerful tools have been established and interesting results have been obtained in this case. In the rest of the paper we will drop the finite rank assumption and investigate to which extent the tools remain applicable in this more general context. As a matter of fact we face the obstacle that several key properties used in the arguments in the finite rank case do not hold when the rank is infinite. For example,

- maximal finite (standard parabolic) subgroups do not necessarily exist;
- accordingly, the definition of the finite continuation may become meaningless;
- the intersection of infinitely many parabolic subgroups is not necessarily a parabolic subgroup.

Moreover, there even exists a known result in the finite rank case which is not true when generalized to arbitrary ranks. Indeed, for the two Coxeter groups of infinite ranks shown in Example 1.4, $W(A_\infty)$ and $W(A_{\infty,\infty})$ are the groups of permutations on the set $\mathbb{N}$ of natural numbers and on the set $\mathbb{Z}$ of integers, respectively, that fix all but finitely many letters in the underlying set. Now a bijection between $\mathbb{N}$ and $\mathbb{Z}$ induces a group isomorphism between $W(A_\infty)$ and $W(A_{\infty,\infty})$ which preserves reflections; therefore the infinite, irreducible and 2-spherical Coxeter group $W(A_\infty)$

is *not* reflection rigid, in spite of the fact mentioned in Sect. 6 that infinite, irreducible and 2-spherical Coxeter groups of finite ranks are reflection rigid (and even strongly rigid).

Based on the structural results on centralizers and normalizers of parabolic subgroups in Coxeter groups (e.g., [6, 7, 40]), the author proposed in [39] an alternative strategy to show that a group isomorphism $f: W \to W'$ between two Coxeter groups maps a generator $s \in S$ into $\mathrm{Ref}_{S'}(W')$, which is also available in the infinite rank case. First, by the result of Richardson [48] on conjugacy classes of involutions, we may assume without loss of generality that $f(s)$ is the longest element of a finite standard parabolic subgroup W'_J of W' ($J \subset S'$) in which $f(s)$ is central. In this case, f induces an isomorphism from $Z_W(s)$ to $Z_{W'}(f(s)) = N_{W'}(W'_J)$ (see e.g., [37] for the latter equality). Now by combining the structural result on $Z_W(s)$ in [6] with the results of [39], it follows that $W'_J \subset f(\langle s \rangle \times (W^{\perp s})_{\mathrm{fin}})$, where $W^{\perp s}$ denotes the subgroup of W generated by the reflections orthogonal to s (which is a Coxeter group due to a general result of Deodhar [14] and Dyer [15], and hence has its own finite part). This inclusion relation implies that, if $(W^{\perp s})_{\mathrm{fin}}$ is trivial, then we have $|J| = 1$, therefore $f(s)$ is a reflection in W' as desired (in fact, the same conclusion holds also in the case that $(W^{\perp s})_{\mathrm{fin}}$ is generated by a single reflection conjugate to s).

In the author's preprint [38], the structure of the group $(W^{\perp s})_{\mathrm{fin}}$ above is completely determined; intuitively speaking, we have $(W^{\perp s})_{\mathrm{fin}} = 1$ in "generic" cases. By this result and the above argument, we obtain the following sufficient conditions for reflection independence in the arbitrary rank case:

Theorem 10.1 ([38]). *For a Coxeter system (W, S) of arbitrary rank, suppose either W is infinite, irreducible and 2-spherical, or W is infinite and S has only one odd component.*[12] *Then W is reflection independent.*

11 Arbitrary Rank Cases: Reduction

As mentioned above, the known results in the finite rank case using finite continuations are based on the properties in Propositions 6.1, 6.2 and 6.3 of parabolic subgroups and maximal finite subgroups, but these properties are not necessarily true or strong enough in the infinite rank case (see Example 1 of [41] for a counterexample for Proposition 6.1 generalized to the case of infinite intersection of parabolic subgroups). To deal with the arbitrary rank case, the author introduced in [41] the notion of *locally parabolic subgroups* of a Coxeter group W: For a reflection subgroup G of W, let $S(G)$ denote the canonical Coxeter generating set of G obtained by the results of [14, 15], which, intuitively, consists of reflections with respect to "indecomposable positive roots of G".

[12] The latter condition is equivalent to that all elements of $\mathrm{Ref}_S(W)$ are conjugate in W.

Definition 11.1 ([41]). *A reflection subgroup G of W is called* locally parabolic, *if any finite subset of $S(G)$ is conjugate in W to a subset of S.*

It is known (Lemma 10 of [41]) that the locally parabolic subgroups of a Coxeter group W of finite rank are nothing but the parabolic subgroups of W; therefore, the notion of locally parabolic subgroups is indeed a natural generalization of parabolic subgroups. For locally parabolic subgroups, the following properties that are analogous to and stronger than Propositions 6.1 and 6.2 are known (we recall that a group is *locally finite*, if its finite subsets always generate finite subgroups):

Proposition 11.2 (**Theorem 1 of** [41]). *Any (not necessarily finite) intersection of locally parabolic subgroups of a Coxeter group is again a locally parabolic subgroup.*

Proposition 11.3 (**Theorem 3 of** [41]). *Let W be a Coxeter group of arbitrary rank. Then any locally finite subgroup of W is contained in a maximal locally finite subgroup of W, which is a locally parabolic subgroup of W.*

Based on those generalized properties, Mühlherr and the author introduced, in an unpublished preprint [35], the following variant of finite continuations, called *locally finite continuations*:

$$\mathrm{LFC}_W(X) := \bigcap\{H \mid H \subset W \text{ is a maximal locally finite subgroup containing } X\}, \quad (7)$$

and studied the properties. For example, if $X \subset W$ generates a locally finite subgroup, then $\mathrm{LFC}_W(X)$ is a locally finite and locally parabolic subgroup of W. In [35], the structure of $\mathrm{LFC}_W(r)$ is completely determined for every reflection r in W; this generalizes the result of Franzsen et al. [19] in Sect. 6.

As mentioned in Sect. 9, Howlett and Mühlherr introduced the notion of reduced Coxeter systems, which are defined as having no pseudo-transpositions, and showed that to solve the isomorphism problem for Coxeter groups of finite ranks, it suffices to consider the case of reduced Coxeter groups. Mühlherr and the author extended the definition of pseudo-transpositions and reduced Coxeter systems to the arbitrary rank case, and proved the following analogous result:

Theorem 11.4 (**Mühlherr–Nuida**). *There is a canonical way to convert the Coxeter generating set S of any Coxeter system (W, S) of arbitrary rank into another Coxeter generating set S' of W for which (W, S') is reduced.*

By using locally finite continuations, we also obtained a characterization of reduced Coxeter systems (W, S) of arbitrary ranks which is "reflection independent among reduced Coxeter systems" (i.e., for any Coxeter generating set $S' \subset W$, if (W, S') is reduced then $\mathrm{Ref}_{S'}(W) = \mathrm{Ref}_S(W)$), and a "nearly characterizing" result on reflection independent Coxeter groups of arbitrary ranks. In particular, we have the following results:

Theorem 11.5 (Mühlherr–Nuida). *Let (W, S) be a Coxeter system of arbitrary rank.*

1. *Suppose that one of the following conditions is satisfied:*
   - *W is bipolar (where bipolarity is defined in the infinite rank case by the characterizing property in Proposition 5.3 rather than the original geometric definition).*
   - *W is infinite and S has only one odd component.*

   Then W is reflection independent.
2. *Suppose that (W, S) is 2-dimensional. Then W is reflection independent if and only if for each vertex s of the presentation diagram $\Delta(W, S)$ of degree 1, the unique edge in $\Delta(W, S)$ adjacent to s does not have label $4k + 2$ with k being an integer.*

See a forthcoming published version of the preprint [35] for the details of the above results.

Acknowledgments This survey paper is based on a two-hour talk by the author at the Workshop and Conference on Groups and Geometries, Indian Statistical Institute Bangalore, India, December 10–21, 2012. The author would like to thank Professor N. S. Narasimha Sastry for inviting him to the conference and giving him an opportunity to give a talk. The author would also like to thank the participants to the conference, especially Professor Luis Paris, who gave precious comments to the talk.

References

1. Bahls, P.: A new class of rigid Coxeter groups. Int. J. Algebra Comput. **13**(1), 87–94 (2003)
2. Bahls, P.: Rigidity of two-dimensional Coxeter groups. Preprint, arXiv:math/0311177v2 (2004)
3. Bahls, P., Mihalik, M.: Reflection independence in even Coxeter groups. Geom. Dedicata **110**, 63–80 (2005)
4. Bannai, E.: Automorphisms of irreducible Weyl groups. J. Fac. Sci. Univ. Tokyo Sect. 1 A **16**, 273–286 (1969)
5. Brady, N., McCammond, J.P., Mühlherr, B., Neumann, W.D.: Rigidity of Coxeter groups and Artin groups. Geom. Dedicata **94**, 91–109 (2002)
6. Brink, B.: On centralizers of reflections in Coxeter groups. Bull. London Math. Soc. **28**(5), 465–470 (1996)
7. Brink, B., Howlett, R.B.: Normalizers of parabolic subgroups in Coxeter groups. Invent. Math. **136**, 323–351 (1999)
8. Caprace, P.-E., Mühlherr, B.: Reflection rigidity of 2-spherical Coxeter groups. Proc. London Math. Soc. **94**(3), 520–542 (2007)
9. Caprace, P.-E., Przytycki, P.: Twist-rigid Coxeter groups. Geom. Topology **14**, 2243–2275 (2010)
10. Caprace, P.-E., Przytycki, P.: Bipolar Coxeter groups. J. Algebra **338**, 35–55 (2011)
11. Castella, A.: Sur les automorphismes et la rigidité des groupes de Coxeter à angles droits. J. Algebra **301**, 642–669 (2006)
12. Charney, R., Davis, M.: When is a Coxeter system determined by its Coxeter groups? J. London Math. Soc. **61**(2), 441–461 (2000)

13. Cohen, A.M.: Coxeter groups and three related topics. In: Barlotti, A., et al. (eds.) Generators and Relations in Groups and Geometries, NATO ASI Series, pp. 235–278. Kluwer Academic Publishers, Dordrecht (1991)
14. Deodhar, V.V.: A note on subgroups generated by reflections in Coxeter groups. Arch. Math. (Basel) **53**, 543–546 (1989)
15. Dyer, M.: Reflection subgroups of Coxeter systems. J. Algebra **135**, 57–73 (1990)
16. Franzsen, W.N.: Automorphisms of Coxeter groups of rank 3 with infinite bonds. J. Algebra **248**, 381–396 (2002)
17. Franzsen, W.N., Howlett, R.B.: Automorphisms of Coxeter groups of rank three. Proc. Amer. Math. Soc. **129**, 2607–2616 (2001)
18. Franzsen, W.N., Howlett, R.B.: Automorphisms of nearly finite Coxeter groups. Adv. Geom. **3**, 301–338 (2003)
19. Franzsen, W.N., Howlett, R.B., Mühlherr, B.: Reflections in abstract Coxeter groups. Comment. Math. Helv. **81**, 665–697 (2006)
20. Hosaka, T.: Coxeter systems with two-dimensional Davis-Vinberg complexes. J. Pure Appl. Algebra **197**, 159–170 (2005)
21. Hosaka, T.: A class of rigid Coxeter groups. Houston J. Math. **32**(4), 1029–1036 (2006)
22. Howlett, R.B., Mühlherr, B.: Isomorphisms of Coxeter groups which do not preserve reflections. Preprint (2004)
23. Howlett, R.B., Rowley, P.J., Taylor, D.E.: On outer automorphism groups of Coxeter groups. Manuscripta Math. **93**, 499–513 (1997)
24. Humphreys, J.E.: Reflection groups and Coxeter groups. Cambridge University Press, Cambridge (1990)
25. James, L.D.: Complexes and Coxeter groups—operations and outer automorphisms. J. Algebra **113**, 339–345 (1988)
26. Kaul, A.: A class of rigid Coxeter groups. J. London Math. Soc. **66**(2), 592–604 (2002)
27. Marquis, T., Mühlherr, B.: Angle-deformations in Coxeter groups. Algebraic Geom. Top. **8**, 2175–2208 (2008)
28. Mihalik, M.: The even isomorphism theorem for Coxeter groups. Trans. Amer. Math. Soc. **359**(9), 4297–4324 (2007)
29. Mihalik, M.: JSJ decompositions of Coxeter groups over virtually abelian splittings. Preprint, arXiv:0804.3963v2 (2009)
30. Mihalik, M., Ratcliffe, J., Tschantz, S.: Matching theorems for systems of a finitely generated Coxeter group. Algebraic Geom. Top. **7**, 919–956 (2007); see also: On the isomorphism problem for finitely generated Coxeter groups. I: Basic matching, arXiv:math/0501075v1 (2005)
31. Mihalik, M., Tschantz, S.: Visual decompositions of Coxeter groups. Groups Geom. Dyn. **3**, 173–198 (2009)
32. Mühlherr, B.: Automorphisms of graph-universal Coxeter groups. J. Algebra **200**, 629–649 (1998)
33. Mühlherr, B.: On isomorphisms between Coxeter groups. Des. Codes Cryptogr. **21**, 189–189 (2000)
34. Mühlherr, B.: The isomorphism problem for Coxeter groups. In: Davis, C., Ellers, E.W. (eds.) The Coxeter Legacy, pp. 1–15. American Mathematical Society, Providence (2006)
35. Mühlherr, B., Nuida, K.: Reflection independent Coxeter groups of arbitrary ranks. Preprint (2013)
36. Mühlherr, B., Weidmann, R.: Rigidity of skew-angled Coxeter groups. Adv. Geom. **2**, 391–415 (2002)
37. Nuida, K.: On the direct indecomposability of infinite irreducible Coxeter groups and the isomorphism problem of Coxeter groups. Commun. Algebra **34**(7), 2559–2595 (2006)
38. Nuida, K.: Centralizers of reflections and reflection-independence of Coxeter groups. Preprint, arXiv:math/0602165v1 (2006)
39. Nuida, K.: Almost central involutions in split extensions of Coxeter groups by graph automorphisms. J. Group Theory **10**, 139–166 (2007)

40. Nuida, K.: On centralizers of parabolic subgroups in Coxeter groups. J. Group Theory **14**(6), 891–930 (2011)
41. Nuida, K.: Locally parabolic subgroups in Coxeter groups of arbitrary ranks. J. Algebra **350**, 207–217 (2012)
42. Paris, L.: Irreducible Coxeter groups. Int. J. Algebra Comput. **17**(3), 427–447 (2007)
43. Radcliffe, D.G.: Rigidity of right-angled Coxeter groups. Preprint, arXiv:math/9901049v1 (1999)
44. Radcliffe, D. G.: Unique presentation of Coxeter groups and related groups. Ph.D. thesis, University of Wisconsin-Milwaukee (2001)
45. Radcliffe, D.G.: Rigidity of graph products of groups. Algebraic Geom. Top. **3**, 1079–1088 (2003)
46. Ratcliffe, J.G., Tschantz, S.T.: Chordal Coxeter groups. Geom. Dedicata **136**, 57–77 (2008)
47. Ratcliffe, J.G., Tschantz, S.T.: JSJ decompositions of Coxeter groups over FA subgroups. Topology Proc. **42**, 57–72 (2013)
48. Richardson, R.W.: Conjugacy classes of involutions in Coxeter groups. Bull. Austral. Math. Soc. **26**, 1–15 (1982)
49. Stillwell, J.: Classical Topology and Combinatorial Group Theory, 2nd edn., GTM, vol. 72. Springer, New York (1993)
50. Tits, J.: Sur le groupe des automorphismes de certains groupes de Coxeter. J. Algebra **113**, 346–357 (1988)

Chapter 13
Lectures on Artin Groups and the $K(\pi, 1)$ Conjecture

Luis Paris

Abstract This paper consists of the notes of a mini-course (3 lectures) on Artin groups that focuses on a central question of the subject, the $K(\pi, 1)$ conjecture.

Keywords Artin group · Coxeter group · Vinberg system · $K(\pi, 1)$ conjecture

Introduction

Artin groups are groups that have presentations with relations of the form $aba \cdots = bab \cdots$, where the left hand side and the right hand side of the equality have the same length. They were introduced by J. Tits [25] as extensions of Coxeter groups, but they win their fame a little later with works by Brieskorn [2–4], Brieskorn, Saito [5], and Deligne [14]. The above cited works by Brieskorn, Saito, and Deligne concern mainly a certain family of Artin groups, those groups associated to finite Coxeter groups and now called spherical type Artin groups. However, some beginnings of results for all Artin groups can be found in [5]. Even now, very few results concerning all Artin groups are known, and some basic questions such as the center and the word problem are still open.

This paper consists of the notes of a mini-course (3 lectures) on Artin groups that focuses on a central question of the subject, the so-called $K(\pi, 1)$ conjecture, but the concepts and ideas introduced here also serve for the study of other questions. In the first lecture (Sect. 1) we introduce the different objects involved in the theory of Artin groups and, more specifically, in the $K(\pi, 1)$ conjecture. In the second lecture (Sect. 2) we state the conjecture and study some examples. Finally, the third lecture

L. Paris (✉)
Institut de Mathématiques de Bourgogne, Université de Bourgogne, UMR 5584 du CNRS,
B.P. 47870, 21078 Dijon cedex, France
e-mail: lparis@u-bourgogne.fr

N. S. N. Sastry (ed.), *Groups of Exceptional Type, Coxeter Groups and Related Geometries*, Springer Proceedings in Mathematics & Statistics 82,
DOI: 10.1007/978-81-322-1814-2_13,

Fig. 1 A Coxeter graph

(Sect. 3) is dedicated to the presentation of some techniques and ideas involved in the study of the $K(\pi, 1)$ conjecture and, more generally, in the study of Artin groups.

1 Lecture 1: Involved Objects

Let S be a finite set. A *Coxeter matrix* over S is defined to be a square matrix $M = (m_{s,t})_{s,t \in S}$ indexed by the elements of S and satisfying

(a) $m_{s,s} = 1$ for all $s \in S$,
(b) $m_{s,t} = m_{t,s} \in \{2, 3, 4, \dots\} \cup \{\infty\}$ for all $s, t \in S$, $s \neq t$.

The *Coxeter graph* which represents the above Coxeter matrix M is the labelled graph $\Gamma = \Gamma(M)$ defined by the following data.

(a) S is the set of vertices of Γ.
(b) Two vertices $s, t \in S$ are linked by an edge if $m_{s,t} \geq 3$.
(c) This edge is labelled by $m_{s,t}$ if $m_{s,t} \geq 4$.

Example. The following is a Coxeter matrix.

$$M = \begin{pmatrix} 1 & 3 & 2 \\ 3 & 1 & 4 \\ 2 & 4 & 1 \end{pmatrix}.$$

Its Coxeter graph is drawn in Fig. 1.

Let Γ be a Coxeter graph. The *Coxeter system* of Γ is the pair $(W, S) = (W_\Gamma, S)$, where S is the set of vertices of Γ, and W is the group defined by the following presentation.

$$W = \left\langle S \;\middle|\; \begin{array}{c} s^2 = 1 \text{ for all } s \in S \\ (st)^{m_{s,t}} = 1 \text{ for all } s, t \in S,\, s \neq t \text{ and } m_{s,t} \neq \infty \end{array} \right\rangle.$$

The group W itself is called the *Coxeter group* of Γ.

Coxeter groups were introduced by J. Tits in his manuscript [24] which was one of the main sources for the seminal book “Groupes et algèbres de Lie, chapitres 4, 5 et 6” [1]. These groups appear in several domains such as finite group theory, Lie theory, geometric group theory, or hyperbolic geometry. We recommend [13] for a complete and detailed study of them.

Fig. 2 The Coxeter graph A_n

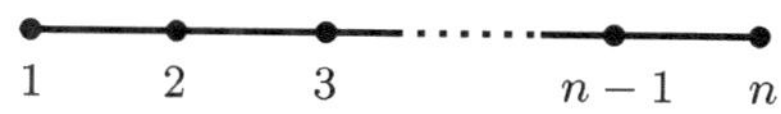

Example. Let $\Gamma = A_n$ be the Coxeter graph depicted in Fig. 2. Then W_Γ has the following presentation.

$$W = \left\langle s_1, \ldots, s_n \;\middle|\; \begin{array}{c} s_i^2 = 1 \text{ for } 1 \le i \le n \\ (s_i s_{i+1})^3 = 1 \text{ for } 1 \le i \le n-1 \\ (s_i s_j)^2 = 1 \text{ for } |i-j| \ge 2 \end{array} \right\rangle.$$

This is the symmetric group $\mathfrak{S}_{n+1}$, where s_i is the transposition $(i, i+1)$ for all $i \in \{1, \ldots, n\}$.

If a, b are two letters and m is an integer ≥ 2, we set

$$\Pi(a, b : m) = \begin{cases} (ab)^{\frac{m}{2}} & \text{if } m \text{ is even} \\ (ab)^{\frac{m-1}{2}} a & \text{if } m \text{ is odd} \end{cases}$$

In other words, $\Pi(a, b : m)$ denotes the word $aba\cdots$ of length m.

Lemma 1.1. *Let Γ be a Coxeter graph. Then W_Γ has the following presentation.*

$$W_\Gamma = \left\langle S \;\middle|\; \begin{array}{c} s^2 = 1 \, for\, all\, s \in S \\ \Pi(s, t : m_{s,t}) = \Pi(t, s : m_{s,t}) \, for\, all\, s, t \in S,\ s \ne t \, and\, m_{s,t} \ne \infty \end{array} \right\rangle.$$

Proof. It suffices to prove that the relation $(st)^m = 1$ is equivalent to the relation $\Pi(s, t : m) = \Pi(t, s : m)$ modulo the relations $s^2 = t^2 = 1$. We show that for $m = 2$ and $m = 3$, and we leave the general case to the reader. Suppose $m = 2$. Then

$$(st)^2 = stst = 1 \Leftrightarrow st = t^{-1}s^{-1} = ts \Leftrightarrow \Pi(s, t : 2) = \Pi(t, s : 2).$$

Suppose $m = 3$. Then

$$(st)^3 = ststst = 1 \Leftrightarrow sts = t^{-1}s^{-1}t^{-1} = tst \Leftrightarrow \Pi(s, t : 3) = \Pi(t, s : 3).$$

□

Let Γ be a Coxeter graph. Let $\Sigma = \{\sigma_s \mid s \in S\}$ be an abstract set in one-to-one correspondence with the set S of vertices of Γ. The *Artin system* of Γ is the pair $(A, \Sigma) = (A_\Gamma, \Sigma)$, where A is the group defined by the following presentation.

$$A = \langle \Sigma \mid \Pi(\sigma_s, \sigma_t : m_{s,t}) = \Pi(\sigma_t, \sigma_s : m_{s,t}) \text{ for all } s, t \in S,\ s \ne t \text{ and } m_{s,t} \ne \infty \rangle.$$

The group A itself is called the *Artin group* of Γ. By Lemma 1.1, the map $\Sigma \to S$, $\sigma_s \mapsto s$, induces an epimorphism $\theta : A_\Gamma \to W_\Gamma$. The kernel of θ is called the *colored Artin group* of Γ. It is denoted by CA_Γ.

Example. The Artin group of the Coxeter graph A_n depicted in Fig. 2 is the braid group $\mathcal{B}_{n+1}$ on $n+1$ strands. It has the following presentation.

$$\mathcal{B}_{n+1} = \left\langle \sigma_1, \dots, \sigma_n \;\middle|\; \begin{array}{r} \sigma_i\sigma_{i+1}\sigma_i = \sigma_{i+1}\sigma_i\sigma_{i+1} \text{ for } 1 \le i \le n-1 \\ \sigma_i\sigma_j = \sigma_j\sigma_i \text{ for } |i-j| \ge 2 \end{array} \right\rangle.$$

The colored Artin group of A_n is the pure braid group $\mathcal{PB}_{n+1}$.

Artin groups were introduced by J. Tits as extensions of Coxeter groups [25]. Although they take part in different areas of mathematics such as the theory of singularities, the theory of mapping class groups, and the geometric group theory, they are badly understood. In particular, the following questions are still open.

(1) Are Artin groups torsion free?
(2) What is the center of an Artin groups?
(3) Do Artin groups have solvable word problem?

We refer to [17] for a presentation on these questions.

Let I be a non-empty open convex cone in a finite dimensional real vector space V. A *hyperplane arrangement* in I is defined to be a set $\mathcal{A}$ of linear hyperplanes of V satisfying the following conditions.

(a) $H \cap I \neq \emptyset$ for all $H \in \mathcal{A}$;
(b) $\mathcal{A}$ is *locally finite*, that is, for all $x \in I$ there exists an open neighborhood U_x of x in I such that $\{H \in \mathcal{A} \mid H \cap U_x \neq \emptyset\}$ is finite.

Example. Let $V = \mathbb{R}^3$, and let $I = \{(x, y, z) \in \mathbb{R}^3 \mid z > 0\}$. For $k \in \mathbb{Z}$ we denote by H_k the plane of equation $x = kz$ and by H'_k the plane of equation $y = kz$. Then $\mathcal{A} = \{H_k, H'_k \mid k \in \mathbb{Z}\}$ is an hyperplane arrangement in I. The trace of $\mathcal{A}$ on the affine plane of equation $z = 1$ is depicted in Fig. 3.

Let V be a finite dimensional real vector space. Recall that a *reflection* in V is a linear transformation of V of order 2 that fixes a hyperplane. Let $\bar{C}_0$ be a closed convex polyhedral cone with non-empty interior, and let C_0 be the interior of $\bar{C}_0$. A *wall* of $\bar{C}_0$ is the support of a face (or 1 codimensional facet) of $\bar{C}_0$, that is, the linear hyperplane spanned by that face. We denote by $H_1, \dots, H_n$ the walls of $\bar{C}_0$, and, for all $i \in \{1, \dots, n\}$, we take a reflection s_i that fixes H_i. We denote by W the subgroup of $\mathrm{GL}(V)$ generated by $S = \{s_1, \dots, s_n\}$. We say that (W, S) is a *Vinberg system* if $wC_0 \cap C_0 = \emptyset$ for all $w \in W \setminus \{1\}$. In that case, W is called a *reflection group* (in Vinberg sense), S is called the *canonical generating system* for W, and C_0 is called the *fundamental chamber* of (W, S).

Fig. 3 Hyperplane arrangement

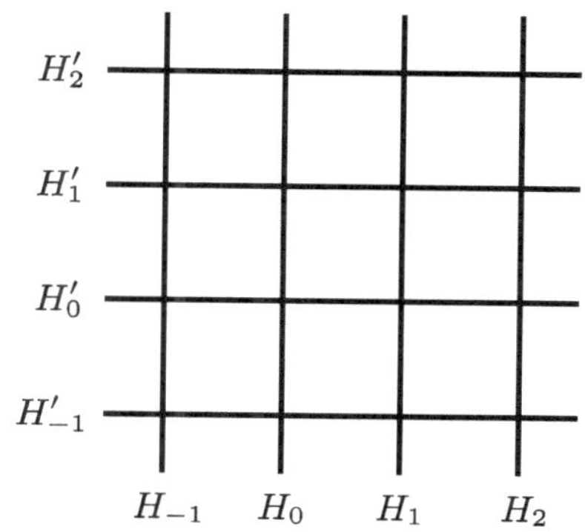

Theorem 1.2 (Vinberg [27]). *Let (W, S) be a Vinberg system. Set*

$$\bar{I} = \bigcup_{w \in W} w\,\bar{C}_0.$$

Then the following claims hold.

(1) *(W, S) is a Coxeter system.*
(2) *$\bar{I}$ is a convex cone with non-empty interior.*
(3) *The interior I of $\bar{I}$ is stable under the action of W, and W acts properly discontinuously on I.*
(4) *Let $x \in I$ such that $W_x = \{w \in W \mid w\,x = x\}$ is different from $\{1\}$. Then there exists a reflection r lying in W such that $r(x) = x$.*

Remark.

(1) There is a difference in Theorem 1.2 (1) between (W, S), viewed as a Vinberg system, and (W, S), viewed as a Coxeter system. In the first case W is a specific subgroup of the linear group $\mathrm{GL}(V)$ while in the second case W is an abstract group given by generators and relations.
(2) Every Coxeter system appears as a Vinberg system (see Sect. 2), but such a "geometric representation" is not unique in general.

The cone I in Theorem 1.2 is called the *Tits cone* of the Vinberg system (W, S). We denote by $\mathcal{R}$ the set of linear reflections lying in W, and, for $r \in \mathcal{R}$, we denote by H_r the hyperplane fixed by r. Set $\mathcal{A} = \{H_r \mid r \in \mathcal{R}\}$. Then, by Theorem 1.2, $\mathcal{A}$ is a hyperplane arrangement in I called the *Coxeter arrangement* of the Vinberg system (W, S).

Example 1. Consider the symmetric group $\mathfrak{S}_{n+1}$ acting on $\mathbb{R}^{n+1}$ by permutations of the coordinates. Let $\bar{C}_0 = \{x \in \mathbb{R}^{n+1} \mid x_1 \le x_2 \le \cdots \le x_{n+1}\}$. For $i, j \in \{1, \ldots, n+1\}$, $i \ne j$, we denote by $H_{i,j}$ the hyperplane of $\mathbb{R}^{n+1}$ of equation $x_i = x_j$.

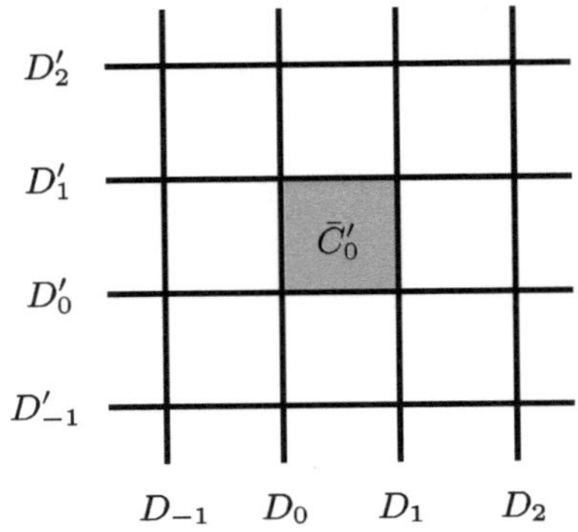

Fig. 4 Line arrangement in the Euclidean affine plane

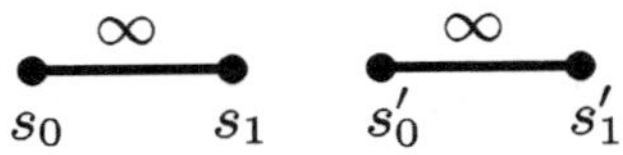

Fig. 5 Coxeter graph

Then $\bar{C}_0$ is a convex polyhedral cone whose walls are $H_{1,2}, H_{2,3}, \dots, H_{n,n+1}$. Note that, for $i \in \{1, \dots, n\}$, the transposition $s_i = (i, i+1)$ is a linear reflection which fixes $H_{i,i+1}$, and $(\mathfrak{S}_{n+1}, \{s_1, \dots, s_n\})$ is a Vinberg system. In this case we have $\bar{I} = I = \mathbb{R}^{n+1}$ and $\mathcal{A} = \{H_{i,j} \mid 1 \le i < j \le n+1\}$.

Example 2. Let $\mathbb{E}^2$ denote the Euclidean affine plane. For $k \in \mathbb{Z}$, we denote by D_k the affine line in $\mathbb{E}^2$ of equation $x = k$, by D'_k the line of equation $y = k$ (see Fig. 4), by s_k the (affine) orthogonal reflection with respect to D_k, and by s'_k the orthogonal reflection with respect to D'_k. Let W the group of isometries of $\mathbb{E}^2$ generated by $\{s_k, s'_k \mid k \in \mathbb{Z}\}$. Then W has the following presentation.

$$W = \langle s_0, s'_0, s_1, s'_1 \mid s_0^2 = s_0'^2 = s_1^2 = s_1'^2 = 1\,,\ (s_0 s'_0)^2 = (s_0 s'_1)^2 \\ = (s_1 s'_0)^2 = (s_1 s'_1)^2 = 1 \rangle.$$

This is the Coxeter group of the Coxeter graph depicted in Fig. 5.

We denote by $\mathrm{Aff}(\mathbb{E}^2)$ the group of affine transformations of the plane $\mathbb{E}^2$. We embed $\mathbb{E}^2$ into $\mathbb{R}^3$ via the map $(x, y) \mapsto (x, y, 1)$, and we embed $\mathrm{Aff}(\mathbb{E}^2)$ into $\mathrm{GL}(\mathbb{R}^3)$ as follows. Let $f \in \mathrm{Aff}(\mathbb{E}^2)$. Let $f_0 \in \mathrm{GL}(\mathbb{R}^2)$ and $u \in \mathbb{R}^2$ such that $f = T_u \circ f_0$, where T_u denotes the translation by vector u. Then the embedding $\mathrm{Aff}(\mathbb{E}^2) \hookrightarrow \mathrm{GL}(\mathbb{R}^3)$ is defined by

$$f \mapsto \begin{pmatrix} f_0 & u \\ 0 & 1 \end{pmatrix}.$$

So, the group W can be viewed through this embedding as a subgroup of $GL(\mathbb{R}^3)$.

For $k \in \mathbb{Z}$, we denote by H_k the linear hyperplane of $\mathbb{R}^3$ spanned by D_k, and by H'_k the hyperplane spanned by D'_k. Then s_k, viewed as an element of $\mathrm{GL}(\mathbb{R}^3)$, is a reflection with fixed hyperplane H_k, and s'_k is a reflection with fixed hyperplane H'_k. Let $\bar{C}'_0 = \{(x, y) \in \mathbb{E}^2 \mid 0 \le x, y \le 1\}$ (see Fig. 4), and let $\bar{C}_0$ be the cone over $\bar{C}'_0$ in $\mathbb{R}^3$. Observe that $\bar{C}_0$ is a convex polyhedral cone whose walls are H_0, H_1, H'_0, H'_1. Moreover, we have $wC_0 \cap C_0 = \emptyset$ for all $w \in W \setminus \{1\}$, thus $(W, \{s_0, s_1, s'_0, s'_1\})$ is a Vinberg system. Here we have

$$\bar{I} = \{(x, y, z) \in \mathbb{R}^3 \mid z > 0\} \cup \{(0, 0, 0)\}, \ I = \{(x, y, z) \in \mathbb{R}^3 \mid z > 0\},$$
$$\text{and } \mathcal{A} = \{H_k, H'_k \mid k \in \mathbb{Z}\}.$$

2 Lecture 2: The $K(\pi, 1)$ Conjecture

For a hyperplane arrangement $\mathcal{A}$ in an open convex cone I we set

$$M(\mathcal{A}) = (I \times I) \setminus \left(\bigcup_{H \in \mathcal{A}} (H \times H) \right).$$

If (W, S) is a Vinberg system and $\mathcal{A}$ is the Coxeter arrangement of (W, S), we set $M(W, S) = M(\mathcal{A})$. Note that, by Theorem 1.2, the group W acts freely and properly discontinuously on $M(W, S)$. Then we set $N(W, S) = M(W, S)/W$.

The starting point of the conjecture that we present in this section is the following.

Theorem 2.1 (Van der Lek [21]). *Let (W, S) be a Vinberg system, and let Γ be the Coxeter graph of (W, S), viewed as a Coxeter system. Then $\pi_1(N(W, S)) = A_\Gamma$, $\pi_1(M(W, S)) = CA_\Gamma$, and the exact sequence associated to the regular covering $M(W, S) \to M(W, S)/W = N(W, S)$ is the following.*

$$1 \longrightarrow CA_\Gamma \longrightarrow A_\Gamma \xrightarrow{\ \theta\ } W \longrightarrow 1.$$

Let X be a connected topological space having the same homotopy type as a CW-complex, and let G be a discrete group. We say that X is an *Eilenberg MacLane space* for G if $\pi_1(X) = G$ and the universal cover of X is contractible. We also say that X is *aspherical*, or that X is a *$K(G, 1)$ space*. Eilenberg MacLane spaces play an important role in the study of the cohomology of groups (see [6], for instance).

The following conjecture is a central question in the theory of Artin groups.

Conjecture 2.2 ($K(\pi, 1)$ conjecture). *Let (W, S) be a Vinberg system, and let Γ be the Coxeter graph of (W, S), viewed as a Coxeter system. Then $N(W, S)$ is an Eilenberg MacLane space for A_Γ.*

This conjecture was firstly stated by Brieskorn [4] for the so-called Artin groups of spherical type. The conjecture as is stated here appeared for the first time in [21]. According to Van der Lek, it is due to Arnold, Pham and Thom. We refer to [23] for more details on this question.

Example. Consider the symmetric group $\mathfrak{S}_n$ acting on $V = \mathbb{R}^n$ by permutations of the coordinates. For all $i, j \in \{1, \dots, n\}$, $i \neq j$, we denote by $H_{i,j}$ the hyperplane of V of equation $x_i = x_j$. Moreover, we denote by s_i the transposition $(i, i+1)$, and we set $S = \{s_1, \dots, s_{n-1}\}$. Recall that $(\mathfrak{S}_n, S)$ is a Vinberg system, where $\bar{I} = I = V$, and $\mathcal{A} = \{H_{i,j} \mid 1 \le i < j \le n\}$ (see Sect. 1). We identify $V \times V = I \times I$ with $\mathbb{C} \otimes V = \mathbb{C}^n$. Then

$$M(\mathfrak{S}_n, S) = \mathbb{C} \otimes V \setminus \left(\bigcup_{i<j} \mathbb{C} \otimes H_{i,j} \right)$$

is the space of ordered configurations of n points in $\mathbb{C}$, and

$$N(\mathfrak{S}_n, S) = M(\mathfrak{S}_n, S)/\mathfrak{S}_n$$

is the space of configurations of n points in $\mathbb{C}$.

Theorem 2.3 (Fox, Neuwirth [16]). *The fundamental group of $N(\mathfrak{S}_n, S)$ is the braid group $\mathcal{B}_n$.*

Let $f, g \in \mathbb{C}[x]$ be two non-constant polynomials. Set

$$f = a_0 x^m + a_1 x^{m-1} + \cdots + a_m\,, \quad a_0 \neq 0$$
$$g = b_0 x^n + b_1 x^{n-1} + \cdots + b_n\,, \quad b_0 \neq 0.$$

The *Sylvester matrix* of f and g is defined to be following $(m+n) \times (m+n)$ matrix.

$$\mathrm{Sylv}(f, g) = \left(\begin{array}{cccc|cccc} a_0 & 0 & \cdots & 0 & b_0 & 0 & \cdots & 0 \\ a_1 & a_0 & \ddots & \vdots & b_1 & b_0 & \ddots & \vdots \\ \vdots & a_1 & \ddots & 0 & \vdots & b_1 & \ddots & 0 \\ a_m & \vdots & \ddots & a_0 & b_n & \vdots & \ddots & b_0 \\ 0 & a_m & & a_1 & 0 & b_n & & b_1 \\ \vdots & \ddots & \ddots & \vdots & \vdots & \ddots & \ddots & \vdots \\ 0 & \cdots & 0 & a_m & 0 & \cdots & 0 & b_n \\ \multicolumn{4}{c}{\underbrace{\hphantom{a_0\ a_0\ a_0\ a_0}}_{n \text{ columns}}} & \multicolumn{4}{c}{\underbrace{\hphantom{b_0\ b_0\ b_0\ b_0}}_{m \text{ columns}}} \end{array} \right).$$

The *resultant* of f and g is

$$\mathrm{Res}(f, g) = \det(\mathrm{Sylv}(f, g)).$$

The following result is classical and well-known (see [10], for instance).

Theorem 2.4. *The polynomials f and g have a common root if and only if* $\mathrm{Res}(f, g) = 0$.

Corollary 2.5. *Let $f \in \mathbb{C}[x]$ be a polynomial of degree ≥ 2. Then f has a multiple root if and only if* $\mathrm{Res}(f, f') = 0$.

Let $f \in \mathbb{C}[x]$ be a polynomial of degree ≥ 2. Then $\mathrm{Res}(f, f')$ is called the *discriminant* of f and is denoted by $\mathrm{Disc}(f)$.

Example. If $f = ax^2 + bx + c$ is of degree 2, then $\mathrm{Disc}(f) = a(-b^2 + 4ac)$.

Let $\mathbb{C}_n[x]$ denote the space of monic polynomials of degree n. Note that $\mathbb{C}_n[x] \simeq \mathbb{C}^n$. Observe that $\mathrm{Disc} : \mathbb{C}_n[x] \to \mathbb{C}$ is a polynomial function, thus

$$\mathcal{D} = \{f \in \mathbb{C}_n[x] \mid f \text{ has a multiple root}\} = \{f \in \mathbb{C}_n[x] \mid \mathrm{Disc}(f) = 0\}$$

is an algebraic set. It is called the *n-th discriminant.*

Proposition 2.6. *We have $N(\mathfrak{S}_n, S) \simeq \mathbb{C}_n[x] \setminus \mathcal{D}$.*

Proof. Let $\phi : M(\mathfrak{S}_n, S) \to \mathbb{C}_n[x] \setminus \mathcal{D}$ be the map defined by

$$\phi(a_1, \dots, a_n) = (x - a_1)(x - a_2) \cdots (x - a_n).$$

Then ϕ is surjective and we have $\phi(a) = \phi(b)$ if and only if there exists $\chi \in \mathfrak{S}_n$ such that $b = \chi(a)$. Il follows that $\mathbb{C}_n[x] \setminus \mathcal{D} = M(\mathfrak{S}_n, S)/\mathfrak{S}_n = N(\mathfrak{S}_n, S)$. □

Theorem 2.7 (Fox, Neuwirth [16]). *The space $N(\mathfrak{S}_n, S)$ is an Eilenber MacLane space for the braid group $\mathcal{B}_n$.*

Proof. We will use the following results. These are classical in algebraic topology. Their proofs can be found for instance in [19].

(1) Let $X \to Y$ be a covering map. Then X is an Eilenberg MacLane space if and only if Y is an Eilenberg MacLane space.
(2) Let $X \to B$ be a locally trivial fibration with connected fiber F. If B and F are both Eilenberg MacLane spaces, then X is also an Eilenberg MacLane space.
(3) Any connected graph is an Eilenberg MacLane space.

By (1), in order to prove that $N(\mathfrak{S}_n, S)$ is an Eilenberg MacLane space, it suffices to prove that $M(\mathfrak{S}_n, S)$ is an Eilenberg MacLane space. We prove that by induction on n.

Suppose $n = 2$. We have

$$M(\mathfrak{S}_2, S) = \{(z_1, z_2) \in \mathbb{C}^2 \mid z_1 \neq z_2\} \simeq \mathbb{C} \times \mathbb{C}^*.$$

The space $\mathbb{C}$ is an Eilenberg MacLane space because it is contractible. On the other hand, the circle is a deformation retract of $\mathbb{C}^*$ (see Fig. 6), and the circle is an Eilenberg

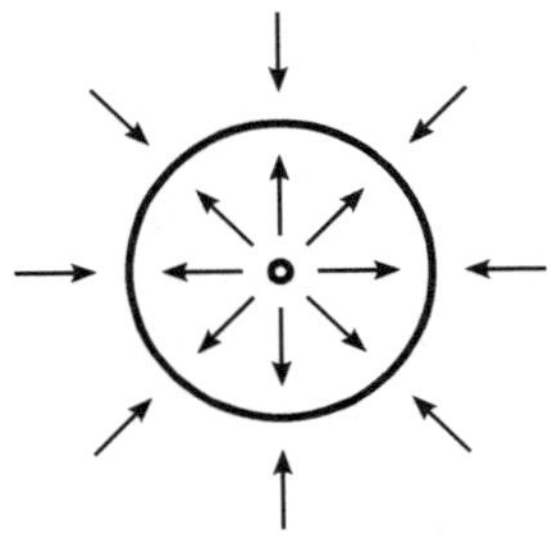

Fig. 6 Retraction of $\mathbb{C}^*$ on a circle

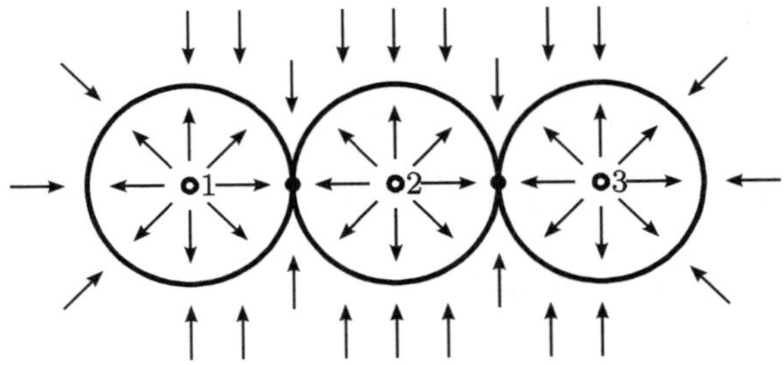

Fig. 7 Retraction of $\mathbb{C} \setminus \{1, 2, 3\}$ on a graph

MacLane space by (3), thus $\mathbb{C}^*$ is an Eilenberg MacLane space. It follows by (2) that $M(\mathfrak{S}_2, S) \simeq \mathbb{C} \times \mathbb{C}^*$ is an Eilenberg MacLane space.

Suppose now that $M(\mathfrak{S}_n, S)$ is an Eilenberg MacLane space. The map

$$\begin{array}{rcl} M(\mathfrak{S}_{n+1}, S) & \to & M(\mathfrak{S}_n, S) \\ (z_1, \dots, z_n, z_{n+1}) & \mapsto & (z_1, \dots, z_n) \end{array}$$

is a locally trivial fibration. The fiber of this map over $(1, 2, \dots, n)$ is

$$F = \{(1, \dots, n, z_{n+1}) \mid z_{n+1} \notin \{1, \dots, n\}\} \simeq \mathbb{C} \setminus \{1, \dots, n\}.$$

We can show that there is a connected graph embedded in $\mathbb{C} \setminus \{1, \dots, n\}$ which is a strong deformation retract of $\mathbb{C} \setminus \{1, \dots, n\}$ (see Fig. 7). Hence, by (3), $\mathbb{C} \setminus \{1, \dots, n\}$ is an Eilenberg MacLane space. We conclude by (2) that $M(\mathfrak{S}_{n+1}, S)$ is an Eilenberg MacLane space. □

Let Γ be a Coxeter graph, and let (W, S) be the Coxeter system of Γ. Let $\{e_s \mid s \in S\}$ be a set in one-to-one correspondence with S, and let V be the real vector space having $\{e_s \mid s \in S\}$ as a basis. Define the symmetric bilinear form $B : V \times V \to \mathbb{R}$ by

$$B(e_s, e_t) = \begin{cases} -\cos(\frac{\pi}{m_{s,t}}) & \text{if } m_{s,t} \neq \infty \\ -1 & \text{if } m_{s,t} = \infty \end{cases}$$

For every $s \in S$ define the linear transformation $\rho_s \in \mathrm{GL}(V)$ by

$$\rho_s(x) = x - 2B(x, e_s)e_s\,, \quad x \in V.$$

It is easily checked that ρ_s is a linear reflection for all $s \in S$, and the map $S \to \mathrm{GL}(V)$, $s \mapsto \rho_s$, extends to a linear representation $\rho : W \to \mathrm{GL}(V)$.

Let V^* denote the dual space of V. Recall that any $f \in \mathrm{GL}(V)$ determines a linear transformation $f^t \in \mathrm{GL}(V^*)$ defined by

$$\langle f^t(\alpha), x\rangle = \langle \alpha, f(x)\rangle\,, \quad \alpha \in V^* \text{ and } x \in V.$$

Then the *dual representation* of ρ is the linear representation $\rho^* : W \to \mathrm{GL}(V^*)$ defined by $\rho^*(w) = (\rho(w)^t)^{-1}$.

For every $s \in S$ we set $H_s = \{\alpha \in V^* \mid \langle \alpha, e_s\rangle = 0\}$. It is easily checked that $\rho^*(s)$ is a linear reflection with fixed hyperplane H_s. On the other hand, we set

$$\bar{C}_0 = \{\alpha \in V^* \mid \langle \alpha, e_s\rangle \geq 0 \text{ for all } s \in S\}.$$

Theorem 2.8 (Tits [24], Bourbaki [1]). *Let Γ be a Coxeter graph, and let (W, S) be the Coxeter system of Γ.*

(1) *The linear representations $\rho : W \to \mathrm{GL}(V)$ and $\rho^* : W \to \mathrm{GL}(V^*)$ are faithful.*
(2) *$\bar{C}_0$ is a convex simplicial cone with non-empty interior whose set of walls is $\{H_s \mid s \in S\}$. Let C_0 be the interior of $\bar{C}_0$. Then $\rho^*(w)(C_0) \cap C_0 = \emptyset$ for all $w \in W \setminus \{1\}$.*

In particular, $(\rho^*(W), \rho^*(S))$ is a Vinberg system whose associated Coxeter graph is Γ.

A Coxeter graph Γ is called of *spherical type* if W_Γ is finite. Note that, if $\Gamma_1, \ldots, \Gamma_\ell$ are the connected components of Γ, then

$$W = W_{\Gamma_1} \times \cdots \times W_{\Gamma_\ell}.$$

So, Γ is of spherical type if and only if all its connected components are of spherical type.

Theorem 2.9 (Coxeter [11, 12]).

(1) *A Coxeter graph Γ is of spherical type if and only if the symmetric bilinear form $B : V \times V \to \mathbb{R}$ is positive definite.*
(2) *The connected spherical type Coxeter graphs are precisely those depicted in Fig. 8.*

One of the main results in the subject is the following.

Theorem 2.10 (Deligne [14]). *Let (W, S) be a Vinberg system, and let Γ be the Coxeter graph of (W, S), viewed as a Coxeter system. If W is finite, then $N(W, S)$ is an Eilenberg MacLane space for A_Γ.*

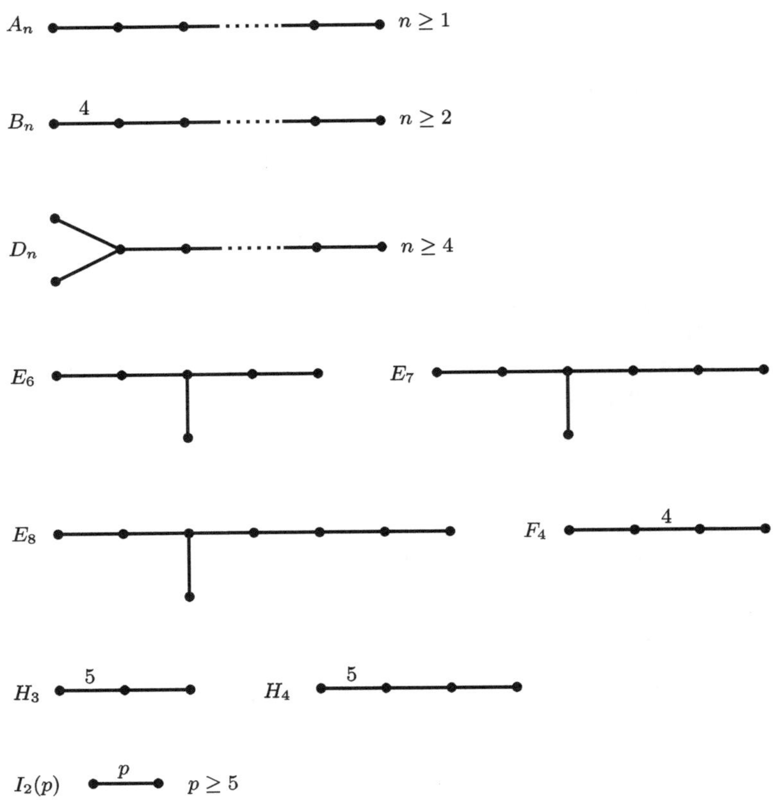

Fig. 8 Connected spherical type Coxeter graphs

Let Γ be a Coxeter graph, and let (W, S) be its associated Coxeter system. For a subset X of S we denote

- by W_X the subgroup of W generated by X,
- by Γ_X the full subgraph of Γ spanned by X.

Two families of subsets of S play an important role in the theory. The first family, denoted by $\mathcal{S}^f$, consists of subsets $X \subset S$ such that W_X is finite. For $X \subset S$, we say that Γ_X is *free of infinity* if $m_{s,t} \neq \infty$ for all $s, t \in X$. The second family, denoted by $\mathcal{S}^{<\infty}$, consists of subsets $X \subset S$ such that Γ_X is free of infinity. Note that $\mathcal{S}^f \subset \mathcal{S}^{<\infty}$.

Since Deligne's paper [14], Conjecture 2.2 has been proved in the following cases.

(1) When $m_{s,t} \geq 3$ for all $s, t \in S$, $s \neq t$ (see [20]). (Such a Coxeter graph is called of *large type*.)
(2) When $|X| \leq 2$ for all $X \in \mathcal{S}^f$ (see [9]). (Such a Coxeter graph is called of *dimension* 2.)
(3) When $\mathcal{S}^f = \mathcal{S}^{<\infty}$ (see [9]). (Such a Coxeter graph is called of *FC type*.)

(4) For the "affine type" Artin groups of type $\tilde{B}$ (see [7]).

Other examples can also be constructed with [15] and [18].

3 Lecture 3: Some Techniques and Proofs

Let Γ be a Coxeter graph, and let (W, S) be its Coxeter system. Let S^* denote the free monoid on S. Let $w \in W$. We say that a word $\mu = s_1 \cdots s_\ell \in S^*$ is an *expression* for w if the equality $w = s_1 \cdots s_\ell$ holds in W. The *length* of w, denoted by $\lg(w)$, is the minimal length of an expression for w. An expression $\mu = s_1 \cdots s_\ell$ for w is said to be *reduced* if $\ell = \lg(w)$.

Let $\mu, \mu' \in S^*$. We say that μ and μ' are related by an *elementary M-operation* if there exist $\nu_1, \nu_2 \in S^*$ and $s, t \in S$ such that $m_{s,t} \neq \infty$,

$$\mu = \nu_1\, \Pi(s, t : m_{s,t})\, \nu_2 \quad \text{and} \quad \mu' = \nu_1\, \Pi(t, s : m_{s,t})\, \nu_2.$$

Theorem 3.1 (Tits [26]). *Let (W, S) be a Coxeter system. Let $w \in W$, and let μ, μ' be two reduced expressions for w. Then μ and μ' are related by a finite sequence of elementary M-operations.*

Let Γ be a Coxeter graph, let (W, S) be its Coxeter system, and let (A, Σ) be its Artin system. Recall that we have an epimorphism $\theta : A \to W$ which sends σ_s to s for all $s \in S$. This epimorphism has a "natural" set-section $\tau : W \to A$ defined as follows. Let $w \in W$. We choose a reduced expression $\mu = s_1 s_2 \cdots s_\ell$ for w, and we set

$$\tau(w) = \sigma_{s_1} \sigma_{s_2} \cdots \sigma_{s_\ell}.$$

By Theorem 3.1, this definition does not depend on the choice of the reduced expression.

Recall that, for a subset X of S, we denote

- by W_X the subgroup of W generated by X,
- by Γ_X the full subgraph of Γ spanned by X.

The subgroup W_X is called *standard parabolic subgroup*.

Theorem 3.2 (Bourbaki [1]). *Let X be a subset of S. Then (W_X, X) is the Coxeter system of Γ_X.*

Example. Let $\Gamma = B_4$ be the Coxeter graph drawn in the left hand side of Fig. 9. We set $S = \{s_1, s_2, s_3, s_4\}$, where the elements of S are numbered according to the figure. Let $X = \{s_1, s_2, s_4\}$. Then Γ_X is the graph $B_2 \sqcup A_1$ drawn in the right hand side of Fig. 9, and

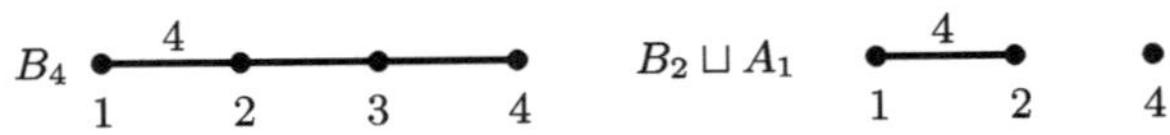

Fig. 9 Coxeter subgraph

$$W_X = \langle s_1, s_2, s_4 \mid s_1^2 = s_2^2 = s_4^2 = 1 \text{ and } (s_1 s_2)^4 = (s_1 s_4)^2 = (s_2 s_4)^2 = 1 \rangle.$$

Let Γ be a Coxeter graph, and let (W, S) be its Coxeter system. Let X, Y be two subsets of S, and let w be an element of W. We say that w is *(X, Y)-minimal* if it is of minimal length among the elements of the double coset $W_X w W_Y$.

Proposition 3.3 (Bourbaki [1]).

(1) *Let $X, Y \subset S$, and let $w \in W$. There is a unique (X, Y)-minimal element lying in $W_X w W_Y$.*
(2) *Let $X \subset S$, and let $w \in W$. Then w is $(\emptyset, X)$-minimal if and only if $\lg(ws) > \lg(w)$ for all $s \in X$.*
(3) *Let $X \subset S$, and let $w \in W$. Then w is $(\emptyset, X)$-minimal if and only if $\lg(wu) = \lg(w) + \lg(u)$ for all $u \in W_X$.*

Recall that an (abstract) *simplicial complex* is a pair $\Upsilon = (S, A)$, where S is a set called set of *vertices*, and A is a set of subsets of S, called set of *simplices*, satisfying the following properties.

(a) Every simplex is non-empty and finite.
(b) Every singleton is a simplex.
(c) Every nonempty subset of a simplex is a simplex.

Let $\Upsilon = (S, A)$ be a simplicial complex. Let $B = \{e_s \mid s \in S\}$ be a set in one-to-one correspondence with S, and let V be the real vector space having B as a basis. For every simplex $\Delta = \{s_0, s_1, \dots, s_p\}$ in A we set

$$|\Delta| = \{t_0 e_{s_0} + t_1 e_{s_1} + \dots + t_p e_{s_p} \mid 0 \le t_0, t_1, \dots, t_p \le 1 \text{ and } \sum_{i=0}^{p} t_i = 1\}.$$

Note that $|\Delta|$ is a geometric simplex of dimension p. The *geometric realization* of Υ is the set

$$|\Upsilon| = \bigcup_{\Delta \in A} |\Delta|,$$

endowed with the so-called *weak topology*.

In this paper the simplicial complexes will often appear in the following way. Let $(E, \le)$ be a partially ordered set. Then the set of non-empty finite chains of E forms an abstract simplicial complex called the *derived complex* of $(E, \le)$. This is denoted by E' or $(E, \le)'$.

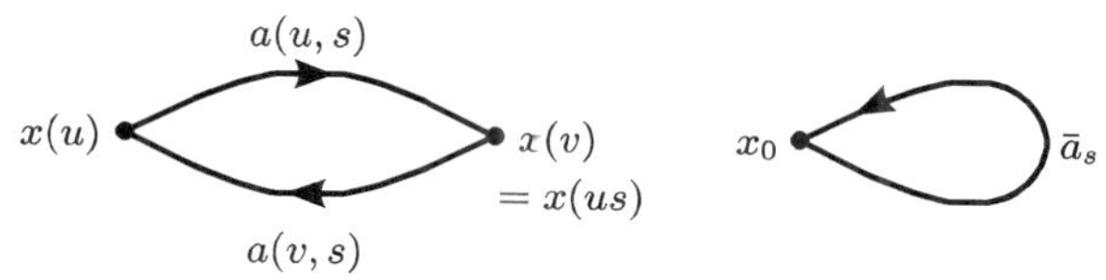

Fig. 10 1-skeletons of $\mathrm{Sal}(\Gamma)$ and $\overline{\mathrm{Sal}}(\Gamma)$

Now, take a Coxeter system (W, S). Recall that $\mathcal{S}^f = \{X \subset S \mid W_X \text{ is finite}\}$. The following can be easily proved by means of the classical theory of Coxeter groups (see [23]).

Lemma 3.4. *Let $\preceq$ be the relation on $W \times \mathcal{S}^f$ defined as follows. We set $(u, X) \preceq (v, Y)$ if*

(1) $X \subset Y$, (2) $v^{-1}u \in W_Y$, (3) $v^{-1}u$ is $(\emptyset, X)$*-minimal.*

Then $\preceq$ is a (partial) order relation on $W \times \mathcal{S}^f$.

Let Γ be a Coxeter graph, and let (W, S) be the Coxeter system of Γ. The *Salvetti complex* of Γ, denoted by $\mathrm{Sal}(\Gamma)$, is defined to be the geometric realization of the derived complex of $(W \times \mathcal{S}^f, \preceq)$. It is easily seen that the action of W on $W \times \mathcal{S}^f$ defined by $w \cdot (u, X) = (wu, X)$ leaves invariant the order relation $\preceq$, hence it induces a "natural" action of W on $\mathrm{Sal}(\Gamma)$.

Theorem 3.5 (Charney, Davis [8]). *Let (W, S) be a Vinberg system. Let Γ be the Coxeter graph of (W, S), viewed as a Coxeter system. Then there exists a homotopy equivalence $f : \mathrm{Sal}(\Gamma) \to M(W, S)$ which is equivariant under the actions of W, and which induces a homotopy equivalence $\bar{f} : \mathrm{Sal}(\Gamma)/W \to M(W, S)/W = N(W, S)$.*

Corollary 3.6. *Let (W, S) be a Vinberg system, and let Γ be the Coxeter graph of (W, S), viewed as a Coxeter system. Then the homotopy type of $N(W, S)$ depends only on the Coxeter graph Γ.*

The spaces $\mathrm{Sal}(\Gamma)$ and $\overline{\mathrm{Sal}}(\Gamma) = \mathrm{Sal}(\Gamma)/W$ have some "cellular decompositions" whose k-skeletons for $k = 0, 1, 2$ can be described as follows.

0-skeletons. The 0-skeleton of $\mathrm{Sal}(\Gamma)$ is a set $\{x(w) \mid w \in W\}$ in one-to-one correspondence with W. The 0-skeleton of $\overline{\mathrm{Sal}}(\Gamma)$ is reduced to a single point that we denote by x_0.

1-skeletons. For every $(u, s) \in W \times S$ there is an edge $a(u, s)$ of the 1-skeleton of $\mathrm{Sal}(\Gamma)$ that connects $x(u)$ with $x(us)$ (see Fig. 10). We orient this edge from $x(u)$ to $x(us)$. So, if $v = us$, then there are an edge oriented from $x(u)$ to $x(v)$ and an edge oriented from $x(v)$ to $x(u)$. On the other hand, there is no edge between $x(u)$ and $x(v)$ if $v \neq us$ for all $s \in S$. For every $s \in S$ there is an edge $\bar{a}_s$ of the 1-skeleton of $\overline{\mathrm{Sal}}(\Gamma)$ whose extremities are x_0 (see Fig. 10).

2-skeletons. Let $u \in W$ and $s, t \in S$ such that $m = m_{s,t} \neq \infty$. With u and $\{s, t\}$ is associated a 2-cell of $\mathrm{Sal}(\Gamma)$, denoted by $\mathbb{B}(u, \{s, t\})$, whose boundary is

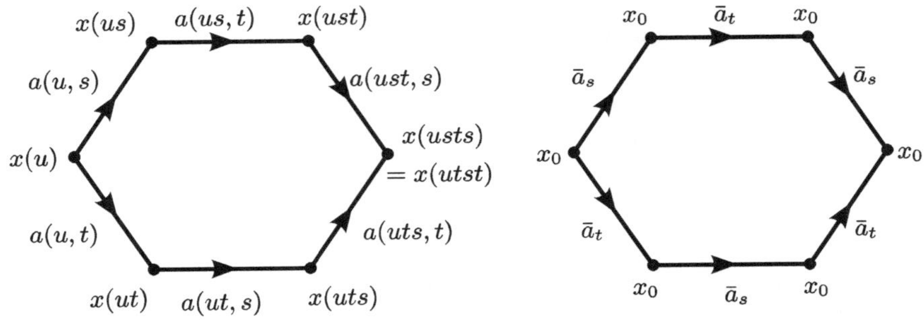

Fig. 11 2-skeletons of Sal(Γ) and $\overline{\text{Sal}}(\Gamma)$

$$a(u,s)\,a(us,t)\cdots a(u\Pi(s,t:m-1),r)\,a(u\Pi(t,s:m-1),r')^{-1}$$
$$\cdots a(ut,s)^{-1}\,a(u,t)^{-1},$$

where $r=t$ and $r'=s$ if m is even, and $r=s$ and $r'=t$ if m is odd (see Fig. 11). All 2-cells of Sal(Γ) are of this form. With $\{s,t\}$ is associated a 2-cell of $\overline{\text{Sal}}(\Gamma)$, denoted by $\bar{\mathbb{B}}(\{s,t\})$, whose boundary is

$$\bar{a}_s\,\bar{a}_t\cdots\bar{a}_r\,\bar{a}_{r'}^{-1}\cdots\bar{a}_s^{-1}\,\bar{a}_t^{-1}=\Pi(\bar{a}_s,\bar{a}_t:m)\,\Pi(\bar{a}_t,\bar{a}_s:m)^{-1}$$

(see Fig. 11). All 2-cells of $\overline{\text{Sal}}(\Gamma)$ are of this form.

The following result follows from these descriptions.

Theorem 3.7. *Let Γ be a Coxeter graph. We have $\pi_1(\overline{\text{Sal}}(\Gamma),x_0)=A_\Gamma$, $\pi_1(\text{Sal}(\Gamma),x(1))=CA_\Gamma$, and the exact sequence associated with the regular covering* Sal(Γ) $\to$ $\overline{\text{Sal}}(\Gamma)$ *is*

$$1\longrightarrow CA_\Gamma\longrightarrow A_\Gamma\xrightarrow{\theta}W\longrightarrow 1$$

Combining Theorem 3.5 and Theorem 3.7 we get the following.

Theorem 3.8 (Van der Lek [21]). *Let (W,S) be a Vinberg system, and let Γ be the Coxeter graph of (W,S), viewed as a Coxeter system. We have $\pi_1(N(W,S))=A_\Gamma$, $\pi_1(M(W,S))=CA_\Gamma$, and the exact sequence associated with the regular covering $M(W,S)\to N(W,S)$ is*

$$1\longrightarrow CA_\Gamma\longrightarrow A_\Gamma\xrightarrow{\theta}W\longrightarrow 1$$

Let Γ be a Coxeter graph. The *Artin monoid* of Γ is the monoid A_Γ^+ defined by the following monoid presentation.

$$A_\Gamma^+=\langle\Sigma\mid\Pi(\sigma_s,\sigma_t:m_{s,t})=\Pi(\sigma_t,\sigma_s:m_{s,t})\text{ for all }s,t\in S,\ s\neq t$$
$$\text{and }m_{s,t}\neq\infty\rangle^+.$$

By [22], the natural homomorphism $A_\Gamma^+ \to A_\Gamma$ is injective.

Let Γ be a Coxeter graph. We extend the epimorphism $\theta : A_\Gamma \to W_\Gamma$ to a (set) map $\tilde{\theta} : A_\Gamma \times \mathcal{S}^f \to W_\Gamma \times \mathcal{S}^f$, $(\alpha, X) \mapsto (\theta(\alpha), X)$. In [23] a (partial) order relation $\preceq$ is defined on $A_\Gamma \times \mathcal{S}^f$ and the following result is proved.

Proposition 3.9 (Paris [23]).

(1) *The map $\tilde{\theta}$ is a homomorphism of ordered sets from $(A_\Gamma \times \mathcal{S}^f, \preceq)$ to $(W_\Gamma \times \mathcal{S}^f, \preceq)$.*
(2) *Let $\widetilde{\mathrm{Sal}}(\Gamma)$ denote the geometric realization of the derived complex of $(A_\Gamma \times \mathcal{S}^f, \preceq)$, and let $\bar{\theta} : \widetilde{\mathrm{Sal}}(\Gamma) \to \mathrm{Sal}(\Gamma)$ denote the map induced by $\tilde{\theta}$. Then $\bar{\theta} : \widetilde{\mathrm{Sal}}(\Gamma) \to \mathrm{Sal}(\Gamma)$ is the universal covering of $\mathrm{Sal}(\Gamma)$.*

We denote by $\widetilde{\mathrm{Sal}}^+(\Gamma)$ the subcomplex of $\widetilde{\mathrm{Sal}}(\Gamma)$ spanned by A_Γ^+.

Theorem 3.10 (Paris [23]). *The complex $\widetilde{\mathrm{Sal}}^+(\Gamma)$ is contractible.*

The following result can be easily proved with Theorem 3.10 and some classical theory of spherical type Artin groups (see [23]).

Theorem 3.11 (Deligne [14]). *Let (W, S) be a Vinberg system, and let Γ be the Coxeter graph of (W, S), viewed as a Coxeter system. If W is finite, then $N(W, S)$ is an Eilenberg MacLane space for A_Γ.*

Let Γ be a Coxeter graph, let S be its set of vertices, and let X be a subset of S. Recall that Γ_X denotes the full subgraph of Γ spanned by X, and W_X denotes the subgroup of W generated by X. We set $\mathcal{S}_X^f = \{Y \subset X \mid W_Y \text{ is finite}\}$, and we observe that the embedding $W_X \times \mathcal{S}_X^f \hookrightarrow W \times \mathcal{S}^f$ induces an embedding $\iota_X : \mathrm{Sal}(\Gamma_X) \hookrightarrow \mathrm{Sal}(\Gamma)$.

Theorem 3.12 (Godelle, Paris [18]). *The embedding $\iota_X : \mathrm{Sal}(\Gamma_X) \to \mathrm{Sal}(\Gamma)$ has a "canonical" retraction $\pi_X : \mathrm{Sal}(\Gamma) \to \mathrm{Sal}(\Gamma_X)$.*

The retraction $\pi_X : \mathrm{Sal}(\Gamma) \to \mathrm{Sal}(\Gamma_X)$ can be described as follows. Let $(u, Y) \in W \times \mathcal{S}^f$. Write $u = u_0 u_1$, where $u_0 \in W_X$ and u_1 is $(X, \emptyset)$-minimal. Set $Y_0 = X \cap (u_1 W_Y u_1^{-1})$. We have $W_{Y_0} \subset u_1 W_Y u_1^{-1}$ and W_Y is finite, thus W_{Y_0} is finite, hence $Y_0 \in \mathcal{S}_X^f$. We set $\pi_X(u, Y) = (u_0, Y_0)$. It is shown in [18] that the map $\pi_X : W \times \mathcal{S}^f \to W_X \times \mathcal{S}_X^f$ induces a continuous map $\pi_X : \mathrm{Sal}(\Gamma) \to \mathrm{Sal}(\Gamma_X)$, and that the latter is a retraction of ι_X.

A straightforward consequence of Theorem 3.12 is the following.

Corollary 3.13. *Let Γ be a Coxeter graph, let S be its set of vertices, and let X be a subset of S. If $\mathrm{Sal}(\Gamma)$ is an Eilenberg MacLane space, then $\mathrm{Sal}(\Gamma_X)$ is also an Eilenberg MacLane space.*

Finally, combining Theorem 3.12 with the fact that W_X embeds into W plus the five lemma, we get the following.

Theorem 3.14 (Van der Lek [21]). *Let Γ be a Coxeter graph, let S be its set of vertices, and let X be a subset of S. Set $\Sigma_X = \{\sigma_s \mid s \in X\}$. Then the inclusion $\Sigma_X \hookrightarrow \Sigma$ induces an injective homomorphism $A_{\Gamma_X} \hookrightarrow A_\Gamma$.*

References

1. Bourbaki, N.: Eléments de mathématique, Fasc. XXXIV: Groupes et algèbres de Lie. Chapitre IV: Groupes de Coxeter et systèmes de Tits. Chapitre V: Groupes engendrés par des réflexions. Chapitre VI: Systèmes de racines. Actualités Scientifiques et Industrielles, No. 1337, Hermann, Paris (1968)
2. Brieskorn, E.: Singular elements of semi-simple algebraic groups. Actes du Congrès International des Mathématiciens (Nice, 1970), Tome 2, pp. 279–284. Gauthier-Villars, Paris (1971)
3. Brieskorn, E.: Die Fundamentalgruppe des Raumes der regulären Orbits einer endlichen komplexen Spiegelungsgruppe. Invent. Math. **12**, 57–61 (1971)
4. Brieskorn, E.: Sur les groupes de tresses [d'après V. I. Arnol'd]. Séminaire Bourbaki, 24ème année (1971/1972), Exp. No. 401. Lecture Notes in Mathematics, vol. 317, pp. 21–44. Springer, Berlin (1973)
5. Brieskorn, E., Saito, K.: Artin-Gruppen und Coxeter-Gruppen. Invent. Math. **17**, 245–271 (1972)
6. Brown, K.S.: Cohomology of Groups: Graduate Texts in Mathematics, 87. Springer-Verlag, New York (1982)
7. Callegaro, F., Moroni, D., Salvetti, M.: The $K(\pi, 1)$ problem for the affine Artin group of type $\tilde{B}_n$ and its cohomology. J. Eur. Math. Soc. (JEMS) **12**(1), 1–22 (2010)
8. Charney, R., Davis, M.W.: Finite $K(\pi, 1)$'s for Artin groups. Prospects in Topology (Princeton, NJ, 1994), Annals of Mathematics Studies, 138, pp. 110–124. Princeton University Press, Princeton (1995)
9. Charney, R., Davis, M.W.: The $K(\pi, 1)$-problem for hyperplane complements associated to infinite reflection groups. J. Amer. Math. Soc. **8**(3), 597–627 (1995)
10. Cox, D., Little, J., O'Shea, D.: Ideals, varieties, and algorithms: an introduction to computational algebraic geometry and commutative algebra (third edition). Undergraduate Texts in Mathematics. Springer, New York (2007)
11. Coxeter, H.S.M.: Discrete groups generated by reflections. Ann. of Math. **35**(2), no. 3, 588–621 (1934)
12. Coxeter, H.S.M.: The complete enumeration of finite groups of the form $R_i^2 = (R_i R_j)^{k_{i,j}} = 1$. J. London Math. Soc. **10**, 21–25 (1935)
13. Davis, M.W.: The geometry and topology of Coxeter groups. London Mathematical Society Monographs Series, 32. Princeton University Press, Princeton (2008)
14. Deligne, P.: Les immeubles des groupes de tresses généralisés. Invent. Math. **17**, 273–302 (1972)
15. Ellis, G., Sköldberg, E.: The $K(\pi, 1)$ conjecture for a class of Artin groups. Comment. Math. Helv. **85**(2), 409–415 (2010)
16. Fox, R., Neuwirth, L.: The braid groups. Math. Scand. **10**, 119–126 (1962)
17. Godelle E., Paris, L.: Basic questions on Artin-Tits groups. Configuration Spaces: Geometry, Combinatorics and Topology, pp. 299–311. Edizione della Normale Scuola Normale Superiore, Pisa (2012)
18. Godelle, E., Paris, L.: $K(\pi, 1)$ and word problems for infinite type Artin-Tits groups, and applications to virtual braid groups. Math. Z. **272**(3), 1339–1364 (2012)
19. Hatcher, A.: Algebraic topology. Cambridge University Press, Cambridge (2002)
20. Hendriks, H.: Hyperplane complements of large type. Invent. Math. **79**(2), 375–381 (1985)
21. van der Lek, H.: The homotopy type of complex hyperplane complements. Ph.D. thesis, University of Nijmegen (1983)

22. Paris, L.: Artin monoids inject in their groups. Comment. Math. Helv. **77**(3), 609–637 (2002)
23. Paris, L.: $K(\pi, 1)$ conjecture for Artin groups. Ann. Fac. Sci. Toulouse Math. (to appear)
24. Tits, J.: Groupes et géométries de Coxeter. Institut des Hautes Etudes Scientifiques, Paris (1961)
25. Tits, J.: Normalisateurs de tores. I. Groupes de Coxeter étendus. J. Algebra **4**, 96–116 (1966)
26. Tits, J.: Le problème des mots dans les groupes de Coxeter. Symposia Mathematica (INDAM, Rome, 1967/68), vol. 1, pp. 175–185. Academic Press, London (1969)
27. Vinberg, È.B.: Discrete linear groups that are generated by reflections. Izv. Akad. Nauk SSSR Ser. Mat. **35**, 1072–1112 (1971)

Chapter 14
Algebraic Codes and Geometry of Some Classical Generalized Polygons

N. S. Narasimha Sastry

Abstract Some results about the geometry of, and the q-ary codes associated with the finite generalized polygons $W(q), q = 2^n$; $H(q), q = 3^n$; and $\mathcal{O}(q), q = 2^{2m+1}$, are surveyed to put in context a few simple observations and state a few related questions. In Sect. 4, using the polarity in the case of $W(q), q = 2^{2m+1}$, and that of $H(q), q = 3^{2m+1}$, we present a nondegenerate symmetric bilinear form on, and a polarity of, the q-ary code associated with each of these geometries which is stabilized by the centralizer of the polarity in the automorphism group of the geometry.

Keywords Generalized polygons · Ovoids · Polarity · Algebraic codes

1 Introduction

(A) In this article, we consider the q-ary codes, $q = p^n$, p a prime and $n \geq 1$, related to the finite generalized polygons $X = (P, L)$: mostly the regular (q, q)-generalized quadrangle (GQ, for short) $W(q), q = 2^n$ and a few remarks about the regular (q, q)-generalized hexagon (GH, for short) $H(q), q = 3^n$, and the (q, q^2)-generalized octagon (GO, for short) $\mathcal{O}(q), q = 2^{2m+1}$. The automorphism groups of these geometries contain subgroups G isomorphic to $PSp(4, 2^n)$ if $X = W(2^n)$, to $G_2(3^n)$ if $X = H(3^n)$ and to the Ree group $G = {}^2F_4(2^{2m+1})$ if $X = \mathcal{O}(2^{2m+1})$. In each case, G acts transitively on the point set P, the line set L and the set $\mathcal{F} = \{(x, l) \in P \times L : x \in l\}$ of all flags of X. Recall that a *generalized n-gon* is a bipartite undirected graph (with no loops and multiple edges) of diameter n and girth $2n$. For all unexplained terminology and results, see the standard references [32, 48, 60]. These codes are of interest because of their close connection to the rich geometry

N. S. N. Sastry (✉)
Division of Theoretical Statistics and Mathematics, Indian Statistical Institute,
8th Mile, Mysore Road, R.V. College Post, Bangalore, India
e-mail: nsastry@isibang.ac.in

N. S. N. Sastry (ed.), *Groups of Exceptional Type, Coxeter Groups and Related Geometries*, Springer Proceedings in Mathematics & Statistics 82,
DOI: 10.1007/978-81-322-1814-2_14, © Springer India 2014

defining these groups (see [5, 7, 8] for an application to study ovoids in the projective 3-space over $\mathbb{F}_{2^n}$) and recent surge of interest in LDPC codes introduced by Gallagher (see [2, 35, 43, 44] for an introduction and some recent developments). These codes are also of interest because they are interesting submodules of some permutation kH-modules with large indecomposable sections, where k is a field and H is a finite group of Lie type defined over a field of the same characteristic as k:

- the k-row span of the incidence matrix of a Lie incidence system of H (whose 'point set' and the 'line set' are objects of two given distinct types of the Lie geometry for H with respect to a H-invariant incidence relation); more generally
- the image and the kernel of a kH-homomorphism between permutation kH-modules.

In this note, we record a few simple observations and put together some related questions.

(B) Let k be an algebraically closed field of characteristic p. For $M \in \{P, L\}$, let k^M denote the kG-module of all functions from M to k with the action of G defined by $(g \cdot f)(x) = f(g^{-1}x)$ for $g \in G$, $f \in k^M$ and $x \in M$. For $A \subseteq M$, we denote by $\mathbf{1}_A \in k^M$ the function taking $x \in M$ to 1 if $x \in A$ and zero otherwise. For a collection Δ of subsets of M, by abuse of notation, we denote by $\langle \Delta \rangle$ the k-subspace of k^M generated by $\{\mathbf{1}_A : A \in \Delta\}$. We write $\mathcal{C}$ for $\langle L \rangle \leq k^P$. Since gcd $(|M|, p) = 1$, the map c from k^M to k taking $f \in k^M$ to its content $\Sigma_{m \in M} f(m)$ has a splitting taking $\lambda \in k$ to $\lambda \mathbf{1}_A \in k^M$. So, we have a kG-module decomposition $k^M = k\mathbf{1}_M \oplus Y$, where $Y = \ker c$. For any subset A of k^P, we write $A \cap Y$ as A^0. Since the number of lines in X containing any given point of X is 1 (mod p), $\mathbf{1}_P \in \mathcal{C}$ and $\mathcal{C} = k\mathbf{1}_P \oplus \mathcal{C}^0$. Being a permutation module, k^P and so Y, are self-dual. For any subset A of P, we denote by $A^{\perp}$ the orthogonal of A with respect to the inner product f on k^P for which the characteristic functions of singleton subsets of P form an orthonormal basis.

(C) In the generalized n-gons X above, n even, each pair of points distance n apart is contained in, and determines, a unique $(1, t)$-subgon of X. There is a natural partition $A \cup B$ of the point set P_1 of a $(1, q)$-subgon of X, where A and B are the equivalence classes in P_1 with respect to the equivalence relation: $x \sim y$ if, and only if, $d(x, y)$ is a multiple of 4. The incidence system (A, B), with $a \in A$ incident with $b \in B$ if a and b are collinear in X, is a complete bipartite graph with $|A| = |B| = q + 1$ and A and B are lines of $PG(3, q)$, but not of $W(q)$, forming a dual grid described in Sect. 2.B.(iv) if $X = W(q)$; A and B are projective planes of order q if $X = H(q)$ (see [51], Sect. 6 for this in terms of the "ideal line condition" for $H(q)$) and the GQ $W(q)$ if $X = \mathcal{O}(q)$. Further, each word of minimum weight in $\mathcal{C}^{\perp}$ appears as a nonzero scalar multiple of $\mathbf{1}_A - \mathbf{1}_B$, where $A \cup B$ is a partition as above of the point set of a $(1, t)$-subgon of X (see, for example, [52]). Our first question is the following:

Problem 1.1. Determine the kG-module structure of:

(i) the subspace of k^P generated by $\{1_A\}$, where the sets $A \subset P$ are as above; and

(ii) the subcode of $\mathcal{C}^{\perp}$ generated by its words of minimum weight.

The subcode of $\mathcal{C}^{\perp}$ in (ii) is proper if $X = W(q)$, $q > 2$ (see Sect. 3D).

2 The GQ $W(q)$, $q = 2^n$

(A) **Geometry of** $W(q)$**:** A *duality* of $PG(3, q)$ is an inclusion reversing bijection of the set of all subspaces of $PG(3, q)$. It interchanges the set P of points and the set Π of planes of $PG(3, q)$, and fixes the set L_0 of all lines of $PG(3, q)$. A duality of order 2 is a *polarity*. Given a polarity τ of $PG(3, q)$, a subspace A of $PG(3, q)$ is said to be *absolute* (relative to τ) if it is incident with $\tau(A)$. We say that τ is a *null* polarity if each point (and so, each plane) is absolute. The set of absolute lines of $PG(3, q)$ with respect to a null polarity is called a *general linear complex*. We denote by $\mathcal{L}$ the set of all general linear complexes in $PG(3, q)$. A polarity of $PG(3, q)$ is induced by a nondegenerate sesquilinear form f on $\mathbb{F}_q^4$. It is a null polarity if $f(x, x) = 0$ for each $x \in \mathbb{F}_q^4$ ([32], p. 42–43). We note that the natural action of $PGL(4, q)$ on the set of all null polarities on $PG(3, q)$, and so on $\mathcal{L}$, is transitive. Further, the stabilizer of a member of $\mathcal{L}$ is isomorphic to $PSp(4, q)$. If τ is a null polarity of $PG(3, q)$ and $L \in \mathcal{L}$ is the set of absolute lines with respect to τ, then the incidence system $X = (P, L)$ is a (q, q)-GQ (see [48], 3.1.1), denoted by $W_L(q)$ (or $W(q)$ if L in question is clear). Recall that an (s, t)*-generalized quadrangle* is an incidence system $X_1 = (P_1, L_1)$, where P_1 is the point set and L_1 is the line set with any two points incident with at most one line, $s + 1$ points incident with each line, $t + 1$ lines incident with each point and, for each nonincident point line pair (p, l), there is exactly one line containing p and sharing a point with l. Then, $|P| = (s + 1)(st + 1)$ and $|L| = (t + 1)(st + 1)$. In the GQ $W_L(q) = (P, L)$ defined above by a polarity τ of $PG(3, q)$, for $x \in P$, $\pi_x = \cup_{x \in l \in L} l = \tau(x) \in \Pi$.

A *duality* of an incidence system $X_1 = (P_1, L_1)$ is an incidence preserving bijection of $P_1 \cup L_1$ which interchanges P_1 and L_1. Thus it defines an isomorphism between X_1 and its dual incidence system $X_1^* = (L_1, \{L_p : p \in P\})$, where $L_p = \{l \in L : p \in l\}$. A *polarity* of X_1 is a duality of order 2. For a polarity τ, an element of $P_1 \cup L_1$ incident with its image under τ is said to be *absolute*. For each n, $W(2^n)$ admits a duality (see [48], 3.2.1). It admits a polarity if, and only if, $n \geq 1$ is odd [59]. To understand this geometry, the Klein correspondence (see [39], Chap. 15) and the fact that the tangent lines of a nondegenerate quadric in $PG(2n, 2^t)$, $t, n \geq 1$ meet at a point (called the *nucleus* of the quadric) ([40], Corollary 2, p. 10). We have: (α) an isomorphism from the dual of $W(q)$ (and so, from $W(q)$) to the (q, q)-GQ $Q(4, q)$ of points and lines contained in a nondegenerate quadric $E \simeq Q(4, q)$ in $PG(4, q)$; and (β) an embedding of $Q(4, q)$ (and so, of $W(q)$) as a subGQ of the (q, q^2)-GQ $Q^-(5, q)$ of points and lines of an elliptic quadric $\widetilde{E}$ in $PG(5, q)$ by realizing $E \cong Q(4, q)$ as $\widetilde{E} \cap \pi$ for a hyperplane π of $PG(5, q)$ not tangent to $\widetilde{E}$. We note that each line of $Q^-(5, q)$ is either a line of $Q(4, q)$ or contains exactly one point of $Q(4, q)$.

We briefly explain (see [61], p.3,4 and [27], p.28, 29) for a beautiful discussion on this). Under the Klein correspondence, lines of $PG(3, q)$ are mapped bijectively to the points of a hyperbolic quadric K of $PG(5, q)$; for $L \in \mathcal{L}$, members of L are mapped to the points of a nondegenerate quadric $E = \pi \cap K$ of π, where π is a hyperplane in $PG(5, q)$ not tangent to K; and elements of L through a point of $PG(3, q)$ are mapped to points of a line contained in $K \cap \pi$. This gives an isomorphism δ from the dual of $W_L(q)$ to the GQ $Q(4, q)$ mentioned in (α). For (β), we choose an elliptic quadric $\widetilde{E}$ in $PG(5, q)$ with $\widetilde{E} \cap K = E$.

To briefly describe the polarity of $W(q)$, we fix an embedding of $PG(3, q)$ as a hyperplane A of $PG(5, q)$, such that the nucleus n of E is not in A. Let π be the projection, through n, of the points and lines E to $PG(3, q)$. This yields the points and lines of a GQ which is isomorphic to $W_L(q)$ and embedded in $PG(3, q)$. Composing $\pi \circ \delta$ with a suitable collineation of $PG(3, q)$ to ensure that the image of the point set of $W_L(q)$ is L, we see that $\pi \circ \delta$ is a duality of $W_L(q)$. Since any duality of $W_L(q)$ is of the form $\psi \circ \pi \circ \delta$ for some collineation ψ of $PG(3, q)$, we obtain a polarity of $W_L(q)$ if we can choose ψ such that $\psi \circ \pi \circ \delta$ is of order two. This is possible if the field $\mathbb{F}_{2^n}$ admits an automorphism τ such that $\tau^2(x^2) = x$ for each x in $\mathbb{F}_{2^n}$ (equivalently, n is odd).

(B) **Ovoids and spreads in** $PG(3, q)$ **and in** $W(q)$**:** (i) An *ovoid* of $PG(3, q)$, $q > 2$, is a set of $q^2 + 1$ points no three collinear. An *ovoid* of $PG(3, 2)$ is a set of five points, no four coplanar [32]. By definition, an elliptic quadric in $PG(3, q)$ is an *ovoid* of $PG(3, q)$. We denote by $\mathcal{O}$ (respectively, $\mathcal{E}$) the set of all ovoids (respectively, all elliptic ovoids) of $PG(3, q)$. An *ovoid* of $W(q)$ is a set of $q^2 + 1$ pairwise noncollinear points. An ovoid of $W(q)$ is also an ovoid of $PG(3, q)$ ([48], 1.8.2). For q odd, elliptic quadrics are the only ovoids in $PG(3, q)$ (independently due to Barlotti and Panella, see [13] and [47]). Further, for a point p, all tangent lines to an elliptic quadric θ through p are coplanar if, and only if, p belongs to θ. Thus, an elliptic ovoid is determined by its set of tangent lines; in other words, distinct ovoids have distinct sets of tangent lines. In contrast, $PG(3, 2^n)$ has ovoids which are not quadrics and distinct ovoids may have the same set of tangent lines. We explain this below assuming q to be even for the rest of this section.

By a fundamental Theorem due to Segre (see [45], 25.10, p. 127), if q is even, the set L of all tangents to an element θ of $\mathcal{O}$ is a general linear complex and θ is an ovoid of the GQ $W_L(q)$. So, the classification of ovoids of $PG(3, q)$ is equivalent to the classification of ovoids of $W(q)$. For $\mathcal{A} \subseteq \mathcal{O}$ and $L \in \mathcal{L}$ as above, We denote by $\mathcal{A}_L$ the set of all members of $\mathcal{A}$ with L as the set of its tangent lines, thus for instance, $\mathcal{E}_L = \mathcal{E} \cap \mathcal{O}_L$. The theorem of Segre mentioned above implies that $\{\mathcal{O}_L\}_{L \in \mathcal{L}}$ partition $\mathcal{O}$.

A collection of $q^2 + 1$ lines of $PG(3, q)$ partitioning P is called a *spread* of $PG(3, q)$. It is called a *symplectic spread of* $PG(3, q)$ or *a spread of* $W_L(q)$ if each of these lines belongs to a given general linear complex L. We recall the important concept of a regular spread. Given any three pairwise skew lines ℓ_1, ℓ_2, ℓ_3 in

$PG(3,q)$, there is a unique transversal to ℓ_1, ℓ_2, ℓ_3 (that is, a line of $PG(3,q)$ meeting each ℓ_i) through any point on any one of the lines l_i, $1 \le i \le 3$. If $\mathcal{R}'$ is the set of the $q+1$ transversals of ℓ_1, ℓ_2, ℓ_3, then a transversal to any three members of $\mathcal{R}'$ is a transversal to all members of $\mathcal{R}'$. The set $\mathcal{R}$ of all transversals of members of $\mathcal{R}'$ is called the *regulus in $PG(3,q)$ determined by ℓ_1, ℓ_2, ℓ_3*. A spread $\mathcal{S}$ in $PG(3,q)$, $q > 2$, is *regular* if it contains the regulus in $PG(3,q)$ determined by each triple of its members ([32], 5.1); equivalently [50], if, for some $\ell_1, \ell_2 \in \mathcal{S}$, $\ell_1 \ne \ell_2$, $\mathcal{S}$ contains the regulus determined by ℓ_1, ℓ_2, ℓ_3 for each $\ell_3 \in \mathcal{S}\backslash\{\ell_1, \ell_2\}$. Any regular spread is symplectic. An ovoid in $W(q)$ is elliptic if, and only if, its image under a duality of $W(q)$ is a regular spread of $W(q)$. Under the Klein correspondence, a regular spread $\mathcal{S}$ of $PG(3,q)$ is mapped onto an elliptic quadric $K \cap \ell^{\perp} \simeq Q^-(3,q)$ in the 3-subspace $\ell^{\perp}$ of $PG(5,q)$, where K is the Klein quadric, ℓ is a line of $PG(5,q)$ disjoint from K and $\perp$ is the orthogonal relation defined by K. Each of the $q+1$ hyperplanes of $PG(5,q)$ containing $\ell^{\perp}$ intersects K in a quadric isomorphic to $Q(4,q)$. Under the Klein correspondence, these are precisely the hyperplane sections of K corresponding to the $q+1$ members of $\mathcal{L}$ containing $\mathcal{S}$.

(ii) *Classical ovoids: (a) Elliptic ovoids:* An elliptic quadric in $PG(3,q)$ is an ovoid of $W_L(q)$, $L \in \mathcal{L}$, if, and only if, its defining quadratic form polarizes to a nondegenerate symplectic bilinear form on $PG(3,q)$ such that L is the set of its isotropic lines. Each elliptic ovoid of $W_L(q)$ can be viewed in the following two useful ways:

(α) as (exactly) one of the $q+1$ orbits for the semi-regular action on P of a cyclic subgroup of $Aut(W_L(q))$ of order q^2+1 (see [33], B.vi.a)

(β) as the set $E_x \subset E$ of all points of $Q(4,q)$ (isomorphic to $W(q)$, see (A.β) and the notation there) collinear to a given point $x \in \widetilde{E}$ of the (q,q^2)-GQ $Q^-(5,q)$ containing $Q(4,q)$ as a hyperplane intersection, but x not a point of $Q(4,q)$.

An elliptic quadric θ of $Q(4,q)$ is E_x for $x \in \widetilde{E}\backslash E$ if, and only if, x belongs to the secant line $\theta^{\perp}$ of $\widetilde{E}$. We note that $E_x = E_y$ for $x, y \in \widetilde{E}\backslash E$ if, and only if, x, y and the nucleus n of E in the hyperplane $\langle E\rangle$ of $PG(5,q)$ are collinear. Thus, there is a 2-to-1 map from $\widetilde{E}\backslash E$ onto the set of all elliptic ovoids in $Q(4,q)$ and $|\mathcal{E}_L| = |\widetilde{E}\backslash E|/2 = q^2\left(q^2-1\right)/2$. Any two elliptic ovoids of $Q(4,q)$ intersect either in a point or in a conic of E.

(γ) If l is a line of $Q^-(5,q)$ meeting E at a point p, then the set $E_l = \{E_x : p \ne x \in \ell\}$ of q elliptic ovoids of $Q(4,q)$, together with the lines of $Q(4,q)$ incident with p, define a partition of $P\backslash\{p\}$, a fact used below and in the last paragraph of Sect. 3D(i). Further, $E_l \subset \mathcal{E}_L$ is maximal in $\mathcal{E}_L$ with respect to pairwise intersection at p. Also, if l and m are lines on $\widetilde{E}$ meeting E at the same point, then $E_l = E_m$ if, and only if, l, m and the nucleus n of E in $PG(5,q)$ are coplanar.

(δ) Let $y \in \widetilde{E}\backslash(E \cup l)$. If y is collinear to p in $\widetilde{E}$, then either $y \in \theta^{\perp}$ for some $\theta \in E_l$ and so $E_y = \theta$; or $\theta \cap E_y$ is a conic in $Q(4,q)$ containing p for each $\theta \in E_l$. If y is not collinear to p, then $p \notin E_y$, $q-2$ members of E_l intersect E_y in mutually disjoint conics, remaining two members of E_l intersect E_y at distinct points and the remaining $q+1$ points of E_y are the points of $Q(4,q)$ collinear to both y and p. Thus, given an elliptic ovoid θ $(= E_x)$ in $Q(4,q)$ and $p \in \theta$, θ is in a unique set $C_{\theta,p}$ (namely, E_{px}) of q elliptic ovoids of $Q(4,q)$ which is maximal with respect to pairwise intersection at p. Further, $\cup\{C_{\theta,p}\backslash\{\theta\}\}_{p\in\theta}$ is the set (having $(q-1)\left(q^2+1\right)$ elements) of all elliptic ovoids of $Q(4,q)$ which intersect θ at a point. For some of this and other related results, see ([14], particularly Sect. 5).

Let θ_1 and θ_2 be elliptic ovoids of $Q(4,q)$. If $\theta_1\cap\theta_2=\{p\}$, then $C_{\theta_1,p}=C_{\theta_2,p}$ and each of the $q-2$ members of $C_{\theta_1,p}\backslash\{\theta_1,\theta_2\}$ intersects both θ_1 and θ_2 at p. Any elliptic ovoid θ of $Q(4,q)$ intersecting θ_1 at a point $x\neq p$ belongs to $C_{\theta_1,x}\backslash\{\theta_1\}$. Restricting to θ_2 the partition of $P\backslash\{x\}$ defined by $C_{\theta_1,x}$ and the $q+1$ lines ℓ, $x\in\ell\subset Q(4,q)$ described in (γ), we see that $q-2$ members of $C_{\theta_1,x}\backslash\{\theta_1\}$ intersect θ_2 in mutually disjoint conics and each line m of $Q(4,q)$ through x intersects θ_2 at a point. So, there is a unique member of $C_{\theta_1,x}\backslash\{\theta_1\}$ intersecting θ_2 at a point different from p. Thus, there are $q^2+(q-2)$ elliptic ovoids of $Q(4,q)$ intersecting each of θ_1 and θ_2 at only one point. If $\theta_1\cap\theta_2$ is a conic, then, by the uniqueness stated above, $C_{\theta_1,p}$ and $C_{\theta_2,p}$ are disjoint. Further, if $x\in\theta_1\backslash\theta_2$, again by (γ), there are exactly two members of $C_{\theta_1,x}$ intersecting (θ_1 at x and) θ_2 at distinct points of $\theta_2\backslash\theta_1$. Thus, there are $2\left(q^2-q\right)$ elliptic ovoids of $Q(4,q)$ intersecting each θ_i, $i\in\{1,2\}$, at a point.

(b) Tits ovoids: If $q=2^{2m+1}$, $m\geq 1$, the set of all absolute points (respectively, all absolute lines) of a polarity of $W_L(q)$ is an ovoid (respectively, spread) of $W_L(q)$ (see [48], 1.8.2), called a *Tits ovoid* (respectively, *Lüneburg spread,* see [45]). Since composition of two polarities of $W_L(q)$ is an automorphism of $W_L(q)$, $Aut\,(W_L(q))\backsimeq P\Gamma Sp(4,q)$ acts transitively on the set $\mathcal{T}_L$ of all Tits ovoids of $W_L\,(q)$. The permutation action of $PS_p(4,q)$ on $\mathcal{T}_L$ is multiplicity free ([8], Theorem 6a, p.150) and the parameters of the association scheme thus defined is conjectured in [11]. Its action on the set $\mathcal{E}_L$ is also transitive. Tits ovoids and elliptic ovoids in $PG(3,q)$ are distinct for $q\geq 8$ because their stabilizers in the subgroup of $Aut(W_L(q))$ isomorphic to $PSp(4,q)$, being respectively isomorphic to ${}^2B_2\,(q)$ [59] and $PSL_2(q^2)\cdot 2$ [39], are not isomorphic.

Elliptic ovoids (which exist for all q) and Tits ovoids (which exist only for all odd powers of 2) are the only known ovoids of $PG(3,q)$. They are collectively called *classical ovoids.* An ovoid of $W\,(q)$ admits a transitive group of automorphisms if, and only if, it is classical ([5], Theorem 1, see also [30]). For $q\in\{2,4,16\}$, each ovoid in $PG(3,q)$ is elliptic. For $q\in\{8,32\}$, each ovoid in $PG(3,q)$ is classical (see [17]).

(iii) In view of the connection of ovoids of $PG(3,q)$ to several other combinatorial structures (for example, inversive planes, see [45], 25.6, p. 126; translation planes [58]; maximal arcs [58]; unitals [22, 46]; association schemes; group divisible designs [21]; semi-biplanes [6]; Tits $\left(q,q^2\right)$-generalized quadrangles [48], etc), *study of various structures on* $\mathcal{O}$ *and their embeddings in* $PG(3,q)$ *may be fruitful.* Questions like classification of ovoids, intersection pattern of members of $\mathcal{O}$ and structure of association schemes on $\mathcal{O}$ and on its suitable subsets; partitions of the point set of $PG(3,q)$ by ovoids or by $q-1$ hyperbolic quadrics and 2 lines, etc., could be of interest. Identifying more differences between elliptic and Tits ovoids (like, for example, their secant plane sections [17], existence of special tangent lines ([32], Sect. 1.(4.58), p. 53), the subcodes of k^P generated by their characteristic functions [53], etc.) may yield pointers towards the nonexistence (conjecturally) of nonclassical ovoids.

(iv) *Dual grids in* $W_L(q)$: We denote by $\mathcal{H}$ the set of all hyperbolic quadrics in $PG(3, q)$. For a general linear complex $L \in \mathcal{L}$, let $\mathcal{H}_L$ denote the set of the members H of $\mathcal{H}$ such that each of the $2\,(q+1)$ lines in H is in L. For each $H \in \mathcal{H}_L$, the duals in $W_L\,(q)$ of the two parallel classes of lines in H (containing $q+1$ elements each) are the lines $m, m^{\perp}$ of $PG(3, q)$ not in L. Here, for any subset A of P, $A^{\perp} \subseteq P$ denotes the set of elements of P collinear in $W_L\,(q)$ to each element of A. The subset $m \cup m^{\perp}$ of P is called a *dual grid* of $W_L(q)$. It has the following equivalent descriptions: (α) the set of all points of a $(1, q)$-subGQ of $W_L\,(q)$; $(\beta)\, m^{\perp} = x^{\perp} \cap y^{\perp}$ and $m = \cap_{z \in x^{\perp} \cap y^{\perp}} z^{\perp}$ for any two distinct points x, y of m; and (γ) if θ is any ovoid of $W_L\,(q)$, then one of $m, m^{\perp}$ is a secant to θ meeting θ at, say, $\{x, y\}$ and the other is $x^{\perp} \cap y^{\perp}$, the intersection of the tangent planes to θ at x and y (See [48], 1.8.4).

(v) *Ovoidal partitions of* P: For a dual grid $m \cup m^{\perp}$, by (iv.γ), m and $m^{\perp}$ meet each ovoid of $W_L\,(q)$ in 0 or 2 points. Since q is even, P cannot be partitioned by ovoids of $W_L\,(q)$.

(a) *Existence of ovoidal partitions:* Subgroups of A, $A \in \{PSL(4, q), G\}$, $PSp(4, q) \simeq G < PSL(4, q)$, of order $q^2 + 1$ form a conjugacy class in A and the centralizer in $PSL(4, q)$ of such a subgroup T is a Singer cycle $\widehat{T} = TK$, where K is a subgroup of order $q + 1$. So, T and K act on P as well as on the set of lines of $PG(3, q)$ semi-regularly. The orbits $\theta_0, \cdots, \theta_q$ of T in P are elliptic quadrics in $PG(3, q)$, defining a partition of P ([33], Theorem 3, p. 1167). Any two such subgroups of $PSL(4, q)$ have at most one orbit in common ([1], Lemma 4.1). If $T < Aut\,(W_L\,(q))$ for some $L \in \mathcal{L}$, then only one of the θ_i's, say θ_0, is an ovoid of $W_L\,(q)$. One of the line orbits of T is a regular spread $\mathcal{S} \subseteq L$ and is the set of all lines of $PG(3, q)$ tangent to each $\theta_i, i \in \{0, \cdots, q\}$. The orbits of K in P are the elements of $\mathcal{S}$.

(b) Let $\{\theta_i\}_{0 \le i \le q}$ be a set of ovoids of $PG(3, q)$ partitioning P and let $L_i \in \mathcal{L}$ denote the set of tangent lines to θ_i. Then, by a simple counting argument, each plane is tangent to exactly one ovoid θ_i; and a line l of $PG(3, q)$ is tangent to each θ_i or it is tangent to a unique ovoid θ_l, secant to $q/2$ ovoids and passant to the $q/2$ remaining ovoids. The Klein correspondence implies that there is a regular spread $\mathcal{S}$ of lines of $PG(3, q)$ such that $L_i \cap L_j = \mathcal{S}$ for $i \neq j$ and $\{L_0, \cdots, L_q\}$ are the only general linear complexes in $PG(3, q)$ containing $\mathcal{S}$ (see [23], Sect. 2 and Lemmas 2.4 and 2.5). On the other hand, any regular spread $\mathcal{S}$ of $PG\,(3, q)$ is contained in precisely $q + 1$ general linear complexes and the cyclic group K of collineations of $PG\,(3, q)$ fixing each member of $\mathcal{S}$ acts regularly on these general linear complexes. Further, if θ is an ovoid of $PG(3, q)$ such that each member of $\mathcal{S}$ is tangent to θ, then $\{g\,(\theta) : g \in K\}$ is a partition of P (see [23], Lemma 2.5 and Theorem 3.1). In particular, there is a partition of P by Tits ovoids (see [29], Theorem 7).

Problem 2.1. Do all partitions of the point set of $PG(3, q)$ *appear in this way? Equivalently, does a cyclic group of* $PSp\,(4, q)$ *of order* $q + 1$ *acts regularly on the ovoids of a partition?*

Problem 2.2. Does there exist partitions of the pointset of $PG(3, q)$ by ovoids of different types? See [29], *Sect. 2.3 for some remarks.*

Problem 2.3. Does there exist a partition of the set of all ovoids (respectively, elliptic ovoids, Tits ovoids) of PG(3, q) such that ovoids in each part determine a partition of the pointset of PG(3, q)?

(vi) We present a few facts about $\mathcal{O}$. For $\theta \in \mathcal{O}$, let θ^* denote the set of all planes tangent to θ and $L(\theta) \in \mathcal{L}$ denote the set of all lines tangent to θ.

Proposition 2.3. *Let $\theta_1, \theta_2 \in \mathcal{O}$. Then,*
(α) (Butler [24], *Theorem 12) $|\theta_1 \cap \theta_2| \leq \frac{1}{2}q^2 + 1$;*
(β) (Bruen and Hirschfeld [19], *Theorem 5.1) $|\theta_1^* \cap \theta_2^*| = |\theta_1 \cap \theta_2|$; and*
(γ) (Govaert and Maldeghem [38], *p. 322; see also Butler* [24], *Lemma 2.2) if $L(\theta_1) \neq L(\theta_2)$, then $L(\theta_1) \cap L(\theta_2)$ is either a regular spread in $PG(3, q)$ or a set of $q^2 + q + 1$ lines meeting a given line l (including l itself) or a set of $(q + 1)^2$ lines of a common $(1, q)$-sub GQ (equivalently, lines of a common dual grid).*

Penttila has found two Tits ovoids of $PG(3, 8)$ with 33 points in common (private communication). However, the bound in (α) seems to be excessive for general q. In (β), the points of tangency in θ_1 and θ_2 of an element of $\theta_1^* \cap \theta_2^*$ need not be in $\theta_1 \cap \theta_2$ (see [9] for a discussion).

Problem 2.4. Does there exists a 'natural' bijection between $\theta_1 \cap \theta_2$ and $\theta_1^* \cap \theta_2^*$?

More generally, the number of points common to two nonsingular quadrics of the same type in a finite projective space of odd dimension is equal to the number of their common tangent hyperplanes. This is not true for finite projective spaces of even dimension, even for projective planes (there are disjoint ovals with common tangent lines (see [19], p. 218 and [9], Lemma 2.1). Though there exist ovals in a projective plane intersecting at a point with distinct tangent lines at the point of intersection, I do not know the answer to the

Problem 2.5. Does there exist a pair of ovoids in $PG(3, q)$ intersecting at a point with distinct tangent planes at the point of intersection?

To see (γ), we note that the Klein correspondence τ maps $L(\theta_1) \cap L(\theta_2)$ onto the intersection of the projective 3-subspace $B = \langle L(\theta_1)\rangle \cap \langle L(\theta_2)\rangle$ of the projective 4−space $\langle L(\theta_1)\rangle$ with the nondegenerate quadric $L(\theta_2)$. The possibilities for the later are : an elliptic quadric in B, a hyperbolic quadric in B and a quadratic cone ([39], chap. 15). They correspond under τ to the possibilities in (γ).

In ([38], Application 1, p. 322 and 321), it is shown that the intersection of the line sets of two distinct generalized hexagons $H(K)$, K a commutative field, on the same quadric $Q(6, K)$ is one of the following: (i) the set of lines at a distance at most 3 from a given point; (ii) the set of lines of a subhexagon with two points on each line and such that each line of $H(K)$ through a point of the subhexagon is a line of the subhexagon; (iii) a set of lines mutually at a distance three from each other such that every other line meets one line of this set.

(vii) *Intersection of classical ovoids in* $W_L(q) \simeq W(q)$, $q = 2^{2m+1} \geq 8$, $L \in \mathcal{L}$: We first describe the subsets of a Tits ovoid θ of $W_L(q)$, each of which appears as the intersection of θ with some Tits ovoid $\neq \theta$ of $W_L(q)$. If $H \simeq {}^2B_2(q)$ is the stabilizer of θ in $PSL(4,q)$, then $H < Aut(W_L(q))$ and $|H| = q^2(q^2+1)(q-1)$. The stabilizer in H of a point x of θ is PK, where $P \in Syl_2(H)$, $K \simeq C_{q-1}$ and the action of P on $\theta \setminus \{x\}$ is regular (see [42], p. 90). Union of $\{x\}$ and an orbit of the centre of P in $\theta \setminus \{x\}$ is an oval and is called a *pseudo-circle in* θ. Union of two distinct pseudo-circles in θ through x is called a *figure of eight* in θ. H contains a unique conjugacy class $\mathcal{A}^+$ (respectively, $\mathcal{A}^-$) of cyclic subgroups of orders $q+r+1$ (respectively, $q-r+1$), where $r^2 = 2q$. The normalizer in H of each of these groups is maximal in H and is of order $4(q+r+1)$ (respectively, $4(q-r+1)$). Each member of $\mathcal{A}^+ \cup \mathcal{A}^-$ is self-centralizing in H and its centralizer in $PSL(4,q)$ (respectively, in $Aut(W_L(q))$) is a cyclic subgroup $\widehat{T}$ of order $|P|$ (respectively, a subgroup T of the Singer cycle $\widehat{T}$ of order q^2+1) (see [42], Theorem 3.10, p. 190). As noted in (B.ii.α), the T-orbits in P are elliptic quadrics and only one of them is an ovoid of $W_L(q)$. Since this accounts for $|H|/4(q+r+1) + |H|/4(q-r+1) = q^2(q^2-1)/2$ elliptic ovoids in $W(q)$ ($\equiv$ subgroups of order q^2+1 of $Aut(W_L(q))$), the map $T \mapsto T \cap H$ is a bijection from the set of subgroups of $Aut(W_L(q))$ of order q^2+1 and $\mathcal{A}^+ \cup \mathcal{A}^-$. Thus, H has two orbits in the set $\mathcal{E}_L$ of all elliptic ovoids of $W_L(q)$ (and two orbits on the set of regular spreads in L, in view of the polarity of $W_L(q)$ centralized by H) with $4(q+r+1)$ and $4(q-r+1)$ elements. An A-orbit in θ is called a *cap* or a *cup* according as $A \in \mathcal{A}^+$ or $\mathcal{A}^-$. Each A-orbit in θ is the intersection of θ with the T-orbit E in P containing it. For use in (Sect. 3C), we note that $\mathbf{1}_E = \Sigma_{x \in T} \mathbf{1}_{x(\theta)}$ ([53], Lemma 14).

Theorem 2.6 *Let L be a general linear complex in $PG(3,q)$ and $\theta_1, \theta_2 \in \mathcal{O}_L$ be distinct. Then, the following hold:*

(i) (Glynn [36]*)* $|\theta_1 \cap \theta_2| \leq q(q-1)/2$.
(ii) (Butler [23]*)* $|\theta_1 \cap \theta_2|$ *is odd. Further,* $|\theta_1 \cap \theta_2| = 1$ *(mod 4) if* θ_1 *is an elliptic ovoid.*
(iii) (Glynn [37]*) If θ_1 and θ_2 are both elliptic, then $\theta_1 \cap \theta_2$ is either a point or a conic (that is, the intersection of θ_i with a nontangent plane).*
(iv) (Bagchi and Sastry [8]*) If θ_1 is elliptic and θ_2 is a Tits ovoid, then $\theta_1 \cap \theta_2$ is either a cup or a cap.*
(v) (Bagchi and Sastry [8]*) If θ_1 and θ_2 are both Tits ovoids, then $\theta_1 \cap \theta_2$ is one of the following: a point, a pseudo-circle, a figure of eight, a cup or a cap.*

(See also [10]; [60], p. 341). Thus any two ovoids (dually, any two spreads) of $W_L(q)$ have at least one element in common. Since an ovoid of $W_L(q)$ with a conic plane section is elliptic (Brown [16], Theorem 4.2, p. 576), any ovoid of $W_L(q)$ intersecting with each elliptic ovoid of $W_L(q)$ in either a point or in a conic is elliptic. I thank H. Van Maldeghem for pointing out this remark to me.

Problem 2.7. If an ovoid of $W_L(q)$ intersects each elliptic ovoid of $W_L(q)$ in a cup or a cap, then is it necessarily a Tits ovoid?

Other possibilities for intersection of ovoids with different sets of tangent lines exist (see [34]).

Example: Let $\ell \cup m$ be a dual grid in the GQ $W_L(q)$. The group $Aut(W_L(q))$ contains a cyclic subgroup A of order $q-1$ which fixes each point of $\ell \cup m$ and acts regularly on the $(q+1)^2$ sets $xy \setminus \{x, y\}$, where $x \in \ell$, $y \in m$ and the line $xy \in L$. Fix $x_0 \in \ell$, $y_0 \in m$ and an ovoid θ of $PG(3,q)$ containing x_0 and y_0. Then, $a\theta$ is an ovoid of $PG(3,q)$ for each $a \in A$ and the $q-1$ ovoids $\{a\theta : a \in A\}$ intersect pairwise at $\{x_0, y_0\}$.

Problem 2.8. Determine the intersection pattern of classical ovoids in $PG(3,q)$ with different sets of tangent lines.

Analysis in ([34], III.C) using the classification of pencils of quadrics in [20] shows that any two elliptic ovoids of $PG(3,q)$ meet in at most $2(q+1)$ points.

(C) **The graph** $\Gamma_{\mathcal{O}}$: Let $\Gamma_{\mathcal{O}}$ denote the graph with vertex set $\mathcal{O}$ and $\theta_1, \theta_2 \in \mathcal{O}$ defined to be adjacent if $|\theta_1 \cap \theta_2| = 1$. For $\mathcal{A} \subseteq \mathcal{O}$, we denote by $\Gamma_{\mathcal{A}}$ the subgraph of $\Gamma_{\mathcal{O}}$ induced on $\mathcal{A}$. Properties of $\Gamma_{\mathcal{A}}$ like connectedness, regularity, diameter etc., are of interest for $A \in \{\mathcal{O}, \mathcal{E}, \mathcal{T}, \mathcal{O}_L, \mathcal{E}_L, \mathcal{T}_L$ for $L \in \mathcal{L}\}$, where $\mathcal{O}_L$, $\mathcal{E}_L$, $\mathcal{T}_L$ are as defined on (B.i). We now record some elementary remarks about the cliques in $\Gamma_{\mathcal{O}}$.

Lemma 2.9 *If $q \geq 4$, a clique in $\Gamma_{\mathcal{O}}$ has at most $q+1$ vertices.*

Proof. If there were $q+2$ pairwise adjacent vertices in $\Gamma_{\mathcal{O}}$, then their union would have at least $(q+2)\left(q^2+1\right) - \binom{q+2}{2}$ points. This exceeds $|P|$ by $\frac{q(q-3)}{2}$, a positive number if $q \geq 4$.

$W(2)$ has 6 ovoids (all elliptic) pairwise adjacent. If $q \geq 4$, $\Gamma_{\mathcal{O}_L}$ can not have a clique of size $q+1$.

Problem 2.10. Does $\Gamma_{\mathcal{O}}$ contains a clique of size $q+1$?

For $x \in P$ and a plane π of $PG(3,q)$ containing x, let $\mathcal{O}_{x,\pi} \subset \mathcal{O}$ denote the set of all ovoids of $PG(3,q)$ containing x with π as their common tangent plane at x. Each clique in $\Gamma_{\mathcal{O}_{x,\pi}}$ of size q (in particular, the cliques E_l in $\Gamma_{\mathcal{E}_L}$ described in B(ii. a. γ) is maximal in $\Gamma_{\mathcal{O}}$. We now show that there is an elementary abelian subgroup of $PSL(4,q)$ of order q acting semi regularly on $\mathcal{O}_{x,\pi}$ whose orbits partition $\mathcal{O}_{x,\pi}$ by maximal cliques of size q.

Proposition 2.11. *Let x, π and $\mathcal{O}_{x,\pi}$ be as above and $L \in \mathcal{L}$. Assume that L contains each line in π incident with x. Let $G < Aut(W_L(q))$ with $G \simeq PSp(4,q)$ and M be the subgroup of G consisting of all elements of G fixing each point of π. Then, the following hold:*

(i) M is isomorphic to the additive group of $\mathbb{F}_q$.
(ii) If $m \in M$ fixes an element of $P \setminus \pi$, then m is trivial.
(iii) M stabilizes each line $\ell \nsubseteq \pi$ of $PG(3,q)$ containing x and acts regularly on $\ell \setminus \{x\}$.

(iv) *For* $\theta \in \mathcal{O}_{x,\pi}$ *and* $m \in M$, $m(\theta) \in \mathcal{O}_{x,\pi}$ *and* $\{m(\theta)\}_{m\in M}$ *is a maximal clique in* $\Gamma_{\mathcal{O}_{x,\pi}}$.

Proof. With an appropriate choice of the symplectic basis for $\mathbb{F}_q^4$ for the standard representation of $Sp(4, q)$ on $\mathbb{F}_q^4$, M is seen to be conjugate to $\{I_4 + \lambda E_{1,4} : \lambda \in \mathbb{F}_q\}$, where $E_{1,4}$ is the 4×4-matrix over $\mathbb{F}_q$ whose $(1, 4)$ th -entry is 1 and the rest are zero. So, (i) follows. If $m \in M$ fixes $z \in P \setminus \pi$ and l is a line incident with z, then m fixes $\ell \cap \pi$. Since the order of m is even, m fixes a third point of ℓ not in $\{z, \ell \cap \pi\}$ and so is the identity on ℓ. Thus m is the identity on each line through z and so on P, proving (ii). If ℓ is as in (iii), then $\ell^{\perp} \subset x^{\perp} = \pi$ and, so $\ell = (\ell^{\perp})^{\perp}$ is fixed by each $m \in M$. Here, the perpendicularity '$\perp$' is in $W_L(q)$. Now, (ii) completes the proof of (iii). For θ and m as in (iv), π is tangent to $m(\theta)$ at x. So, $m(\theta) \in \mathcal{O}_{x,\pi}$. Each of the q^2 lines $l \not\subseteq \pi$ incident with x is secant to $m(\theta)$ and, by (iii), $l \cap m(\theta) \neq l \cap m'(\theta)$ for distinct $m, m' \in M$. So, the ovoids $\{m(\theta)\}_{m\in M}$ are distinct and pairwise intersect at $\{x\}$. Maximality of the clique follows because the complement of their union is $\pi \setminus \{x\}$.

Proposition 2.12. *Let* $L \in \mathcal{L}$ *and* $q > 2$.

(i) *(Hubaut, Metz [41])* $\Gamma_{\mathcal{E}_L}$ *is a strongly regular graph with parameters* $v = \binom{q^2}{2}$, $k = (q-1)(q^2+1)$, $\lambda = q^2 + (q-2)$ *and* $\mu = 2(q^2 - q)$.
(ii) *Any clique in* $\Gamma_{\mathcal{E}_L}$ *containing three members intersecting mutually at a common point is a subset of a maximal clique of the type* E_l *described in B(ii. a.* γ*).*
(iii) *(Van Maldeghem and Sastry) Let* $\mathcal{A}$ *be a clique in* $\Gamma_{\mathcal{E}_L}$ *whose members meet mutually at distinct points. Then,* $\mathcal{A}$ *has at most 6 elements.*

Proof. (i) is well-known. (ii) follows from the discussion in (B.ii.a.δ). For (iii), we use the set up outlined in the last two paragraphs of B(ii. a. γ). It also contains a proof of (i). Let $\mathcal{A}$ be a clique in $\Gamma_{\mathcal{E}_L}$ with at least 3 elements $\theta_1, \theta_2, \theta_3$. Let $\theta_i^{\perp} \cap \widetilde{E} = \{a_i, b_i\} \subset \widetilde{E} \setminus E$, $i = 1, 2, 3$. Then, $a_i \nsim b_i$ and $\theta_i = E_{a_i} = E_{b_i}$. With proper indexing, we can take $\{a_1, a_2, a_3\}$ and $\{b_1, b_2, b_3\}$ to be triads (that is, triples of mutually noncollinear points) in $Q^-(5, q)$ and their traces are conics C and D, respectively. If $\theta_4 \in \mathcal{A}$ and $\theta_4 \neq \theta_i$ for $i = 1, 2, 3$, then $\theta_4 = E_{a_4} = E_{b_4}$ for non collinear points a_4, b_4 in $\widetilde{E} \setminus E$ and, each of them is collinear to a_i or b_i (and not both) for $i = 1, 2, 3$. Let $b_4 \sim a_1$, say. Then, as $a_1 \sim b_2$, $b_4 \nsim b_2$. So, $b_4 \sim a_2$. Similarly, $b_4 \sim a_3$ and $b_4 \in C$. If $\mathcal{A}$ has a fifth element $\theta_5 = E_{a_5} = E_{b_5}$, then C contains an element b_5 collinear to a_4. Since the plane π containing C is not contained in $a_4^{\perp}$ (since $b_4 \in a_4^{\perp}$) and $a_4^{\perp} \cap \pi$ is a line, there are at most two possibilities for b_5 and (iii) follows.

A more complete analysis of the cliques in $\Gamma_{\mathcal{E}_L}$ will be presented in [25]. If each ovoid of $W_L(q)$ is classical, then $\Gamma_{\mathcal{E}_L}$ and Γ_{θ_L} are the connected components of Γ_{θ_L} . There are two G-orbits of edges in $\Gamma_{\mathcal{J}_L}$. In fact, if $\theta_1, \theta_2 \in \mathcal{J}_L$ and $H_i = Stab_G(\theta_i)$, $i = 1, 2$, then $|\theta_1 \cap \theta_2| = 1$ if either $|H_1 \cap H_2|$ is q^2or $2q$ (see 3.2 (b), p.156). We note that if $|H_1 \cap H_2| = 2q$. $|\theta_1 \cap \theta_2| = 1$ could also be $2q + 1$. The proof of ([15], Lemma 5.1, p.304) shows the connectivity of the graph with vertex set $\mathcal{J}_L$ with two verticies declared to be adjacent if their stabilizers in G intersect in a subgroup of order q^2.

3 The Code $\mathcal{C}$

Let $X = (P, L)$ be the GQ $W(q)$ and let the notation be as in Sect. 1(i, ii).

(A) *Socle of k^P and of $\mathcal{C}$*: For $x \in P$, let π_x denote the union of the lines of $W(q)$ incident with x. Since q and the characteristic of k are even and since the number of lines as well as the number of planes of $W(q)$ incident with each $x \in P$ is odd, $\mathbf{1}_{\pi_x} \in \mathcal{C}$ and $\mathbf{1}_P = \Sigma_{x \in P} \mathbf{1}_{\pi_x} \in \langle \Pi \rangle \subseteq \mathcal{C}$. We note that $\langle \Pi \rangle^0$ is a simple kG-module; in fact, the Steinberg module for G. If θ is an ovoid of $W(q)$, then the characteristic functions of its tangent planes $\{\pi_x\colon x \in \theta\}$ form a basis for $\langle \Pi \rangle$ ([5], Lemma 6, p. 144): in fact, for a secant plane π of θ, a point $p \in P$ is on $q+1, 1, 2$ or 0 tangent planes to θ at the points of the oval $\pi \cap \theta$ according as p is the nucleus n of $\pi \cap \theta$; $n \neq p \in \pi$; $p \notin \pi$ but p lies on, or does not lie on, the tangent planes considered. So, $\mathbf{1}_\pi = \Sigma_{x \in \theta \cap \pi} \mathbf{1}_{\pi_x}$. For any subset A of θ, the restriction of $\Sigma_{x \in A} \lambda_x \mathbf{1}_{\pi_x}$ to θ is $\Sigma_{x \in A} \lambda_x x$. This is zero if, and only if, $\lambda_x = 0$ for each $x \in A$. Thus, dimension of $\langle \Pi \rangle$ is $q^2 + 1$. Since $\mathbf{1}_P \in \langle \Pi \rangle$, $\langle \Pi \rangle^0$ is of codimension one in $\langle \Pi \rangle$. Recall that the *socle* of a module M is its largest semisimple submodule. It is written as $Soc(M)$. The *radical* of M is the smallest submodule of M such that the quotient is semisimple. It is written as $rad\,(M)$.

Proposition 3.1. *(i) $Soc(k^P) = \langle \Pi \rangle \simeq k\mathbf{1}_P \oplus \langle \Pi \rangle^0 \simeq k^P/rad\,(k^P)$ and is of dimension q^2+1.*

(ii) Socle $(Y) = \langle \Pi \rangle^0 \simeq Y/rad\,(Y)$. In particular, Y is indecomposable.

Proof. Let S be a Sylow 2-subgroup of G. Then, S fixes a flag (p, ℓ) in $W(q)$ and so, also π_p. Further, it acts transitively on the sets of points $\ell \setminus \{p\}$, $\pi_p \setminus \ell$ and $P \setminus \pi_p$. So, $\mathbf{1}_{\{p\}}$, $\mathbf{1}_\ell$,$\mathbf{1}_{\pi_p}$ and $\mathbf{1}_P$ generate the space of S-fixed points in k^P. Hence, $k\mathbf{1}_P$ and $\langle \mathbf{1}_P - \mathbf{1}_{\pi_p} : p \in P \rangle = \langle \Pi \rangle^0$ are the only simple kG-submodules of k^P and the socle of k^P is $\langle \Pi \rangle$. From the preceeding remarks and the self-duality of k^P and Y, the rest follows.

In particular, socle of $\mathcal{C}$ is $\langle \Pi \rangle$, that of $\mathcal{C}^\perp$ is $\langle \Pi \rangle^0$ and that of any kG-submodule of k^P is either $k\mathbf{1}_P$, $\langle \Pi \rangle^0$ or $\langle \Pi \rangle$. Given a duality δ of $W(q)$, the map from P to $\mathcal{C}$ taking $x \in P$ to $\mathbf{1}_{\delta(x)}$ defines a linear map $\widehat{\delta}$ from k^P onto $\mathcal{C}$ with kernel $\mathcal{C}^\perp$. Further, for $\ell \in L, \widehat{\delta}(\mathbf{1}_\ell) = \mathbf{1}_{\pi_{\delta(\ell)}} \in \langle \Pi \rangle$ and δ is a bijection between P and L. So, $\mathcal{C} \simeq k^P/\mathcal{C}^\perp$and $\mathcal{C}/\mathcal{C} \cap \mathcal{C}^\perp \simeq \langle \Pi \rangle$.

We note that, if G is a finite group admitting a split (B, N)-pair and defined over a field of the same characteristic p as that of k and A is the set of cosets of any maximal parabolic subgroup of G, then the module k^A for the permutation action of G on A has a kG-module decomposition $k^A = k\mathbf{1} \oplus Y$ (as in Sect. 1A). Further, $Soc(Y)$ and $Y/rad\,(Y)$ are simple kG-modules. For details, see ([3], Sect. 11 and Remark 11.4).

(B) *Structure of* $\mathcal{C}$: The dimension of $\mathcal{C}$ is

$$1 + 2^{-2n}[(1 + \sqrt{17})^{2n} + (1 - \sqrt{17})^{2n}] = \Sigma_{I \in \mathcal{N}} 4^{|I|}, \qquad (*)$$

where $\mathcal{N}$ is the collection of all subsets of $\mathbb{Z}/2n\mathbb{Z}$ which do not have consecutive elements ([53], Theorem 1, p. 485; for the last equality, see [28], p. 34–35). This formula is obtained by identifying the composition factors appearing in each socle layer of $\mathcal{C}^0$. Note that (*) is not a rational function of q. The lattice of submodules of $\mathcal{C}$ is also known ([53], p. 491). As each kG-composition factor of $\mathcal{C}^0$ appears precisely once, each submodule of $\mathcal{C}^0$ is determined by the isomorphism type of its quotient by its radical. Further, the number of submodules of $\mathcal{C}^0$ is finite and is independent of k. The composition factors of $\mathcal{C}^0$ and of κ^P are the same ([56], Theorem 1, p. 238). From a geometric point of view, the following may be interesting:

Question: *For each socle layer of* $C_2 = \langle \mathbf{1}_\ell : \ell \in L \rangle$ *considered as an* $\mathbb{F}_2G$-*module, find 'canonical' subset of* P *'defined geometrically' whose characteristic functions define a basis for the layer.*

In a significant development ([27] for even characteristic and [26] for odd characteristic), Chandler, Sin and Xiang obtained a formula for the dimension of the q-ary code generated by the characteristic functions of isotropic subspaces of a fixed dimension l in a symplectic space over $\mathbb{F}_q$ of projective dimension n. They also identify the composition factors of the module for the corresponding symplectic group. Their method is very similar to the representation theoretic approach in [12] for the famous Hamada formula for the dimension of the q-ary code generated by the characteristic functions of the subspaces of fixed dimension l in a finite projective space over $\mathbb{F}_q$. If $n = 3$ and $q = p^n$, p any prime, $\mathbb{F}_q$-rank of points $\times$ isotropic lines incidence matrix reduces to $1+\alpha_1^n+\alpha_2^n$, where $\alpha_1, \alpha_2 = \frac{p(p+1)^2}{4} \pm \frac{p(p+1)(p-1)}{12}\sqrt{17}$. It agrees with the expression above for $p = 2$. However, the methods in [27] and in [53] are different.

The words of minimum nonzero weight in $\mathcal{C}$ are nonzero scalar multiples of $\mathbf{1}_\ell, \ell \in L$ ([4], Corollary 5.7.5, p. 186).

(C) *Subcodes of* $\mathcal{C}$ *generated by ovoids in* $W(q)$: For a subgroup A of G, let σ_A denote the k-endomorphism of k^P taking $\mathbf{1}_{\{x\}} \in k^P, x \in P$, to $\Sigma_{a \in A} \mathbf{1}_{\{a(x)\}} \in k^P$. If T is a subgroup of $G \simeq PSp(4, q)$ of order $q^2 + 1$ and the spread $\mathcal{S}$ and the ovoids $\{\theta_i\}_{i=0}^q$ are defined by T as in Sect. 2B(v a), then $\sigma_T(l) = \mathbf{1}_P$ or $\mathbf{1}_{\theta_i}$ according as $l \in \mathcal{S}$ or l is tangent to θ_i for some i, $0 \le i \le q$ (see Sect 2B(ii) α and v.b). Thus, $\sigma_T(k^P) = \langle \mathbf{1}_{\theta_0}, \cdots, \mathbf{1}_{\theta_q} \rangle$ and $\sigma_T(\mathcal{C}) = \langle \mathbf{1}_{\theta_0}, \mathbf{1}_P \rangle \subset \mathcal{C}$. Since G is transitive on $\mathcal{E}_L$ and $\mathcal{C}$ is a kG-module, $\langle \mathcal{E}_L \rangle \subseteq \mathcal{C}$. The dimension of $\langle \mathcal{E}_L \rangle$ is 5^n. The composition factors of each of its socle layers is known ([53], Theorem 13, p. 493). Further, $\langle \mathcal{E}_L \rangle = \langle \mathcal{H}_L \rangle$ ([54], Theorem 3.1, p. 5).

Problem 3.2. Determine the words of the minimum weight of $\langle \mathcal{E}_L \rangle$.

If $W(q)$ admits a polarity τ, then $\mathcal{C}$ contains the characteristic function $\mathbf{1}_\theta$ of the set θ of all absolute points of τ, a Tits ovoid of $W(q)$. This follows because $\mathbf{1}_\theta$ is the diagonal of the incidence matrix of $W(q)$ (written as $\left(a_{x,y}\right)_{x,y\in P}$ with $a_{x,y} = 1$ or 0 according as x is, or is not, in y^τ) and the $\mathbb{F}_2$-row span of a symmetric (0, 1)-matrix contains the diagonal ([5], Theorem 3, p. 143; see [18], Theorem 4.10, p. 369 for a very elegant proof). Since G is transitive on the set $\mathcal{T}_L$ of Tits ovoids in $W(q)$ and $\mathcal{C}$ is a kG-module, it follows that $\langle \mathcal{J}_L \rangle \subseteq \mathcal{C}$. As noted at the end of the first paragraph of (§2.B.vii), the characteristic function of an elliptic ovoid of $W(q)$ is a sum of the characteristic functions of some Tits ovoids of $W(q)$. So, $\langle \mathcal{E}_L \rangle \subseteq \langle \mathcal{J}_L \rangle \subseteq \mathcal{C}$. For a proof of $\mathbf{1}_P \notin \langle \mathcal{J}_L \rangle$ and more on $\langle \mathcal{J}_L \rangle / \langle \mathcal{E}_L \rangle$, see [53], Theorem 15, p. 495.

Problem 3.3. Determine the dimension and the words of minimum weight of $\langle \mathcal{J}_L \rangle$.

(D) *On $\mathcal{C}^\perp$* : (i) *The subcode $\langle \mathcal{D} \rangle$ of $\mathcal{C}^\perp$*: Let $\mathcal{D}$ denote the set of all dual grids in $W(q)$ (see Sect. 2B(iv). Each word of $\mathcal{C}^\perp$ of minimum weight is of the form $\mathbf{1}_{m\cup m^\perp}$, where $m \cup m^\perp \in \mathcal{D}$ and m is a line of $PG(3, q)$ not in L ([52]). An element of $\mathcal{D}$ meets each element of $L \cup \mathcal{D}$ in either 0, 2 or $2(q + 1)$ points. So, $\langle \mathcal{D} \rangle \subseteq \mathcal{C}^\perp \cap \mathcal{D}^\perp$. Let T and $\theta_0, \cdots, \theta_q$ be as in (Sect 2B(v. a) and $m \cup m^\perp \in \mathcal{D}$. Then, by a result with R. P. Shukla (unpublished), if $\theta_m, \theta_{m^\perp} \in \{\theta_1, \cdots, \theta_q\}$ are the unique ovoids the lines m and $m^\perp$ are tangent to, then $\theta_m \neq \theta_{m^\perp}$. Consequently, for σ_T defined in (C), $\sigma_T\left(m \cup m^\perp\right) = \mathbf{1}_{\theta_m} + \mathbf{1}_{\theta_{m^\perp}}$ which is not in $\sigma_T(\mathcal{C})$ if $q > 2$. If $\langle \mathcal{D} \rangle = \mathcal{C}^\perp$, then $\mathcal{C} = \langle \mathcal{D} \rangle^\perp \supseteq \langle \mathcal{D} \rangle$, a contradiction if $q > 2$. Thus, $\langle \mathcal{D} \rangle \neq \mathcal{C}^\perp$ if $q > 2$. This situation is similar to the case of the code orthogonal to the q-ary code generated by the lines of a projective 3-space over $\mathbb{F}_q$. Since $\langle \mathcal{D} \rangle$ has only words of even weight, $\langle \mathcal{D} \rangle \neq \langle \mathcal{E}_L \rangle$.

Problem 3.4. Determine the dimension of $\langle \mathcal{D} \rangle$? Does $\langle \mathcal{D} \rangle$ contains $\langle 1_{P\setminus\theta} : \theta \in \mathcal{E}_L \rangle$?

For each ovoid $\theta \in W_L(q)$, the set of lines of $PG(3, q)$ not in L is partitioned into the set $S(\theta)$ of secants of θ and the set $E(\theta)$ of external lines of θ. The map $l \longleftrightarrow l^\perp$ is a bijection between $S(\theta)$ and $E(\theta)$ (which does not extend linearly!) and $\mathcal{D} = \{l \cup l^\perp : l \in S(\theta)\}$.

Problem 3.5. Let H be the stabilizer of θ in $Aut(W(q))$. Determine the structure of the kH-modules $\langle S(\theta) \rangle$ and $\langle E(\theta) \rangle$ of k^P in the case when θ is a classical ovoid of $W(q)$.

See Problem 1. [56] studies similar codes from ovals of Desarguesian projective planes of odd order.

By (γ) in (Sect 2B(ii. a), $\mathcal{C}^\perp$ is a one-step-completely orthogonalizable code. This means that, for each coordinate position x, its dual (that is, $\mathcal{C}$) has $2(q + 1) - 1 = q + (q + 1)$ vectors whose supports intersect pairwise at $\{x\}$ [6].

(ii) *The subcode M of $\mathcal{C}^\perp$*: For a subgroup T of G of order $q^2 + 1$, let M_T denote the subspace of k^P of dimension q spanned by the characteristic functions of unions of even number of the T-orbits $\theta_1, \cdots, \theta_q$ in P (see Sect. 2B(v. a)). Then, $M = \Sigma\{M_T : T \text{ a subgroup of } G \text{ of order } q^2+1\} \leq k^P$ is a kG-submodule of $\mathcal{C}^{\perp}$ (see Sect 2B(v. b).

Problem 3.6. Determine the dimension and the words of minimum weight of M. Is $\langle \mathcal{D} \rangle \subseteq M$?

(iii) The support of a word of maximum weight in $\mathcal{C}^{\perp}$ is the complement of an ovoid of $W(q)$ ([5, 49]). Thus, determination of the weight enumerator of $\mathcal{C}^{\perp}$ (more to the point, the number of words of $\mathcal{C}^{\perp}$ of maximum weight) settles the question of the existence of ovoids in $W(q)$ other than the classical ovoids. We mention that there are no code words of $\mathcal{C}^{\perp}$ whose weight is in the interval $(q^3 + (5q-4)/6, q^3 + q)$ ([49], Theorem 16, p. 31–37). Another approach to understand the nonexistence of nonclassical ovoids in $W_L(q)$ is given by the result ([15], Corollary 5.6): if each ovoid of $W_L(q)$ is classical, then the $GQQ^-(5, q)$ defined in (§ 2.$A(\beta)$) is the only $(q, q^2) - GQ$ containing a hyperplane isomorphic to $W_L(q)$.

Study of the codes $\langle \mathcal{O}_L \rangle$ and $\langle P \backslash \theta : \theta \in \mathcal{O}_L \rangle$ may be instructive.

From the structure of $\mathcal{C}$ in ([53] ,Theorem 1, p. 485), $\mathcal{C} \cap \mathcal{C}^{\perp}$ is the radical of $\mathcal{C}$ and its radical series is known. If $k = \mathbb{F}_2$ and $q > 2$, then the weight of each element of $\mathcal{C} \cap \mathcal{C}^{\perp}$ is a multiple of 4, because if $w \in \mathcal{C} \cap \mathcal{C}^{\perp}$ has weight congruent to 2 (mod 4), then so does $\sigma_T(w)$, but $\sigma_T(\mathcal{C})$ has no such element. Considering the image under σ_T of sums of even number of members of $\{\theta_1, \cdots, \theta_q\}$, we conclude that dim $\left(\mathcal{C}^{\perp}/\mathcal{C} \cap \mathcal{C}^{\perp}\right) \geq q - 2$.

Problem 3.7. Determine the composition factors of $\mathcal{C}^{\perp}/\mathcal{C} \cap \mathcal{C}^{\perp}$.

4 A Bilinear Form and a Polarity on $\mathcal{C}$

The structure of the code $\mathcal{C}$ discussed in Sect. 3 as a $k[M]$-module, where M is the stabilizer in $PSp(4, q)$ of a quadratic form (either hyperbolic or elliptic) polarizing to a symplectic form on $PG(3, q)$ defining $PSp(4, q)$, is determined in ([55], Theorem 1). We hope that the bilinear form and a polarity on $\mathcal{C}$ introduced in this section (which exist when the incidence system admits a polarity) will be helpful in understanding $\mathcal{C}$.

Let $X = (P, L)$ be a finite connected partial linear space with $s + 1$ points on each line and $s+1$ lines through each point and $G = Aut(X)$. Assume that X admits a polarity τ, $H = \{g \in G : g \circ \tau = \tau \circ g \text{ on } P \cup L\}$, $\mathcal{O} = \{x \in P : x \in x^{\tau}\}$ and $\mathcal{S} = \{\ell \in L : \ell^{\tau} \in \ell\}$. The sets $\mathcal{O}$ and $\mathcal{S}$ are H- invariant. The important examples here are $W(q)$ and $H(q)$ with $q = p^{2m+1} (= s)$, $p = 2$ if $X = W(q)$ and $p = 3$ if $X = H(q)$. We note that $H = {}^2B_2(2^{2n+1})$ if $X = W(2^{2n+1})$ and $H = {}^2G_2(3^{2n+1})$ if $X = H(3^{2n+1})$, $n \geq 1$.

(A) **A nondegenerate bilinear form on $\mathcal{C}$:** Let k be a field of characteristic p; k^P, k^L and $\mathcal{C}$ be as in §1.B and $\eta : k^L \to \mathcal{C}$ be the surjective kG-morphism taking $\ell \in L$ to $\mathbf{1}_\ell = \ell \in \mathcal{C}$. Let B' be the symmetric k-bilinear form on k^L defined by setting, for ℓ and $m \in L$, $B'(\ell, m)$ to be 1 if $\ell^\tau \in m$ and zero otherwise.

Proposition 4.1. *(i) B' is H-invariant, (ii) radical of $B' = \ker \eta$ and (iii) B' induces a symmetric nondegenerate H-invariant bilinear form B on $\mathcal{C}$.*

Proof. (i) follows from the equivalence of the following statements for $\ell, m \in L$ and $h \in H$: $B'(h\ell, hm) = 1$; $(h\ell)^\tau \in hm$; $h(\ell^\tau) \in hm$; $\ell^\tau \in m$; $B'(\ell, m) = 1$.

(ii) follows from the equivalence of the following statements for $\alpha = \Sigma_{l\in L}\lambda_\ell \ell$, $\lambda_\ell \in k$: $\alpha \in Rad(B')$; $B'(\alpha, m) = 0$ for each $m \in L$; $\Sigma_{m^\tau \in \ell \in L}\lambda_\ell = 0$ for each $m \in L$; $\Sigma_{p\in \ell \in L}\lambda_\ell = 0$ for each $p \in P$; $\alpha \in \ker \eta$.

(iii) If $v, v', w, w' \in k^L$ are such that $\eta(v) = \eta(v')$ and $\eta(w) = \eta(w')$, then $x = v - v'$ and $y = w - w'$ are in $\ker \eta$ and $B'(v, w) = B'(v' + x, w' + y) = B'(v', w')$, by (ii). So, B' induces a symmetric H-invariant bilinear form B on $\mathcal{C}$. Further, if $v = \Sigma_{\ell\in L}\lambda_\ell \ell \in k^L$ and $w = \eta(v) \in \mathcal{C}$, then, for each $m \in L$,

$$B(\mathbf{1}_m, w) = B'(m, \Sigma\lambda_\ell \ell) = \Sigma\lambda_\ell B'(m, \ell) = \Sigma_{m^\tau \in \ell}\lambda_\ell = w_{m_\tau} \qquad (**)$$

where, for $p \in P$, w_p denotes the 'pth-coordinate of w'. As τ is a bijection between P and L, this implies that $B(\mathbf{1}_m, w) = 0$ for each $m \in L$ if, and only if, $w = 0$. Thus, B is nondegenerate. □

Note that the radical of the restriction of f (defined in Sect. 1B) to $\mathcal{C}$ is $\mathcal{C} \cap \mathcal{C}^\perp$, which unlike radical of B, may be nonzero. For any subspace D of $\mathcal{C}$, we denote by $D^\hbar$ the orthogonal complement of D with respect to B. If $(p, s+1) = 1$, then $\mathbf{1}_P = \Sigma_{\ell\in L}\mathbf{1}_\ell \in \mathcal{C}$ and, for $w \in \mathcal{C}$, by (**),

$$B(w, \mathbf{1}_P) = B(w, \Sigma_{\ell\in L}\mathbf{1}_\ell) = \Sigma_{\ell\in L} w_{\ell^\tau} = \Sigma_{p\in P} w_p = f(w, \mathbf{1}_P).$$

So, $\mathbf{1}_P^\hbar = \mathbf{1}_P^\perp = \mathcal{C}^0$. For $x \in P$, let $\widetilde{\pi}_x = \Sigma_{x\in m\in L} m \in k^L$ and $\pi_x = \eta(\widetilde{\pi}_x) \in \mathcal{C}$. Then, for $w \in \mathcal{C}$, $B(\pi_x, w) = \Sigma_{p\in x^\tau} w_p$ (by (**)) $= f(w, x^\tau)$. So, for $\Pi = \langle \pi_x : x \in P\rangle \leq \mathcal{C}$, $\Pi^\hbar = \mathcal{C} \cap \mathcal{C}^\perp$.

The subspaces $\mathcal{M} = \{v \in \mathcal{C} : B(v, \mathbf{1}_m) = f(v, \mathbf{1}_m) \text{ for each } m \in L\}$, $U = \langle \mathbf{1}_\ell : \ell \in \mathcal{S}\rangle$ and $W = \langle \mathbf{1}_\ell : \ell \in L \setminus \mathcal{S}\rangle$ are kH-submodules of $\mathcal{C}$. An element $w \in \mathcal{C}$ is in $\mathcal{M}$ if, and only if, $w_{m^\tau} = \Sigma_{p\in m^\tau} w_p$ for each $m \in L$. So, $\mathbf{1}_P \in \mathcal{M}$.

Let $p = 2$ and write $w \in \mathcal{C}$ as $w = \Sigma_{\ell\in\mathcal{S}}\lambda_\ell \mathbf{1}_l + \Sigma_{\ell\in L\setminus\mathcal{S}}\mu_\ell \mathbf{1}_l$, $\lambda_\ell, \mu_\ell \in k$. Then, $B(w, w) = \Sigma_{\ell\in\mathcal{S}}\lambda_\ell^2 = (\Sigma_{\ell\in\mathcal{S}}\lambda_\ell)^2$. So, the set of all isotropic elements of $\mathcal{C}$ with respect to B is $U^0 + W$. Using ([5], Theorem 3, p. 143) again, we see that $\mathbf{1}_\mathcal{O} \in \mathcal{C}$. Since $x \in \mathcal{O}$ if, and only if, $x^\tau \in \mathcal{S}$, by (**), $B(\mathbf{1}_m, \mathbf{1}_\mathcal{O}) = 1$ if $m^\tau \in \mathcal{O}$ and zero otherwise. So, $\mathbf{1}_\mathcal{O} \in \left(U^0 + W\right)^\hbar$. Since $U^0 + W$ is of codimension zero in $\mathcal{C}$ if $\mathcal{O}$ is empty and 1 otherwise, $\left(U^0 + W\right)^\hbar = k\mathbf{1}_\mathcal{O}$.

(B) **A polarity of $\mathcal{C}$:** Let $\kappa : k^P \longrightarrow \mathcal{C}$ be the surjective kH-module homomorphism taking $x \in P$ to $\tau(x)$. Recall $A^\perp$ defined for $A \subseteq k^P$ in §1.B.

Proposition 4.2. *(i)* $0 \to \mathcal{C}^{\perp} \to k^P \xrightarrow{\kappa} \mathcal{C} \to 0$ *is an exact sequence of* kH*-modules.*

(ii) For any subspace A *of* $\mathcal{C}$, $\dim A + \dim \kappa(A^{\perp}) = \dim \mathcal{C}$.

Proof. (i) By definition, κ is onto. (i) follows from the equivalence of the following statements for $v = \Sigma_{x\in P} v_x x \in k^P$: $v \in \ker \kappa$; $\Sigma_{x\in P} v_x \mathbf{1}_{x^\tau} = 0$; $\Sigma\{v_x : x \in P$ such that $p \in x^\tau\} = 0$ for each $p \in P$; $\Sigma_{x\in p^\tau} v_x = 0$ for each $p \in P$; $\Sigma_{x\in \ell} v_x = 0$ for each $\ell \in L$; $\Sigma_{x\in P} v_x x \in \mathcal{C}^{\perp}$.

(ii) $\dim A + \dim A^{\perp} = |P| = \dim \mathcal{C} + \dim \mathcal{C}^{\perp}$. Therefore, $\dim A + \dim A^{\perp} - \dim \mathcal{C}^{\perp} = \dim \mathcal{C}$. But $\mathcal{C}^{\perp} \subseteq A^{\perp}$ and $A^{\perp}/\mathcal{C}^{\perp} \simeq \kappa(A^{\perp})$. So, (ii) holds. □

Let $\mathcal{P}(V)$ denote the set of all subspaces of a vector space V, partially ordered by inclusion. Consider the maps

$$\mathcal{P}(k^P) \xrightarrow{\perp} \mathcal{P}(k^P) \xrightarrow{\widetilde{\kappa}} \mathcal{P}(\mathcal{C}),$$

where, for $A \subseteq k^P$, '$\perp$' takes A to its orthogonal complement $A^{\perp}$ with respect to f and, if $A \in \mathcal{P}(k^P)$, $\widetilde{\kappa}(A) = \kappa(A)$. We write $A^{\diamond} = \kappa\left(A^{\perp}\right)$. Clearly, if A is a kH-submodule of k^P, then so is $A^{\diamond}$.

Proposition 4.3. *Let* $A \in \mathcal{P}(k^P)$.

(a) $\mathcal{C} \subseteq A$ *if, and only if,* $A^{\diamond} = 0$.

(b) If $A \in \mathcal{P}(\mathcal{C})$, *then*

(i) $\dim A + \dim A^{\diamond} = \dim \mathcal{C}$; *and (ii)* $A = A^{\diamond\diamond}$.

(c) The map $A \to A^{\diamond}$, $A \in \mathcal{P}(\mathcal{C})$, *is an inclusion reversing involutory permutation of* $\mathcal{P}(\mathcal{C})$.

(d) If $(p, s+1) = 1$, *then* $\mathbf{1}_P \in \mathcal{C}$, $(k\mathbf{1}_P)^{\diamond} = \mathcal{C}^0$ *and* $(A^{\diamond})^0 = (A + k\mathbf{1}_P)^{\diamond}$ *for each subspace* A *of* $\mathcal{C}$. *In particular, the map in (c) is an inclusion reversing bijection between the set of all subspaces of* $\mathcal{C}$ *containing* $\mathbf{1}_P$ *and the set of all subspaces of* $\mathcal{C}^0$.

Proof. Since a subspace A contains $\mathcal{C}$ if, and only if, $A^{\perp} \leq \mathcal{C}^{\perp}$, (a) follows from Proposition 4.2 (i). Proposition 4.2 (ii) is restated as (b.i).

Since the dimensions of A and $A^{\diamond\diamond}$ are equal (by (b.i)), we need only to show that $A \subseteq A^{\diamond\diamond}$. Note that

$$A^{\diamond} = \left\{\Sigma_{p\in P}\left(\Sigma_{x\in p^\tau}\beta_x\right) p : \Sigma_{p\in P}\beta_p p \in A^{\perp}\right\}.$$

Consider $w = \eta(\Sigma_{l\in L}\lambda_\ell \ell) \in \mathcal{C}$. Then, $w = \Sigma_{x\in P}\left(\Sigma_{x\in\ell\in L}\lambda_\ell\right) x$. Now, $\Sigma_{l\in L}\lambda_\ell \ell^\tau \in \left(A^{\diamond}\right)^{\perp}$ if, and only if, for each $w' = \Sigma \beta_x x \in A^{\perp}$,

$$0 = \Sigma_{\ell^\tau\in P}\lambda_\ell\left(\Sigma_{\ell^\tau\in x^\tau}\beta_x\right) = \Sigma_{\ell^\tau\in P}\lambda_\ell\left(\Sigma_{x\in\ell}\beta_x\right) = \Sigma_{x\in P}\left(\Sigma_{x\in\ell\in L}\lambda_\ell\right)\beta_x = f(w, w').$$

So, $\Sigma_{l\in L}\lambda_\ell \ell^\tau \in \left(A^{\diamond}\right)^{\perp}$ if, and only if, $w \in A^{\perp\perp} = A$. Since $w = \kappa\left(\Sigma_{l\in L}\lambda_\ell \ell^\tau\right)$, it follows that $A = A^{\diamond\diamond}$. So, (b.ii) follows. (c) now follows from (b).

For $w = \Sigma_{x\in P}\lambda_x x \in k^P$, $\kappa(w) = \eta(\Sigma_{x\in P}\lambda_x x^\tau)$ and $c(\kappa(w)) = (s+1)c(w)$. Since $\Sigma_{l\in L}\mathbf{1}_l = (s+1)\mathbf{1}_P \in \mathcal{C}$ and $(p, s+1) = 1$, $\mathbf{1}_P \in \mathcal{C}$. For any subspace A of k^P, $\kappa(A \cap \mathbf{1}_P^\perp) = \kappa(A) \cap \mathbf{1}_P^\perp \subseteq \mathcal{C}^0$. So, $(k\mathbf{1}_P)^\diamondsuit = \kappa(k\mathbf{1}_P^\perp) \subseteq \mathcal{C}^0$. Since $\mathcal{C}^0$ and $(k\mathbf{1}_P)^\diamondsuit$ are both of codimension one in $\mathcal{C}$ (see Proposition 4.3 (b. i)), $(k\mathbf{1}_P)^\diamondsuit = \mathcal{C}^0$. For $A \in \mathcal{P}(\mathcal{C})$, $A^\perp \cap \mathbf{1}_P^\perp = (A + k\mathbf{1}_P)^\perp$. So,

$$(A^\diamondsuit)^\circ = \kappa\left(A^\perp\right) \cap \mathbf{1}_P^\perp = \kappa(A^\perp \cap \mathbf{1}_P^\perp) = \kappa((A + k\mathbf{1}_P)^\perp) = (A + k\mathbf{1}_P)^\diamondsuit.$$

So, a subspace A of k^Pcontains $\mathbf{1}_P$ if, and only if, $A^\diamondsuit \subseteq \mathcal{C}^0$. So, (d) follows.

Let $l \in L$. If $w = \Sigma_{p\in P}\beta_p p \in k^P$, then $w \in l^\perp$ if, and only if, $\Sigma_{p\in l}\beta_p = 0$. In this case, l^τ-th coordinate of $\kappa(w)$ is zero. So, $\kappa(k\mathbf{1}_l^\perp) \leq \{w \in \mathcal{C} : w_{\ell}\tau = 0\}$. Since both subspaces are of codimension one in $\mathcal{C}$, equality holds. As noted above, $(k\mathbf{1}_P)^\diamondsuit = \mathcal{C}^0$.

When $(p, s+1) = 1$, the polarity of $\mathcal{P}(\mathcal{C})$ taking $A \in \mathcal{P}(\mathcal{C})$ to $A^\diamondsuit$ is not symplectic, because $\mathbf{1}_P \notin (k\mathbf{1}_P)^\diamondsuit = \mathcal{C}^0$.

Acknowledgments I thank H. Van Maldeghem for several very useful remarks and his permission to include Proposition 2.12 (iii) here Ilaria Cardinali for discussions on the material in Sect 2B(i. a) and T. Penttila for some useful remarks. Finally, I dedicate this article to the memory of two recently departed, very important maternal persons in my life Kanakammagaru and Venkatalakshammagaru.

References

1. Alderson, T.L., Mellinger, K.E.: Partitions in finite geometry and related constant composition codes. Innov. Incid. Geom. **8**, 49–71 (2008)
2. Arslan, O.: The dimension of $LU(3, q)$ codes. J. Comb. Th. Ser. A **116**, 1073–1079 (2009)
3. Arslan, O., Sin, P.: Some simple modules for classical groups and $p-$ ranks of orthogonal and Hermitian geometries. J. Algebra **327**, 141–169 (2011)
4. Assmus, E.F., Key, J.D.: Designs and Their Codes. Cambridge University Press, Cambridge (1993)
5. Bagchi, B., Sastry, N.S.N.: Even order inversive planes, generalized quadrangles and codes. Geom. Dedicata **22**: 137–147 (1987)(see the footnote in [52], p. 49 for a correction)
6. Bagchi, B., Sastry, N.S.N.: One step completely orthogonalizable codes from generalized quadrangles. Inf. Comput. **77**, 123–130 (1988)
7. Bagchi, B., Sastry, N.S.N.: Codes associated with generalized polygons. Geom. Dedicata **27**, 1–8 (1988)
8. Bagchi, B., Sastry, N.S.N.: Intersection pattern of the classical ovoids in symplectic 3-space of even order. J. Algebra **126**, 147–160 (1989)
9. Bagchi, B., Sastry, N.S.N.: Ovoidal packings of $PG(3, q)$ for even q. Discrete Math. **313**, 2213–2217 (2013)
10. Ball, S.: On ovoids of $O(5, q)$. Adv. Geom. **4**, 1–7 (2004)
11. Bannai, E., Song, S.Y.: On the character table of the associate scheme $S_p(4, q)/S_z(q)$, Graphs and combinatorics 5, 291–293 (1989)
12. Bardoe, M., Sin, P.: The permutation modules for the action of $GL\left(n+1, \mathbb{F}_q\right)$ acting on $\left(\mathbb{F}_q^{n+1}\right)$. J. London Math. Soc. **61**, 58–80 (2000)

13. Barlotti, A.: Some Topics in Finite Geometrical Structures, Institute of Statistics Mimeo Series 439. University of North Carolina, North Carolina (1965)
14. Brouns, L., Thas, J.A., Van Maldeghem, H.: A characterization of $Q(5,q)$ using one subquadrangle $Q(4,q)$. European J. Combin. **23**, 153–177 (2002)
15. Brown, M.R.: Generalized quadrangle of order (q,q^2), q even, containing $W(q)$ as a subgeneralized quadrangle, Geom. Dedicata **56**, 299–306 (1995)
16. Brown, M.R.: Ovoids of $PG(3,q)$, q even, with a conic section. J. London Math. Soc **62**, 569–582 (2000)
17. Brown, M.R.: (Hyper) Ovals and Ovoids in Projective Spaces. Ghent University, Socrates Intensive course (2000)
18. Brouwer, A.E., Wilbrink, H.A.: Block designs. In: Buekenhout, F. (ed.) Hand Book of Incidence Geometry. Elsevier Science, Amsterdam (1995)
19. Bruen, A.A., Hirschfeld, J.W.P.: Intersection in projective spaces, I: combinatorics. Math. Zeit. **193**, 215–225 (1986)
20. Bruen, A.A., Hirschfeld, J.W.P.: Intersection in projective spaces, II: pencils of quadrics. European J. Combin. **9**, 255–270 (1988)
21. Bruen, A.A., Wehlau, D.L.: Partitioning quadrics. Symmetric group divisible designs and caps, Design Code Cryptogr. **10**, 145–155 (1997)
22. Buekenhout, F.: Existance of unitals in finite translation planes of order q^2 with kernel of order q. Geom. Dedicata **5**, 189–104 (1976)
23. Butler, D.: On the intersection of ovoids sharing a polarity. Geom. Dedicata **135**, 157–165 (2008)
24. Butler, D.: The maximum size of the intersections of two ovoids. J. Combin. Th. Ser. A **116**, 242–245 (2009)
25. Cardinali, I, Sastry, N.S.N.: On elliptic ovoids and their rosettes in a classical generalized quadrangle of even order, submitted (26 pages)
26. Chandler, D.B., Sin, P., Xiang, Q.: The permutation action of finite symplectic groups of odd characteristic on their standard modules. J. Algebra **318**, 871–892 (2007)
27. Chandler, D.B., Sin, P., Xiang, Q.: Incidence modules for symplectic spaces in characteristic two. J. Algebra **323**, 3157–3181 (2010)
28. L. Chastkofsky and W. Feit, On the projective characters in characteristic 2 of the groups $Sz(2^m)$ and $Sp_4(2^n)$, Inst. Hautes Études Sci. Publ. Math. No. 51 (1980) 9–35.
29. Cossidente, A., Vereecke, S.J.K.: Some geometry of the isomorphism $Sp(4,q) \simeq O(5,q)$, q even. J. Geom. **70**, 28–37 (2001)
30. Cossidente, A., King, O.H.: Group-theoretic characterizations of classical ovoids. In: Finite Geometries, vol.3, pp. 121–131. Kluwer Academic Publishers, Dordrecht (2001)
31. Debroy, I., Thas, J.A.: On semi-partial geometries. J. Comb. Th. Ser. A **25**, 242–250 (1978)
32. Dembowski, P.: Finite Geometries. Springer, Berlin (1968)
33. Ebert, G.L.: Partitioning projective geometries into caps. Canad. J. Math. **37**, 1163–1175 (1985)
34. Edoukou, F.A.B.: Codes defined by forms of degree 2 on quadratic surfaces. IEEE Tans. Inform. Theory **54**(2), 860–864 (2008)
35. Gallagher, R.G.: Low density parity check codes. IRE Trans. Inform. Theory **8**, 21–28 (1962)
36. Glynn, D.: Projective planes and related combinatorial systems, Ph.D. Thesis, University of Adelaide (1978)
37. Glynn, D.: On a set of lines of $PG(3,q)$ corresponding to a maximal cap contained in the Klein quadric of $PG(5,q)$. Geom. Dedicata **26**, 273–280 (1988)
38. Govert, E., Van Maldeghem, H.: Distance-preserving maps in generalized polygons, part II: maps on points and/or lines. Beitr. Alg. Geom. **43**, 303–324 (2002)
39. Hirschfeld, J.W.P.: Finite Projective Spaces of Three Dimensions. Oxford University Press, Oxford (1986)
40. Hirschfeld, J.W.P., Thas, J.A.: General Galois Geometries. Oxford University Press, Oxford/New york (1991)
41. Hubaut, X., Metz, R.: A class of strongly regular graphs related to orthogonal groups. In: Barlotti, A., et al. (eds.) Combinatorics 81. North-Holland, Amsterdam/New York/Oxford, (1982)

42. Huppert, B., Blackburn, N.: Finite Groups III. Springer, Berlin, Heidelberg, New York, Tokyo (1982)
43. Kim, J.-L., Miller, K.E., Storme, L.: Small weight codewards in LDPC codes defined by (dual) classical generalized quadrangles. Des. Codes and Cryptogr. **43**, 73–92 (2007)
44. Liu, Z., Pados, D.A.: LDPC codes from generalized polygons. IEEE Trans. Inform. Theory **51**(11), 3890–3898 (2005)
45. Lüneburg, H.: Translation Planes. Springer, Berlin, Heidelberg, New York (1980)
46. Metz, R.: On a class of unitals. Geom. Dedicata **8**, 125–126 (1979)
47. Panella, G.: Caratterizzazione delle quadriche di uno spazio (tridimensionale) lineare sopra un corpo finito. Bull. Un. Mat. Ital. **10**, 507–513 (1955)
48. Payne, S., Thas, J.A.: Finite generalized quadrangles. Second Edition, EMS Series of Lectures in Mathematics (2009)
49. Pepe, V., Storme, L., Voorde, G.V.: On codewords in the dual code of classical generalized quadrangles and classical polar spaces. Discrete Math. **310**, 3132–3148 (2010)
50. Prohaska, O., Walker, M.: A note on Hering type inversive planes of even order. Arch. Math. **28**, 431–432 (1977)
51. Ronan, M.: A geometric characterization of Moufang hexagons. Invent. Math. **57**, 227–262 (1980)
52. Sastry, N.S.N.: Codes and generalized polygons. In: Combinatorics and Applications (Calcutta : Volume in honour of Professor S. S. Shrikhande, Indian Statistical Institute) **1984**, 303–310 (1982)
53. Sastry, N.S.N., Sin, P.: The code of a regular generalized quadrangle of even order. In: Proceedings of the Symposium in Pure Mathematics. Am. Math. Soc **63**, 485–496 (1998)
54. Sastry, N.S.N., Sin, P.: The binary code associated with nondegenerate quadrics of a symplectic space of even order. J. Comb. Th. Ser. A **94**, 1–14 (2001)
55. Sastry, N.S.N., Shukla, R.P.: Structure of a code related to $Sp(4, q)$, q even. In: Proceedings of Indian Academy of Sciences-Math. Sciences **117**, 457–470 (2007)
56. Sastry, N.S.N., Shukla, R.P.: Multiplicities of simple modules in $Sp(4, q)$ permutation module on $P(3, q)$, q even. Arch. Math. **97**, 237–245 (2011)
57. Sin, P., Wu, J., Xiang, Q.: Dimensions of some binary codes arising from a conic in $PG(2, q)$. J. Comb. Th. Ser. A **118**, 853–878 (2011)
58. Thas, J.A.: Construction of maximal arcs and partial geometries. Geom. Dedicata **3**, 61–64 (1974)
59. Tits, J.: Ovoides et groupes de Suzuki. Arch. Math. **13**, 187–192 (1962)
60. Van Maldeghem, H.: Generalized Polygons, Monographis in Mathematics 93. Birkhäuser , Basel, Boston, Berlin (1998)
61. Todd, J.A.: As it might have been, Bull. London Math. Soc. 2, 1–4 (1970).

Chapter 15
Some Weyl Modules of the Algebraic Groups of Type E_6

Peter Sin

Abstract Let G be a simple algebraic group of type E_6 over an algebraically closed field of characteristic $p > 0$. We determine the submodule structure of the Weyl modules with highest weight $r\omega_1$ for $0 \le r \le p-1$, where ω_1 is the fundamental weight of the standard 27-dimensional module. In the process, the structures of other Weyl modules with highest weights linked to $r\omega_1$ are also found.

Keywords Exceptional groups · Weyl modules · Irreducible modular representations

1 Introduction

In this note we study certain Weyl modules for a simple, simply connected algebraic group G of type E_6 over an algebraically closed field of characteristic $p > 0$. The modules we consider are for highest weights which are of the form $r\omega_1$, $0 \le r \le p-1$, where ω_1 is the highest weight of the "standard" 27-dimensional module, and we will give a full description of their G-submodules. If P is the maximal parabolic subgroup stabilizing the highest weight vector in the 27-dimensional module $H^0(\omega_1)^*$, then the embedding of the projective variety G/P for the associated line bundle is projectively normal [3], so the homogeneous coordinate ring is $\bigoplus_{r\ge 0} H^0(r\omega_1)$.

Peter Sin—This work was partially supported by a grant from the Simons Foundation (#204181 to Peter Sin).

P. Sin (✉)
Department of Mathematics, University of Florida, PO Box 118105,
Gainesville, FL 32611, USA
e-mail: sin@ufl.edu

N. S. N. Sastry (ed.), *Groups of Exceptional Type, Coxeter Groups and Related Geometries*, Springer Proceedings in Mathematics & Statistics 82,
DOI: 10.1007/978-81-322-1814-2_15,

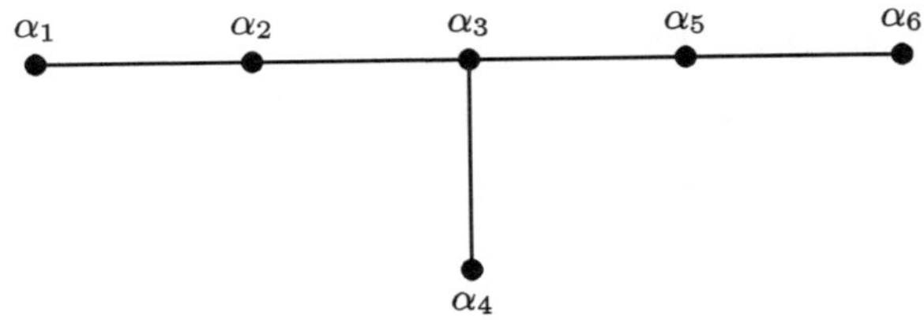

Fig. 1 Dynkin diagram of type E_6

As a consequence of Steinberg's Tensor Product Theorem [5], our results also describe the simple G-socles of the modules $H^0(r\omega_1)$ for all $r \geq 0$. Analogous results for classical groups were obtained in [1].

Our labelling of the fundamental roots and weights is according to Fig. 1. We describe the E_6 root system as follows. Let e_i, $i = 1, \ldots, 8$ be an orthonomal basis of an 8-dimensional Euclidean space. Then, in coordinates, our root system R is the union of the set

$$\{\pm e_i \pm e_j \mid 4 \leq i < j \leq 8\}$$

with the set

$$\{\pm\frac{1}{2}[(e_1 - e_2 - e_3) + \sum_{i=5}^{8} \pm e_i] \mid \text{number of minus signs is even}\}.$$

A set of fundamental roots is

$$S = \{\alpha_1 = e_4 - e_5, \alpha_2 = e_5 - e_6, \alpha_3 = e_6 - e_7, \alpha_4 = e_7 + e_8,$$
$$\alpha_5 = e_7 - e_8, \alpha_6 = \frac{1}{2}(e_1 - e_2 - e_3 - e_4 - e_5 - e_6 - e_7 + e_8)\}.$$

The fundamental dominant weights have coordinates

$$\omega_1 = \frac{1}{3}(1, -1, -1, 3, 0, 0, 0, 0), \qquad \omega_2 = \frac{1}{3}(2, -2, -2, 3, 3, 0, 0, 0),$$
$$\omega_3 = (1, -1, -1, 1, 1, 1, 0, 0), \qquad \omega_4 = \frac{1}{2}(1, -1, -1, 1, 1, 1, 1, 1),$$
$$\omega_5 = \frac{1}{6}(5, -5, -5, 3, 3, 3, 3, -3), \qquad \omega_6 = \frac{1}{3}(2, -2, -2, 0, 0, 0, 0, 0).$$

Our notation will be standard, following [2]. In particular we denote the Weyl module with highest weight λ by $V(\lambda)$ and its simple quotient by $L(\lambda)$. By definition, $V(\lambda) = H^0(-w_0\lambda)^*$, where w_0 is the longest element of the Weyl group [2, II.2.13]. Also, $V(\lambda) \cong {}^{\tau}H^0(\lambda)$, for a certain anti-automorphism τ of G that induces the identity map on characters [2, II. 2.12]. As $-w_0\omega_1 = \omega_6$, the submodule structure of $V(r\omega_1)$

will yield the submodule structures of $V(r\omega_6)$, $H^0(r\omega_1)$ and $H^0(r\omega_6)$ by applying τ and duality.

Theorem 1.1. *Let G be a simply connected, semisimple algebraic group of type E_6 over an algebraically closed field of characteristic p. The following statements give a complete description of the submodule structure of the module $V(r\omega_1)$, for $0 \le r \le p-1$.*

(a) For $0 \le r \le p-4$ the Weyl module $V(r\omega_1)$ is simple.
(b) $(r = p-3)$

(i) If $p = 3$, then $V((p-3)\omega_1) = V(0)$ is simple.
(ii) If $p = 5$, there is an exact sequence

$$0 \to V(\omega_6) \to V(2\omega_1) \to L(2\omega_1) \to 0.$$

(iii) If $p = 7$, there is an exact sequence

$$0 \to V(\omega_1 + \omega_4) \to V(2\omega_1 + \omega_6) \to V(4\omega_1) \to L(4\omega_1) \to 0.$$

(iv) For $p \ge 11$, there is an exact sequence

$$\begin{aligned} 0 \to V((p-9)\omega_1) &\to V((p-8)\omega_1 + \omega_6) \\ &\to V((p-8)\omega_1 + \omega_2) \to V((p-6)\omega_1 + \omega_4) \\ &\to V((p-5)\omega_1 + \omega_6) \to V((p-3)\omega_1) \to L((p-3)\omega_1) \to 0 \end{aligned}$$

(c) $(r = p-2)$

(i) If $p = 2$ or $p = 3$ the Weyl module $V((p-2)\omega_1)$ is simple.
(ii) If $p = 5$ there is an exact sequence

$$0 \to V(0) \to V(3\omega_1) \to L(3\omega_1) \to 0.$$

(iii) If $p = 7$, there is an exact sequence

$$0 \to V(\omega_4 + \omega_6) \to V(\omega_1 + 2\omega_6) \to V(5\omega_1) \to L(5\omega_1) \to 0.$$

(iv) For $p \ge 11$ there is an exact sequence

$$\begin{aligned} 0 \to V((p-10)\omega_1 + \omega_2) &\to V((p-9)\omega_1 + \omega_5) \\ &\to V((p-8)\omega_1 + \omega_3) \to V((p-7)\omega_1 + \omega_4 + \omega_6) \\ &\to V((p-6)\omega_1 + 2\omega_6) \to V((p-2)\omega_1) \to L((p-2)\omega_1) \to 0 \end{aligned}$$

(d) $(r = p-1)$

(i) If $p \le 5$ the the Weyl module $V((p-1)\omega_1)$ is simple.

(ii) *If $p = 7$, there is an exact sequence*

$$0 \to V(3\omega_6) \to V(6\omega_1) \to L(6\omega_1) \to 0.$$

(iii) *For $p \geq 11$ there is an exact sequence*

$$\begin{aligned} 0 \to V((p-11)\omega_1 + 2\omega_2) &\to V((p-10)\omega_1 + \omega_2 + \omega_5) \\ &\to V((p-9)\omega_1 + \omega_3 + \omega_6) \to V((p-8)\omega_1 + \omega_4 + 2\omega_6) \\ &\to V((p-7)\omega_1 + 3\omega_6) \to V((p-1)\omega_1) \to L((p-1)\omega_1) \to 0 \end{aligned}$$

(e) *In each of the above sequences the first and last nonzero terms are simple modules and the other terms have two composition factors.*

We shall apply the *Jantzen Sum Formula* [2, II.8.19] [1] : The Weyl module $V(\lambda)$ has a descending filtration, of submodules $V(\lambda)^i$, $i > 0$, such that

$$V(\lambda)^1 = \mathrm{rad}(V(\lambda)), \quad (\text{so that} V(\lambda)/V(\lambda)^1 \cong L(\lambda))$$

and

$$J(\lambda) := \sum_{i>0} \mathrm{Ch}(V(\lambda)^i) = -\sum_{\alpha>0} \sum_{\{m:0<mp<\langle\lambda+\rho\alpha^\vee\rangle\}} v_p(mp)\chi(\lambda - mp\alpha)$$

We shall refer to the quantity $J(\lambda)$ as the Jantzen sum for λ (or for $V(\lambda)$). We recall that the weight ρ is the half-sum of the positive roots and $v_p(m)$ denotes the exponent of p in the prime factorization of m. Finally, the formal character $\chi(\mu)$ is the so-called Weyl character, defined in [2, II.5.7], which has the following concrete description. There is a unique dominant weight of the form $w(\mu+\rho)$, where $w \in W$. Let $\mu' = w(\mu + \rho) - \rho$. Then $\chi(\mu)$ is equal to $\mathrm{sign}(w)\mathrm{Ch}(V(\mu'))$ if μ' is dominant, and zero otherwise. In particular $\chi(\mu) = \mathrm{Ch}(V(\mu))$ if μ is dominant and $\chi(\mu) = 0$ if and only if $\lambda + \rho - mp\alpha$ is orthogonal to some root.

To aid our computation, when p and $\lambda \in X_+$ have been fixed, we shall say that a multiple $m\alpha$ of a positive root is *relevant* if $0 < mp < \langle\lambda + \rho\alpha^\vee\rangle\}$ and that $m\alpha$ is a *contributor* if $m\alpha$ is relevant and $\chi(\lambda - mp\alpha) \neq 0$. We will call the quantity $v_p(mp)\chi(\lambda - mp\alpha) = \mathrm{sign}(w)\chi(\mu')$ a *contribution* and the dominant weight $w(\lambda + \rho - mp\alpha) - \rho$ a *contributing weight*.

Thus, in computing the Jantzen sums we can begin by determining the relevant root multiples, then determine which among them is a contributor and finally add up the contributions.

We note that if α_0 is the highest root, then $\langle r\omega_1 + \rho, \alpha_0^\vee\rangle = r + 11 < 2p$, when $p \geq 11$, so the only relevant root multiples are actually roots. For the primes $p = 2$, 3, 5 and 7, we have to take into account higher multiples.

[1] The validity of the sum formula for all p was proved by Andersen.

1.1 Discussion of the Proof of Theorem 1.1

The proof is by means of computations, whose results are compiled in the Tables 1, 2, 3, 4, 5, 6, 7, 8, 9, 10, 11, 12, 13, 14, 15, 16, 17 and 18. The tables all have the same form. In the first column are dominant weights λ. In the second column, for each λ we list all the relevant root multiples. A root multiple is given by the tuple of coefficients in its expression as a sum of fundamental roots. Since we are dealing throughout with a single root system E_6, the relevant root multiples for any given weight are easily computed. The relevant root multiples for each weight λ are divided into non-contributors and contributors. For those root multiples $m\alpha$ which we claim to be noncontributors, we must exhibit a (co)root β that is orthogonal to $\lambda + \rho - pm\alpha$. Note that β is not necessarily unique up to a sign. The third column gives the coordinate tuple of the weight $\lambda+\rho-pm\alpha$ with respect to the fundamental weights and the fourth columns gives the coordinate tuple of a suitable β with respect to the fundamental (co)roots. The reader can immediately verify that $\lambda + \rho - pm\alpha$ and β are orthogonal, hence that $m\alpha$ is indeed a non-contributor.

For a contributing root multiple $m\alpha$, the third column has the Weyl group element w such that $w(\lambda + \rho - pm\alpha)$ is dominant and fourth column has the contributing weight $w(\lambda + \rho - pm\alpha) - \rho$. The element w is given as a tuple of indices $[i_1, i_2, \ldots, i_r]$, where $w = w_{i_1} w_{i_2} \cdots w_{i_r}$ as a word in the fundamental reflections. The weight $w(\lambda + \rho - pm\alpha) - \rho$ is given by its tuple of coefficients with respect to the fundamental weights. It is visually obvious, from the fact that all entries in the fourth column tuples are nonnegative, that $m\alpha$ is a contributor. The sign of the contribution is given by the length of w.

We have discussed the immediate verifiability of the tables except for checking that for each w entry the weight $w(\lambda+\rho-pm\alpha)-\rho$ is as given. This can be carried out by longer but routine computations (which can easily be automated).

In order to prove the theorem for a particular weight $r\omega_1$ and in characteristic p, we first find all the relevant root multiples and contributions for this weight, which may depend on p. Then we repeat the procedure for all contributing weights in a second iteration of the sum formula. In principle, further iterations of this process might be expected, but for the weights being considered here they turn out not to be necessary; two iterations provide enough information to deduce Theorem 1.1.

1.2 Proof of Theorem 1.1 in Detail

(a). We may assume that $p \geq 5$. Now $V(0)$ is trivially simple, and it is well known that $V(\omega_1)$ is simple for all p. (This can also be can be checked from our tables.) For $p = 7$, we have to check also that $V(2\omega_1)$, and $V(3\omega_1)$ are simple. From Table 14 we can see that no relevant root multiples are contributors. Therefore the Jantzen sums are zero. Assume then that $p \geq 11$. For $r < p - 10$, there are no relevant roots for $r\omega_1$, so $V(r\omega_1)$ is simple. For $r = p - 10,\ldots,p - 4$, Table 1 lists the relevant roots and shows that none is a contributor, so $V(r\omega_1)$ is simple.

Table 1 $r\omega_1$, $p-10 \le r \le p-4$, $p \ge 11$

λ	$m\alpha$	$\lambda+\rho-pm\alpha$	β
$(p-10)\omega_1$	[1, 2, 3, 2, 2, 1]	$[p-9, 1, 1, -p+1, 1, 1]$	[1, 2, 3, 1, 2, 1]
$(p-9)\omega_1$	[1, 2, 3, 2, 2, 1]	$[p-8, 1, 1, -p+1, 1, 1]$	[1, 2, 2, 1, 2, 1]
	[1, 2, 3, 1, 2, 1]	$[p-8, 1, -p+1, p+1, 1, 1]$	[1, 2, 2, 1, 2, 1]
$(p-8)\omega_1$	[1, 2, 3, 2, 2, 1]	$[p-7, 1, 1, -p+1, 1, 1]$	[1, 2, 2, 1, 1, 1]
	[1, 2, 2, 1, 2, 1]	$[p-7, -p+1, p+1, 1, -p+1, 1]$	[1, 2, 2, 1, 1, 1]
	[1, 2, 3, 1, 2, 1]	$[p-7, 1, -p+1, p+1, 1, 1]$	[1, 2, 2, 1, 1, 1]
$(p-7)\omega_1$	[1, 2, 3, 2, 2, 1]	$[p-6, 1, 1, -p+1, 1, 1]$	[1, 1, 2, 1, 1, 1]
	[1, 2, 2, 1, 1, 1]	$[p-6, -p+1, 1, 1, p+1, -p+1]$	[1, 1, 2, 1, 1, 1]
	[1, 1, 2, 1, 2, 1]	$[-6, p+1, 1, 1, -p+1, 1]$	[1, 1, 2, 1, 1, 1]
	[1, 2, 2, 1, 2, 1]	$[p-6, -p+1, p+1, 1, -p+1, 1]$	[1, 2, 2, 1, 1, 0]
	[1, 2, 3, 1, 2, 1]	$[p-6, 1, -p+1, p+1, 1, 1]$	[1, 1, 2, 1, 1, 1]
$(p-6)\omega_1$	[1, 2, 3, 2, 2, 1]	$[p-5, 1, 1, -p+1, 1, 1]$	[1, 1, 1, 1, 1, 1]
	[1, 1, 2, 1, 1, 1]	$[-5, p+1, -p+1, 1, p+1, -p+1]$	[1, 1, 1, 1, 1, 1]
	[1, 2, 2, 1, 1, 1]	$[p-5, -p+1, 1, 1, p+1, -p+1]$	[1, 1, 1, 1, 1, 1]
	[1, 1, 2, 1, 2, 1]	$[-5, p+1, 1, 1, -p+1, 1]$	[1, 1, 1, 1, 1, 1]
	[1, 2, 2, 1, 2, 1]	$[p-5, -p+1, p+1, 1, -p+1, 1]$	[1, 1, 1, 1, 1, 1]
	[1, 2, 3, 1, 2, 1]	$[p-5, 1, -p+1, p+1, 1, 1]$	[1, 1, 2, 1, 1, 0]
	[1, 2, 2, 1, 1, 0]	$[p-5, -p+1, 1, 1, 1, p+1]$	[1, 1, 2, 1, 1, 0]
$(p-5)\omega_1$	[1, 2, 3, 2, 2, 1]	$[p-4, 1, 1, -p+1, 1, 1]$	[1, 1, 1, 1, 1, 0]
	[1, 1, 1, 1, 1, 1]	$[-4, 1, p+1, -p+1, 1, -p+1]$	[1, 1, 1, 0, 1, 1]
	[1, 1, 2, 1, 1, 1]	$[-4, p+1, -p+1, 1, p+1, -p+1]$	[1, 1, 1, 0, 1, 1]
	[1, 2, 2, 1, 1, 1]	$[p-4, -p+1, 1, 1, p+1, -p+1]$	[1, 1, 1, 0, 1, 1]
	[1, 1, 2, 1, 2, 1]	$[-4, p+1, 1, 1, -p+1, 1]$	[1, 1, 1, 0, 1, 1]
	[1, 2, 2, 1, 2, 1]	$[p-4, -p+1, p+1, 1, -p+1, 1]$	[1, 1, 1, 0, 1, 1]
	[1, 2, 3, 1, 2, 1]	$[p-4, 1, -p+1, p+1, 1, 1]$	[1, 1, 1, 0, 1, 1]
	[1, 2, 2, 1, 1, 0]	$[p-4, -p+1, 1, 1, 1, p+1]$	[1, 1, 1, 1, 1, 0]
	[1, 1, 2, 1, 1, 0]	$[-4, p+1, -p+1, 1, 1, p+1]$	[1, 1, 1, 1, 1, 0]
$(p-4)\omega_1$	[1, 2, 3, 2, 2, 1]	$[p-3, 1, 1, -p+1, 1, 1]$	[1, 1, 1, 1, 0, 0]
	[1, 1, 1, 1, 1, 1]	$[-3, 1, p+1, -p+1, 1, -p+1]$	[1, 1, 1, 1, 0, 0]
	[1, 1, 2, 1, 1, 1]	$[-3, p+1, -p+1, 1, p+1, -p+1]$	[1, 1, 1, 1, 0, 0]
	[1, 2, 2, 1, 1, 1]	$[p-3, -p+1, 1, 1, p+1, -p+1]$	[1, 1, 1, 1, 0, 0]
	[1, 1, 2, 1, 2, 1]	$[-3, p+1, 1, 1, -p+1, 1]$	[1, 1, 1, 0, 1, 0]
	[1, 2, 2, 1, 2, 1]	$[p-3, -p+1, p+1, 1, -p+1, 1]$	[1, 1, 1, 0, 1, 0]
	[1, 2, 3, 1, 2, 1]	$[p-3, 1, -p+1, p+1, 1, 1]$	[1, 1, 1, 0, 1, 0]
	[1, 1, 1, 0, 1, 1]	$[-3, 1, 1, p+1, 1, -p+1]$	[1, 1, 1, 0, 1, 0]
	[1, 2, 2, 1, 1, 0]	$[p-3, -p+1, 1, 1, 1, p+1]$	[1, 1, 1, 0, 1, 0]
	[1, 1, 2, 1, 1, 0]	$[-3, p+1, -p+1, 1, 1, p+1]$	[1, 1, 1, 0, 1, 0]
	[1, 1, 1, 1, 1, 0]	$[-3, 1, p+1, -p+1, -p+1, p+1]$	[1, 1, 1, 0, 1, 0]

Table 2 $(p-3)\omega_1$, $p \geq 11$

λ	$m\alpha$	$\lambda+\rho-pm\alpha$	β
$(p-3)\omega_1$	[1, 1, 2, 1, 1, 1]	$[-2, p+1, -p+1, 1, p+1, -p+1]$	[1, 1, 1, 0, 0, 0]
	[1, 2, 2, 1, 1, 1]	$[p-2, -p+1, 1, 1, p+1, -p+1]$	[1, 1, 1, 0, 0, 0]
	[1, 2, 3, 1, 2, 1]	$[p-2, 1, -p+1, p+1, 1, 1]$	[1, 1, 1, 0, 0, 0]
	[1, 1, 1, 0, 1, 1]	$[-2, 1, 1, p+1, 1, -p+1]$	[1, 1, 1, 0, 0, 0]
	[1, 2, 2, 1, 1, 0]	$[p-2, -p+1, 1, 1, 1, p+1]$	[1, 1, 1, 0, 0, 0]
	[1, 1, 2, 1, 1, 0]	$[-2, p+1, -p+1, 1, 1, p+1]$	[1, 1, 1, 0, 0, 0]
	[1, 1, 1, 0, 1, 0]	$[-2, 1, 1, p+1, -p+1, p+1]$	[1, 1, 1, 0, 0, 0]
	[1, 1, 1, 1, 0, 0]	$[-2, 1, 1, -p+1, p+1, 1]$	[1, 1, 1, 0, 0, 0]
		w	$w(\lambda+\rho-pm\alpha)-\rho$
	[1, 2, 3, 2, 2, 1]	[1, 2, 3, 5, 4, 3, 2, 6, 5, 3, 4]	$[p-9, 0, 0, 0, 0, 0]$
	[1, 1, 1, 1, 1, 1]	[1, 2, 3, 5, 6, 4, 2, 1]	$[p-6, 0, 0, 1, 0, 0]$
	[1, 1, 2, 1, 2, 1]	[1, 2, 3, 5, 4, 3, 6, 5, 1]	$[p-8, 1, 0, 0, 0, 0]$
	[1, 2, 2, 1, 2, 1]	[1, 2, 3, 5, 4, 3, 1, 6, 5, 2]	$[p-8, 0, 0, 0, 0, 1]$
	[1, 1, 1, 1, 1, 0]	[1, 2, 3, 5, 4, 2, 1]	$[p-5, 0, 0, 0, 0, 1]$

(b). Part (i) is obvious. For (ii), the starting point is the simplicity of $V(\omega_1)$. By the graph automorphism from the symmetry of the Dynkin diagram this implies that $V(\omega_6)$ is also simple. Also, in Table 12 the only contributor for $2\omega_1$ is $\alpha = \alpha = \sum_{i=1}^{5}\alpha_i$, and the contribution is $-\chi(\omega_6)$. Hence, $\text{rad}(V(2\omega_1)) \cong L(\omega_6)$. To prove (iii), we examine the rows of Table 15 corresponding to $4\omega_1$, We see that the only two contributions are $\chi(\omega_1+\omega_4)$ (from $\alpha = \sum_{i=1}^{6}\alpha_i$) and $-\chi(2\omega_1+\omega_6)$ (from $\alpha = \sum_{i=1}^{5}\alpha_i$). We then consider the Jantzen sums for the highest weights of these two contributions, by looking at Table 17. There, we see that $V(\omega_1+\omega_4)$ is simple. Also, the only contribution to the Jantzen sum for $2\omega_1+\omega_6$ is $-\chi(\omega_1+\omega_4)$, and this means that $\text{rad}(V(2\omega_1+\omega_6)) \cong V(\omega_1+\omega_4)$, and the proof of (iii) is complete. (iv) When $p \geq 11$, the relevant roots and contributions for $(p-3)\omega_1$ are given by Table 2. The contributing weights are $(p-9)\omega_1$, $(p-6)\omega_1+\omega_4$, $(p-8)\omega_1+\omega_2$, $(p-8)\omega_1+\omega_6$, and $(p-5)\omega_1+\omega_6$. The data for the Jantzen sums for these highest weights is given in Table 3. From Table 3 we see first that $V((p-9)\omega_1)$ is simple. Then we see that $J((p-8)\omega_1+\omega_6) = \chi((p-9)\omega_1)$, which implies that $\text{rad}(V((p-8)\omega_1+\omega_6)) \cong L((p-9)\omega_1)$. Next, we have

$$\begin{aligned} J((p-8)\omega_1+\omega_2) &= \chi((p-8)\omega_1+\omega_6) - \chi((p-9)\omega_1) \\ &= \chi((p-8)\omega_1+\omega_6) - \text{Ch}(\text{rad}(V((p-8)\omega_1+\omega_6))). \end{aligned} \tag{1}$$

Hence $\text{rad}(V((p-8)\omega_1+\omega_2))$ is a simple module isomorphic to $L((p-8)\omega_1+\omega_6)$. Next, we have

Table 3 $(p-3)\omega_1$, $p \geq 11$, second iteration.

λ	$m\alpha$		
		$\lambda+\rho-pm\alpha$	β
$(p-9)\omega_1$	[1, 2, 3, 2, 2, 1]	$[p-8, 1, 1, -p+1, 1, 1]$	[1, 2, 2, 1, 2, 1]
	[1, 2, 3, 1, 2, 1]	$[p-8, 1, -p+1, p+1, 1, 1]$	[1, 2, 2, 1, 2, 1]
		$\lambda+\rho-pm\alpha$	β
$(p-8)\omega_1+\omega_6$	[1, 2, 3, 2, 2, 1]	$[p-7, 1, 1, -p+1, 1, 2]$	[1, 1, 2, 1, 1, 1]
	[1, 2, 2, 1, 1, 1]	$[p-7, -p+1, 1, 1, p+1, -p+2]$	[1, 1, 2, 1, 1, 1]
	[1, 1, 2, 1, 2, 1]	$[-7, p+1, 1, 1, -p+1, 2]$	[1, 1, 2, 1, 1, 1]
	[1, 2, 3, 1, 2, 1]	$[p-7, 1, -p+1, p+1, 1, 2]$	[1, 1, 2, 1, 1, 1]
		w	$w(\lambda+\rho-pm\alpha)-\rho$
	[1, 2, 2, 1, 2, 1]	[1, 3, 2, 6, 5, 4, 3, 2, 5, 4, 3, 1, 6, 5, 2]	$[p-9, 0, 0, 0, 0, 0]$
		$\lambda+\rho-pm\alpha$	β
$(p-8)\omega_1+\omega_2$	[1, 2, 3, 2, 2, 1]	$[p-7, 2, 1, -p+1, 1, 1]$	[1, 1, 2, 1, 1, 1]
	[1, 2, 2, 1, 1, 1]	$[p-7, -p+2, 1, 1, p+1, -p+1]$	[1, 1, 2, 1, 1, 1]
	[1, 1, 2, 1, 2, 1]	$[-7, p+2, 1, 1, -p+1, 1]$	[1, 1, 2, 1, 1, 1]
	[1, 2, 3, 1, 2, 1]	$[p-7, 2, -p+1, p+1, 1, 1]$	[1, 1, 2, 1, 1, 1]
		w	$w(\lambda+\rho-pm\alpha)-\rho$
	[1, 2, 2, 1, 2, 1]	[1, 3, 2, 5, 4, 3, 2, 5, 4, 3, 1, 6, 5, 2]	$[p-9, 0, 0, 0, 0, 0]$
	[1, 2, 2, 1, 1, 0]	[2, 1, 3, 2, 5, 4, 3, 2, 1, 5, 4, 3, 2]	$[p-8, 0, 0, 0, 0, 1]$
		$\lambda+\rho-pm\alpha$	β
$(p-6)\omega_1+\omega_4$	[1, 2, 3, 2, 2, 1]	$[p-5, 1, 1, -p+2, 1, 1]$	[1, 1, 1, 1, 1, 0]
	[1, 1, 1, 1, 1, 1]	$[-5, 1, p+1, -p+2, 1, -p+1]$	[1, 1, 1, 1, 1, 0]
	[1, 1, 2, 1, 2, 1]	$[-5, p+1, 1, 2, -p+1, 1]$	[1, 1, 1, 1, 1, 0]
	[1, 2, 2, 1, 2, 1]	$[p-5, -p+1, p+1, 2, -p+1, 1]$	[1, 1, 1, 1, 1, 0]
	[1, 2, 2, 1, 1, 0]	$[p-5, -p+1, 1, 2, 1, p+1]$	[1, 1, 1, 1, 1, 0]
	[1, 1, 2, 1, 1, 0]	$[-5, p+1, -p+1, 2, 1, p+1]$	[1, 1, 1, 1, 1, 0]
		w	$w(\lambda+\rho-pm\alpha)-\rho$
	[1, 1, 2, 1, 1, 1]	[1, 2, 4, 3, 5, 3, 2, 6, 4, 3, 1]	$[p-8, 1, 0, 0, 0, 0]$
	[1, 2, 2, 1, 1, 1]	[1, 2, 4, 3, 5, 3, 2, 1, 6, 4, 3, 2]	$[p-8, 0, 0, 0, 0, 1]$
	[1, 2, 3, 1, 2, 1]	[1, 2, 4, 3, 2, 1, 5, 4, 3, 2, 6, 5, 3]	$[p-9, 0, 0, 0, 0, 0]$
		$\lambda+\rho-pm\alpha$	β
$(p-5)\omega_1+\omega_6$	[1, 2, 3, 2, 2, 1]	$[p-4, 1, 1, -p+1, 1, 2]$	[1, 1, 1, 1, 1, 0]
	[1, 1, 1, 1, 1, 1]	$[-4, 1, p+1, -p+1, 1, -p+2]$	[1, 1, 1, 1, 1, 0]
	[1, 1, 2, 1, 2, 1]	$[-4, p+1, 1, 1, -p+1, 2]$	[1, 1, 1, 1, 1, 0]
	[1, 2, 2, 1, 2, 1]	$[p-4, -p+1, p+1, 1, -p+1, 2]$	[1, 1, 1, 1, 1, 0]
	[1, 2, 2, 1, 1, 0]	$[p-4, -p+1, 1, 1, 1, p+2]$	[1, 1, 1, 1, 1, 0]
	[1, 1, 2, 1, 1, 0]	$[-4, p+1, -p+1, 1, 1, p+2]$	[1, 1, 1, 1, 1, 0]
		w	$w(\lambda+\rho-pm\alpha)-\rho$
	[1, 1, 2, 1, 1, 1]	[1, 2, 3, 5, 3, 2, 6, 4, 3, 1]	$[p-8, 1, 0, 0, 0, 0]$
	[1, 2, 2, 1, 1, 1]	[1, 2, 3, 5, 3, 2, 1, 6, 4, 3, 2]	$[p-8, 0, 0, 0, 0, 1]$
	[1, 2, 3, 1, 2, 1]	[1, 2, 3, 2, 1, 5, 4, 3, 2, 6, 5, 3]	$[p-9, 0, 0, 0, 0, 0]$
	[1, 1, 1, 0, 1, 1]	[1, 2, 3, 5, 6, 5, 3, 2, 1]	$[p-6, 0, 0, 1, 0, 0]$

Table 4 $(p-2)\omega_1$, $p \geq 11$

λ	$m\alpha$		
		$\lambda+\rho-pm\alpha$	β
$(p-2)\omega_1$	$[1, 1, 1, 1, 1, 1]$	$[-1, 1, p+1, -p+1, 1, -p+1]$	$[1, 1, 0, 0, 0, 0]$
	$[1, 2, 2, 1, 1, 1]$	$[p-1, -p+1, 1, 1, p+1, -p+1]$	$[1, 1, 0, 0, 0, 0]$
	$[1, 2, 2, 1, 2, 1]$	$[p-1, -p+1, p+1, 1, -p+1, 1]$	$[1, 1, 0, 0, 0, 0]$
	$[1, 1, 1, 0, 1, 1]$	$[-1, 1, 1, p+1, 1, -p+1]$	$[1, 1, 0, 0, 0, 0]$
	$[1, 2, 2, 1, 1, 0]$	$[p-1, -p+1, 1, 1, 1, p+1]$	$[1, 1, 0, 0, 0, 0]$
	$[1, 1, 1, 0, 0, 0]$	$[-1, 1, -p+1, p+1, p+1, 1]$	$[1, 1, 0, 0, 0, 0]$
	$[1, 1, 1, 1, 1, 0]$	$[-1, 1, p+1, -p+1, -p+1, p+1]$	$[1, 1, 0, 0, 0, 0]$
	$[1, 1, 1, 0, 1, 0]$	$[-1, 1, 1, p-1, -p+1, p+1]$	$[1, 1, 0, 0, 0, 0]$
	$[1, 1, 1, 1, 0, 0]$	$[-1, 1, 1, -p-1, p+1, 1]$	$[1, 1, 0, 0, 0, 0]$
		w	$w(\lambda+\rho-pm\alpha)-\rho$
	$[1, 2, 3, 2, 2, 1]$	$[1, 2, 3, 5, 4, 3, 2, 6, 5, 3, 4]$	$[p-10, 1, 0, 0, 0, 0]$
	$[1, 1, 2, 1, 1, 1]$	$[1, 2, 3, 5, 6, 4, 3, 1]$	$[p-7, 0, 0, 1, 0, 1]$
	$[1, 1, 2, 1, 2, 1]$	$[1, 2, 3, 5, 4, 3, 6, 5, 1]$	$[p-8, 0, 1, 0, 0, 0]$
	$[1, 2, 3, 1, 2, 1]$	$[1, 2, 3, 5, 4, 3, 2, 6, 5, 3]$	$[p-9, 0, 0, 0, 1, 0]$
	$[1, 1, 2, 1, 1, 0]$	$[1, 2, 3, 5, 4, 3, 1]$	$[p-6, 0, 0, 0, 0, 2]$

$$
\begin{aligned}
J((p-6)\omega_1+\omega_4) &= \chi((p-8)\omega_1+\omega_2) - \chi((p-8)\omega_1+\omega_6) + \chi((p-9)\omega_1) \\
&= \chi((p-8)\omega_1+\omega_2) - J((p-8)\omega_1+\omega_2) \\
&= \chi((p-8)\omega_1+\omega_2) - \mathrm{Ch}(\mathrm{rad}(V((p-8)\omega_1+\omega_2))),
\end{aligned} \tag{2}
$$

which implies that $\mathrm{rad}(V((p-6)\omega_1+\omega_4)) \cong L((p-8)\omega_1+\omega_2)$. Continuing in the same way, we see that $\mathrm{rad}(V((p-5)\omega_1+\omega_6)) \cong L((p-6)\omega_1+\omega_4)$ and that $\mathrm{rad}(V((p-3)\omega_1)) \cong L((p-5)\omega_1+\omega_6)$. This completes the proof of (iv) and of (b).
(c). (i) The result is clear for $p=2$ and $p=3$ (see Table 10). (ii) For $p=5$, it is immediate from the $3\omega_1$ rows of Table 13 that the only contribution is $-\chi(0)$, so $\mathrm{rad}(V(\omega_3))$ is a one-dimensional trivial module. (iii) When $p=7$, the contributions for $5\omega_1$ can be found from Table 15. The contributions are $\chi(\omega_4+\omega_6)$ and $-\chi(\omega_1+2\omega_6)$. The data for a second iteration of the sum formula, applied to the contributing weights are given in Table 18. We see that $V(\omega_4+\omega_6)$ is simple and that $\mathrm{rad}(V(\omega_1+2\omega_6)) \cong V(\omega_4+\omega_6)$. Thus, (iii) holds.

(iv) Table 4 gives the contributions for $(p-2)\omega_1$. They are $-\chi((p-10)\omega_1+\omega_2)$, $\chi((p-9)\omega_1+\omega_5)$, $-\chi((p-8)\omega_1+\omega_3)$, $\chi((p-7)\omega_1+\omega_4+\omega_6)$ and $-\chi((p-6)\omega_1+2\omega_6)$. For each of the contributing weights, the data for a second iteration of the sum formula are given in Table 5. From Table 5 we see first that $V((p-10)\omega_1+\omega_2)$ is simple. Then we see that $J((p-9)\omega_1+\omega_5) = \chi((p-10)\omega_1+\omega_2)$, which implies that $\mathrm{rad}(V((p-9)\omega_1+\omega_5)) \cong V((p-10)\omega_1+\omega_2)$. Next, we have

$$
\begin{aligned}
J((p-8)\omega_1+\omega_3) &= \chi((p-9)\omega_1+\omega_5) - \chi((p-10)\omega_1+\omega_2) \\
&= \chi((p-9)\omega_1+\omega_5) - \mathrm{Ch}(\mathrm{rad}(V((p-9)\omega_1+\omega_5))).
\end{aligned} \tag{3}
$$

Table 5 $(p-2)\omega_1$, $p \geq 11$, second iteration

λ	$m\alpha$		
		$\lambda+\rho-pm\alpha$	β
$(p-10)\omega_1+\omega_2$	[1, 2, 3, 2, 2, 1]	$[p-9, 2, 1, -p+1, 1, 1]$	[1, 2, 2, 1, 1, 1]
	[1, 2, 2, 1, 2, 1]	$[p-9, -p+2, p+1, 1, -p+1, 1]$	[1, 2, 2, 1, 1, 1]
	[1, 2, 3, 1, 2, 1]	$[p-9, 2, -p+1, p+1, 1, 1]$	[1, 2, 2, 1, 1, 1]
		$\lambda+\rho-pm\alpha$	β
$(p-9)\omega_1+\omega_5$	[1, 2, 3, 2, 2, 1]	$[p-8, 1, 1, -p+1, 2, 1]$	[1, 2, 2, 1, 1, 1]
	[1, 2, 2, 1, 2, 1]	$[p-8, -p+1, p+1, 1, -p+2, 1]$	[1, 2, 2, 1, 1, 1]
	[1, 2, 3, 1, 2, 1]	$[p-8, 1, -p+1, p+1, 2, 1]$	[1, 2, 2, 1, 1, 1]
		w	$w(\lambda+\rho-pm\alpha)-\rho$
	[1, 1, 2, 1, 2, 1]	[1, 5, 3, 2, 6, 5, 4, 3, 2, 5, 4, 3, 6, 5, 1]	$[p-10, 1, 0, 0, 0, 0]$
		$\lambda+\rho-pm\alpha$	β
$(p-8)\omega_1+\omega_3$	[1, 2, 3, 2, 2, 1]	$[p-7, 1, 2, -p+1, 1, 1]$	[1, 1, 2, 1, 1, 0]
	[1, 1, 2, 1, 1, 1]	$[-7, p+1, -p+2, 1, p+1, -p+1]$	[1, 1, 2, 1, 1, 0]
	[1, 1, 2, 1, 2, 1]	$[-7, p+1, 2, 1, -p+1, 1]$	[1, 1, 2, 1, 1, 0]
	[1, 2, 3, 1, 2, 1]	$[p-7, 1, -p+2, p+1, 1, 1]$	[1, 1, 2, 1, 1, 0]
	[1, 2, 2, 1, 1, 0]	$[p-7, -p+1, 2, 1, 1, p+1]$	[1, 1, 2, 1, 1, 0]
		w	$w(\lambda+\rho-pm\alpha)-\rho$
	[1, 2, 2, 1, 1, 1]	[1, 3, 2, 4, 3, 5, 3, 2, 1, 6, 4, 3, 2]	$[p-9, 0, 0, 0, 1, 0]$
	[1, 2, 2, 1, 2, 1]	[1, 3, 2, 5, 4, 3, 2, 5, 4, 3, 1, 6, 5, 2]	$[p-10, 1, 0, 0, 0, 0]$
		$\lambda+\rho-pm\alpha$	β
$(p-7)\omega_1+\omega_4+\omega_6$	[1, 2, 3, 2, 2, 1]	$[p-6, 1, 1, -p+2, 1, 2]$	[1, 1, 2, 1, 1, 0]
	[1, 1, 2, 1, 1, 1]	$[-6, p+1, -p+1, 2, p+1, -p+2]$	[1, 1, 2, 1, 1, 0]
	[1, 1, 2, 1, 2, 1]	$[-6, p+1, 1, 2, -p+1, 2]$	[1, 1, 2, 1, 1, 0]
	[1, 2, 3, 1, 2, 1]	$[p-6, 1, -p+1, p+2, 1, 2]$	[1, 1, 2, 1, 1, 0]
	[1, 2, 2, 1, 1, 0]	$[p-6, -p+1, 1, 2, 1, p+2]$	[1, 1, 2, 1, 1, 0]
		w	$w(\lambda+\rho-pm\alpha)-\rho$
	[1, 1, 1, 1, 1, 1]	[1, 2, 4, 3, 6, 5, 3, 6, 4, 2, 1]	$[p-8, 0, 1, 0, 0, 0]$
	[1, 2, 2, 1, 1, 1]	[1, 2, 4, 3, 5, 3, 2, 1, 6, 4, 3, 2]	$[p-9, 0, 0, 0, 1, 0]$
	[1, 2, 2, 1, 2, 1]	[1, 2, 5, 4, 3, 2, 5, 4, 3, 1, 6, 5, 2]	$[p-10, 1, 0, 0, 0, 0]$
		$\lambda+\rho-pm\alpha$	β
$(p-6)\omega_1+2\omega_6$	[1, 2, 3, 2, 2, 1]	$[p-5, 1, 1, -p+1, 1, 3]$	[1, 1, 2, 1, 1, 0]
	[1, 1, 2, 1, 1, 1]	$[-5, p+1, -p+1, 1, p+1, -p+3]$	[1, 1, 2, 1, 1, 0]
	[1, 1, 2, 1, 2, 1]	$[-5, p+1, 1, 1, -p+1, 3]$	[1, 1, 2, 1, 1, 0]
	[1, 2, 3, 1, 2, 1]	$[p-5, 1, -p+1, p+1, 1, 3]$	[1, 1, 2, 1, 1, 0]
	[1, 2, 2, 1, 1, 0]	$[p-5, -p+1, 1, 1, 1, p+3]$	[1, 1, 2, 1, 1, 0]
		w	$w(\lambda+\rho-pm\alpha)-\rho$
	[1, 1, 1, 1, 1, 1]	[1, 2, 3, 6, 5, 3, 6, 4, 2, 1]	$[p-8, 0, 1, 0, 0, 0]$
	[1, 2, 2, 1, 1, 1]	[1, 2, 3, 5, 3, 2, 1, 6, 4, 3, 2]	$[p-9, 0, 0, 0, 1, 0]$
	[1, 2, 2, 1, 2, 1]	[1, 2, 5, 3, 2, 5, 4, 3, 1, 6, 5, 2]	$[p-10, 1, 0, 0, 0, 0]$
	[1, 1, 1, 0, 1, 1]	[1, 2, 3, 5, 6, 5, 3, 2, 1]	$[p-7, 0, 0, 1, 0, 1]$

Hence $\mathrm{rad}(V((p-8)\omega_1+\omega_3))$ is a simple module isomorphic to $L((p-9)\omega_1+\omega_5)$. Next, we have

Table 6 $(p-1)\omega_1,\ p \geq 11$

λ	$m\alpha$	$\lambda+\rho-pm\alpha$	β
$(p-1)\omega_1$	[1, 1, 1, 1, 1, 1]	$[0, 1, p+1, -p+1, 1, -p+1]$	[1, 0, 0, 0, 0, 0]
	[1, 1, 2, 1, 1, 1]	$[0, p+1, -p+1, 1, p+1, -p+1]$	[1, 0, 0, 0, 0, 0]
	[1, 1, 2, 1, 2, 1]	$[0, p+1, 1, 1, -p+1, 1]$	[1, 0, 0, 0, 0, 0]
	[1, 1, 1, 0, 1, 1]	$[0, 1, 1, p+1, 1, -p+1]$	[1, 0, 0, 0, 0, 0]
	[1, 1, 0, 0, 0, 0]	$[0, -p+1, p+1, 1, 1, 1]$	[1, 0, 0, 0, 0, 0]
	[1, 1, 2, 1, 1, 0]	$[0, p+1, -p+1, 1, 1, p+1]$	[1, 0, 0, 0, 0, 0]
	[1, 1, 1, 0, 0, 0]	$[0, 1, -p+1, p+1, p+1, 1]$	[1, 0, 0, 0, 0, 0]
	[1, 1, 1, 1, 1, 0]	$[0, 1, p+1, -p+1, -p+1, p+1]$	[1, 0, 0, 0, 0, 0]
	[1, 1, 1, 0, 1, 0]	$[0, 1, 1, p+1, -p+1, p+1]$	[1, 0, 0, 0, 0, 0]
	[1, 1, 1, 1, 0, 0]	$[0, 1, 1, -p+1, p+1, 1]$	[1, 0, 0, 0, 0, 0]
		w	$w(\lambda+\rho-pm\alpha)-\rho$
	[1, 2, 3, 2, 2, 1]	[1, 2, 3, 5, 4, 3, 2, 6, 5, 3, 4]	$[p-11, 2, 0, 0, 0, 0]$
	[1, 2, 2, 1, 1, 1]	[1, 2, 3, 5, 6, 4, 3, 2]	$[p-8, 0, 0, 1, 0, 2]$
	[1, 2, 2, 1, 2, 1]	[1, 2, 3, 5, 4, 3, 6, 5, 2]	$[p-9, 0, 1, 0, 0, 1]$
	[1, 2, 3, 1, 2, 1]	[1, 2, 3, 5, 4, 3, 2, 6, 5, 3]	$[p-10, 1, 0, 0, 1, 0]$
	[1, 2, 2, 1, 1, 0]	[1, 2, 3, 5, 4, 3, 2]	$[p-7, 0, 0, 0, 0, 3]$

$$
\begin{aligned}
J((p-7)\omega_1+\omega_4+\omega_6) &= \chi((p-8)\omega_1+\omega_3) - \chi((p-9)\omega_1+\omega_5) \\
&\quad + \chi((p-10)\omega_1+\omega_2) \\
&= \chi((p-8)\omega_1+\omega_3) - J((p-8)\omega_1+\omega_3) \\
&= \chi((p-8)\omega_1+\omega_3) - \mathrm{Ch}(\mathrm{rad}(V((p-8)\omega_1+\omega_3))),
\end{aligned} \tag{4}
$$

which implies that $\mathrm{rad}(V((p-7)\omega_1+\omega_4+\omega_6)) \cong L((p-8)\omega_1+\omega_3)$. Continuing in the same way, we see that $\mathrm{rad}(V((p-6)\omega_1+2\omega_6)) \cong L((p-7)\omega_1+\omega_4+\omega_6)$ and that $\mathrm{rad}(V((p-2)\omega_1)) \cong L((p-6)\omega_1+2\omega_6)$, and the proof of (iv), and of (c) is complete.

(d). The procedure for verifying (d) follows the same pattern as in (b) and (c), and was sketched in [4]. The steps are as follows. Tables 8, 9, 11 and 13 yield (i). For part (ii) Table 16 shows that $J(6\omega_1) = \chi(3\omega_6)$. The simplicity of $V(3\omega_6)$ can be checked by a further sum formula calculation or by observing that $V(3\omega_1)$ is simple by Table 14 and the two modules are conjugate under the graph automorphism of G. Finally, to prove (iii) we find the contributions for $V((p-1)\omega_1)$ in Table 6 and consider a second iteration the sum formula for each of the Weyl modules for contributing weights. The necessary data appear in Table 7. In a manner completely analogous to the proof of (c) (iv), we can make the the following sequence of inferences from Table 7.

1. $V((p-11)\omega_1+2\omega_2)$ is simple.
2. $\mathrm{rad}(V((p-10)\omega_1+\omega_2+\omega_5)) \cong L((p-11)\omega_1+2\omega_2)$.
3. $\mathrm{rad}(V((p-9)\omega_1+\omega_3+\omega_6)) \cong L((p-10)\omega_1+\omega_2+\omega_5)$.
4. $\mathrm{rad}(V((p-8)\omega_1+\omega_4+2\omega_6)) \cong L((p-9)\omega_1+\omega_3+\omega_6)$.
5. $\mathrm{rad}(V((p-7)\omega_1+3\omega_6)) \cong L((p-8)\omega_1+\omega_4+2\omega_6)$.

Table 7 $(p-1)\omega_1$, $p \geq 11$, second iteration

λ	$m\alpha$		
		$\lambda+\rho-pm\alpha$	β
$(p-11)\omega_1+2\omega_2$	$[1,2,3,2,2,1]$	$[p-10,3,1,-p+1,1,1]$	$[1,2,2,1,1,0]$
	$[1,2,2,1,1,1]$	$[p-10,-p+3,1,1,p+1,-p+1]$	$[1,2,2,1,1,0]$
	$[1,2,2,1,2,1]$	$[p-10,-p+3,p+1,1,-p+1,1]$	$[1,2,2,1,1,0]$
	$[1,2,3,1,2,1]$	$[p-10,3,-p+1,p+1,1,1]$	$[1,2,2,1,1,0]$
		$\lambda+\rho-pm\alpha$	β
$(p-10)\omega_1+\omega_2+\omega_5$	$[1,2,3,2,2,1]$	$[p-9,2,1,-p+1,2,1]$	$[1,2,2,1,1,0]$
	$[1,2,2,1,1,1]$	$[p-9,-p+2,1,1,p+2,-p+1]$	$[1,2,2,1,1,0]$
	$[1,2,2,1,2,1]$	$[p-9,-p+2,p+1,1,-p+2,1]$	$[1,2,2,1,1,0]$
	$[1,2,3,1,2,1]$	$[p-9,2,-p+1,p+1,2,1]$	$[1,2,2,1,1,0]$
		w	$w(\lambda+\rho-pm\alpha)-\rho$
	$[1,1,2,1,2,1]$	$[1,5,3,2,6,5,4,3,2,5,4,3,6,5,1]$	$[p-11,2,0,0,0,0]$
		$\lambda+\rho-pm\alpha$	β
$(p-9)\omega_1+\omega_3+\omega_6$	$[1,2,3,2,2,1]$	$[p-8,1,2,-p+1,1,2]$	$[1,2,2,1,1,0]$
	$[1,2,2,1,1,1]$	$[p-8,-p+1,2,1,p+1,-p+2]$	$[1,2,2,1,1,0]$
	$[1,2,2,1,2,1]$	$[p-8,-p+1,p+2,1,-p+1,2]$	$[1,2,2,1,1,0]$
	$[1,2,3,1,2,1]$	$[p-8,1,-p+2,p+1,1,2]$	$[1,2,2,1,1,0]$
		w	$w(\lambda+\rho-pm\alpha)-\rho$
	$[1,1,2,1,1,1]$	$[1,3,2,4,3,6,5,3,2,6,4,3,1]$	$[p-10,1,0,0,1,0]$
	$[1,1,2,1,2,1]$	$[1,3,2,6,5,4,3,2,5,4,3,6,5,1]$	$[p-11,2,0,0,0,0]$
		$\lambda+\rho-pm\alpha$	β
$(p-8)\omega_1+\omega_4+2\omega_6$	$[1,2,3,2,2,1]$	$[p-7,1,1,-p+2,1,3]$	$[1,2,2,1,1,0]$
	$[1,2,2,1,1,1]$	$[p-7,-p+1,1,2,p+1,-p+3]$	$[1,2,2,1,1,0]$
	$[1,2,2,1,2,1]$	$[p-7,-p+1,p+1,2,-p+1,3]$	$[1,2,2,1,1,0]$
	$[1,2,3,1,2,1]$	$[p-7,1,-p+1,p+2,1,3]$	$[1,2,2,1,1,0]$
		w	$w(\lambda+\rho-pm\alpha)-\rho$
	$[1,1,1,1,1,1]$	$[1,2,4,3,6,5,3,6,4,2,1]$	$[p-9,0,1,0,0,1]$
	$[1,1,2,1,1,1]$	$[1,2,4,3,6,5,3,2,6,4,3,1]$	$[p-10,1,0,0,1,0]$
	$[1,1,2,1,2,1]$	$[1,2,6,5,4,3,2,5,4,3,6,5,1]$	$[p-11,2,0,0,0,0]$
		$\lambda+\rho-pm\alpha$	β
$(p-7)\omega_1+3\omega_6$	$[1,2,3,2,2,1]$	$[p-6,1,1,-p+1,1,4]$	$[1,2,2,1,1,0]$
	$[1,2,2,1,1,1]$	$[p-6,-p+1,1,1,p+1,-p+4]$	$[1,2,2,1,1,0]$
	$[1,2,2,1,2,1]$	$[p-6,-p+1,p+1,1,-p+1,4]$	$[1,2,2,1,1,0]$
	$[1,2,3,1,2,1]$	$[p-6,1,-p+1,p+1,1,4]$	$[1,2,2,1,1,0]$
		w	$w(\lambda+\rho-pm\alpha)-\rho$
	$[1,1,1,1,1,1]$	$[1,2,3,6,5,3,6,4,2,1]$	$[p-9,0,1,0,0,1]$
	$[1,1,2,1,1,1]$	$[1,2,3,6,5,3,2,6,4,3,1]$	$[p-10,1,0,0,1,0]$
	$[1,1,2,1,2,1]$	$[1,2,6,5,3,2,5,4,3,6,5,1]$	$[p-11,2,0,0,0,0]$
	$[1,1,1,0,1,1]$	$[1,2,3,5,6,5,3,2,1]$	$[p-8,0,0,1,0,2]$

Table 8 $p = 2$, relevant root multiples for ω_1, Part 1

λ	$m\alpha$	$\lambda+\rho-pm\alpha$	β
ω_1	[1, 2, 3, 2, 2, 1]	[2, 1, 1, −1, 1, 1]	[0, 0, 1, 1, 0, 0]
	[2, 4, 6, 4, 4, 2]	[2, 1, 1, −3, 1, 1]	[0, 0, 1, 1, 1, 1]
	[3, 6, 9, 6, 6, 3]	[2, 1, 1, −5, 1, 1]	[1, 2, 3, 2, 2, 1]
	[4, 8, 12, 8, 8, 4]	[2, 1, 1, −7, 1, 1]	[1, 1, 2, 1, 1, 1]
	[5, 10, 15, 10, 10, 5]	[2, 1, 1, −9, 1, 1]	[1, 2, 2, 1, 2, 1]
	[0, 0, 1, 1, 1, 1]	[2, 3, 1, −1, 1, −1]	[0, 0, 1, 1, 1, 1]
	[0, 1, 1, 1, 1, 1]	[4, −1, 3, −1, 1, −1]	[0, 0, 0, 0, 1, 1]
	[0, 2, 2, Z, 2, 2]	[6, −3, 5, −3, 1, −3]	[0, 0, 1, 1, 1, 1]
	[1, 1, 1, 1, 1, 1]	[0, 1, 3, −1, 1, −1]	[0, 0, 0, 0, 1, 1]
	[2, 2, 2, 2, 2, 2]	[−2, 1, 5, −3, 1, −3]	[0, 0, 1, 1, 1, 1]
	[3, 3, 3, 3, 3, 3]	[−4, 1, 7, −5, 1, −5]	[1, 1, 1, 0, 1, 1]
	[0, 1, 2, 1, 1, 1]	[4, 1, −1, 1, 3, −1]	[0, 1, 1, 0, 0, 0]
	[0, 2, 4, 2, 2, 2]	[6, 1, −3, 1, 5, −3]	[0, 0, 1, 1, 1, 1]
	[1, 1, 2, 1, 1, 1]	[0, 3, −1, 1, 3, −1]	[1, 0, 0, 0, 0, 0]
	[2, 2, 4, 2, 2, 2]	[−2, 5, −3, 1, 5, −3]	[0, 0, 1, 1, 1, 1]
	[3, 3, 6, 3, 3, 3]	[−4, 7, −5, 1, 7, −5]	[0, 1, 2, 1, 1, 1]
	[1, 2, 2, 1, 1, 1]	[2, −1, 1, 1, 3, −1]	[0, 1, 1, 0, 0, 0]
	[2, 4, 4, 2, 2, 2]	[2, −3, 1, 1, 5, −3]	[0, 1, 1, 0, 1, 1]
	[3, 6, 6, 3, 3, 3]	[2, −5, 1, 1, 7, −5]	[0, 1, 2, 1, 1, 1]
	[4, 8, 8, 4, 4, 4]	[2, −7, 1, 1, 9, −7]	[1, 1, 2, 1, 1, 1]
	[0, 1, 2, 1, 2, 1]	[4, 1, 1, 1, −1, 1]	[0, 0, 0, 0, 1, 1]
	[0, 2, 4, 2, 4, 2]	[6, 1, 1, 1, −3, 1]	[0, 0, 1, 1, 1, 1]
	[0, 3, 6, 3, 6, 3]	[8, 1, 1, 1, −5, 1]	[0, 1, 2, 1, 1, 1]
	[1, 1, 2, 1, 2, 1]	[0, 3, 1, 1, −1, 1]	[0, 0, 0, 0, 1, 1]
	[2, 2, 4, 2, 4, 2]	[−2, 5, 1, 1, −3, 1]	[0, 0, 1, 1, 1, 1]
	[3, 3, 6, 3, 6, 3]	[−4, 7, 1, 1, −5, 1]	[1, 1, 1, 0, 1, 1]
	[4, 4, 8, 4, 8, 4]	[−6, 9, 1, 1, −7, 1]	[1, 1, 2, 1, 1, 1]
	[1, 2, 2, 1, 2, 1]	[2, −1, 3, 1, −1, 1]	[0, 0, 0, 0, 1, 1]
	[2, 4, 4, 2, 4, 2]	[2, −3, 5, 1, −3, 1]	[0, 1, 1, 0, 1, 1]
	[3, 6, 6, 3, 6, 3]	[2, −5, 7, 1, −5, 1]	[1, 1, 1, 0, 1, 1]
	[4, 8, 8, 4, 8, 4]	[2, −7, 9, 1, −7, 1]	[1, 2, 2, 1, 1, 0]
	[1, 2, 3, 1, 2, 1]	[2, 1, −1, 3, 1, 1]	[0, 1, 1, 0, 0, 0]
	[2, 4, 6, 2, 4, 2]	[2, 1, −3, 5, 1, 1]	[0, 1, 1, 0, 1, 1]
	[3, 6, 9, 3, 6, 3]	[2, 1, −5, 7, 1, 1]	[0, 1, 2, 1, 1, 1]
	[4, 8, 12, 4, 8, 4]	[2, 1, −7, 9, 1, 1]	[1, 1, 2, 1, 1, 1]
	[5, 10, 15, 5, 10, 5]	[2, 1, −9, 11, 1, 1]	[1, 2, 2, 1, 2, 1]

Table 9 $p = 2$, relevant root multiples for ω_1, Part 2.

λ	$m\alpha$	$\lambda+\rho-pm\alpha$	β
ω_1	[0, 0, 1, 0, 1, 1]	[2, 3, −1, 3, 1, −1]	[0, 0, 0, 0, 1, 1]
	[0, 1, 1, 0, 1, 1]	[4, −1, 1, 3, 1, −1]	[0, 0, 0, 0, 1, 1]
	[1, 1, 1, 0, 1, 1]	[0, 1, 1, 3, 1, −1]	[0, 0, 0, 0, 1, 1]
	[2, 2, 2, 0, 2, 2]	[−2, 1, 1, 5, 1, −3]	[0, 1, 1, 0, 1, 1]
	[1, 2, 2, 1, 1, 0]	[2, −1, 1, 1, 1, 3]	[0, 1, 1, 0, 0, 0]
	[2, 4, 4, 2, 2, 0]	[2, −3, 1, 1, 1, 5]	[1, 2, 2, 1, 1, 0]
	[3, 6, 6, 3, 3, 0]	[2, −5, 1, 1, 1, 7]	[1, 1, 1, 1, 1, 0]
	[1, 1, 0, 0, 0, 0]	[0, −1, 3, 1, 1, 1]	[1, 0, 0, 0, 0, 0]
	[1, 1, 2, 1, 1, 0]	[0, 3, −1, 1, 1, 3]	[1, 0, 0, 0, 0, 0]
	[2, 2, 4, 2, 2, 0]	[−2, 5, −3, 1, 1, 5]	[1, 1, 1, 0, 0, 0]
	[3, 3, 6, 3, 3, 0]	[−4, 7, −5, 1, 1, 7]	[1, 1, 1, 1, 1, 0]
	[1, 1, 1, 0, 0, 0]	[0, 1, −1, 3, 3, 1]	[1, 0, 0, 0, 0, 0]
	[1, 1, 1, 1, 1, 0]	[0, 1, 3, −1, −1, 3]	[1, 0, 0, 0, 0, 0]
	[2, 2, 2, 2, 2, 0]	[−2, 1, 5, −3, −3, 5]	[0, 1, 1, 1, 1, 0]
	[1, 1, 1, 0, 1, 0]	[0, 1, 1, 3, −1, 3]	[1, 0, 0, 0, 0, 0]
	[2, 2, 2, 0, 2, 0]	[−2, 1, 1, 5, −3, 5]	[1, 1, 1, 0, 0, 0]
	[1, 1, 1, 1, 0, 0]	[0, 1, 1, −1, 3, 1]	[1, 0, 0, 0, 0, 0]
	[2, 2, 2, 2, 0, 0]	[−2, 1, 1, −3, 5, 1]	[1, 1, 1, 0, 0, 0]
	[0, 1, 2, 1, 1, 0]	[4, 1, −1, 1, 1, 3]	[0, 1, 1, 0, 0, 0]
	[0, 2, 4, 2, 2, 0]	[6, 1, −3, 1, 1, 5]	[0, 1, 1, 1, 1, 0]
	[0, 1, 1, 1, 1, 0]	[4, −1, 3, −1, −1, 3]	[0, 1, 1, 1, 1, 0]
	[0, 1, 1, 0, 1, 0]	[4, −1, 1, 3, −1, 3]	[0, 1, 1, 0, 0, 0]
	[0, 1, 1, 1, 0, 0]	[4, −1, 1, −1, 3, 1]	[0, 1, 1, 0, 0, 0]
	[0, 0, 1, 1, 1, 0]	[2, 3, 1, −1, −1, 3]	[0, 0, 1, 0, 1, 0]

Table 10 $p = 3$, relevant root multiples for ω_1

λ	$m\alpha$	$\lambda+\rho-pm\alpha$	β
ω_1	[1, 2, 3, 2, 2, 1]	[2, 1, 1, −2, 1, 1]	[0, 1, 1, 1, 0, 0]
	[2, 4, 6, 4, 4, 2]	[2, 1, 1, −5, 1, 1]	[1, 2, 3, 2, 2, 1]
	[3, 6, 9, 6, 6, 3]	[2, 1, 1, −8, 1, 1]	[1, 2, 2, 1, 1, 1]
	[0, 0, 1, 1, 1, 1]	[2, 4, 1, −2, 1, −2]	[0, 0, 1, 0, 1, 1]
	[0, 1, 1, 1, 1, 1]	[5, −2, 4, −2, 1, −2]	[0, 1, 1, 1, 0, 0]
	[1, 1, 1, 1, 1, 1]	[−1, 1, 4, −2, 1, −2]	[1, 1, 0, 0, 0, 0]
	[2, 2, 2, 2, 2, 2]	[−4, 1, 7, −5, 1, −5]	[1, 1, 1, 0, 1, 1]
	[0, 1, 2, 1, 1, 1]	[5, 1, −2, 1, 4, −2]	[0, 1, 2, 1, 1, 1]
	[1, 1, 2, 1, 1, 1]	[−1, 4, −2, 1, 4, −2]	[0, 0, 1, 0, 1, 1]
	[2, 2, 4, 2, 2, 2]	[−4, 7, −5, 1, 7, −5]	[0, 1, 2, 1, 1, 1]
	[1, 2, 2, 1, 1, 1]	[2, −2, 1, 1, 4, −2]	[1, 1, 0, 0, 0, 0]
	[2, 4, 4, 2, 2, 2]	[2, −5, 1, 1, 7, −5]	[0, 1, 2, 1, 1, 1]
	[0, 1, 2, 1, 2, 1]	[5, 1, 1, 1, −2, 1]	[0, 0, 1, 0, 1, 1]
	[0, 2, 4, 2, 4, 2]	[8, 1, 1, 1, −5, 1]	[0, 1, 2, 1, 1, 1]

(Continued)

Table 10 (Continued)

λ	$m\alpha$	$\lambda+\rho-pm\alpha$	β
	[1, 1, 2, 1, 2, 1]	[−1, 4, 1, 1, −2, 1]	[0, 0, 1, 0, 1, 1]
	[2, 2, 4, 2, 4, 2]	[−4, 7, 1, 1, −5, 1]	[1, 1, 1, 0, 1, 1]
	[1, 2, 2, 1, 2, 1]	[2, −2, 4, 1, −2, 1]	[1, 1, 0, 0, 0, 0]
	[2, 4, 4, 2, 4, 2]	[2, −5, 7, 1, −5, 1]	[1, 1, 1, 0, 1, 1]
	[3, 6, 6, 3, 6, 3]	[2, −8, 10, 1, −8, 1]	[1, 2, 2, 1, 1, 1]
	[1, 2, 3, 1, 2, 1]	[2, 1, −2, 4, 1, 1]	[0, 0, 1, 0, 1, 1]
	[2, 4, 6, 2, 4, 2]	[2, 1, −5, 7, 1, 1]	[0, 1, 2, 1, 1, 1]
	[3, 6, 9, 3, 6, 3]	[2, 1, −8, 10, 1, 1]	[1, 2, 2, 1, 1, 1]
	[0, 1, 1, 0, 1, 1]	[5, −2, 1, 4, 1, −2]	[0, 0, 1, 0, 1, 1]
	[1, 1, 1, 0, 1, 1]	[−1, 1, 1, 4, 1, −2]	[0, 0, 1, 0, 1, 1]
	[1, 2, 2, 1, 1, 0]	[2, −2, 1, 1, 1, 4]	[1, 1, 0, 0, 0, 0]
	[2, 4, 4, 2, 2, 0]	[2, −5, 1, 1, 1, 7]	[1, 1, 1, 1, 1, 0]
	[1, 1, 2, 1, 1, 0]	[−1, 4, −2, 1, 1, 4]	[0, 0, 1, 1, 1, 0]
	[2, 2, 4, 2, 2, 0]	[−4, 7, −5, 1, 1, 7]	[1, 1, 1, 1, 1, 0]
	[1, 1, 1, 0, 0, 0]	[−1, 1, −2, 4, 4, 1]	[1, 1, 0, 0, 0, 0]
	[1, 1, 1, 1, 1, 0]	[−1, 1, 4, −2, −2, 4]	[1, 1, 0, 0, 0, 0]
	[1, 1, 1, 0, 1, 0]	[−1, 1, 1, 4, −2, 4]	[1, 1, 0, 0, 0, 0]
	[1, 1, 1, 1, 0, 0]	[−1, 1, 1, −2, 4, 1]	[1, 1, 0, 0, 0, 0]
	[0, 1, 2, 1, 1, 0]	[5, 1, −2, 1, 1, 4]	[0, 1, 1, 0, 1, 0]
	[0, 1, 1, 1, 1, 0]	[5, −2, 4, −2, −2, 4]	[0, 1, 1, 0, 1, 0]

Table 11 $p = 3$, relevant root multiples for $2\omega_1$

λ	$m\alpha$	$\lambda+\rho-pm\alpha$	β
$2\omega_1$	[1, 2, 3, 2, 2, 1]	[3, 1, 1, −2, 1, 1]	[0, 1, 1, 1, 0, 0]
	[2, 4, 6, 4, 4, 2]	[3, 1, 1, −5, 1, 1]	[0, 1, 2, 1, 1, 1]
	[3, 6, 9, 6, 6, 3]	[3, 1, 1, −8, 1, 1]	[1, 1, 2, 1, 1, 1]
	[4, 8, 12, 8, 8, 4]	[3, 1, 1, −11, 1, 1]	[1, 2, 3, 1, 2, 1]
	[0, 0, 1, 1, 1, 1]	[3, 4, 1, −2, 1, −2]	[0, 0, 1, 0, 1, 1]
	[0, 1, 1, 1, 1, 1]	[6, −2, 4, −2, 1, −2]	[0, 1, 1, 1, 0, 0]
	[1, 1, 1, 1, 1, 1]	[0, 1, 4, −2, 1, −2]	[1, 0, 0, 0, 0, 0]
	[2, 2, 2, 2, 2, 2]	[−3, 1, 7, −5, 1, −5]	[1, 1, 1, 1, 0, 0]
	[0, 1, 2, 1, 1, 1]	[6, 1, −2, 1, 4, −2]	[0, 1, 2, 1, 1, 1]
	[1, 1, 2, 1, 1, 1]	[0, 4, −2, 1, 4, −2]	[0, 0, 1, 0, 1, 1]
	[2, 2, 4, 2, 2, 2]	[−3, 7, −5, 1, 7, −5]	[0, 1, 2, 1, 1, 1]
	[1, 2, 2, 1, 1, 1]	[3, −2, 1, 1, 4, −2]	[0, 1, 1, 1, 0, 0]
	[2, 4, 4, 2, 2, 2]	[3, −5, 1, 1, 7, −5]	[0, 1, 2, 1, 1, 1]
	[3, 6, 6, 3, 3, 3]	[3, −8, 1, 1, 10, −8]	[1, 1, 2, 1, 1, 1]
	[0, 1, 2, 1, 2, 1]	[6, 1, 1, 1, −2, 1]	[0, 0, 1, 0, 1, 1]
	[0, 2, 4, 2, 4, 2]	[9, 1, 1, 1, −5, 1]	[0, 1, 2, 1, 1, 1]
	[1, 1, 2, 1, 2, 1]	[0, 4, 1, 1, −2, 1]	[0, 0, 1, 0, 1, 1]
	[2, 2, 4, 2, 4, 2]	[−3, 7, 1, 1, −5, 1]	[1, 1, 1, 0, 1, 0]

(Continued)

Table 11 (Continued)

λ	$m\alpha$	$\lambda+\rho-pm\alpha$	β
	[3, 3, 6, 3, 6, 3]	[−6, 10, 1, 1, −8, 1]	[1, 1, 2, 1, 1, 1]
	[1, 2, 2, 1, 2, 1]	[3, −2, 4, 1, −2, 1]	[0, 1, 1, 0, 1, 0]
	[2, 4, 4, 2, 4, 2]	[3, −5, 7, 1, −5, 1]	[1, 1, 1, 0, 1, 0]
	[3, 6, 6, 3, 6, 3]	[3, −8, 10, 1, −8, 1]	[1, 2, 2, 1, 1, 0]
	[1, 2, 3, 1, 2, 1]	[3, 1, −2, 4, 1, 1]	[0, 0, 1, 0, 1, 1]
	[2, 4, 6, 2, 4, 2]	[3, 1, −5, 7, 1, 1]	[0, 1, 2, 1, 1, 1]
	[3, 6, 9, 3, 6, 3]	[3, 1, −8, 10, 1, 1]	[1, 1, 2, 1, 1, 1]
	[0, 1, 1, 0, 1, 1]	[6, −2, 1, 4, 1, −2]	[0, 0, 1, 0, 1, 1]
	[1, 1, 1, 0, 1, 1]	[0, 1, 1, 4, 1, −2]	[0, 0, 1, 0, 1, 1]
	[2, 2, 2, 0, 2, 2]	[−3, 1, 1, 7, 1, −5]	[1, 1, 1, 0, 1, 0]
	[1, 2, 2, 1, 1, 0]	[3, −2, 1, 1, 1, 4]	[0, 1, 1, 0, 1, 0]
	[2, 4, 4, 2, 2, 0]	[3, −5, 1, 1, 1, 7]	[1, 1, 1, 0, 1, 0]
	[1, 1, 0, 0, 0, 0]	[0, −2, 4, 1, 1, 1]	[1, 0, 0, 0, 0, 0]
	[1, 1, 2, 1, 1, 0]	[0, 4, −2, 1, 1, 4]	[1, 0, 0, 0, 0, 0]
	[2, 2, 4, 2, 2, 0]	[−3, 7, −5, 1, 1, 7]	[1, 1, 1, 0, 1, 0]
	[1, 1, 1, 0, 0, 0]	[0, 1, −2, 4, 4, 1]	[1, 0, 0, 0, 0, 0]
	[1, 1, 1, 1, 1, 0]	[0, 1, 4, −2, −2, 4]	[1, 0, 0, 0, 0, 0]
	[2, 2, 2, 2, 2, 0]	[−3, 1, 7, −5, −5, 7]	[1, 1, 1, 0, 1, 0]
	[1, 1, 1, 0, 1, 0]	[0, 1, 1, 4, −2, 4]	[1, 0, 0, 0, 0, 0]
	[1, 1, 1, 1, 0, 0]	[0, 1, 1, −2, 4, 1]	[1, 0, 0, 0, 0, 0]
	[0, 1, 2, 1, 1, 0]	[6, 1, −2, 1, 1, 4]	[0, 1, 1, 0, 1, 0]
	[0, 1, 1, 1, 1, 0]	[6, −2, 4, −2, −2, 4]	[0, 1, 1, 0, 1, 0]

Table 12 $p=5$, relevant root multiples for ω_1 and $2\omega_1$

λ	$m\alpha$	$\lambda+\rho-pm\alpha$	β
ω_1	[1, 2, 3, 2, 2, 1]	[2, 1, 1, −4, 1, 1]	[0, 1, 1, 1, 1, 1]
	[2, 4, 6, 4, 4, 2]	[2, 1, 1, −9, 1, 1]	[1, 2, 2, 1, 2, 1]
	[1, 1, 1, 1, 1, 1]	[−3, 1, 6, −4, 1, −4]	[0, 1, 1, 1, 1, 1]
	[0, 1, 2, 1, 1, 1]	[7, 1, −4, 1, 6, −4]	[0, 1, 1, 1, 1, 1]
	[1, 1, 2, 1, 1, 1]	[−3, 6, −4, 1, 6, −4]	[1, 1, 1, 1, 0, 0]
	[1, 2, 2, 1, 1, 1]	[2, −4, 1, 1, 6, −4]	[0, 1, 1, 1, 1, 1]
	[0, 1, 2, 1, 2, 1]	[7, 1, 1, 1, −4, 1]	[0, 1, 1, 1, 1, 1]
	[1, 1, 2, 1, 2, 1]	[−3, 6, 1, 1, −4, 1]	[1, 1, 1, 0, 1, 0]
	[1, 2, 2, 1, 2, 1]	[2, −4, 6, 1, −4, 1]	[0, 1, 1, 1, 1, 1]
	[1, 2, 3, 1, 2, 1]	[2, 1, −4, 6, 1, 1]	[1, 1, 1, 0, 1, 0]
	[2, 4, 6, 2, 4, 2]	[2, 1, −9, 11, 1, 1]	[1, 2, 2, 1, 2, 1]
	[1, 1, 1, 0, 1, 1]	[−3, 1, 1, 6, 1, −4]	[1, 1, 1, 0, 1, 0]
	[1, 2, 2, 1, 1, 0]	[2, −4, 1, 1, 1, 6]	[1, 1, 1, 0, 1, 0]
	[1, 1, 2, 1, 1, 0]	[−3, 6, −4, 1, 1, 6]	[1, 1, 1, 0, 1, 0]
	[1, 1, 1, 1, 1, 0]	[−3, 1, 6, −4, −4, 6]	[1, 1, 1, 0, 1, 0]

(Continued)

Table 12 (Continued)

λ	$m\alpha$	$\lambda+\rho-pm\alpha$	β
$2\omega_1$	[1, 2, 3, 2, 2, 1]	[3, 1, 1, −4, 1, 1]	[0, 1, 1, 1, 1, 1]
	[2, 4, 6, 4, 4, 2]	[3, 1, 1, −9, 1, 1]	[1, 2, 2, 1, 1, 1]
	[1, 1, 1, 1, 1, 1]	[−2, 1, 6, −4, 1, −4]	[0, 1, 1, 1, 1, 1]
	[0, 1, 2, 1, 1, 1]	[8, 1, −4, 1, 6, −4]	[0, 1, 1, 1, 1, 1]
	[1, 1, 2, 1, 1, 1]	[−2, 6, −4, 1, 5, −4]	[1, 1, 1, 0, 0, 0]
	[1, 2, 2, 1, 1, 1]	[3, −4, 1, 1, 6, −4]	[0, 1, 1, 1, 1, 1]
	[0, 1, 2, 1, 2, 1]	[8, 1, 1, 1, −4, 1]	[0, 1, 1, 1, 1, 1]
	[1, 1, 2, 1, 2, 1]	[−2, 6, 1, 1, −4, 1]	[1, 1, 2, 1, 2, 1]
	[1, 2, 2, 1, 2, 1]	[3, −4, 6, 1, −4, 1]	[0, 1, 1, 1, 1, 1]
	[2, 4, 4, 2, 4, 2]	[3, −9, 11, 1, −9, 1]	[1, 2, 2, 1, 1, 1]
	[1, 2, 3, 1, 2, 1]	[3, 1, −4, 6, 1, 1]	[1, 1, 1, 0, 0, 0]
	[2, 4, 6, 2, 4, 2]	[3, 1, −9, 11, 1, 1]	[1, 2, 2, 1, 1, 1]
	[1, 1, 1, 0, 1, 1]	[−2, 1, 1, 6, 1, −4]	[1, 1, 1, 0, 0, 0]
	[1, 2, 2, 1, 1, 0]	[3, −4, 1, 1, 1, 6]	[1, 1, 1, 0, 0, 0]
	[1, 1, 2, 1, 1, 0]	[−2, 6, −4, 1, 1, 6]	[1, 1, 1, 0, 0, 0]
	[1, 1, 1, 0, 1, 0]	[−2, 1, 1, 6, −4, 6]	[1, 1, 1, 0, 0, 0]
	[1, 1, 1, 1, 0, 0]	[−2, 1, 1, −4, 6, 1]	[1, 1, 1, 0, 0, 0]
		w	$w(\lambda+\rho-pm\alpha)-\rho$
	[1, 1, 1, 1, 1, 0]	[1, 2, 3, 5, 4, 2, 1]	[0, 0, 0, 0, 0, 1]

Table 13 $p=5$, relevant root multiples for $3\omega_1$ and $4\omega_1$

λ	$m\alpha$	$\lambda+\rho-pm\alpha$	β
$3\omega_1$	[1, 2, 3, 2, 2, 1]	[4, 1, 1, −4, 1, 1]	[0, 1, 1, 1, 1, 1]
	[2, 4, 6, 4, 4, 2]	[4, 1, 1, −9, 1, 1]	[1, 1, 2, 1, 1, 1]
	[1, 1, 1, 1, 1, 1]	[−1, 1, 5, −4, 1, −4]	[0, 1, 1, 1, 1, 1]
	[0, 1, 2, 1, 1, 1]	[9, 1, −4, 1, 6, −4]	[0, 1, 1, 1, 1, 1]
	[1, 1, 2, 1, 1, 1]	[−1, 6, −4, 1, 6, −4]	[1, 1, 2, 1, 1, 1]
	[1, 2, 2, 1, 1, 1]	[4, −4, 1, 1, 6, −4]	[0, 1, 1, 1, 1, 1]
	[2, 4, 4, 2, 2, 2]	[4, −9, 1, 1, 11, −9]	[1, 1, 2, 1, 1, 1]
	[0, 1, 2, 1, 2, 1]	[9, 1, 1, 1, −4, 1]	[0, 1, 1, 1, 1, 1]
	[2, 2, 4, 2, 4, 2]	[−6, 11, 1, 1, −9, 1]	[1, 1, 2, 1, 1, 1]
	[1, 2, 2, 1, 2, 1]	[4, −4, 6, 1, −4, 1]	[0, 1, 1, 1, 1, 1]
	[2, 4, 4, 2, 4, 2]	[4, −9, 11, 1, −9, 1]	[1, 2, 2, 1, 1, 0]
	[1, 2, 3, 1, 2, 1]	[4, 1, −4, 6, 1, 1]	[0, 1, 2, 1, 1, 0]
	[2, 4, 6, 2, 4, 2]	[4, 1, −9, 11, 1, 1]	[1, 1, 2, 1, 1, 1]
	[1, 1, 1, 0, 1, 1]	[−1, 1, 1, 6, 1, −4]	[1, 1, 0, 0, 0, 0]
	[1, 2, 2, 1, 1, 0]	[4, −4, 1, 1, 1, 6]	[1, 2, 2, 1, 1, 0]
	[1, 1, 2, 1, 1, 0]	[−1, 6, −4, 1, 1, 6]	[0, 1, 2, 1, 1, 0]
	[1, 1, 1, 0, 0, 0]	[−1, 1, −4, 6, 6, 1]	[1, 1, 0, 0, 0, 0]
	[1, 1, 1, 1, 1, 0]	[−1, 1, 6, −4, −4, 6]	[1, 1, 0, 0, 0, 0]
	[1, 1, 1, 0, 1, 0]	[−1, 1, 1, 6 −4, 6]	[1, 1, 0, 0, 0, 0]
	[1, 1, 1, 1, 0, 0]	[−1, 1, 1, −4, 6, 1]	[1, 1, 0, 0, 0, 0]

(Continued)

Table 13 (Continued)

λ	$m\alpha$	$\lambda+\rho-pm\alpha$	β
		w	$w(\lambda+\rho-pm\alpha)-\rho$
	[1, 1, 2, 1, 2, 1]	[3, 5, 4, 3, 6, 5, 1]	[0, 0, 0, 0, 0, 0]
$4\omega_1$	[1, 2, 3, 2, 2, 1]	[5, 1, 1, −4, 1, 1]	[0, 1, 1, 1, 1, 1]
	[2, 4, 6, 4, 4, 2]	[5, 1, 1, −9, 1, 1]	[1, 1, 1, 1, 1, 1]
	[1, 1, 1, 1, 1, 1]	[0, 1, 6, −4, 1, −4]	[0, 1, 1, 1, 1, 1]
	[0, 1, 2, 1, 1, 1]	[10, 1, −4, 1, 6, −4]	[0, 1, 1, 1, 1, 1]
	[1, 1, 2, 1, 1, 1]	[0, 6, −4, 1, 6, −4]	[1, 0, 0, 0, 0, 0]
	[2, 2, 4, 2, 2, 2]	[−5, 11, −9, 1, 11, −9]	[1, 1, 1, 1, 1, 1]
	[1, 2, 2, 1, 1, 1]	[5, −4, 1, 1, 6, −4]	[0, 1, 1, 1, 1, 1]
	[2, 4, 4, 2, 2, 2]	[5, −9, 1, 1, 11, −9]	[1, 1, 1, 1, 1, 1]
	[0, 1, 2, 1, 2, 1]	[10, 1, 1, 1, −4, 1]	[0, 1, 1, 1, 1, 1]
	[1, 1, 2, 1, 2, 1]	[0, 6, 1, 1, −4, 1]	[1, 0, 0, 0, 0, 0]
	[2, 2, 4, 2, 4, 2]	[−5, 11, 1, 1, −9, 1]	[1, 1, 1, 1, 1, 1]
	[1, 2, 2, 1, 2, 1]	[5, −4, 6, 1, −4, 1]	[0, 1, 1, 1, 1, 1]
	[2, 4, 4, 2, 4, 2]	[5, −9, 11, 1, −9, 1]	[1, 1, 1, 1, 1, 1]
	[1, 2, 3, 1, 2, 1]	[5, 1, −4, 6, 1, 1]	[0, 1, 2, 1, 1, 0]
	[2, 4, 6, 2, 4, 2]	[5, 1, −9, 11, 1, 1]	[1, 1, 2, 1, 1, 0]
	[1, 1, 1, 0, 1, 1]	[0, 1, 1, 6, 1, −4]	[1, 0, 0, 0, 0, 0]
	[1, 2, 2, 1, 1, 0]	[5, −4, 1, 1, 1, 6]	[0, 1, 2, 1, 1, 0]
	[2, 4, 4, 2, 2, 0]	[5, −9, 1, 1, 1, 11]	[1, 1, 2, 1, 1, 0]
	[1, 1, 0, 0, 0, 0]	[0, −4, 6, 1, 1, 1]	[1, 0, 0, 0, 0, 0]
	[1, 1, 2, 1, 1, 0]	[0, 6, −4, 1, 1, 6]	[1, 0, 0, 0, 0, 0]
	[1, 1, 1, 0, 0, 0]	[0, 1, −4, 6, 6, 1]	[1, 0, 0, 0, 0, 0]
	[1, 1, 1, 1, 1, 0]	[0, 1, 6, −4, −4, 6]	[1, 0, 0, 0, 0, 0]
	[1, 1, 1, 0, 1, 0]	[0, 1, 1, 6, −4, 6]	[1, 0, 0, 0, 0, 0]
	[1, 1, 1, 1, 0, 0]	[0, 1, 1, −4, 6, 1]	[1, 0, 0, 0, 0, 0]

Table 14 $p = 7$, relevant root multiples for $r\omega_1$, $1 \leq r \leq 3$

λ	$m\alpha$	$\lambda+\rho-pm\alpha$	β
ω_1	[1, 2, 3, 2, 2, 1]	[2, 1, 1, −6, 1, 1]	[1, 1, 1, 1, 1, 1]
	[1, 1, 2, 1, 1, 1]	[−5, 8, −6, 1, 8, −6]	[1, 1, 1, 1, 1, 1]
	[1, 2, 2, 1, 1, 1]	[2, −6, 1, 1, 8, −6]	[1, 1, 1, 1, 1, 1]
	[1, 1, 2, 1, 2, 1]	[−5, 8, 1, 1, −6, 1]	[1, 1, 1, 1, 1, 1]
	[1, 2, 2, 1, 2, 1]	[2, −6, 8, 1, −6, 1]	[1, 1, 1, 1, 1, 1]
	[1, 2, 3, 1, 2, 1]	[2, 1, −6, 8, 1, 1]	[0, 1, 2, 1, 2, 1]
	[1, 2, 2, 1, 1, 0]	[2, −6, 1, 1, 1, 8]	[1, 1, 2, 1, 1, 0]
		$\lambda+\rho-pm\alpha$	β
$2\omega_1$	[1, 2, 3, 2, 2, 1]	[3, 1, 1, −6, 1, 1]	[0, 1, 2, 1, 2, 1]
	[1, 1, 1, 1, 1, 1]	[−4, 1, 8, −6, 1, −6]	[1, 1, 1, 0, 1, 1]
	[1, 1, 2, 1, 1, 1]	[−4, 8, −6, 1, 8, −6]	[1, 1, 1, 0, 1, 1]
	[1, 2, 2, 1, 1, 1]	[3, −6, 1, 1, 8, −6]	[1, 1, 1, 0, 1, 1]
	[1, 1, 2, 1, 2, 1]	[−4, 8, 1, 1, −6, 1]	[0, 1, 2, 1, 2, 1]

(Continued)

Table 14 (Continued)

λ	$m\alpha$		
		$\lambda+\rho-pm\alpha$	β
$3\omega_1$	[1, 2, 2, 1, 2, 1]	[3, −6, 8, 1, −6, 1]	[0, 1, 2, 1, 2, 1]
	[1, 2, 3, 1, 2, 1]	[3, 1, −6, 8, 1, 1]	[0, 1, 2, 1, 2, 1]
	[1, 2, 2, 1, 1, 0]	[3, −6, 1, 1, 1, 8]	[1, 1, 1, 1, 1, 0]
	[1, 1, 2, 1, 1, 0]	[−4, 8, −6, 1, 1, 8]	[1, 1, 1, 1, 1, 0]
	[1, 2, 3, 2, 2, 1]	[4, 1, 1, −6, 1, 1]	[1, 2, 3, 2, 2, 1]
	[1, 1, 1, 1, 1, 1]	[−3, 1, 8, −6, 1, −6]	[1, 1, 1, 1, 0, 0]
	[1, 1, 2, 1, 1, 1]	[−3, 8, −6, 1, 8, −6]	[1, 1, 1, 1, 0, 0]
	[1, 2, 2, 1, 1, 1]	[4, −6, 1, 1, 8, −6]	[1, 1, 1, 1, 0, 0]
	[1, 1, 2, 1, 2, 1]	[−3, 8, 1, 1, −6, 1]	[0, 1, 2, 1, 2, 1]
	[1, 2, 2, 1, 2, 1]	[4, −6, 8, 1, −6, 1]	[0, 1, 2, 1, 2, 1]
	[1, 2, 3, 1, 2, 1]	[4, 1, −6, 8, 1, 1]	[0, 1, 2, 1, 2, 1]
	[1, 1, 1, 0, 1, 1]	[−3, 1, 1, 8, 1, −6]	[1, 1, 1, 0, 1, 0]
	[1, 2, 2, 1, 1, 0]	[4, −6, 1, 1, 1, 8]	[1, 1, 1, 0, 1, 0]
	[1, 1, 2, 1, 1, 0]	[−3, 8, −6, 1, 1, 8]	[1, 1, 1, 0, 1, 0]
	[1, 1, 1, 1, 1, 0]	[−3, 1, 8, −6, −6, 8]	[1, 1, 1, 0, 1, 0]

Table 15 $p = 7$, relevant root multiples for $r\omega_1, 4 \leq r \leq 5$

λ	$m\alpha$		
		$\lambda+\rho-pm\alpha$	β
$4\omega_1$	[1, 2, 3, 2, 2, 1]	[5, 1, 1, −6, 1, 1]	[0, 1, 2, 1, 2, 1]
	[2, 4, 6, 4, 4, 2]	[5, 1, 1, −13, 1, 1]	[1, 2, 3, 1, 2, 1]
	[1, 1, 2, 1, 1, 1]	[−2, 8, −6, 1, 8, −6]	[1, 1, 1, 0, 0, 0]
	[1, 2, 2, 1, 1, 1]	[5, −6, 1, 1, 8, −6]	[1, 1, 1, 0, 0, 0]
	[1, 1, 2, 1, 2, 1]	[−2, 8, 1, 1, −6, 1]	[0, 1, 2, 1, 2, 1]
	[1, 2, 2, 1, 2, 1]	[5, −6, 8, 1, −6, 1]	[0, 1, 2, 1, 2, 1]
	[1, 2, 3, 1, 2, 1]	[5, 1, −6, 8, 1, 1]	[0, 1, 2, 1, 2, 1]
	[1, 1, 1, 0, 1, 1]	[−2, 1, 1, 8, 1, −6]	[1, 1, 1, 0, 0, 0]
	[1, 2, 2, 1, 1, 0]	[5, −6, 1, 1, 1, 8]	[1, 1, 1, 0, 0, 0]
	[1, 1, 2, 1, 1, 0]	[−2, 8, −6, 1, 1, 8]	[1, 1, 1, 0, 0, 0]
	[1, 1, 1, 0, 1, 0]	[−2, 1, 1, 8, −6, 8]	[1, 1, 1, 0, 0, 0]
	[1, 1, 1, 1, 0, 0]	[−2, 1, 1, −6, 8, 1]	[1, 1, 1, 0, 0, 0]
		w	$w(\lambda+\rho-pm\alpha)-\rho$
	[1, 1, 1, 1, 1, 1]	[1, 2, 3, 5, 6, 4, 2, 1]	[1, 0, 0, 1, 0, 0]
	[1, 1, 1, 1, 1, 0]	[1, 2, 3, 5, 4, 2, 1]	[2, 0, 0, 0, 0, 1]
		$\lambda+\rho-pm\alpha$	β
$5\omega_1$	[1, 2, 3, 2, 2, 1]	[6, 1, 1, −6, 1, 1]	[0, 1, 2, 1, 2, 1]
	[2, 4, 6, 4, 4, 2]	[6, 1, 1, −13, 1, 1]	[1, 2, 2, 1, 2, 1]
	[1, 1, 1, 1, 1, 1]	[−1, 1, 8, −6, 1, −6]	[1, 1, 0, 0, 0, 0]
	[1, 2, 2, 1, 1, 1]	[6, −6, 1, 1, 8, −6]	[1, 1, 0, 0, 0, 0]
	[1, 1, 2, 1, 2, 1]	[−1, 8, 1, 1, −6, 1]	[0, 1, 2, 1, 2, 1]
	[1, 2, 2, 1, 2, 1]	[6, −6, 8, 1, −6, 1]	[0, 1, 2, 1, 2, 1]
	[1, 2, 3, 1, 2, 1]	[6, 1, −6, 8, 1, 1]	[0, 1, 2, 1, 2, 1]
	[2, 4, 6, 2, 4, 2]	[6, 1, −13, 15, 1, 1]	[1, 2, 2, 1, 2, 1]

(Continued)

Table 15 (Continued)

λ	$m\alpha$	$\lambda+\rho-pm\alpha$ / w	β / $w(\lambda+\rho-pm\alpha)-\rho$
	[1, 1, 1, 0, 1, 1]	[−1, 1, 1, 8, 1, −6]	[1, 1, 0, 0, 0, 0]
	[1, 2, 2, 1, 1, 0]	[6, −6, 1, 1, 1, 8]	[1, 1, 0, 0, 0, 0]
	[1, 1, 1, 0, 0, 0]	[−1, 1, −6, 8, 8, 1]	[1, 1, 0, 0, 0, 0]
	[1, 1, 1, 1, 1, 0]	[−1, 1, 8, −6, −6, 8]	[1, 1, 0, 0, 0, 0]
	[1, 1, 1, 0, 1, 0]	[−1, 1, 1, 8, −6, 8]	[1, 1, 0, 0, 0, 0]
	[1, 1, 1, 1, 0, 0]	[−1, 1, 1, −6, 8, 1]	[1, 1, 0, 0, 0, 0]
	[1, 1, 2, 1, 1, 1]	[1, 2, 3, 5, 6, 4, 3, 1]	[0, 0, 0, 1, 0, 1]
	[1, 1, 2, 1, 1, 0]	[1, 2, 3, 5, 4, 3, 1]	[1, 0, 0, 0, 0, 2]

Table 16 $p=7$, relevant root multiples for $6\omega_1$

λ	$m\alpha$	$\lambda+\rho-pm\alpha$	β
$6\omega_1$	[1, 2, 3, 2, 2, 1]	[7, 1, 1, −6, 1, 1]	[0, 1, 2, 1, 2, 1]
	[2, 4, 6, 4, 4, 2]	[7, 1, 1, −13, 1, 1]	[1, 2, 2, 1, 1, 1]
	[1, 1, 1, 1, 1, 1]	[0, 1, 8, −6, 1, −6]	[1, 0, 0, 0, 0, 0]
	[1, 1, 2, 1, 1, 1]	[0, 8, −6, 1, 8, −6]	[1, 0, 0, 0, 0, 0]
	[1, 2, 2, 1, 1, 1]	[7, −6, 1, 1, 8, −6]	[1, 2, 2, 1, 1, 1]
	[1, 1, 2, 1, 2, 1]	[0, 8, 1, 1, −6, 1]	[0, 1, 2, 1, 2, 1]
	[1, 2, 2, 1, 2, 1]	[7, −6, 8, 1, −6, 1]	[0, 1, 2, 1, 2, 1]
	[2, 4, 4, 2, 4, 2]	[7, −13, 15, 1, −13, 1]	[1, 2, 2, 1, 1, 1]
	[1, 2, 3, 1, 2, 1]	[7, 1, −6, 8, 1, 1]	[0, 1, 2, 1, 2, 1]
	[2, 4, 6, 2, 4, 2]	[7, 1, −13, 15, 1, 1]	[1, 2, 2, 1, 1, 1]
	[1, 1, 1, 0, 1, 1]	[0, 1, 1, 8, 1, −6]	[1, 0, 0, 0, 0, 0]
	[1, 1, 0, 0, 0, 0]	[0, −6, 8, 1, 1, 1]	[1, 0, 0, 0, 0, 0]
	[1, 1, 2, 1, 1, 0]	[0, 8, −6, 1, 1, 8]	[1, 0, 0, 0, 0, 0]
	[1, 1, 1, 0, 0, 0]	[0, 1, −6, 8, 8, 1]	[1, 0, 0, 0, 0, 0]
	[1, 1, 1, 1, 1, 0]	[0, 1, 8, −6, −6, 8]	[1, 0, 0, 0, 0, 0]
	[1, 1, 1, 0, 1, 0]	[0, 1, 1, 8, −6, 8]	[1, 0, 0, 0, 0, 0]
	[1, 1, 1, 1, 0, 0]	[0, 1, 1, −6, 8, 1]	[1, 0, 0, 0, 0, 0]
		w	$w(\lambda+\rho-pm\alpha)-\rho$
	[1, 2, 2, 1, 1, 0]	[1, 2, 3, 5, 4, 3, 2]	[0, 0, 0, 0, 0, 3]

Table 17 $p=7$, second iteration relevant root multiples for $4\omega_1$

λ	$m\alpha$	$\lambda+\rho-pm\alpha$	β
$2\omega_1+\omega_6$	[1, 2, 3, 2, 2, 1]	[3, 1, 1, −6, 1, 2]	[1, 2, 3, 2, 2, 1]
	[1, 1, 1, 1, 1, 1]	[−4, 1, 8, −6, 1, −5]	[1, 1, 1, 1, 1, 0]
	[1, 1, 2, 1, 1, 1]	[−4, 8, −6, 1, 8, −5]	[0, 1, 2, 1, 1, 1]
	[1, 2, 2, 1, 1, 1]	[3, −6, 1, 1, 8, −5]	[0, 1, 2, 1, 1, 1]
	[0, 1, 2, 1, 2, 1]	[10, 1, 1, 1, −6, 2]	[0, 1, 2, 1, 1, 1]
	[1, 1, 2, 1, 2, 1]	[−4, 8, 1, 1, −6, 2]	[1, 1, 1, 1, 1, 0]
	[1, 2, 2, 1, 2, 1]	[3, −6, 8, 1, −6, 2]	[1, 1, 1, 1, 1, 0]

(Continued)

Table 17 (Continued)

λ	$m\alpha$		
		$\lambda+\rho-pm\alpha$	β
$\omega_1+\omega_4$	[1, 2, 3, 1, 2, 1]	[3, 1, −6, 8, 1, 2]	[0, 1, 2, 1, 1, 1]
	[1, 2, 2, 1, 1, 0]	[3, −6, 1, 1, 1, 9]	[1, 1, 1, 1, 1, 0]
	[1, 1, 2, 1, 1, 0]	[−4, 8, −6, 1, 1, 9]	[1, 1, 1, 1, 1, 0]
	[1, 1, 1, 0, 1, 1]	[1, 2, 3, 5, 6, 5, 3, 2, 1]	[1, 0, 0, 1, 0, 0]
	[1, 2, 3, 2, 2, 1]	[2, 1, 1, −5, 1, 1]	[1, 2, 3, 2, 2, 1]
	[1, 1, 1, 1, 1, 1]	[−5, 1, 8, −5, 1, −6]	[1, 1, 1, 1, 1, 0]
	[1, 1, 2, 1, 1, 1]	[−5, 8, −6, 2, 8, −6]	[0, 1, 2, 1, 1, 1]
	[1, 2, 2, 1, 1, 1]	[2, −6, 1, 2, 8, −6]	[0, 1, 2, 1, 1, 1]
	[0, 1, 2, 1, 2, 1]	[9, 1, 1, 2, −6, 1]	[0, 1, 2, 1, 1, 1]
	[1, 1, 2, 1, 2, 1]	[−5, 8, 1, 2, −6, 1]	[1, 1, 1, 1, 1, 0]
	[1, 2, 2, 1, 2, 1]	[2, −6, 8, 2, −6, 1]	[1, 1, 1, 1, 1, 0]
	[1, 2, 3, 1, 2, 1]	[2, 1, −6, 9, 1, 1]	[0, 1, 2, 1, 1, 1]
	[1, 2, 2, 1, 1, 0]	[2, −6, 1, 2, 1, 8]	[1, 1, 1, 1, 1, 0]
	[1, 1, 2, 1, 1, 0]	[−5, 8, −6, 2, 1, 8]	[1, 1, 1, 1, 1, 0]

Table 18 $p=7$, second iteration relevant root multiples for $5\omega_1$

λ	$m\alpha$		
		$\lambda+\rho-pm\alpha$	β
$\omega_1+2\omega_6$	[1, 2, 3, 2, 2, 1]	[2, 1, 1, −5, 1, 3]	[1, 2, 3, 2, 2, 1]
	[1, 1, 1, 1, 1, 1]	[−5, 1, 8, −5, 1, −4]	[0, 1, 1, 1, 1, 1]
	[0, 1, 2, 1, 1, 1]	[9, 1, −6, 1, 8, −4]	[0, 1, 1, 1, 1, 1]
	[1, 1, 2, 1, 1, 1]	[−5, 8, −6, 1, 8, −4]	[1, 1, 2, 1, 1, 0]
	[1, 2, 2, 1, 1, 1]	[2, −6, 1, 1, 8, −4]	[0, 1, 1, 1, 1, 1]
	[0, 1, 2, 1, 2, 1]	[9, 1, 1, 1, −6, 3]	[0, 1, 1, 1, 1, 1]
	[1, 1, 2, 1, 2, 1]	[−5, 8, 1, 1, −6, 3]	[1, 1, 2, 1, 1, 0]
	[1, 2, 2, 1, 2, 1]	[2, −6, 8, 1, −6, 3]	[0, 1, 1, 1, 1, 1]
	[1, 2, 3, 1, 2, 1]	[2, 1, −6, 8, 1, 3]	[1, 1, 2, 1, 1, 0]
	[1, 2, 2, 1, 1, 0]	[2, −6, 1, 1, 1, 10]	[1, 1, 2, 1, 1, 0]
		w	$w(\lambda+\rho-pm\alpha)-\rho$
	[1, 1, 1, 0, 1, 1]	[1, 2, 3, 5, 6, 5, 3, 2, 1]	[0, 0, 0, 1, 0, 1]
		$\lambda+\rho-pm\alpha$	β
$\omega_4+\omega_6$	[1, 2, 3, 2, 2, 1]	[1, 1, 1, −5, 1, 2]	[1, 2, 3, 2, 2, 1]
	[1, 1, 1, 1, 1, 1]	[−6, 1, 8, −5, 1, −5]	[0, 1, 1, 1, 1, 1]
	[0, 1, 2, 1, 1, 1]	[8, 1, −6, 2, 8, −5]	[0, 1, 1, 1, 1, 1]
	[1, 1, 2, 1, 1, 1]	[−6, 8, −6, 2, 8, −5]	[1, 1, 2, 1, 1, 0]
	[1, 2, 2, 1, 1, 1]	[1, −6, 1, 2, 8, −5]	[0, 1, 1, 1, 1, 1]
	[0, 1, 2, 1, 2, 1]	[8, 1, 1, 2, −6, 2]	[0, 1, 1, 1, 1, 1]
	[1, 1, 2, 1, 2, 1]	[−6, 8, 1, 2, −6, 2]	[1, 1, 2, 1, 1, 0]
	[1, 2, 2, 1, 2, 1]	[1, −6, 8, 2, −6, 2]	[0, 1, 1, 1, 1, 1]
	[1, 2, 3, 1, 2, 1]	[1, 1, −6, 9, 1, 2]	[1, 1, 2, 1, 1, 0]
	[1, 2, 2, 1, 1, 0]	[1, −6, 1, 2, 1, 9]	[1, 1, 2, 1, 1, 0]

6. $\mathrm{rad}(V((p-1)\omega_1)) \cong L((p-7)\omega_1 + 3\omega_6)$.

(e). It is clear from the above discussion that (e) holds.

Acknowledgments I wish to thank the referees for their careful reading and helpful suggestions, which have improved the exposition.

References

1. Arslan, O., Sin, P.: Some simple modules for classical groups and p-ranks of orthogonal and Hermitian geometries. J. Algebra **327**, 141–169 (2011)
2. Jantzen, J.C.: Representations of Algebraic Groups. Academic Press, London (1987)
3. Ramanan, S., Ramanathan, A.: Projective normality of flag varieties and Schubert varieties. Inventiones Mathematicae **79**, 217–224 (1985)
4. Sin, P.: Oppositeness in buildings and simple modules for finite groups of Lie type. Buildings, finite geometries and groups. In: Springer Proceedings in Mathematics, vol. 10, pp. 273–286 (2011)
5. Steinberg, R.: Representations of algebraic groups. Nagoya Math. J. **22**, 33–56 (1963)

About the Editor

N.S. NARASIMHA SASTRY is a Professor of Mathematics and Head, Indian Statistical Institute, Bengaluru Centre, India. He received his Ph.D. from University of Pittsburgh, Pennsylvania, USA in 1978 and has held visiting positions at Tata Institute of Fundamental Research, Mumbai, India; Michigan State University, East Lansing, USA; University of Western Australia, Perth, Australia; Rutgers University, New Brunswick, USA; University of Florida, Gainsville, USA and University of Ghent, Belgium. He has served as referee for journals such as *Journal of Algebra; Codes, Designs and Cryptography* and *Journal of Pure and Applied Algebra* and has published in journals such as *Journal of Algebra, Journal of Combinatorial Theory (Series A), Geometriae Dedicata, Journal of Functional Analysis, Journal of Operator Theory, Proceedings in Indian Academy of Sciences, Journal of Algebraic Combinatorics* and *Archiv der Mathematik*. He has edited two books: *Groups, Finite Geometries and Buildings,* published by Springer and *Essays in Geometric Group Theory,* published by the Ramanujam Mathematical Society. He has co-edited *Perspectives in Mathematical Sciences,* (Volumes 1 and 2), published by World Scientific. His research interests are finite groups including finite simple groups, geometries related to finite simple groups, algebraic codes, Coxeter groups and the Monster.

N. S. N. Sastry (ed.), *Groups of Exceptional Type, Coxeter Groups and Related Geometries,* Springer Proceedings in Mathematics & Statistics 82,
DOI: 10.1007/978-81-322-1814-2,

Problem Set

A Geometric Description of Hoffman Singleton?

Bhaskar Bagchi

A strongly regular graph with parameters (v, k, λ, μ) is a simple graph on v vertices which is regular of degree k such that (i) any two adjacent vertices have exactly λ common neighbours and (ii) any two non-adjacent vertices have exactly μ common neighbours.

It is well known that, up to isomorphism, there is a unique strongly regular graph with parameters $(50, 7, 0, 1)$. This is known as Ho-Si (after its co-discoverers Hoffman and Singleton). It is also known that the (full) automorphism group of Ho-Si is $U(3, 5) : 2$. In my opinion, no satisfactory proof of this fact is known. (The known proof may be called a "proof by coincidence" : correct, but not satisfactory!) So, my question, in a nutshell, is:

Question: Explain the equality Aut (Ho-Si) = $U(3, 5) : 2$.

Let's elaborate. Take any srg $(50, 7, 0, 1)$, say Γ. (We do not assume the uniqueness theorem.) Define a *Robertson system* in Γ to be a partition of the fifty vertices of Γ into ten pentagons in Γ. The following Lemma is well known, and easily proved by counting.

Lemma 1 *(a) There are exactly 126 Robertson systems in Γ.*

(b) The ten pentagons of any Robertson system are naturally split into two classes of five pentagons such that (i) no edge of Γ joins two pentagons from the same class, (ii) any two pentagons from different classes are joined by five edges of Γ, forming a copy of the Petersen graph in Γ.

From Lemma 1, it is more or less immediate that there are exactly 525 copies of Petersen in Γ, and each Robertson system in Γ yields $5 \times 5 = 25$ of these Petersens. The following may again be proved by (a little more difficult) counting:

Lemma 2 *Any two distinct Robertson systems in Γ correspond to a unique common Petersen in Γ. Thus, the incidence system with the Robertson systems of Γ as points and the Petersens in Γ as the blocks (with obvious incidence) is a $2-(5^3+1, 5+1, 1)$ design.*

N. S. N. Sastry (ed.), *Groups of Exceptional Type, Coxeter Groups and Related Geometries,* Springer Proceedings in Mathematics & Statistics 82,
DOI: 10.1007/978-81-322-1814-2, © Springer India 2014

Now, there is a "natural" $2-(5^3+1, 5+1, 1)$ design coming from geometry, namely, the "unital" admitting $U(3,5):2$ as its automorphism group (acting doubly transitively on the points of the design). It consists of the totally isotropic points and lines of $PG(2, \mathbb{F}_{25})$ with respect to a given non-degenerate hermitian form. The determination of the automorphism group (as well as uniqueness) of the srg (50, 7, 0, 1) proceeds by identifying the design of Lemma 2 with this unital. This is hard.

This line of argument shows that there MUST BE 50 geometric objects sitting inside $PG(2, \mathbb{F}_{25})$ with a given hermitian form, on which the associated group $U(3,5):2$ acts. These 50 objects should serve as the vertices of Ho-Si. So, an elaborated (and more concrete) version of my question is:

Question: Find these fifty objects explicitly!

Let me mention one observation which can be the beginning of an answer to my question.

Recall that a *symmetric net* of order n is the incidence system obtained from an affine plane of order n by deleting the lines of a fixed parallel class. Any symmetric net of order n can be completed uniquely to an affine plane of order n. Since the affine plane of order 5 is unique (and its group acts transitively on the parallel classes), it follows that the symmetric net of order five is also unique.

Now, the following may easily be proved by counting.

Lemma 3 *Let $\tilde{\Gamma}$ be the subgraph of Γ obtained by deleting the 5×10 edges of the pentagons in a fixed Robertson system. Then $\tilde{\Gamma}$ is the incidence graph of a symmetric net of order five.*

Given the lemma, one may use the uniqueness of the net to deduce uniqueness of the graph. More to the point, from Lemma 3, one sees the emergence of the (affine, hence) projective plane of order five from the s.r.g. Γ. Can one go on and complete it to a projective plane of order 25 inside Γ?